# The Analysis of Time Series

The field of time series analysis has undergone a remarkable transformation since the publication of the seventh edition of this book. While classical statistical models such as autoregressive integrated moving average (ARIMA), state-space models, and spectral methods remain essential, the rise of artificial intelligence (AI) has introduced groundbreaking approaches to modelling, forecasting, and generating time-dependent data. This eighth edition of *The Analysis of Time Series: An Introduction with R* reflects these advancements with the addition of two new chapters: Predictive AI for Time Series and Generative AI for Time Series. These chapters bridge the gap between traditional time series methods and cutting-edge AI techniques, offering readers a comprehensive and integrated perspective on the field.

**Features**

- Comprehensive coverage of classical time series models including ARIMA, state-space models, and spectral methods
- Two new chapters on predictive and generative AI, introducing cutting-edge methods like transformers, variational autoencoders, and diffusion models
- Practical examples and illustrations using R, demonstrating the application of both classical and AI-based approaches to real-world time series data
- Emphasis on the integration of classical statistical rigor with the flexibility and scalability of AI methods
- Clear explanations and intuitive insights, making advanced concepts accessible to a broad audience
- Updated content reflecting the latest developments in time series analysis, with a focus on modern, high-dimensional, and nonlinear data challenges

*The Analysis of Time Series: An Introduction with R, Eighth Edition* is designed for students, researchers, and practitioners in statistics, as well as in finance, economics, climate science, health, and engineering. It serves as both a foundational text for those new to time series analysis and a valuable resource for experienced analysts seeking to engage with the rapidly evolving landscape of predictive and generative AI. With its balance of theory, practical implementation, and real-world examples, the book is ideal for use in academic courses, professional training, and self-study.

**Haipeng Xing** is a professor of applied mathematics and statistics at the State University of New York, Stony Brook, USA, and the author of three books and numerous research papers. His research interests include quantitative finance and risk management, econometrics, applied stochastic control, and sequential statistical methodology.

**Chris Chatfield** is a retired reader in statistics at the University of Bath, UK, the author of five books and numerous research papers, and an elected senior fellow of the International Institute of Forecasters.

**CHAPMAN & HALL/CRC**

**Texts in Statistical Science Series**



**Recently Published Titles**

**Statistics in Survey Sampling**
*Jae Kwang Kim*

**Time Series: A Data Analysis Approach Using R**
*Robert H. Shumway and David S. Stoffer*

**Bayesian Statistical Methods: With Applications to Machine Learning**
*By Brian J. Reich, Sujit K. Ghosh*

**Experimental Design for Data Science and Engineering**
*By V. Roshan Joseph*

**Survival Analysis: Principles and Applications in Clinical Trials and Beyond**
*Song Yang*

**Foundations of Bayesian Statistics for Data Scientists: With R and Python**
*Alan Agresti, Maria Kateri, Ranjini Grove, and Antonietta Mira*

**A Course in Regression and Smoothing Methods**
*Zhiqiang Tan*

**A Course in Large-sample and High-dimensional Theory**
*Zhiqiang Tan*

**Markov Chain Monte Carlo**
Stochastic Simulation for Bayesian Inference
*Dani Gamerman, Hedibert Lopes, and Flávio Bambirra Gonçalves*

**Introduction to Modern Randomization-Based Design and Analysis for Causal Inference**
*Tirthankar Dasgupta and Donald B. Rubin*

**The Analysis of Time Series**
An Introduction with R, Eighth Edition
*Haipeng Xing and Chris Chatfield*

**Regressions in Covariances, Dependencies and Graphs**
*Mohsen Pourahmadi and Aramayis Dallakyan*

For more information about this series, please visit: https://www.routledge.com/Chapman--HallCRC-Texts-in-Statistical-Science/book-series/CHTEXSTASCI

# The Analysis of Time Series

## An Introduction with R

### Eighth Edition

Haipeng Xing and Chris Chatfield

CRC Press
Taylor & Francis Group
Boca Raton London New York

CRC Press is an imprint of the
Taylor & Francis Group, an **informa** business
A CHAPMAN & HALL BOOK

Eighth edition published 2027
by CRC Press
2385 NW Executive Center Drive, Suite 320, Boca Raton, FL 33431

and by CRC Press
4 Park Square, Milton Park, Abingdon, Oxon, OX14 4RN

*CRC Press is an imprint of Taylor & Francis Group, LLC*



Sixth edition published by CRC Press 2003
Seventh edition published by CRC Press 2019



ISBN: 978-1-041-08586-7 (hbk)
ISBN: 978-1-041-02633-4 (pbk)
ISBN: 978-1-003-64611-2 (ebk)

DOI: 10.1201/9781003646112

Typeset in Nimbus Roman
by KnowledgeWorks Global Ltd.

Publisher's note: This book has been prepared from camera-ready copy provided by the authors.

**To Ying and Liz**

Alice sighed wearily. 'I think you might do something better with the time,' she said, 'than waste it in asking riddles that have no answers.'

'If you knew time as well as I do,' said the Hatter, 'you wouldn't talk about wasting *it*. It's *him*.'

'I don't know what you mean,' said Alice.

'Of course you don't!' the Hatter said, tossing his head contemptuously. 'I dare say you never even spoke to Time!'

'Perhaps not,' Alice cautiously replied, 'but I know I have to beat time when I learn music.'

'Ah! that accounts for it,' said the Hatter. 'He won't stand beating.'

Lewis Carroll, *Alice's Adventures in Wonderland*

# Contents

# Preface to the Eighth Edition

Since the publication of the seventh edition of this book, this field of time series analysis has undergone a profound transformation. While classical statistical models—such as ARIMA, state-space models, and spectral methods—remain foundational and indispensable, recent advances in AI have reshaped how time-dependent data are modelled, predicted, and generated. This revised edition reflects these developments by introducing two new chapters: Predictive AI for Time Series and Generative AI for Time Series.

The primary motivation for adding these chapters is twofold. First, prediction has become increasingly data-driven and high-dimensional. Modern applications in finance, economics, climate science, health, and engineering often involve large-scale, nonlinear, and nonstationary time series that challenge the assumptions underlying classical models. Predictive AI methods—such as recurrent neural networks, transformers, and other deep learning architectures—offer powerful alternatives that complement traditional approaches. These methods emphasize out-of-sample forecasting performance, representation learning, and scalability, making them particularly well suited for contemporary data environments. The new chapter on predictive AI is designed to introduce these ideas in a principled yet accessible way, highlighting both their connections to and departures from classical time series theory.

Second, the scope of time series analysis has expanded beyond forecasting to data generation and simulation. Generative AI models—including variational autoencoders, diffusion models, and generative adversarial networks—enable the synthesis of realistic time series that preserve complex temporal dependence structures. These models play an increasingly important role in scenario analysis, stress testing, data augmentation, and uncertainty quantification. The chapter on generative AI introduces these emerging tools and explains how they can be understood as extensions of probabilistic modelling ideas long familiar to time series analysts.

The addition of these chapters is not intended to replace classical methods with AI-based approaches. Rather, the goal of this revised edition is integration. Classical time series models provide interpretability, statistical rigor, and theoretical insight, while AI methods offer flexibility and expressive power for complex data environments. Understanding both perspectives—and how they complement one another—allows readers to make informed modelling choices in practice.

I would like to thank many students, instructors, and anonymous readers of the previous edition who generously shared comments and pointed out typographical errors and minor inaccuracies. In particular, I am grateful to Coad Alex, Adrian Keister, and John Miller for pointing out some errors in the seventh edition. Their thoughtful feedback has been invaluable in improving the clarity and accuracy of this revised edition. Any remaining errors are, of course, my own responsibility.

This revised edition is intended for students, researchers, and practitioners who seek a solid foundation in time series analysis while engaging with the rapidly evolving landscape of predictive and generative AI. As in the original edition, this book emphasizes clarity, intuition, and practical implementation. Examples and illustrations using R are provided throughout, including in the new chapters, to demonstrate how both classical and AI-based methods can be applied to time series data. Specifically, examples in Chapters 1–13 and 14–15 are implemented using both `R` or `Python`. Data and `R`/`Python` scripts for these examples are available at

`https://www.ams.sunysb.edu/~xing/tsRbook/index.html`.

Haipeng Xing
*Department of Applied Mathematics and Statistics*
*State University of New York, Stony Brook*
*Stony Brook, NY, USA*
haipeng.xing@stonybrook.edu
December 2025

# Abbreviations and Notations

| | |
|---|---|
| AR | Autoregressive |
| ARIMA | Autoregressive integrated moving average |
| ARMA | Autoregressive moving average |
| GARCH | Generalized autoregressive conditionally heteroscedastic |
| MA | Moving average |
| SARIMA | Seasonal ARIMA |
| TAR | Threshold autoregressive |
| SWN | Strict white noise |
| WN | White noise |
| | |
| FFT | Fast Fourier transform |
| MMSE | Minimum mean square error |
| P.I. | Prediction interval |
| | |
| ac.f. | Autocorrelation function |
| acv.f. | Autocovariance function |
| | |
| N | Sample size, or length of observed time series |
| $\widehat{x}_N(h)$ | Forecast of $x_{N+h}$ made at time $N$ |
| $B$ | Backward shift operator such that $BX_t = X_{t-1}$ |
| $B^d$ | $B^d X_t = X_{t-d}$ |
| $\nabla$ | First differencing operator, $(1-B)$, such that $\nabla X_t = X_t - X_{t-1}$ |
| $\nabla_d$ | $\nabla_d = 1 - B^d$ |
| $\nabla_d^r$ | $\nabla_d^r = (1-B^d)^r$ |
| $E$ | Expectation or expected value |
| Var | Variance |
| $I$ | Identity matrix – a square matrix with ones on the diagonal and zeros otherwise |
| $N(\mu, \sigma^2)$ | Normal distribution with mean $\mu$ and variance $\sigma^2$ |
| $\chi_\nu^2$ | Chi-square distribution with $\nu$ degrees of freedom |
| $\{Z_t\}$ or $\{\varepsilon_t\}$ | Purely random process of independent random variables, usually $N(0, \sigma^2)$-distributed |
| $A^T$ or $X^T$ | Transpose of a matrix A or vector $X$ — vectors are indicated by boldface, but not scalars or matrices |

# Chapter 1

# Introduction

A time series is a collection of observations made sequentially through time. Examples occur in a variety of fields, ranging from economics to engineering, and methods of analysing time series constitute an important area of statistics.

## 1.1 Some Representative Time Series

We begin with some examples of the sort of time series that arise in practice.

*Economic and financial time series*

Many time series are routinely recorded in economics and finance. Examples include share prices on successive days, export totals in successive months, average incomes in successive months, company profits in successive years, and so on.

The classic Beveridge wheat price index series consists of the average wheat price in nearly 50 places in various countries measured in successive years from 1500 to 1869 (Berveridge, 1921). This series is of particular interest to economics historians and is available in many places (e.g. in the `tseries` package of `R`). Figure 1.1 shows this series, and some apparent cyclic behaviour can be seen. The trend of the series will be studied in Sections 2.5 and 2.6.

To plot the data using the `R` statistical package, you can load the data `bev` in the `tseries` package and plot the time series (the > below are prompts):

```
> library(tseries)     # load the library
> data(bev)     # load the dataset
> plot(bev, xlab="Year", ylab="Wheat price index", xaxt="n")
> x.pos<-c(1500, 1560, 1620, 1680, 1740, 1800, 1869)
          # define x-axis labels
> axis(1, x.pos, x.pos)
```

As an example of financial time series, Figure 1.2 shows the daily returns (or percentage change) of the adjusted closing prices of the Standard & Poor's 500 (S&P500) index from January 4, 1995, to December 30, 2016. The data shown in Figure 1.2 are typical of return data. The mean of the return series seems to be stable with an average return of approximately zero, but the volatility of data changes over time. This series will be analysed in Chapter 12.

DOI: 10.1201/9781003646112-1

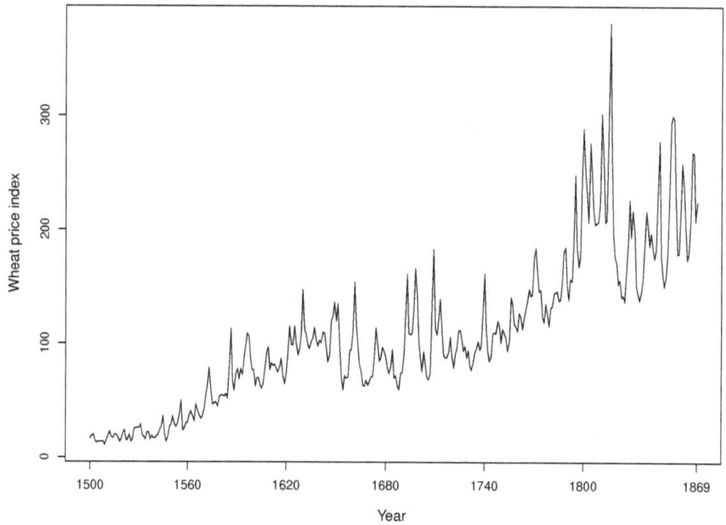


Figure 1.1 *The Beveridge wheat price annual index series from 1500 to 1869.*

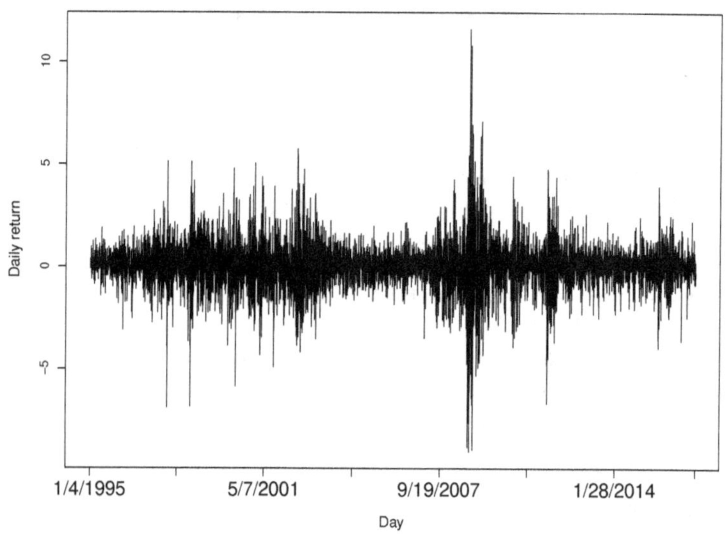


Figure 1.2 *Daily returns of the adjusted closing prices of the S&P500 index from January 4, 1995, to December 30, 2016.*

To reproduce Figure 1.2 in R, suppose you save the data as sp500_ret_1995-2016.csv in the directory mydata. Then you can use the following command to read the data and plot the time series.

```
> sp500<-read.csv("mydata/sp500_ret_1995-2016.csv")
> n<-nrow(sp500)
> x.pos<-c(seq(1,n,800),n)
> plot(sp500$Return, type="l", xlab="Day",
    ylab="Daily return", xaxt="n")
> axis(1, x.pos, sp500$Date[x.pos])
```

*Physical time series*

Many types of time series occur in the physical sciences, particularly in meteorology, marine science, and geophysics. Examples are rainfall on successive days and air temperature measured in successive hours, days, or months. Figure 1.3 shows the average air temperature in Anchorage, Alaska in the United States in successive months over a 16-year period. The series can be downloaded from the U.S. National Centers for Environmental Information (https://www.ncdc.noaa.gov/cag/). Seasonal fluctuations can be clearly seen in the series.

Some mechanical recorders take measurements continuously and produce a continuous trace rather than observations at discrete intervals of time. For example, in some laboratories it is important to keep temperature and humidity as constant as possible, and so devices are installed to measure these variables continuously. Action may be taken when the trace goes outside pre-specified limits. Visual examination of the trace may be adequate for many purposes, but for more detailed analysis, it is customary to convert the continuous trace to a series in discrete time by sampling the trace at appropriate equal intervals of time. The resulting analysis is more straightforward and can readily be handled by standard time series software.

*Marketing time series*

The analysis of time series arising in marketing is an important problem in commerce. Observed variables could include sales figures in successive weeks or months, monetary receipts, advertising costs, and so on. As an example, Figure 1.4 shows the domestic sales of Australian fortified wine by winemakers in successive quarters over a 30-year period, which are available at the Australian Bureau of Statistics (http://www.abs.gov.au/AUSSTATS/). This series will be analyzed in Sections 4.8 and 4.9. Note the trend and seasonal variation which is typical of sales data. It is often important to forecast future sales so as to plan production. It may also be of interest to examine the relationship between sales and other time series such as advertising expenditure.

*Demographic time series*

Various time series occur in the study of population change. Examples include the total population of Canada measured annually, and monthly birth totals in England.

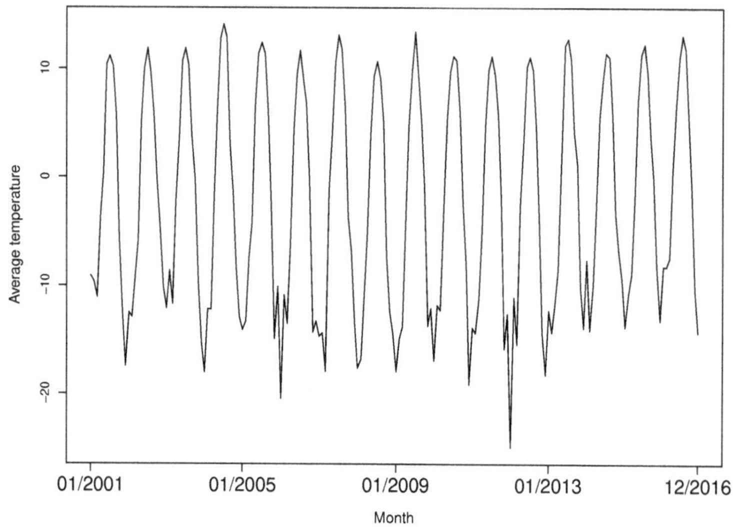


Figure 1.3 *Monthly average air temperature (deg C) in Anchorage, Alaska, the United States, in successive months from 2001 to 2016.*

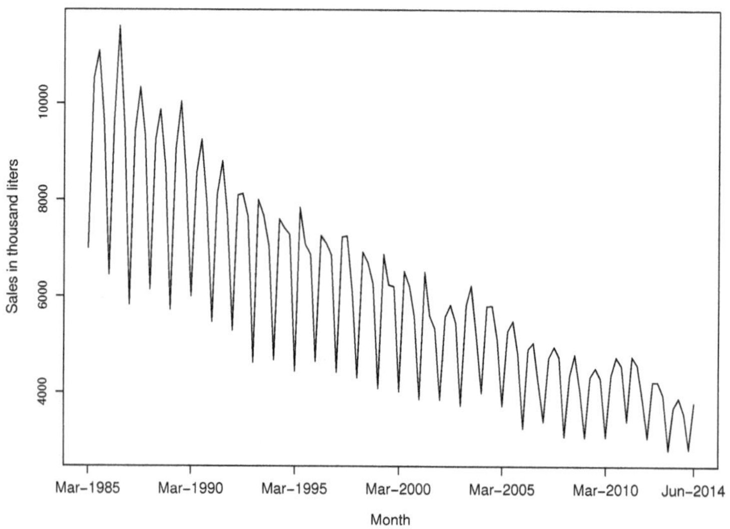


Figure 1.4 *Domestic sales (unit: thousand litres) of Australian fortified wine by winemakers in successive quarters from March 1985 to June 2014.*

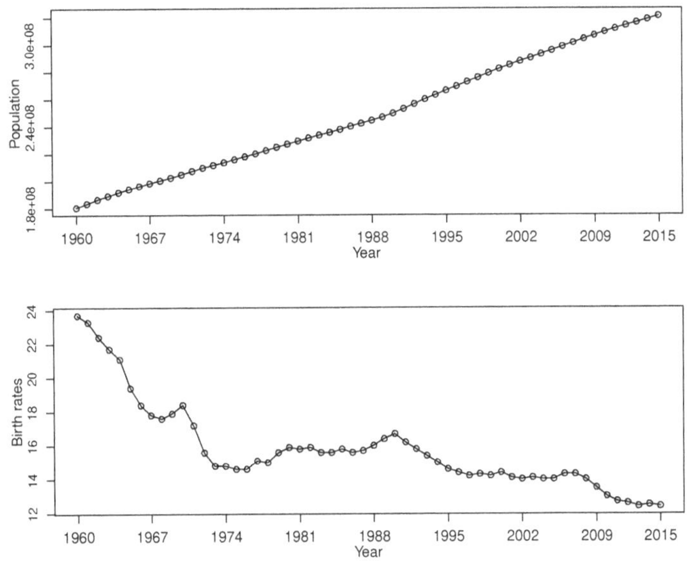


Figure 1.5 *Total population and birth rate (per 1,000 people) for the United States from 1965 to 2015.*

Figure 1.5 shows the total population and crude birth rate (per 1,000 people) for the United States from 1965 to 2015. The data are available at the U.S. Federal Reserve Bank of St. Louis (`https://fred.stlouisfed.org/`). Demographers want to predict changes in population for as long as 10 or 20 years into the future and are helped by the slowly changing structure of a human population. Standard time series methods can be applied to study this problem.

To reproduce Figure 1.5 in R, you can use the following commands to read the data and plot the time series.

```
> pop<-read.csv("mydata/US_pop_birthrate.csv", header=T)
> x.pos<-c(seq(1, 56, 7), 56)
> x.label<-c(seq(1960, 2009, by=7), 2015)

> par(mfrow=c(2,1), mar=c(3,4,3,4))
> plot(pop[,2], type="l", xlab="", ylab="", xaxt="n")
> points(pop[,2])
> axis(1, x.pos, x.label, cex.axis=1.2)
> title(xlab="Year", ylab="Population", line=2, cex.lab=1.2)
> plot(pop[,3], type="l", xlab="", ylab="", xaxt="n")
> points(pop[,3])
> axis(1, x.pos, x.label, cex.axis=1.2)
> title(xlab="Year", ylab="Birth rates", line=2, cex.lab=1.2)
```

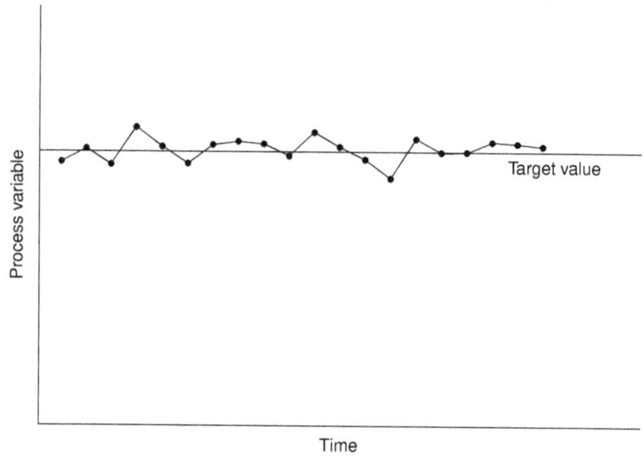


Figure 1.6 *A process control chart.*

*Process control data*

In process control, the problem is to detect changes in the performance of a manufacturing process by measuring a variable which shows the quality of the process. These measurements can be plotted against time as in Figure 1.6. When the measurements stray too far from some target value, appropriate corrective action should be taken to control the process. Special techniques have been developed for this type of time series problem, and the reader is referred to a book on statistical quality control, e.g. Montgomery (1996).

*Binary processes*

A special type of time series arises when observations can take one of only two values, usually denoted by 0 and 1 (see Figure 1.7). For example, in computer science, the position of a switch, either 'on' or 'off', could be recorded as one or zero, respectively. Time series of this type, called **binary processes**, occur in many situations, including the study of communication theory. A problem here is to predict when the process will take a different value. One way to solve this problem is to use regime-switching models, which will be discussed in Chapter 11 (Section 11.6).

*Point processes*

A completely different type of time series occurs when we consider a series of events occurring 'randomly' through time. For example, we could record the dates of major railway disasters. A series of events of this type is usually called a **point process**. As an example, Figure 1.8 shows the intraday transaction data of the International Business Machines Corporation (IBM) from 9:35:00 to 9:38:00 on January 4, 2010. When a trade event occurs, the corresponding trading price and trading volume are

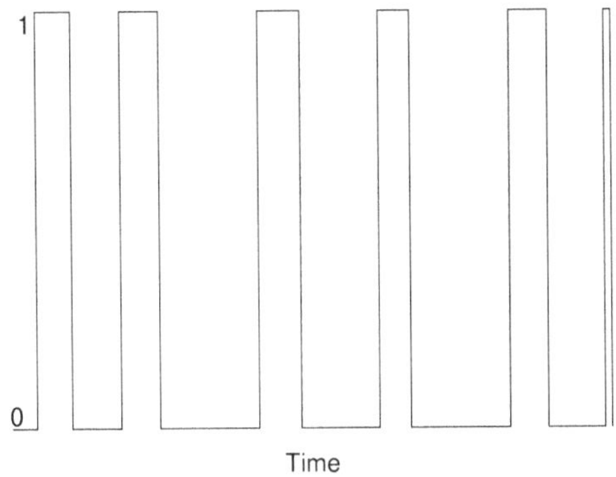


Figure 1.7 *A realization of a binary process.*

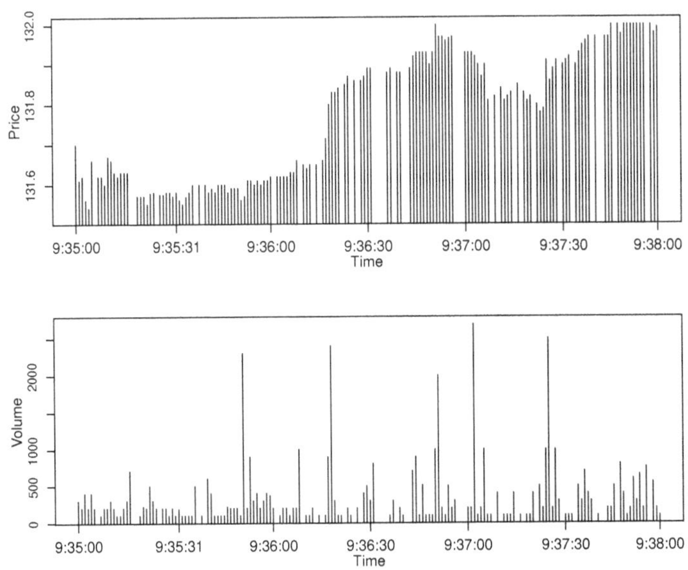


Figure 1.8 *Transaction prices and volumes of IBM stocks from 9:35:00 to 9:38:00 on January 4, 2010.*

observed. However, trades do not occur equally spaced in time; hence, time intervals between trades (or trade durations) are considered as random variables. For observations of this type, we are interested in such quantities as the distribution of the number of events occurring in a given time period and the distribution of time intervals between events. Methods of analysing point process data are generally very different from those used for analysing standard time series data, and the reader is referred, for example, to Cox and Isham (1980).

To reproduce Figure 1.8 in R, you can use the following commands to read the data and plot the time series.

```
> ibm<-read.table("mydata/taq_trade_ibm_100104.txt",
    header=T, sep="\t")
> ibm.new<-ibm[,c(1,2,7)]
> ibm[,2]<-as.numeric(as.character(ibm[,2]))

> ### take 9:35:00-9:37:59am trading record
> data<-ibm.new[1458:2371,]
> newtime<-rep(0, nrow(data))
> for (i in 1:nrow(data)){
   min<-as.numeric(substr(as.character(data$TIME[i]),3,4))
   sec<-as.numeric(substr(as.character(data$TIME[i]),6,7))
   newtime[i]<- (min-30)*60+sec
  }

> x.label<-c("9:35:00", "9:35:31", "9:36:00", "9:36:30",
  "9:37:00", "9:37:30", "9:38:00")
> x.pos<-c(1, 139, 249, 485, 619, 776, 914)

> par(mfrow=c(2,1), mar=c(2,4,2,4))
> plot(newtime, data[,2],xlab="",ylab="",xaxt="n",type="h")
> axis(1, newtime[x.pos], x.label, cex.axis=1.2)
> title(xlab="Time", ylab="Price", line=2, cex.lab=1.2)
> plot(newtime, data[,3],xlab="",ylab="",xaxt="n",type="h")
> axis(1, newtime[x.pos], x.label, cex.axis=1.2)
> title(xlab="Time", ylab="Volume", line=2, cex.lab=1.2)
```

## 1.2 Terminology

A time series is said to be **continuous** when observations are made continuously through time as in Figure 1.7. The adjective ‘continuous’ is used for series of this type even when the measured variable can only take a discrete set of values, as in Figure 1.7. A time series is said to be **discrete** when observations are taken only at specific times, usually equally spaced. The term ‘discrete’ is used for series of this type even when the measured variable is a continuous variable.

This book is mainly concerned with discrete time series, where the observations are taken at equal intervals. Discrete time series can arise in several ways. Given a continuous time series, we could read off (or digitize) the values at equal intervals of time to give a discrete time series, sometimes called a **sampled** series. The sampling interval between successive readings must be carefully chosen so as to lose little information (see Section 7.7). A different type of discrete series arises when a variable does not have an instantaneous value, but we can **aggregate** (or accumulate) the values over equal intervals of time. Examples of this type are monthly exports and daily rainfalls. Finally, some time series are inherently discrete, an example being the dividend paid by a company to shareholders in successive years.

Much statistical theory is concerned with random samples of independent observations. The special feature of time series analysis is the fact that successive observations are usually *not* independent and that the analysis must take into account the *time order* of the observations. When successive observations are dependent, future values may be predicted from past observations. If a time series can be predicted exactly, it is said to be **deterministic**. However, most time series are **stochastic** in that the future is only partly determined by past values, so that exact predictions are impossible and must be replaced by the idea that future values have a probability distribution which is conditioned on a knowledge of past values.

## 1.3 Objectives of Time Series Analysis

There are several possible objectives in analysing a time series. These objectives may be classified as description, explanation, prediction, and control and will be considered in turn.

(i) *Description*

When presented with a time series, the first step in the analysis is usually to plot the observations against time to give what is called a **time plot**, and then to obtain simple descriptive measures of the main properties of the series. This is described in detail in Chapter 2. The power of the time plot as a descriptive tool is illustrated by Figure 1.4, which clearly shows that there is a regular seasonal effect with wine sales 'high' in the third quarter and 'low' in the first quarter of each year. The time plot also shows that annual sales are decreasing (i.e. there is a downward trend). For some series, the variation is dominated by such 'obvious' features, and a fairly simple model, which only attempts to describe trend and seasonal variation, may be perfectly adequate to describe the variation in the time series. For other series, more sophisticated techniques will be required to provide an adequate analysis. Then a more complex model will be constructed, such as the various types of stochastic processes described in Chapter 3.

This book devotes a greater amount of space to the more advanced techniques, but this does not mean that elementary descriptive techniques are unimportant. Anyone who tries to analyze a time series without plotting it first is asking for trouble. A graph will not only show up trend and seasonal variation but will also reveal missing observations and any 'wild' observations or **outliers** that do not appear to be consistent with the rest of the data. The treatment of outliers is a complex subject in which

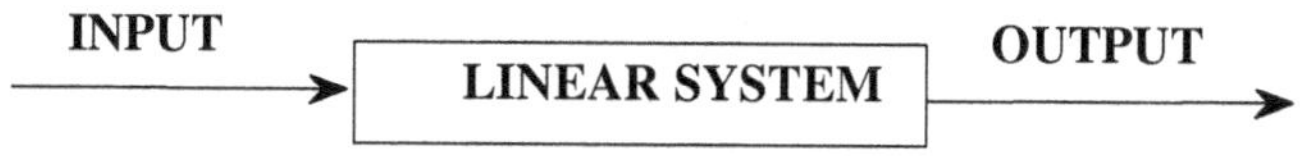


Figure 1.9 *Schematic representation of a linear system.*

common sense is as important as theory. An outlier may be a perfectly valid, but extreme, observation, which could, for example, indicate that the data are not normally distributed. Alternatively, an outlier may be a freak observation arising, for example, when a recording device goes wrong or when a strike severely affects sales. In the latter case, the outlier needs to be adjusted in some way before further analysis of the data. Robust methods are designed to be insensitive to outliers.

Other features to look for in a time plot include sudden or gradual changes in the properties of the series. For example, a step change in the level of the series would be very important to notice, if one exists. Any changes in the seasonal pattern should also be noted. The analyst should also look out for the possible presence of turning points where, for example, an upward trend suddenly changes to a downward trend. If there is some sort of discontinuity in the series, then different models may need to be fitted to different parts of the series.

(ii) *Explanation*

When observations are taken on two or more variables, it may be possible to use the variation in one time series to explain the variation in another series. This may lead to a deeper understanding of the mechanism that generated a given time series.

Although regression models are occasionally helpful here, they are not really designed to handle time series data with all the correlations inherent therein, and so we will see that alternative classes of models should be considered. Chapter 9 considers the analysis of what are called **linear systems**. A linear system converts an input series to an output series by a linear operation. Given observations on the input and output to a linear system (see Figure 1.9), the analyst wants to assess the properties of the linear system. For example, it is of interest to see how sea level is affected by temperature and pressure, and to see how sales are affected by price and economic conditions. A class of models, called transfer function models, enables us to model time series data in an appropriate way.

(iii) *Prediction*

Given an observed time series, one may want to predict the future values of the series. This is an important task in sales forecasting and in the analysis of economic and industrial time series. Many writers, including ourselves, use the terms 'prediction' and 'forecasting' interchangeably, but note that some authors do not. For example, Brown (1963) uses 'prediction' to describe subjective methods and 'forecasting' to describe objective methods.

(iv) *Control*

Time series are sometimes collected or analyzed so as to improve control over some physical or economic system. For example, when a time series is generated that measures the 'quality' of a manufacturing process, the aim of the analysis may be to keep the process operating at a 'high' level. Control problems are closely related to prediction in many situations. For example, if one can predict that a manufacturing process is going to move off target, then appropriate corrective action can be taken.

Control procedures vary considerably in style and sophistication. In statistical quality control, the observations are plotted on control charts and the controller takes action as a result of studying the charts. A more complicated type of approach is based on modelling the data and using the model to work out an 'optimal' control strategy — see, for example, Box et al. (1994). In this approach, a stochastic model is fitted to the series, future values of the series are predicted, and then the input process variables are adjusted so as to keep the process on target.

Many other contributions to control theory have been made by control engineers and mathematicians rather than statisticians. This topic is rather outside the scope of this book and will not be covered.

## 1.4 Approaches to Time Series Analysis

This book will describe various approaches to time series analysis. Chapter 2 describes **simple descriptive techniques**, which include plotting the data and looking for trends, seasonal fluctuations, and so on. Chapter 3 introduces a variety of probability models for time series, while Chapter 4 discusses ways of fitting these models to time series. The major diagnostic tool that is used in Chapter 4 is the **autocorrelation** function which helps to describe the evolution of a process through time. Inference based on this function is often called an **analysis in the time domain**. Chapter 5 goes on to discuss a variety of forecasting procedures, but this chapter is not a prerequisite for the rest of the book.

Chapter 6 introduces the **spectral density** function which describes how the variation in a time series may be accounted for by cyclic components at different frequencies. Chapter 7 shows how to estimate the spectral density function by means of **spectral analysis**. Inference based on the spectral density function is often called an **analysis in the frequency domain**.

Chapter 8 discusses the analysis of bivariate time series, while Chapter 9 extends this work by considering **linear systems** in which one series is regarded as the input while the other series is regarded as the output. Chapter 10 introduces an important class of models, called **state-space models**. It also describes the **Kalman filter**, which is a general method of updating the best estimate of the 'signal' in a time series in the presence of noise. Chapters 11 and 12 introduce **non-linear** and **volatility** time series models, respectively. Chapter 13 introduces **multivariate** time series models, while Chapters 14 and 15 introduce predictive and generative AI methods for time series analysis.

## 1.5 Review of Books on Time Series

This section gives a brief review of some alternative books on time series that may be helpful to the reader. The literature has expanded considerably in recent years and so a selective approach is necessary.

Alternative general introductory texts include Brockwell and Davis (2002), Harvey (1993), Kendall and Ord (1990), and Wei (1990). Diggle (1990) is aimed primarily at biostatisticians, while Enders (1995), Mills (1990, 1999), and Harris and Sollis (2003) are aimed at economists.

There are many more advanced texts including Anderson (1971), Brockwell and Davis (1991), Fuller (1996), and Priestley (1981). The latter is particularly strong on spectral analysis and multivariate time series modelling. Brillinger (2001) is a classic reprint of a book concentrating on the frequency domain. Kendall et al. (1983), now in the fourth edition, is a valuable reference source, but note that earlier editions are somewhat dated. Hamilton (1994) is aimed at econometricians. Tsay (2010), now in the third edition, provides a comprehensive introduction to financial econometric models and their applications to modelling and prediction of financial time series data.

The classic book by Box and Jenkins (1970) describes an approach to time series analysis, forecasting, and control that is based on autoregressive integrated moving average (ARIMA) models. This important book is not really suitable for the beginner, who is recommended to read Chapters 3–5 in this book, Vandaele (1983), or the relevant chapters of Wei (1990). The 1976 revised edition of Box and Jenkins (1970) was virtually unchanged, but the third edition (Box et al., 1994), with G. Reinsel as third author, has changed substantially.

Some time series books introduce the subject with statistical computing languages. For example, Cowpertwit and Metcalfe (2009) shows how to use `R` to apply time series methods in some practical applications. Shumway and Stoffer (2010) introduces time series models in time and frequency domains with `R` codes provided for some examples. Tsay (2010) demonstrates the application of time series models in financial econometrics with examples in `R` and `S-Plus`.

In this edition of the book, the first 13 chapters of the new edition have a very similar structure to the original edition. We added two new chapters (Chapters 14 and 15) to introduce predictive and generative AI methods for time series. We added more examples and exercises in this edition. For illustration purposes, we also present `R` code for data examples in the book.

Additional books will be referenced, as appropriate, in later chapters.

Chapter 2

# Basic Descriptive Techniques

Statistical techniques for analysing time series range from relatively straightforward descriptive methods to sophisticated inferential techniques. This chapter introduces the former, which will often clarify the main properties of a given series. Descriptive methods should generally be tried before attempting more complicated procedures, because they can be vital in 'cleaning' the data, and then getting a 'feel' for them, before trying to generate ideas for a suitable model.

Before doing anything, the analyst should make sure that the practical problem being tackled is properly understood. In other words, the **context** of a given problem is crucial in time series analysis, as in all areas of statistics. If necessary, the analyst should ask questions so as to get appropriate background information and clarify the objectives. These preliminary questions should not be rushed. In particular, make sure that appropriate data have been, or will be, collected. If the series are too short, or the wrong variables have been measured, it may not be possible to solve the given problem.

For a chapter on descriptive techniques, the reader may be expecting the first section to deal with **summary statistics**. Indeed, in most areas of statistics, a typical analysis begins by computing the sample mean (or median or mode) and the standard deviation (or interquartile range) to measure 'location' and 'dispersion'. However, *time-series analysis is different!* If a time series contains trend, seasonality, or some other systematic component, the usual summary statistics can be seriously misleading and should not be calculated. Moreover, even when a series does *not* contain any systematic components, the summary statistics do not have their usual properties (see Section 4.1.2). Thus, this chapter focuses on ways of understanding typical time-series effects, such as trend, seasonality, and correlations between successive observations.

## 2.1 Types of Variation

Traditional methods of time-series analysis are mainly concerned with decomposing the variation in a series into components representing trend, seasonal variation, and other cyclic changes. Any remaining variation is attributed to 'irregular' fluctuations. This approach is not always the best but is particularly valuable when the variation is dominated by trend and seasonality. However, it is worth noting that a decomposition into trend and seasonal variation is generally not unique unless certain assumptions

DOI: 10.1201/9781003646112-2

are made. Thus some sort of modelling, either explicit or implicit, may be involved in carrying out these descriptive techniques, and this demonstrates the blurred borderline that always exists between descriptive and inferential techniques in statistics.

The different sources of variation will now be described in more detail.

*Seasonal variation*

Many time series, such as sales figures and temperature readings, exhibit variation that is annual in period. For example, unemployment is typically 'high' in winter but lower in summer. This yearly variation is easy to understand and can readily be estimated if seasonality is of direct interest. Alternatively, seasonal variation can be removed from the data to give deseasonalized data, if seasonality is not of direct interest.

*Other cyclic variation*

Apart from seasonal effects, some time series exhibit variation at a fixed period due to some other physical cause. An example is daily variation in temperature. In addition, some time series exhibit oscillations which do not have a fixed period but which are predictable to some extent. For example, economic data are sometimes thought to be affected by business cycles with a period varying from about 3 or 4 years to more than 10 years, depending on the variable measured. However, the existence of such business cycles is the subject of some controversy, and there is increasing evidence that any such cycles are not symmetric. An economy usually behaves differently when going into recession rather than emerging from recession.

*Trend*

This may be loosely defined as 'long-term change in the mean level'. A difficulty with this definition is deciding what is meant by 'long term'. For example, climatic variables sometimes exhibit cyclic variation over a very long time period such as 50 years. If one just had 20 years of data, this long-term oscillation may look like a trend, but if several hundred years of data were available, then the long-term cyclic variation would be visible. Nevertheless in the short term it may still be more meaningful to think of such a long-term oscillation as a trend. Thus in speaking of a 'trend', we must take into account the number of observations available and make a subjective assessment of what is meant by the phrase 'long term'. As for seasonality, methods are available either for estimating the trend or for removing it so that the analyst can look more closely at other sources of variation.

*Other irregular fluctuations*

After trend and cyclic variations have been removed from a set of data, we are left with a series of residuals that may or may not be 'random'. In due course, we will examine various techniques for analysing series of this type, either to see whether any cyclic variation is still left in the residuals, or whether apparently irregular variation may be explained in terms of probability models such as moving average (MA) or autoregressive (AR) models, introduced in Chapter 3.

## 2.2 Stationary Time Series

A mathematical definition of a stationary time series model will be given in Section 3.2. However, it may be helpful to introduce here the idea of stationarity from an intuitive point of view. Broadly speaking a time series is said to be **stationary** if there is no systematic change in mean (no trend), if there is no systematic change in variance, and if strictly periodic variations have been removed. In other words, the properties of one section of the data are much like those of any other section. Strictly speaking, it is very often that time series data violate the stationarity property. However, the phrase is often used for time series data meaning that they exhibit characteristics that suggest a stationary model can sensibly be fitted.

Much of the probability theory of time series is concerned with stationary time series, and for this reason time series analysis often requires one to transform a non-stationary series into a stationary one so as to use this theory. For example, it may be of interest to remove the trend and seasonal variation from a set of data and then try to model the variation in the residuals by means of a stationary stochastic process. However, it is also worth stressing that the non-stationary components, such as the trend, may be of more interest than the stationary residuals.

## 2.3 The Time Plot

The first, and most important, step in any time-series analysis is to plot the observations against time. This graph, called a **time plot**, will show important features of the series such as trend, seasonality, outliers, and discontinuities. The plot is vital, both to describe the data and to help in formulating a sensible model; several examples have already been given in Chapter 1.

Plotting a time series is not as easy as it sounds. The choice of scales, the size of the intercept, and the way that the points are plotted (e.g. as a continuous line or as separate dots or crosses) may substantially affect the way the plot 'looks'; the analyst must exercise care and judgement. In addition, the usual rules for drawing 'good' graphs should be followed: a clear title must be given, units of measurement should be stated and axes should be properly labelled.

Nowadays, most graphs are produced by computers. Although many are well designed, software packages may also generate poor graphs. Users should be prepared to revise these figures or, preferably, give precise instructions in advance to produce clear and effective graphs. For instance, since the software simply prints the title it is given, it is the user's responsibility to supply a clear and meaningful one.

## 2.4 Transformations

Plotting the data may suggest that it is sensible to consider transforming them, for example, by taking logarithms or square roots. The three main reasons for making a transformation are as follows.

*(i) To stabilize the variance*

If there is a trend in the series and the variance appears to increase with the mean, then it may be advisable to transform the data. In particular, if the standard deviation is directly proportional to the mean, a logarithmic transformation is indicated. On the other hand, if the variance changes through time *without* a trend being present, then a transformation will not help. Instead, a model that allows for changing variance should be considered.

*(ii) To make the seasonal effect additive*

If there is a trend in the series and the size of the seasonal effect appears to increase with the mean, then it may be advisable to transform the data so as to make the seasonal effect constant from year to year. The seasonal effect is then said to be **additive**. In particular, if the size of the seasonal effect is directly proportional to the mean, then the seasonal effect is said to be **multiplicative** and a logarithmic transformation is appropriate to make the effect additive. However, this transformation will only stabilize the variance if the error term is also thought to be multiplicative (see Section 2.6), a point that is sometimes overlooked.

*(iii) To make the data normally distributed*

Model building and forecasting are usually carried out on the assumption that the data are normally distributed. In practice this is not necessarily the case; there may, for example, be evidence of skewness in that there tend to be 'spikes' in the time plot that are all in the same direction (either up or down). This effect can be difficult to eliminate with a transformation and it may be necessary to model the data using a different 'error' distribution.

The logarithmic and square-root transformations, mentioned above, are special cases of a general class of transformations called the Box–Cox transformation. Given an observed time series $\{x_t\}$ and a transformation parameter $\lambda$, the transformed series is given by

$$y_t = \begin{cases} (x_t^{\lambda} - 1)/\lambda & \lambda \neq 0 \\ \log x_t & \lambda = 0. \end{cases}$$

This is effectively just a power transformation when $\lambda \neq 0$, as the constants are introduced to make $y_t$ a continuous function of $\lambda$ at the value $\lambda = 0$. The 'best' value of $\lambda$ can be 'guesstimated', or alternatively estimated by a proper inferential procedure such as maximum likelihood.

It is instructive to note that Nelson and Granger (1979) found little improvement in forecast performance when a general Box–Cox transformation was tried on a number of series. There are problems in practice with transformations in that a transformation which makes the seasonal effect additive, for example, may fail to stabilize the variance. Thus it may be impossible to achieve all the above requirements at the same time. In any case a model constructed for the transformed data may be less than helpful. It is more difficult to interpret, and forecasts produced by the transformed model may have to be 'transformed back' in order to be of use. This can introduce biasing effects. Usually, transformations should be avoided wherever possible except where the transformed variable has a direct physical interpretation. For example, when percentage increases are of interest, then taking logarithms makes sense.

Further general remarks on transformations are given by Granger and Newbold (1986, Section 10.5).

## 2.5 Analysing Series That Contain a Trend and No Seasonal Variation

In Section 2.1, we loosely defined trend as a 'long-term change in the mean level'. It is much more difficult to give a precise definition of trend and different authors may use the term in different ways. The simplest type of trend is the familiar 'linear trend + noise', for which the observation at time $t$ is a random variable $X_t$ given by

$$X_t = \alpha + \beta t + \varepsilon_t, \tag{2.1}$$

where $\alpha, \beta$ are constants and $\varepsilon_t$ denotes a random error term with zero mean. The mean level at time $t$ is given by $m_t = (\alpha + \beta t)$; this is sometimes called 'the trend term'. Other writers prefer to describe the slope $\beta$ as the trend, so that trend is the *change* in the mean level per unit time. It is usually clear from the context what is meant by 'trend'.

The trend in Equation (2.1) is a deterministic function of time and is sometimes called a **global** linear trend. In practice, this generally provides an unrealistic model, and nowadays there is more emphasis on models that allow for **local** linear trends. One possibility is to fit a **piecewise linear** model where the trend line is locally linear but with change points where the slope and intercept change (abruptly). It is usually arranged that the lines join up at the change points, but even so, the sudden changes in slope often seem unnatural. Thus, it often seems more sensible to look at models that allow a smooth transition between the different submodels. Extending this idea, it seems even more natural to allow the parameters $\alpha$ and $\beta$ in Equation (2.1) to evolve through time. This could be done deterministically, but it is more common to assume that $\alpha$ and $\beta$ evolve stochastically, giving rise to what is called a **stochastic trend**. Some examples of suitable models under the general heading of state-space models are given in Chapter 10. Another possibility, depending on how the data look, is that the trend has a *non-linear* form such as quadratic growth. Exponential growth can be particularly difficult to handle, even if logarithms are taken to transform the trend to a linear form. Even with present-day computing aids, it can still be difficult to decide what form of trend is appropriate in a given context (see Ball and Wood (1996) and the discussion that followed).

The analysis of a time series that exhibits trend depends on whether one wants to (1) measure the trend and/or (2) remove the trend in order to analyze local fluctuations. It also depends on whether the data exhibit seasonality (see Section 2.6). With seasonal data, it is a good idea to start by calculating successive yearly averages, as these will provide a simple description of the underlying trend. An approach of this type is sometimes perfectly adequate, particularly if the trend is fairly small; but sometimes a more sophisticated approach is desired.

We now describe some different general approaches to describing trend.

### 2.5.1 *Curve fitting*

A traditional method of dealing with non-seasonal data that contain a trend, particularly yearly data, is to fit a simple function of time such as a polynomial curve (linear, quadratic, etc.), a Gompertz curve, or a logistic curve (e.g. see Meade, 1984; Franses, 1998, Chapter 4). The global linear trend in Equation (2.1) is the simplest type of polynomial curve. The Gompertz curve can be written in the form

$$\ln(x_t) = a + b\, r^t,$$

where $a, b, r$ are parameters with $0 < r < 1$, or in the alternative form of

$$x_t = \alpha \exp[\beta \exp(-\gamma t)],$$

which looks quite different but is actually equivalent provided $\gamma > 0$. The logistic curve is given by

$$x_t = a/(1 + b\, e^{-ct}).$$

Both these curves are S-shaped and approach an asymptotic value as $t \to \infty$, with the Gompertz curve generally converging slower than the logistic. Fitting the curves to data may lead to non-linear simultaneous equations.

For all curves of this type, the fitted function provides a measure of the trend; the residuals provide an estimate of local fluctuations, where the residuals are the differences between the observations and the corresponding values of the fitted curve.

### 2.5.2 *Filtering*

A second procedure for dealing with a trend is to use a **linear filter**, which converts one time series, $\{x_t\}$, into another, $\{y_t\}$, by the linear operation

$$y_t = \sum_{r=-q}^{+s} a_r x_{t+r},$$

where $\{a_r\}$ is a set of weights. In order to smooth out local fluctuations and estimate the local mean, we should clearly choose the weights so that $\sum a_r = 1$, and then the operation is often referred to as a **moving average**. Moving averages are discussed in detail by Kendall et al. (1983, Chapter 46), and we will only provide a brief introduction.

Moving averages are often symmetric with $s = q$ and $a_j = a_{-j}$. The simplest example of a symmetric smoothing filter is the simple moving average, for which $a_r = 1/(2q+1)$ for $r = -q, \ldots, +q$, and the smoothed value of $x_t$ is given by

$$\mathrm{Sm}(x_t) = \frac{1}{2q+1} \sum_{r=-q}^{+q} x_{t+r}. \tag{2.2}$$

The simple moving average is not generally recommended by itself for measuring trend, although it can be useful for removing seasonal variation. Another symmetric example is provided by the case where the $\{a_r\}$ are successive terms in the expansion of $(\frac{1}{2}+\frac{1}{2})^{2q}$. Thus when $q=1$, the weights are $a_{-1}=a_{+1}=\frac{1}{4}, a_0=\frac{1}{2}$. As $q$ gets large, the weights approximate a normal curve.

A third example is **Spencer's 15-point moving average**, which is used for smoothing mortality statistics to get life tables. This covers 15 consecutive points with $q=7$, and the symmetric weights are

$$\frac{1}{320}[-3,-6,-5,3,21,46,67,74,67,46,21,\ldots].$$

A fourth example, called the **Henderson moving average**, is described by Kenny and Durbin (1982) and is widely used, for example, in the X-11 and X-12 seasonal packages (see Section 2.6). This moving average aims to follow a cubic polynomial trend without distortion, and the choice of $q$ depends on the degree of irregularity. The symmetric nine-term moving average, for example, is given by

$$[-0.041,-0.010,0.119,0.267,0.330,\ldots].$$

The general idea is to fit a polynomial curve, not to the whole series, but to a local set of points. For example, a polynomial fitted to the first $(2q+1)$ data points can be used to determine the interpolated value at the middle of the range where $t=(q+1)$, and the procedure can then be repeated using the data from $t=2$ to $t=(2q+2)$, and so on.

To demonstrate the effect of these moving averages, we use the Beveridge wheat price annual index series from 1500 to 1869 as an example (see Figure 1.1). For comparison purpose, the top panel of Figure 2.1 shows the original Beveridge wheat price annual index series from 1500 to 1869, and the middle and bottom panels show the smoothed series by using the moving average (2.2) with $q=7$ and Spencer's 15-point moving average. Note that unequal weights in Spencer's 15-point moving average allow us to capture more local variation in the original series.

To reproduce Figure 2.1 and the two smoothed series in R, use the following command:

```
> smooth.sym<-function(my.ts, window.q){
    window.size<-2*window.q+1
    my.ts.sm<-rep(0, length(my.ts)-window.size)
    for (i in 1:length(my.ts.sm)) {
      my.ts.sm[i]<-mean(my.ts[i:(i+window.size-1)])
    }
    my.ts.sm
  }     # moving average with equal and symmetric weights
> smooth.spencer<-function(my.ts){
    weight<-c(-3,-6,-5,3,21,46,67,74,67,46,21,3,-5,-6,-3)/320
    my.ts.sm<-rep(0, length(my.ts)-15)
    for (i in 1:length(my.ts.sm)) {
       my.ts.sm[i]<-sum(my.ts[i:(i+14)]*weight)
    }
```

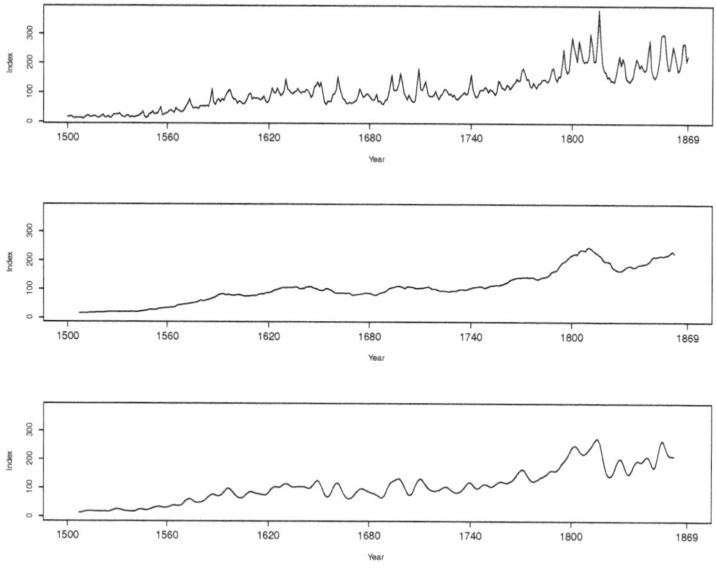


Figure 2.1 *The original and smoothed Beveridge wheat price annual index series from 1500 to 1869 (Top: The raw data; Middle: Smoothed series with filter (2.2) and $q = 7$; Bottom: Smoothed series with Spencer's 15-point moving average).*

```
  my.ts.sm
 }    # Spencer's 15-point moving average

> library(tseries)
> data(bev)
> bev.sm<-smooth.sym(bev, 7)
> bev.spencer<-smooth.spencer(bev)
> x.pos<-c(1500, 1560, 1620, 1680, 1740, 1800, 1869)
> par(mfrow=c(3,1), mar=c(4,4,4,4))
> plot(bev, type="l", xlab="Year", ylab="Index", xaxt="n")
> axis(1, x.pos, x.pos)
> plot(c(1, length(bev)), c(0, max(bev)), type="n", xlab="Year",
                 ylab="Index", xaxt="n")
> lines(seq(8, length(bev)-8), bev.sm)
> axis(1, x.pos-1500+1, x.pos)
> plot(c(1, length(bev)), c(0, max(bev)), type="n", xlab="Year",
                 ylab="Index", xaxt="n")
> lines(seq(8, length(bev)-8), bev.spencer)
> axis(1, x.pos-1500+1, x.pos)
```

Whenever a symmetric filter is chosen, there is likely to be an **end-effects** problem (e.g. Kendall et al., 1983, Section 46.11), since $\mathrm{Sm}(x_t)$ can only be calculated for $t = (q+1)$ to $t = N - q$. In some situations this may not be important, as, for example, in carrying out some retrospective analyses. However, in other situations, such as in forecasting, it is particularly important to get smoothed values right up to $t = N$. The analyst can project the smoothed values by eye or use an asymmetric filter that only involves present and past values of $x_t$. For example, the popular technique known as **exponential smoothing** (see Section 5.2.2) effectively assumes that

$$\mathrm{Sm}(x_t) = \sum_{j=0}^{\infty} \alpha(1-\alpha)^j x_{t-j},$$

where $\alpha$ is a constant such that $0 < \alpha < 1$. Here we note that the weights $a_j = \alpha(1-\alpha)^j$ decrease geometrically with $j$.

Having estimated the trend, we can look at the local fluctuations by examining

$$\begin{aligned} \mathrm{Res}(x_t) &= \text{residual from smoothed value} \\ &= x_t - \mathrm{Sm}(x_t) = \sum_{r=-q}^{+s} b_r x_{t+r}. \end{aligned}$$

This is also a linear filter with $b_0 = 1 - a_0$, and $b_r = -a_r$ for $r \neq 0$. If $\sum a_r = 1$, then $\sum b_r = 0$ and the filter is a trend remover.

How do we choose the appropriate filter? The answer to this question really requires considerable experience plus a knowledge of the frequency aspects of time-series analysis, which will be discussed in later chapters. As the name implies, filters are usually designed to produce an output with emphasis on variation at particular frequencies. For example, to get smoothed values we want to remove the local fluctuations that constitute what is called the high-frequency variation. In other words we want what is called a **low-pass** filter. To get $\mathrm{Res}(x_t)$, we want to remove the long-term fluctuations or the low-frequency variation. In other words we want what is called a **high-pass** filter. The **Slutsky** or **Slutsky–Yule effect** is related to this problem. Slutsky showed that by operating on a completely random series with both averaging and differencing procedures one could induce sinusoidal variation in the data. Slutsky went on to suggest that apparently periodic behaviour in some economic time series might be accounted for by the smoothing procedures used to form the data. We will return to this question later.

*Filters in series*

A smoothing procedure may be carried out in two or more stages.

As an example, two filters in series may be represented as in Figure 2.2. It is easy to show that a series of linear operations is still a linear filter overall. Suppose filter I, with weights $\{a_r\}$, acts on $\{x_t\}$ to produce $\{y_t\}$. Then filter II with weights $\{b_j\}$ acts on $\{y_t\}$ to produce $\{z_t\}$. Now

$$z_t = \sum_j b_j y_{t+j} = \sum_j b_j \sum_r a_r x_{t+j+r} = \sum_k c_k x_{t+k}$$

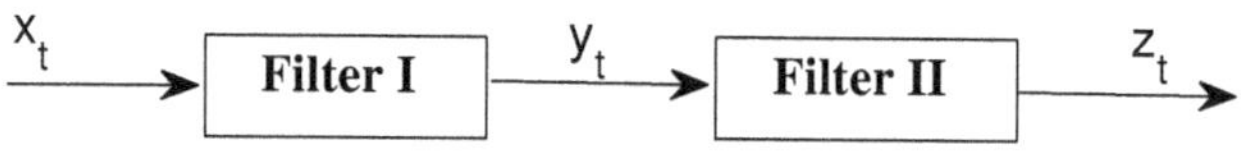


Figure 2.2 *Two filters in series.*

where

$$c_k = \sum_r a_r b_{(k-r)}$$

are the weights for the overall filter. The weights $\{c_k\}$ are obtained by a procedure called **convolution**, and we may write

$$\{c_k\} = \{a_r\} * \{b_j\},$$

where the symbol $*$ represents the convolution operator. For example, the filter $(\frac{1}{4}, \frac{1}{2}, \frac{1}{4})$ may be written as

$$\left(\frac{1}{4}, \frac{1}{2}, \frac{1}{4}\right) = \left(\frac{1}{2}, \frac{1}{2}\right) * \left(\frac{1}{2}, \frac{1}{2}\right).$$

The Spencer 15-point moving average is actually a convolution of four filters, namely

$$\left(\frac{1}{4}, \frac{1}{4}, \frac{1}{4}, \frac{1}{4}\right) * \left(\frac{1}{4}, \frac{1}{4}, \frac{1}{4}, \frac{1}{4}\right) * \left(\frac{1}{5}, \frac{1}{5}, \frac{1}{5}, \frac{1}{5}, \frac{1}{5}\right) * \left(-\frac{3}{4}, \frac{3}{4}, 1, \frac{3}{4}, -\frac{3}{4}\right),$$

and this may be the best way to compute it.

### 2.5.3 *Differencing*

A special type of filtering, which is particularly useful for removing a trend, is simply to difference a given time series until it becomes stationary. We will see that this method is an integral part of the so-called Box-Jenkins procedure. For non-seasonal data, first-order differencing is usually sufficient to attain apparent stationarity. Here a new series, say $\{y_2, \dots, y_N\}$, is formed from the original observed series, say $\{x_1, \dots, x_N\}$, by $y_t = x_t - x_{t-1} = \nabla x_t$ for $t = 2, 3, \dots, N$. Occasionally second-order differencing is required using the operator $\nabla^2$, where

$$\nabla^2 x_t = \nabla x_t - \nabla x_{t-1} = x_t - 2x_{t-1} + x_{t-2}.$$

First differencing is widely used and often works well. For example, Franses and Kleibergen (1996) show that better out-of-sample forecasts are usually obtained with economic data by using first differences rather than fitting a deterministic (or global) trend. *Seasonal differencing*, to remove seasonal variation, will be introduced in the next section.

### 2.5.4 Other approaches

Other more complicated approaches to handling trend will be introduced later in the book. In particular, several state-space models involving trend terms will be introduced in Chapter 10.

## 2.6 Analysing Series that Contain a Trend and Seasonal Variation

In Section 2.1 we introduced seasonal variation, which is generally annual in period, while Section 2.4 distinguished between additive seasonality, which is constant from year to year, and multiplicative seasonality. Three seasonal models in common use are

$$\begin{array}{ll} \text{A} & X_t = m_t + S_t + \varepsilon_t \\ \text{B} & X_t = m_t S_t + \varepsilon_t \\ \text{C} & X_t = m_t S_t \varepsilon_t \end{array}$$

where $m_t$ is the deseasonalized mean level at time $t, S_t$ is the seasonal effect at time $t$, and $\varepsilon_t$ is the random error.

Model A describes the **additive** case, while models B and C both involve **multiplicative** seasonality. In model C the error term is also multiplicative, and a logarithmic transformation will turn this into a (linear) additive model, which may be easier to handle. The time plot should be examined to see which model is likely to give the better description. The seasonal indices $\{S_t\}$ are usually assumed to change slowly through time so that $S_t \simeq S_{t-s}$, where $s$ is the number of observations per year. The indices are usually normalized so that they sum to zero in the additive case, or average to one in the multiplicative case. Difficulties arise in practice if the seasonal and/or error terms are not exactly multiplicative or additive. For example, the seasonal effect may increase with the mean level but not so quickly, so that it is somewhere 'in between' being multiplicative and additive. Durbin and Murphy (1975) describe a mixed additive-multiplicative seasonal model.

The analysis of time series which exhibit seasonal variation depends on whether one wants to (1) measure the seasonal effect and/or (2) eliminate seasonality. For series showing little trend, it is usually adequate to estimate the seasonal effect for a particular period (e.g. January) by finding the average of each January observation minus the corresponding yearly average in the additive case, or the January observation divided by the yearly average in the multiplicative case.

For series that do contain a substantial trend, a more sophisticated approach is required. With monthly data, the most common way of eliminating the seasonal effect is to calculate

$$\text{Sm}(x_t) = \frac{\frac{1}{2}x_{t-6} + x_{t-5} + x_{t-4} + \cdots + x_{t+5} + \frac{1}{2}x_{t+6}}{12}.$$

Note that the two end coefficients are different from the rest but that the coefficients sum to unity. A simple moving average cannot be used, as this would span 12 months and would not be centred on an integer value of $t$. A simple moving average over 13 months cannot be used, as this would give twice as much weight to the month appearing at both ends. For quarterly data, the seasonal effect can be eliminated by calculating

$$\text{Sm}(x_t) = \frac{\frac{1}{2}x_{t-2} + x_{t-1} + x_t + x_{t+1} + \frac{1}{2}x_{t+2}}{4}.$$

For 4-weekly data, one *can* use a simple moving average over 13 successive observations.

These smoothing procedures all effectively estimate the local (deseasonalized) level of the series. The seasonal effect itself can then be estimated by calculating $x_t - \text{Sm}(x_t)$ or $x_t/\text{Sm}(x_t)$ depending on whether the seasonal effect is thought to be additive or multiplicative. A check should be made that the seasonals are reasonably stable, and then the average monthly (or quarterly etc.) effects can be calculated.

As an example, Figure 2.3 decomposes the domestic monthly sales series of Australian fortified wine by winemakers into trend, additive, seasonal, and remainder series. The decomposition is carried out by using the R command `decompose`, which decomposes a time series into trend, (additive or multiplicative) seasonal, and remainder components using moving averages.

```
> wine<-read.csv("../data/aus_wine_sales.csv", header=F)
> wine.ts<-ts(wine[,2], frequency=4, start=c(1985,1))
```

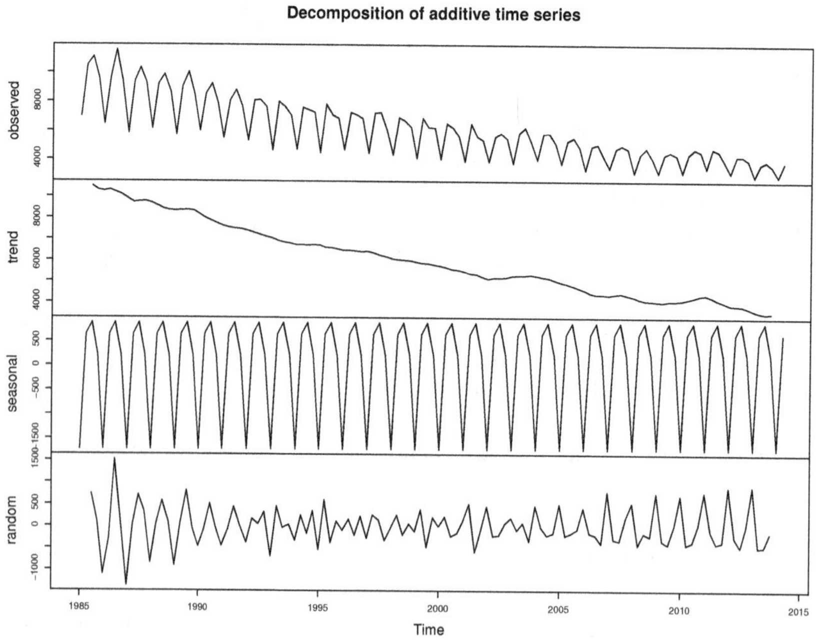


Figure 2.3 *The decomposition of the domestic sales of Australian fortified wine.*

```
        # create a time series object
> wine.de<-decompose(wine.ts, type="additive")
> plot(wine.de)
```

A seasonal effect can also be eliminated by a simple linear filter called **seasonal differencing**. For example, with monthly data one can employ the operator $\nabla_{12}$ where

$$\nabla_{12}x_t = x_t - x_{t-12}.$$

Further details on seasonal differencing will be given in Sections 4.8 and 5.3.

Two general reviews of methods for seasonal adjustment are Butter and Fase (1991) and Hylleberg (1992). Without going into great detail, we should mention the widely used **X-11 method**, now updated as the **X-12 method** (Findley et al., 1998), which is used for estimating or removing both trend and seasonal variation. It is a fairly complicated procedure that employs a series of linear filters and adopts a recursive approach. Preliminary estimates of trend are used to get preliminary estimates of seasonal variation, which in turn are used to get better estimates of trend, and so on. The new software for X-12 gives the user more flexibility in handling outliers as well as providing better diagnostics and an improved user interface. X-12 also allows the user to deal with the possible presence of **calendar effects**, which should always be considered when dealing with seasonal data (e.g. Bell and Hillmer, 1983). For example, if Easter falls in March one year and April the next, then this will alter the seasonal effect on sales for both months. The X-11 or X-12 packages are often combined with ARIMA modelling, as introduced in the next three chapters, to interpolate the values near the end of the series and avoid the end-effects problem arising from using symmetric linear filters alone. The package is called **X-12-ARIMA**. On mainland Europe, many governments use an alternative approach, based on packages called **SEATS** (Signal Extraction in ARIMA Time Series) and **TRAMO** (Time-Series Regression with ARIMA Noise). They are described in Gómez and Maravall (2001).

## 2.7 Autocorrelation and the Correlogram

An important guide to the properties of a time series is provided by a series of quantities called the **sample autocorrelation coefficients**. They measure the correlation, if any, between observations at different distances apart and provide useful descriptive information. In Chapter 4, we will see that they are also an important tool in model building and often provide valuable clues to a suitable probability model for a given set of data.

We assume that the reader is familiar with the ordinary correlation coefficient.[1] Given $N$ pairs of observations on two variables $x$ and $y$, say $\{(x_1, y_1),$

[1] This topic is briefly covered in Appendix C.

$(x_2,y_2),\ldots,(x_N,y_N)\}$, the sample correlation coefficient is given by

$$r = \frac{\sum_{i=1}^{N}(x_i-\bar{x})(y_i-\bar{y})}{\sqrt{\sum_{i=1}^{N}(x_i-\bar{x})^2\sum_{i=1}^{N}(y_i-\bar{y})^2}}. \tag{2.3}$$

This quantity lies in the range $[-1,1]$ and measures the strength of the linear association between the two variables. It can easily be shown that the value does not depend on the units in which the two variables are measured. The correlation is negative if 'high' values of $x$ tend to go with 'low' values of $y$. If the two variables are independent, then the true correlation is zero. We apply an analogous formula to time-series data to measure whether successive observations are correlated.

Given $N$ observations $x_1,\ldots,x_N$, on a time series, we can form $N-1$ pairs of observations, namely, $(x_1,x_2),(x_2,x_3),\ldots,(x_{N-1},x_N)$, where each pair of observations is separated by one time interval. Regarding the first observation in each pair as one variable and the second observation in each pair as a second variable, then, by analogy with Equation (2.3), we can measure the correlation coefficient between adjacent observations, $x_t$ and $x_{t+1}$, using the formula

$$r_1 = \frac{\sum_{t=1}^{N-1}(x_t-\bar{x}_{(1)})(x_{t+1}-\bar{x}_{(2)})}{\sqrt{\sum_{t=1}^{N-1}(x_t-\bar{x}_{(1)})^2\sum_{t=1}^{N-1}(x_{t+1}-\bar{x}_{(2)})^2}}, \tag{2.4}$$

where

$$\bar{x}_{(1)} = \sum_{t=1}^{N-1} x_t/(N-1)$$

is the mean of the first observation in each of the $(N-1)$ pairs and so is the mean of the first $N-1$ observations, while

$$\bar{x}_{(2)} = \sum_{t=2}^{N} x_t/(N-1)$$

is the mean of the last $N-1$ observations. As the coefficient given by Equation (2.4) measures correlation between successive observations, it is called the **sample autocorrelation coefficient** or a **serial correlation coefficient** at lag one.

Equation (2.4) is rather complicated, and so, as $\bar{x}_{(1)} \simeq \bar{x}_{(2)}$, it is usually approximated by

$$r_1 = \frac{\sum_{t=1}^{N-1}(x_t-\bar{x})(x_{t+1}-\bar{x})}{(N-1)\sum_{t=1}^{N}(x_t-\bar{x})^2/N} \tag{2.5}$$

where $\bar{x} = \sum_{t=1}^{N} x_t/N$ is the overall mean. It is often convenient to further simplify this expression by dropping the factor $N/(N-1)$, which is close to one for large $N$. This gives the even simpler formula

$$r_1 = \frac{\sum_{t=1}^{N-1} (x_t - \bar{x})(x_{t+1} - \bar{x})}{\sum_{t=1}^{N} (x_t - \bar{x})^2} \tag{2.6}$$

and this is the form that will be used in this book.

In a similar way, we can find the correlation between observations that are $k$ steps apart, and this is given by

$$r_k = \frac{\sum_{t=1}^{N-k} (x_t - \bar{x})(x_{t+k} - \bar{x})}{\sum_{t=1}^{N} (x_t - \bar{x})^2}, \quad k = 0, 1, 2, \ldots. \tag{2.7}$$

This is called the sample autocorrelation coefficient at lag $k$. The $r_k$-values are always in $[-1, 1]$. Note that $r_0 = 1$.

In practice the autocorrelation coefficients are usually calculated by computing the series of autocovariance coefficients, $\{c_k\}$, which we define by analogy with the usual covariance formula as

$$c_k = \frac{1}{N} \sum_{t=1}^{N-k} (x_t - \bar{x})(x_{t+k} - \bar{x}). \tag{2.8}$$

This is the sample autocovariance coefficient at lag $k$.

We then compute

$$r_k = c_k / c_0 \tag{2.9}$$

for $k = 0, 1, 2, \ldots, M$, where $M < N$.

Note that some authors prefer to use

$$c_k = \frac{1}{N-k} \sum_{t=1}^{N-k} (x_t - \bar{x})(x_{t+k} - \bar{x})$$

rather than Equation (2.8), and this gives Equation (2.5) when $k = 1$, but we prefer to use Equations (2.8) and (2.6) for reasons explained later in Section 4.1. We note, once again, that there are only small differences between the different formulae for large $N$.

### 2.7.1 *The correlogram*

A useful aid in interpreting a set of autocorrelation coefficients is a graph called a **correlogram** in which the sample autocorrelation coefficients $r_k$ are plotted against the lag $k$ for $k = 0, 1, \ldots, M$, where $M$ is usually much less than $N$. For example if

$N = 200$, then the analyst might look at the first 20 or 30 coefficients. Examples are given in Figures 2.4–2.8. A visual inspection of the correlogram is often very helpful. Of course, $r_0$ is always unity but is still worth plotting for comparative purposes. The correlogram may alternatively be called the sample autocorrelation function (ac.f.).

### 2.7.2 *Interpreting the correlogram*

Interpreting the meaning of a set of autocorrelation coefficients is not always easy. Here we offer some general advice.

*Random series*

A time series is said to be completely random (or i.i.d.) if it consists of a series of independent observations having the same distribution. Figure 2.4 plots a completely random series and its correlogram. Then, for large $N$, we expect to find that $r_k \simeq 0$ for all non-zero values of $k$. In fact we will see later that, for a random time series, $r_k$, $k \geq 1$, is approximately Normal$(0, 1/N)$. Thus, if a time series is random, we can expect 19 out of 20 of the values of $r_k$ to lie between $\pm 1.96/\sqrt{N}$. As a result, it is common practice to regard any values of $r_k$ outside these limits as being 'significant'. However, if one plots say the first 20 values of $r_k$, then one can expect to find one 'significant' value on average even when the time series really is random. This spotlights one of the difficulties in interpreting the correlogram, in that a large number of

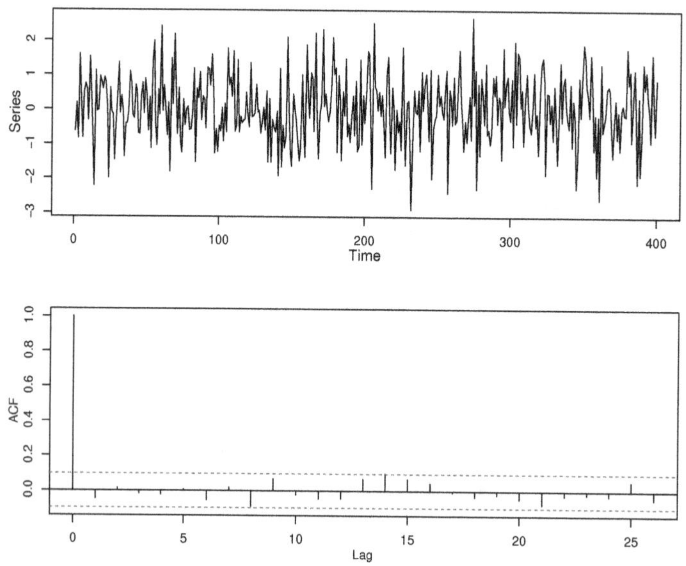


Figure 2.4 *A completely random series together with its correlogram. The dotted lines in the correlogram are at* $\pm 1.96/\sqrt{N}$. *Values outside these lines are said to be significantly different from zero.*

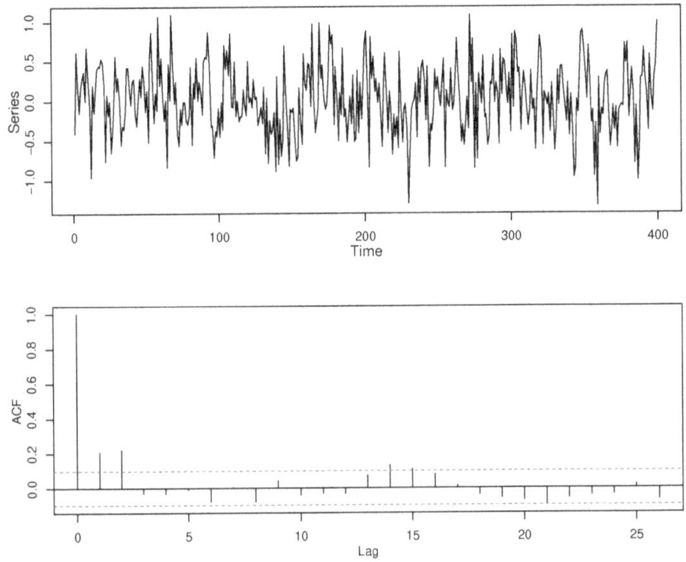


Figure 2.5 *A time series showing short-term correlation together with its correlogram.*

coefficients is quite likely to contain one (or more) 'unusual' results, even when no real effects are present. (See also Section 4.1.)

Figure 2.4 can be reproduced using the following R code.

```
> set.seed(1)
> x<-rnorm(400)
> par(mfrow=c(2,1), mar=c(3,4,3,4))
> plot(x, type="l", xlab="", ylab="")
> title(xlab="Time", ylab="Series", line=2, cex.lab=1.2)
> acf(x, ylab="", main="")
> title(xlab="Lag", ylab="ACF", line=2)
```

*Short-term correlation*

Stationary series often exhibit short-term correlation characterized by a fairly large value of $r_1$ followed by one or two further coefficients which, while greater than zero, tend to get successively smaller. Values of $r_k$ for longer lags tend to be approximately zero. An example of such a correlogram is shown in Figure 2.5. A time series that gives rise to such a correlogram is one for which an observation above the mean tends to be followed by one or more further observations above the mean, and similarly for observations below the mean.

*Alternating series*

If a time series has a tendency to alternate, with successive observations on different sides of the overall mean, then the correlogram also tends to alternate. With

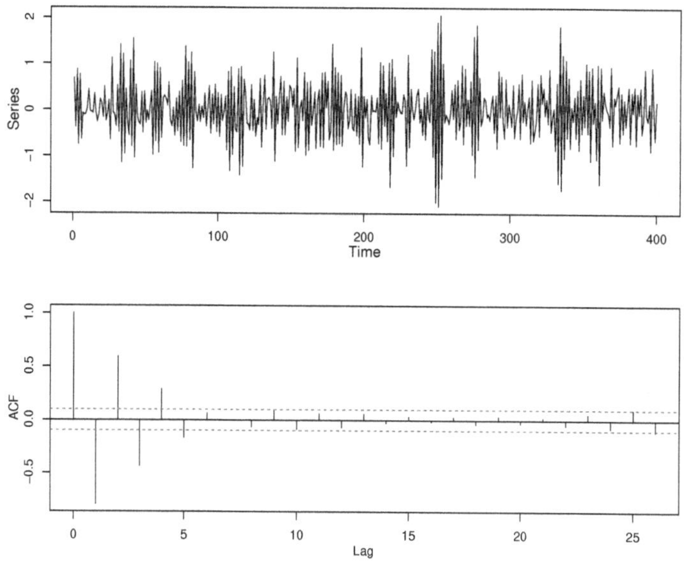


Figure 2.6 *An alternating time series together with its correlogram.*

successive values on opposite sides of the mean, the value of $r_1$ will naturally be negative, but the value of $r_2$ will be positive, as observations at lag 2 will tend to be on the same side of the mean. An alternating time series together with its correlogram is shown in Figure 2.6. The time series plots in Figures 2.5 and 2.6 also suggest that it is not necessarily easy to visually distinguish a series with short-term correlation from the one with alternating correlation.

*Non-stationary series*

If a time series contains a trend, then the values of $r_k$ will not come down to zero except for very large values of the lag. This is because an observation on one side of the overall mean tends to be followed by a large number of further observations on the same side of the mean because of the trend. A typical non-stationary time series together with its correlogram is shown in Figure 2.7. Little can be inferred from a correlogram of this type as the trend dominates all other features. In fact the sample ac.f. $\{r_k\}$ is only meaningful for data from a **stationary** time-series model (see Chapters 3 and 4) and so any trend should be removed before calculating $\{r_k\}$. Of course, if the trend itself is of prime interest, then it should be modelled, rather than removed, and then the correlogram is not helpful.

Figure 2.7 can be reproduced using the following `R` code.

```
> set.seed(1)
> ts.sim3<-cumsum(rnorm(400))
> par(mfrow=c(2,1), mar=c(3,4,3,4))
```

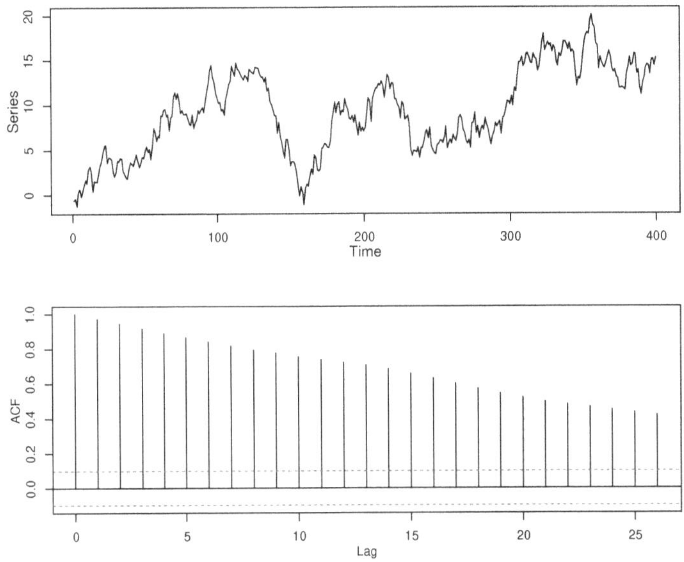


Figure 2.7 *A non-stationary time series together with its correlogram.*

```
> plot(ts.sim3, type="l", xlab="", ylab="")
> title(xlab="Time", ylab="Series", line=2, cex.lab=1.2)
> acf(ts.sim3, ylab="",main="")
> title(xlab="Lag", ylab="ACF", line=2)
```

*Seasonal series*

If a time series contains seasonal variation, then the correlogram will also exhibit oscillation at the same frequency. For example, with monthly observations, we can expect $r_6$ to be 'large' and negative, while $r_{12}$ will be 'large' and positive. In particular if $x_t$ follows a sinusoidal pattern, then so does $r_k$. For example, if

$$x_t = a\cos(\omega t)$$

where $a$ is a constant and the frequency $\omega$ is such that $0 < \omega < \pi$, then it can be shown (see Exercise 2.3) that

$$r_k \simeq \cos(\omega k) \qquad \text{for large } N.$$

The top panel of Figure 2.8 shows the correlogram of the monthly air temperature data shown in Figure 1.3. The sinusoidal pattern of the correlogram is clearly evident; but for seasonal data of this type the correlogram provides little extra information, as the seasonal pattern is usually displayed in the time plot of the data (e.g. Figure 1.3). Note that it is generally wise to look at coefficients covering at least three seasons. For example, coefficients up to lag 36 may be useful for monthly data.

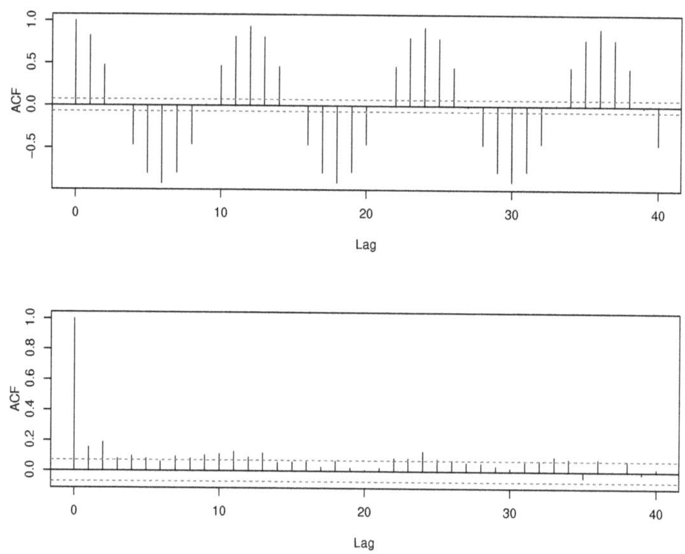


Figure 2.8 *The correlograms of monthly observations on air temperature in Anchorage, Alaska for the raw data (top) and for the seasonally adjusted data (bottom).*

If the seasonal variation is removed from seasonal data, then the correlogram may provide useful information. The seasonal variation was removed from the air temperature data by the simple, but rather crude, procedure of calculating the 12 monthly averages and subtracting the appropriate one from each individual observation. The correlogram of the resulting series (The bottom panel of Figure 2.8) shows that the first three coefficients are significantly different from zero. This indicates short-term correlation in that a month, which is say colder than the average for that month, will tend to be followed by one or two further months that are colder than average.

*Outliers*

If a time series contains one or more outliers, the correlogram may be seriously affected and it may be advisable to adjust outliers in some way before starting the formal analysis. For example, if there is one outlier in the time series at, say, time $t_0$, and if it is not adjusted, then the plot of $x_t$ against $x_{t+k}$ will contain **two** 'extreme' points, namely, $(x_{t_0-k}, x_{t_0})$ and $(x_{t_0}, x_{t_0+k})$. The effect of these two points will be to depress the sample correlation coefficients towards zero. If there are two outliers, this effect becomes even more noticeable, except when the lag equals the distance between the outliers, in which case a spuriously large correlation may occur.

*General remarks*

Considerable experience is required to interpret sample autocorrelation coefficients. In addition, it is necessary to study the probability theory of stationary series and learn about the classes of models that may be appropriate. It is also necessary to investigate the sampling properties of $r_k$. These topics will be covered in the next two chapters and we will then be in a better position to interpret the correlogram of a given time series.

## 2.8 Other Tests of Randomness

In most cases, a visual examination of the graph of a time series is enough to see that the series is **not** random, as, for example, if trend or seasonality is present or there is obvious short-term correlation. However, it is occasionally desirable to assess whether an apparently stationary time series is 'random'. One type of approach is to carry out what is called a **test of randomness** in which one tests whether the observations $x_1, \ldots, x_N$ could have arisen in that order by chance by taking a simple random sample size $N$ from a population assumed to be stationary but with unknown characteristics. Various tests exist for this purpose as described, for example, by Kendall et al. (1983, Section 45.15) and by Kendall and Ord (1990, Chapter 2). It is convenient to briefly mention such tests here.

One type of test is based on counting the number of *turning points*, meaning the number of times there is a local maximum or minimum in the time series. A local maximum is defined to be any observation $x_t$ such that $x_t > x_{t-1}$ and also $x_t > x_{t+1}$. A converse definition applies to local minima. If the series really is random, one can work out the expected number of turning points and compare it with the observed value. An alternative type of test is based on *runs* of observations. For example, the analyst can count the number of runs where successive observations are all greater than the median or all less than the median. This may show short-term correlation. Alternatively, the analyst can count the number of runs where successive observations are (monotonically) increasing or are (monotonically) decreasing. This may show up trend. Under the null hypothesis of randomness, the expected number of such runs can be found and compared with the observed value, giving tests that are non-parametric or distribution-free in character.

Tests of the above types will not be described here, as we have generally found it more convenient to simply examine the correlogram (and possibly the spectral density function) of a given time series to see whether it is random. This can often be done visually, but if a test is required, then the so-called **portmanteau test** can be used (see Section 4.7). The latter test can also be used when assessing models by means of a residual analysis, where the residual of an observation is the difference between the observation and its fitted value from the model. Thus, having fitted a model to a non-random series, one wants to see if the residuals are random, as they should be if the correct model has been fitted. Testing residuals for randomness will be discussed in Section 4.9.

## 2.9 Handling Real Data

We close this chapter with some important comments on how to handle real data. Analysts generally like to think they have 'good' data, meaning that the data have been carefully collected with no outliers or missing values. In reality, this does not always happen, so that an important part of the initial examination of the data is to assess the quality of the data and consider modifying them, if necessary. An even more basic question is whether the most appropriate variables have been measured in the first place, and whether they have been measured to an appropriate accuracy. Assessing the structure and format of the data is a key step. Practitioners will tell you that these types of questions often take longer to sort out than might be expected, especially when data come from a variety of sources. It really is important to avoid being driven to bad conclusions by bad data.

The process of checking through data is often called *cleaning* the data, or *data editing*. It is an essential precursor to attempts at modelling data. Data cleaning could include **modifying outliers**, identifying and correcting obvious **errors** and filling in (or *imputing*) any **missing observations**. This can sometimes be done using fairly crude devices, such as downweighting outliers to the next most extreme value or replacing missing values with an appropriate mean value. However, more sophisticated methods may be needed, requiring a deeper understanding of time-series models. The analyst should also deal with any other known peculiarities, such as a change in the way a variable is defined during the course of the data-collection process. Data cleaning often arises naturally during a simple preliminary descriptive analysis. In particular, in time-series analysis, the construction of a time plot for each variable is the most important tool for revealing any oddities such as outliers and discontinuities.

After cleaning the data, the next step for the time-series analyst is to determine whether trend and seasonality are present. If so, how should such effects be modelled, measured, or removed? In our experience, the treatment of such effects, together with the treatment of outliers and missing values, is often *more* important than any subsequent choices as regards analysing and modelling time-series data.

The *context* of the problem is crucial in deciding how to modify data, if at all, and how to handle trend and seasonality. This explains why it is essential to get background knowledge about the problem, and in particular to clarify the study objectives. A corollary is that it is difficult to make any general remarks or give general recommendations on data cleaning. It is essential to combine statistical theory with sound common sense and knowledge of the particular problem being tackled.

We close by giving the following checklist of possible actions, while noting that the list is not exhaustive and needs to be adapted to the particular problem under consideration.

- Do you understand the context? Have the 'right' variables been measured?
- Have all the time series been plotted?
- Are there any missing values? If so, what should be done about them?
- Are there any outliers? If so, what should be done about them?
- Are there any obvious discontinuities in the data? If so, what does this mean?
- Does it make sense to transform any of the variables?
- Is trend present? If so, what should be done about it?
- Is seasonality present? If so, what should be done about it?

## Exercises

**2.1** The following data show the coded sales of company X in successive 4-week periods over 1995–1998.

| | I | II | III | IV | V | VI | VII | VIII | IX | X | XI | XII | XIII |
|---|---|---|---|---|---|---|---|---|---|---|---|---|---|
| 1995 | 153 | 189 | 221 | 215 | 302 | 223 | 201 | 173 | 121 | 106 | 86 | 87 | 108 |
| 1996 | 133 | 177 | 241 | 228 | 283 | 255 | 238 | 164 | 128 | 108 | 87 | 74 | 95 |
| 1997 | 145 | 200 | 187 | 201 | 292 | 220 | 233 | 172 | 119 | 81 | 65 | 76 | 74 |
| 1998 | 111 | 170 | 243 | 178 | 248 | 202 | 163 | 139 | 120 | 96 | 95 | 53 | 94 |

(a) Plot the data.

(b) Assess the trend and seasonal effects.

**2.2** Sixteen successive observations on a stationary time series are as follows: 1.6, 0.8, 1.2, 0.5, 0.9, 1.1, 1.1, 0.6, 1.5, 0.8, 0.9, 1.2, 0.5, 1.3, 0.8, 1.2

(a) Plot the observations.

(b) Looking at the graph plotted in (a), guess an approximate value for the autocorrelation coefficient at lag 1.

(c) Plot $x_t$ against $x_{t+1}$, and again try to guess the value of $r_1$.

(d) Calculate $r_1$.

**2.3** If $x_t = a\cos(\omega t)$ where $a$ is a constant and $\omega$ is a constant in $(0, \pi)$, show that $r_k \to \cos(\omega k)$ as $N \to \infty$.
(Hint: You will need to use the trigonometrical results listed in Section 7.2. Using Equation (7.2) it can be shown that $\bar{x} \to 0$ as $N \to \infty$, so that $r_k \to \sum\cos(\omega t)\cos[\omega(t+k)]/\sum\cos^2(\omega t)$. Now use the result that $2\cos A\cos B = \cos(A+B)+\cos(A-B)$ together with the result that $\sum\cos^2(\omega t) = N/2$ for a suitably chosen $N$.)

**2.4** A computer generates a series of 400 observations that are supposed to be random. The first 10 sample autocorrelation coefficients of the series are $r_1 = 0.02$, $r_2 = 0.05$, $r_3 = -0.09$, $r_4 = 0.08$, $r_5 = -0.02$, $r_6 = 0.00$, $r_7 = 0.12$, $r_8 = 0.06$, $r_9 = 0.02$, $r_{10} = -0.08$. Is there any evidence of non-randomness?

**2.5** Suppose we have a seasonal series of monthly observations $\{X_t\}$, for which the seasonal factor at time $t$ is denoted by $\{S_t\}$. Further suppose that the seasonal pattern is constant through time so that $S_t = S_{t-12}$ for all $t$. Denote a stationary series of random deviations by $\{\varepsilon_t\}$.

(a) Consider the model $X_t = a + bt + S_t + \varepsilon_t$ having a global linear trend and additive seasonality. Show that the seasonally differenced series $\nabla_{12}X_t := X_t - X_{t-12}$ is stationary.

(b) Consider the model $X_t = (a+bt)S_t + \varepsilon_t$ having a global linear trend and multiplicative seasonality. Does the operator $\nabla_{12}$ transform $X_t$ to stationarity? If not, find a differencing operator that does.

(Note: As stationarity is not formally defined until Chapter 3, you should use heuristic arguments. A stationary process may involve a constant mean value (that could be zero) plus any linear combination of the stationary series $\{\varepsilon_t\}$, but should not include terms such as trend and seasonality.)

**2.6** Consider the process $X_t = t^3 + (1+t+t^2)S_t + \varepsilon_t$, in which $S_t$ is a seasonal component with period $d$ (i.e. $S_{t-d} = S_t$) and $\{\varepsilon_t\}$ is a stationary series of random devations.

(a) Compute the seasonally differenced series $\nabla_d X_t = X_t - X_{t-d}$.

(b) Define the following operator $\nabla_d^r$ for $X_t$ ($d > 1$, $r \geq 2$, $d$ and $r$ are positive integers),

$$\nabla_d^r X_t = \nabla(\nabla_d^{r-1} X_t).$$

For example, when $d = 12$ and $r = 2$, we have

$$\begin{aligned} \nabla_{12}^2 X_t &= \nabla_{12}(\nabla_{12} X_t) = \nabla_{12}(X_t - X_{t-12}) \\ &= (X_t - X_{t-12}) - (X_{t-12} - X_{t-24}) \\ &= X_t - 2X_{t-12} + X_{t-24}. \end{aligned}$$

What is the minimum value of $r$ so that $\nabla_d^r X_t$ does not contain trend or seasonal components?

Chapter 3

# Some Linear Time Series Models

This chapter introduces various probability models for time series. Some tools for describing the properties of such models are specified and the important notion of stationarity is formally defined.

## 3.1 Stochastic Processes and Their Properties

We concentrate on various types of time-series models that collectively come under the general title of 'stochastic processes'. Most physical processes in the real world involve a random element in their structure and a **stochastic process** can be described as 'a statistical phenomenon that evolves in time according to probabilistic laws'. Examples include the length of a queue, the number of accidents in a particular town in successive months and the air temperature at a particular site on successive days. The word 'stochastic', which is of Greek origin, is used to mean 'pertaining to chance', and many writers use 'random process' as a synonym for stochastic process.

Mathematically, a stochastic process may be defined as a collection of random variables that are ordered in time and defined at a set of time points, which may be continuous or discrete. We will denote the random variable at time $t$ by $X(t)$ if time is continuous (usually $-\infty < t < \infty$), and by $X_t$ if time is discrete (usually $t = 0, \pm 1, \pm 2, \ldots$). Some stochastic processes, such as the size of a bacterial colony, are studied using specialist tools, and we concentrate here on models suitable for time-series variables measured at equal intervals of time.

The theory of stochastic processes has been extensively developed and is discussed in many books including (Cox and Miller, 1968, Chapter 7), Grimmett and Stirzaker (2001), Papoulis (1984), written primarily for engineers, and Ross (1997). In this chapter we concentrate on those aspects particularly relevant to time-series analysis.

Most statistical problems are concerned with estimating the properties of a population from a sample. The properties of the latter are typically determined by the investigator, including the sample size and whether randomness is incorporated into the selection procedure. In time-series analysis, there is a rather different situation in that the order of observations is determined by time and it is usually impossible to make more than one observation at any given time. Thus, although it may be possible to increase the sample size by varying the *length* of the observed time series, there will only be a single outcome of the process and a single observation on the random

DOI: 10.1201/9781003646112-3

variable at time $t$. Nevertheless we may regard the observed time series as just one example of the infinite set of time series that might have been observed. This infinite set of time series is sometimes called the **ensemble**. Every member of the ensemble is a possible **realization** of the stochastic process. The observed time series can be thought of as one particular realization and will be denoted by $x(t)$ for $(0 \leq t \leq T)$ if time is continuous, and by $x_t$ for $t = 1, \ldots, N$ if time is discrete. Time-series analysis is essentially concerned with evaluating the properties of the underlying probability model from this observed time series, even though this single realization is the only one we will ever observe.

Many models for stochastic processes are expressed by means of an algebraic formula relating the random variable at time $t$ to past values of the process, together with values of an unobservable 'error' process. From this model, it may be possible to specify the joint probability distribution of $X(t_1), \ldots, X(t_k)$ for any set of times $t_1, \ldots, t_k$ and any value of $k$. However, this is rather complicated and is not usually attempted in practice. A simpler, more useful way of describing a stochastic process is to give the *moments* of the process, particularly the first and second moments that are called the **mean** and **autocovariance function** (acv.f.), respectively. The **variance function** is a special case of the acv.f. These functions will now be defined for continuous time, with similar definitions applying in discrete time.

*Mean.* The mean function $\mu(t)$ is defined for all $t$ by

$$\mu(t) = E[X(t)].$$

*Variance.* The variance function $\sigma^2(t)$ is defined for all $t$ by

$$\sigma^2(t) = \text{Var}[X(t)] = E[(X(t) - \mu(t))^2].$$

*Autocovariance.* The variance function alone is not enough to specify the second moments of a sequence of random variables. More generally, we define the acv.f. $\gamma(t_1, t_2)$ to be the covariance[1] of $X(t_1)$ with $X(t_2)$, namely

$$\gamma(t_1, t_2) = E\{[X(t_1) - \mu(t_1)][X(t_2) - \mu(t_2)]\}.$$

Clearly, the variance function is a special case of the acv.f. when $t_1 = t_2$.

*Higher moments* of a stochastic process may be defined in an obvious way, but are rarely used in practice.

## 3.2 Stationary Processes

An important class of stochastic processes are those that are stationary. A heuristic idea of stationarity was introduced in Section 2.2.

A time series is said to be **strictly stationary** if the joint distribution of $X(t_1), \ldots, X(t_k)$ is the same as the joint distribution of $X(t_1 + \tau), \ldots, X(t_k + \tau)$ for

[1]Readers who are unfamiliar with the term 'covariance' should read Appendix C. When applied to a sequence of random variables, it is called an autocovariance.

all $t_1, \ldots, t_k, \tau$. In other words, shifting the time origin by an amount $\tau$ has no effect on the joint distributions, which must therefore depend only on the intervals between $t_1, t_2, \ldots, t_k$. The above definition holds for any value of $k$. In particular, if $k = 1$, strict stationarity implies that the distribution of $X(t)$ is the same for all $t$, so that, provided the first two moments are finite, we have

$$\mu(t) = \mu, \qquad \sigma^2(t) = \sigma^2$$

are both constants, which do not depend on the value of $t$.

Furthermore, if $k = 2$ the joint distribution of $X(t_1)$ and $X(t_2)$ depends only on the time difference $(t_2 - t_1) = \tau$, which is called the **lag**. Thus the acv.f. $\gamma(t_1, t_2)$ also depends only on $(t_2 - t_1)$ and may be written as $\gamma(\tau)$, where

$$\gamma(\tau) = E\{[X(t) - \mu][X(t+\tau) - \mu]\} = \text{Cov}[X(t), X(t+\tau)]$$

is called the autocovariance coefficient at lag $\tau$.

The size of an autocovariance coefficient depends on the units in which $X(t)$ is measured. Thus, for interpretative purposes, it is helpful to standardize the acv.f. to produce a function called the **autocorrelation function** (ac.f.), which is defined by

$$\rho(\tau) = \gamma(\tau)/\gamma(0).$$

This quantity measures the correlation between $X(t)$ and $X(t+\tau)$. Its empirical counterpart was introduced in Section 2.7. Note that the argument $\tau$ of $\gamma(\tau)$ and $\rho(\tau)$ is discrete if time is discrete, but continuous if time is continuous. We typically use $\gamma(k)$ and $\rho(k)$ to denote these functions in the discrete-time case.

At first sight it may seem surprising to suggest that there are processes for which the distribution of $X(t)$ should be the same for all $t$. However, readers with some knowledge of stochastic processes will know that there are many processes $\{X(t)\}$, which have what is called an **equilibrium** distribution as $t \to \infty$, whereby the probability distribution of $X(t)$ tends to a limit that does *not* depend on the initial conditions. Thus once such a process has been running for some time, the distribution of $X(t)$ will change very little. Indeed if the initial conditions are specified to be identical to the equilibrium distribution, the process is stationary in time and the equilibrium distribution is then the stationary distribution of the process. Of course the **conditional** distribution of $X(t_2)$ given that $X(t_1)$ has taken a particular value, say $x(t_1)$, may be quite different from the stationary distribution, but this is perfectly consistent with the process being stationary.

In practice it is often useful to define stationarity in a less restricted way than that described above. A process is called **second-order stationary** (or **weakly stationary**) if its mean is constant and its acv.f. depends only on the lag, so that

$$E[X(t)] = \mu$$

and

$$\text{Cov}[X(t), X(t+\tau)] = \gamma(\tau).$$

No requirements are placed on moments higher than second order. By letting $\tau = 0$, we note that the form of a stationary acv.f. implies that the variance, as well as the mean, is constant. The definition also implies that both the variance and the mean must be finite.

This weaker definition of stationarity will generally be used from now on, as many of the properties of stationary processes depend only on the structure of the process as specified by its first and second moments. One important class of processes where this is particularly true is the class of **normal** processes where the joint distribution of $X(t_1),\ldots,X(t_k)$ is multivariate normal for all $t_1,\ldots,t_k$. The multivariate normal distribution is completely characterized by its first and second moments, and hence by $\mu(t)$ and $\gamma(t_1,t_2)$, and so it follows that second-order stationarity implies strict stationarity for normal processes. However, $\mu$ and $\gamma(\tau)$ may not adequately describe stationary processes, which are very 'non-normal'.

## 3.3 Properties of the Autocorrelation Function

We have already noted in Section 2.7 that the sample autocorrelation coefficients of an observed time series are an important set of statistics for describing the time series. Similarly the (theoretical) ac.f. of a stationary stochastic process is an important tool for assessing its properties. This section investigates the general properties of the ac.f.

Suppose a stationary stochastic process $X(t)$ has mean $\mu$, variance $\sigma^2$, acv.f. $\gamma(\tau)$ and ac.f. $\rho(\tau)$. Then

$$\rho(\tau) = \gamma(\tau)/\gamma(0) = \gamma(\tau)/\sigma^2.$$

Note that $\rho(0) = 1$.

*Property 1*: The ac.f. is an even function of lag, so that $\rho(\tau) = \rho(-\tau)$.

This property simply says that the correlation between $X(t)$ and $X(t+\tau)$ is the same as that between $X(t)$ and $X(t-\tau)$. The result is easily proved using $\gamma(\tau) = \rho(\tau)\sigma^2$ by

$$\begin{aligned} \gamma(\tau) &= \text{Cov}[X(t), X(t+\tau)] \\ &= \text{Cov}[X(t-\tau), X(t)] \qquad \text{since } X(t) \text{ is stationary.} \\ &= \gamma(-\tau) \end{aligned}$$

*Property 2*: $|\rho(\tau)| \leq 1$.

This is the 'usual' property of correlation, namely, that it lies between $\pm 1$. It is proved by noting that

$$\text{Var}[\lambda_1 X(t) + \lambda_2 X(t+\tau)] \geq 0$$

for any constants $\lambda_1, \lambda_2$, since a variance is always non-negative. The variance is equal to $\lambda_1^2 \text{Var}[X(t)] + \lambda_2^2 \text{Var}[X(t+\tau)] + 2\lambda_1\lambda_2 \text{Cov}[X(t), X(t+\tau)]$

$$= (\lambda_1^2 + \lambda_2^2)\sigma^2 + 2\lambda_1\lambda_2\gamma(\tau)$$

When $\lambda_1 = \lambda_2 = 1$, we find $\gamma(\tau) \geq -\sigma^2$, so that $\rho(\tau) \geq -1$.
When $\lambda_1 = 1, \lambda_2 = -1$, we find $\sigma^2 \geq \gamma(\tau)$, so that $\rho(\tau) \leq +1$.

Thus $|\rho(\tau)| \leq 1$ as required.

The standardized nature of a correlation coefficient means that the value of $\rho(\tau)$ does not depend on the units in which the time series is measured, as can readily be demonstrated by multiplying all values in a series by the same constant and showing that the resulting autocorrelations are unchanged.

*Property 3*: The ac.f. does not uniquely identify the underlying model.

Although a given stochastic process has a unique covariance structure, the converse is not in general true. It is usually possible to find many normal and non-normal processes with the same ac.f. and this creates further difficulty in interpreting sample ac.f.s. Jenkins and Watts (1968, p. 170) give an example of two different stochastic processes, which have exactly the same ac.f. Even for stationary normal processes, which are completely determined by the mean, variance and ac.f., we will see in Section 3.6 that a requirement, called the invertibility condition, is needed to ensure uniqueness.

## 3.4 Purely Random Processes

A discrete-time process is called a purely random process if it consists of a sequence of random variables, $\{Z_t\}$, which are mutually *independent and identically distributed* (i.i.d.). We normally further assume that the random variables are normally distributed with mean zero and variance $\sigma_Z^2$. From the definition it follows that the process has constant mean and variance. Moreover, the independence assumption means that

$$\gamma(k) = \text{Cov}(Z_t, Z_{t+k}) = \begin{cases} \sigma_Z^2 & k = 0 \\ 0 & k = \pm 1, \pm 2, \ldots. \end{cases} \tag{3.1}$$

This means that different values are uncorrelated so that the ac.f. is given by

$$\rho(k) = \begin{cases} 1 & k = 0 \\ 0 & k = \pm 1, \pm 2, \ldots. \end{cases} \tag{3.2}$$

As the mean and acv.f. do not depend on time, the process is second-order stationary. In fact the independence assumption implies that the process is also strictly stationary.

Purely random processes are useful in many situations, particularly as building blocks for more complicated processes such as moving average processes – see Section 3.6. In practice, if all sample ac.f.'s of a series are close to zero, then the series is considered as a realization of a purely random process. Figure 3.1 shows an example of a purely random process, $Z_t \sim N(0,1)$, $1 \leq t \leq 500$, and its correlogram, which can be reproduced by the following command in R:

```
> z<-rnorm(500, 0, 1)
> par(mfrow=c(2,1), mar=c(3,4,3,4))
> plot(z, type="l", xlab="Time", ylab="Series")
> acf(z, xlab="Lag",ylab="ACF", main="")
```

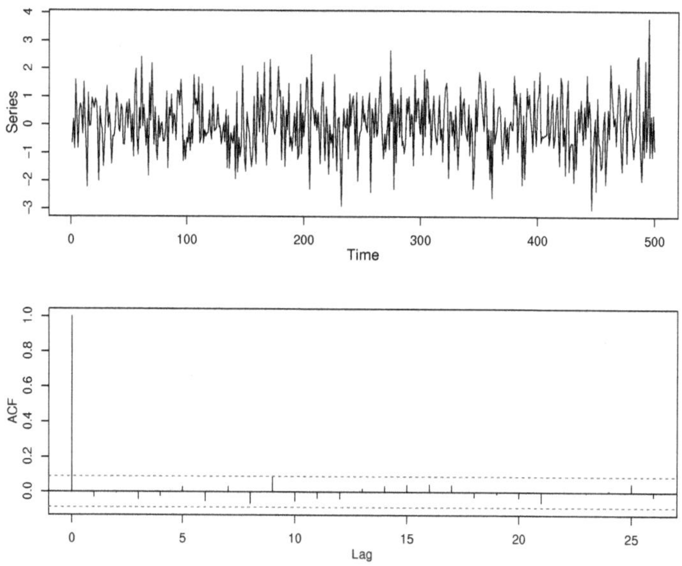


Figure 3.1 *A purely random process with $\sigma_Z^2 = 1$ (top) and its correlogram (bottom).*

Some authors prefer to make the weaker assumption that the $Z_t'$s are mutually uncorrelated, rather than independent. This is adequate for linear, normal processes, but the stronger independence assumption is needed when considering non-linear models (see Chapter 11). Note that a purely random process is sometimes called **white noise**, particularly by engineers.

The possibility of defining a continuous-time purely random process is discussed in Section 3.12.

## 3.5 Random Walks

Suppose that $\{Z_t\}$ is a discrete-time, purely random process with mean $\mu$ and variance $\sigma_Z^2$. A process $\{X_t\}$ is said to be a random walk if

$$X_t = X_{t-1} + Z_t. \tag{3.3}$$

The process is customarily started at zero when $t = 0$, so that

$$X_1 = Z_1$$

and

$$X_t = \sum_{i=1}^{t} Z_i.$$

Then we find that $E(X_t) = t\mu$ and that $\text{Var}(X_t) = t\sigma_Z^2$ since the $Z_i$'s are independent. As the mean and variance change with $t$, the process is non-stationary.

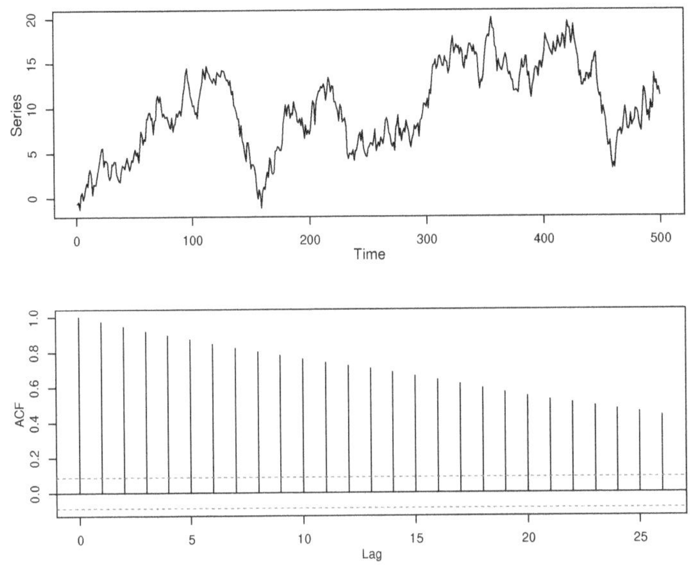


Figure 3.2 *Simulated random walk (top) and its correlogram (bottom). The random walk series is generated from the white noise series in Figure 3.1.*

However, it is interesting to note that the first differences of a random walk, given by

$$\nabla X_t = X_t - X_{t-1} = Z_t$$

form a purely random process, which is therefore stationary. This feature can be used to construct a random walk. For instance, Figure 3.2 shows a random walk series and its correlogram. The random walk series is generated by cumulative sums of a white noise series via the following command in R.

```
> n<-500
> z<-rnorm(n, 0, 1)
> x.rw<-cumsum(z)
> par(mfrow=c(2,1), mar=c(4,4,4,4))
> plot(x.rw, type="l", xlab="Time", ylab="Series")
> acf(x.rw, xlab="Lag", ylab="ACF", main="")
```

The best-known examples of time series, which behave like random walks, are share prices on successive days. A model, which often gives a good approximation to such data, is

$$\text{share price on day } t = \text{share price on day } (t-1) + \text{random error.}$$

## 3.6 Moving Average Processes

Suppose that $\{Z_t\}$ is a purely random process with mean zero and variance $\sigma_Z^2$. Then a process $\{X_t\}$ is said to be a moving average process of order $q$ (abbreviated to a MA$(q)$ process) if

$$X_t = \beta_0 Z_t + \beta_1 Z_{t-1} + \cdots + \beta_q Z_{t-q}, \tag{3.4}$$

where $\{\beta_i\}$ are constants. The $Z$s are usually scaled so that $\beta_0 = 1$.

### *3.6.1 Stationarity and autocorrelation function of a MA process*

We find immediately that

$$\begin{aligned} E(X_t) &= 0, \\ \text{Var}(X_t) &= \sigma_Z^2 \sum_{i=0}^{q} \beta_i^2, \end{aligned}$$

since the $Z$s are independent. We also have

$$\begin{aligned} \gamma(k) &= \text{Cov}(X_t, X_{t+k}) \\ &= \text{Cov}(\beta_0 Z_t + \cdots + \beta_q Z_{t-q}, \beta_0 Z_{t+k} + \cdots + \beta_q Z_{t+k-q}) \\ &= \begin{cases} 0 & k > q \\ \sigma_Z^2 \sum_{i=0}^{q-k} \beta_i \beta_{i+k} & k = 0, 1, \ldots, q \\ \gamma(-k) & k < 0 \end{cases} \end{aligned} \tag{3.5}$$

since

$$\text{Cov}(Z_s, Z_t) = \begin{cases} \sigma_Z^2 & s = t \\ 0 & s \neq t. \end{cases}$$

As $\gamma(k)$ does not depend on $t$, and the mean is constant, the process is second-order stationary for all values of the $\{\beta_i\}$. Furthermore, if the $Z$s are normally distributed, then so are the $X$s, and we have a strictly stationary normal process.

The ac.f. of the above MA$(q)$ process is given by

$$\rho(k) = \begin{cases} 1 & k = 0 \\ \sum_{i=0}^{q-k} \beta_i \beta_{i+k} \Big/ \sum_{i=0}^{q} \beta_i^2 & k = 1, \ldots, q \\ 0 & k > q \\ \rho(-k) & k < 0. \end{cases} \tag{3.6}$$

Note that the ac.f. 'cuts off' at lag $q$, which is a special feature of MA processes. In particular, the MA(1) process with $\beta_0 = 1$ has an ac.f. given by

$$\rho(k) = \begin{cases} 1 & k = 0 \\ \beta_1/(1+\beta_1^2) & k = \pm 1 \\ 0 & \text{otherwise.} \end{cases}$$

Using the definition of the MA($q$) process, one can construct a MA series. For example, we can simulate a white noise series, $Z_t \sim N(0,1)$, and then construct a MA(1) process, $X_t = Z_t - 0.8Z_{t-1}$, and a MA(2) process, $X_t = Z_t + 0.7Z_{t-1} - 0.2Z_{t-2}$ using $Z_t$. The resulting series and their correlograms are shown in Figure 3.3, which can be reproduced by the following command in R. Note that the $r_k$'s are significant only for $k = 0$ and 1, and $k = 0, 1, 2$, respectively.

```
> n<-500
> z<-rnorm(n)
> x.ma1<-z[2:n]-0.8*z[1:(n-1)]
> x.ma2<-z[3:n]+0.7*z[2:(n-1)]-0.2*z[1:(n-2)]
```

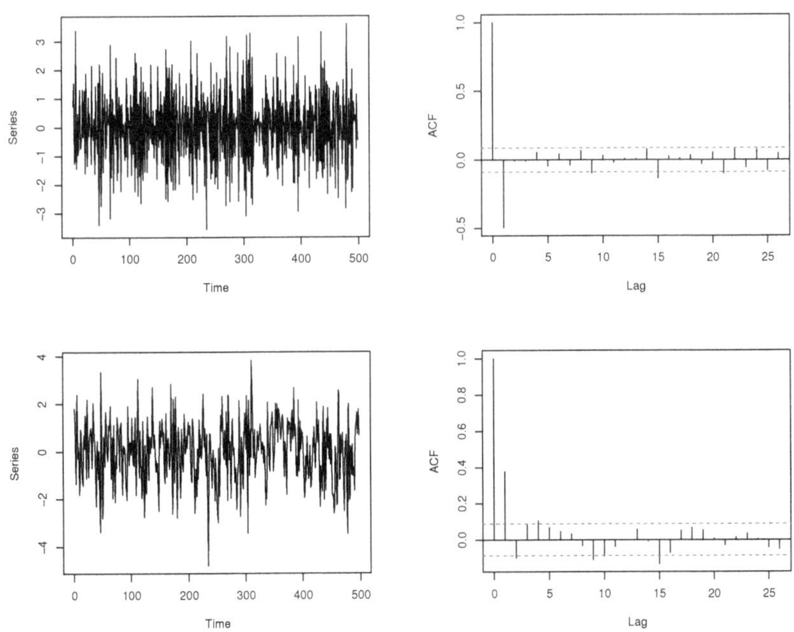


Figure 3.3 *Simulated MA(1) (top left) and MA(2) (bottom left) processes and their corresponding correlograms (top right and bottom right).*

```
> par(mfrow=c(2,2), mar=c(4,4,4,4))
> plot(x.ma1, type="l", xlab="Time", ylab="Series")
> acf(x.ma1, xlab="Lag", ylab="ACF", main="")
> plot(x.ma2, type="l", xlab="Time", ylab="Series")
> acf(x.ma2, xlab="Lag", ylab="ACF", main="")
```

### 3.6.2 *Invertibility of a MA process*

No restrictions on the $\{\beta_i\}$ are required for a (finite-order) MA process to be stationary, but it is generally desirable to impose restrictions on the $\{\beta_i\}$ to ensure that the process satisfies a condition called **invertibility**. This condition may be explained in the following way. Consider the following first-order MA processes:

$$\text{A:} \qquad X_t = Z_t + \theta Z_{t-1}.$$

$$\text{B:} \qquad X_t = Z_t + \frac{1}{\theta} Z_{t-1}.$$

It can easily be shown that these two different processes have exactly the same ac.f. (Are you surprised? Then check $\rho(k)$ for models A and B.) Thus we cannot identify a MA process uniquely from a given ac.f., as we noted in Property 3 in Section 3.3. Now, if we 'invert' models A and B by expressing $Z_t$ in terms of $X_t, X_{t-1}, \ldots$, we find by successive substitution that

$$\text{A:} \qquad Z_t = X_t - \theta X_{t-1} + \theta^2 X_{t-2} - \cdots.$$

$$\text{B:} \qquad Z_t = X_t - \frac{1}{\theta} X_{t-1} + \frac{1}{\theta^2} X_{t-2} - \cdots.$$

If $|\theta| < 1$, the series of coefficients of $X_{t-j}$ for model A converges whereas that of B does not. Thus model B cannot be 'inverted' in this way. More generally, a process, $\{X_t\}$, is said to be invertible if the random disturbance at time $t$, sometimes called the *innovation*, can be expressed, as a convergent sum of present and past values of $X_t$ in the form

$$Z_t = \sum_{j=0}^{\infty} \pi_j X_{t-j}, \tag{3.7}$$

where $\sum |\pi_j| < \infty$. This effectively means that the first-order MA process can be rewritten in the form of an autoregressive process, possibly of infinite order, whose coefficients form a convergent sum. Autoregressive processes are introduced below in Section 3.7. It turns out that an estimation procedure for a MA process, which involves estimating the residuals (see Section 4.3.1), will lead naturally to model A. Thus, if $|\theta| < 1$, model A is said to be invertible whereas model B is not. The imposition of the invertibility condition ensures that there is a unique invertible first-order MA process for a given ac.f.

The invertibility condition for a MA process of any order is best expressed by using the backward shift operator, denoted by $B$, which is defined by

$$B^j X_t = X_{t-j} \qquad \text{for all } j$$

Then Equation (3.4) may be written as

$$X_t = (\beta_0 + \beta_1 B + \cdots + \beta_q B^q) Z_t = \theta(B) Z_t \tag{3.8}$$

where $\theta(B)$ is a polynomial of order $q$ in $B$. It can be shown that a MA($q$) process is invertible if the roots of the equation

$$\theta(B) = \beta_0 + \beta_1 B + \cdots + \beta_q B^q = 0 \tag{3.9}$$

all lie outside the unit circle, where we regard $B$ as a complex variable and not as an operator. This means that the roots, which may be complex, have modulus greater than unity.

How the invertibility condition ensures $X_t$ to be invertible? This can be explained by utilizing the backward shift operator. In particular, Equation (3.8) can be written equivalently as

$$Z_t = \frac{1}{\theta(B)} X_t. \tag{3.10}$$

In the first-order case for model A, we have $\theta(B) = 1 + \theta B$, which has root $B = -1/\theta$. Provided that $|\theta| < 1$, the root $B = -1/\theta$ is real and lies 'outside the unit circle'. So again we see that model A is invertible if $|\theta| < 1$. Furthermore, if we regard $B$ as a complex variable, the operator $1/\theta(B)$ can be expanded as an infinite series, i.e.

$$\frac{1}{1+\theta B} = 1 + \sum_{i=1}^{\infty} (-\theta)^i B^i. \tag{3.11}$$

When $|\theta| < 1$, this infinite series is convergent since

$$|1 + \sum_{i=1}^{\infty} (-\theta)^i| \leq 1 + \sum_{i=1}^{\infty} |\theta|^i = \frac{1}{1-|\theta|}.$$

Hence Equation (3.10) becomes

$$Z_t = \Big(1 + \sum_{i=1}^{\infty} (-\theta)^i B^i\Big) X_t = X_t + \sum_{i=1}^{\infty} (-\theta)^i X_{t-i},$$

and thus $\{X_t\}$ is invertible.

The argument above can be extended to MA($q$) processes. Suppose that $\theta(B)$ can be decomposed as the following form

$$\theta(B) = (1 + \theta_1 B) \cdots (1 + \theta_q B),$$

where $\theta_1, \ldots, \theta_q$ could possibly take complex values. Then the operator $1/\theta(B)$ can be written as

$$\frac{1}{\theta(B)} = \prod_{j=1}^{q} \frac{1}{1+\theta_j B} = \prod_{j=1}^{q} \Big(1 + \sum_{i=1}^{\infty} (-\theta_j)^i B^i\Big). \tag{3.12}$$

When all the roots, $-1/\theta_1, \ldots, -1/\theta_q$, are outside the unit circle, the product of infinite series in (3.12) is convergent, and hence $Z_t = 1/\theta(B)X_t$ can be written in the form of (3.7), and therefore $\{X_t\}$ is invertible.

MA processes have been used in many areas, particularly econometrics. For example, economic indicators are affected by a variety of 'random' events such as strikes, government decisions, shortages of key materials, and so on. Such events will not only have an immediate effect but may also affect economic indicators to a lesser extent in several subsequent periods, and so it is at least plausible that an MA process may be appropriate.

Note that an arbitrary constant, $\mu$ say, may be added to the right-hand side of Equation (3.4) to give a process with mean $\mu$. This does not affect the ac.f. (see Exercise 3.5) and has been omitted for simplicity.

## 3.7 Autoregressive Processes

Suppose that $\{Z_t\}$ is a purely random process with mean zero and variance $\sigma_Z^2$. Then a process $\{X_t\}$ is said to be an autoregressive process of order $p$ (abbreviated to an AR($p$) process) if

$$X_t = \alpha_1 X_{t-1} + \cdots + \alpha_p X_{t-p} + Z_t. \tag{3.13}$$

This is rather like a multiple regression model, but $X_t$ is regressed on past values of $X_t$ rather than on separate predictor variables. This explains the prefix 'auto'.

### *3.7.1 First-order process*

For simplicity, we begin by examining the first-order case, where $p = 1$. Then

$$X_t = \alpha X_{t-1} + Z_t. \tag{3.14}$$

The AR(1) process is sometimes called the Markov process, after the Russian A.A. Markov. By successive substitution into Equation (3.14) we may write

$$\begin{aligned} X_t &= \alpha(\alpha X_{t-2} + Z_{t-1}) + Z_t \\ &= \alpha^2(\alpha X_{t-3} + Z_{t-2}) + \alpha Z_{t-1} + Z_t \end{aligned}$$

and eventually we find that $X_t$ may be expressed as an infinite-order MA process in the form

$$X_t = Z_t + \alpha Z_{t-1} + \alpha^2 Z_{t-2} + \cdots$$

provided $-1 < \alpha < +1$, so that the sum converges.

The possibility that AR processes may be written in MA form, and vice versa — see also Section 3.6 — means that there is a duality between AR and MA processes, which is useful for a variety of purposes. Rather than use successive substitution to explore this duality, it is simpler to use the backward shift operator $B$. Thus Equation (3.14) may be written

$$(1 - \alpha B)X_t = Z_t,$$

so that

$$
\begin{aligned}
X_t &= Z_t/(1-\alpha B) = (1+\alpha B+\alpha^2B^2+\cdots)Z_t \\
&= Z_t+\alpha Z_{t-1}+\alpha^2 Z_{t-2}+\cdots.
\end{aligned}
$$

When expressed in this form it is clear that

$$
\begin{aligned}
E(X_t) &= 0 \\
\text{Var}(X_t) &= \sigma_Z^2(1+\alpha^2+\alpha^4+\cdots)
\end{aligned}
$$

by independence of the $Z$'s. Thus the variance is finite provided that $|\alpha|^2 < 1$, so if $|\alpha| < 1$, in which case

$$\text{Var}(X_t) = \sigma_X^2 = \sigma_Z^2/(1-\alpha^2).$$

The acv.f. is given by

$$
\begin{aligned}
\gamma(k) &= E[X_tX_{t+k}] = E[(\Sigma\alpha^i Z_{t-i})(\Sigma\alpha^j Z_{t+k-j})] \\
&= \sigma_Z^2\sum_{i=0}^{\infty}\alpha^i\alpha^{k+i} \qquad \text{for } k \geq 0, \\
&= \alpha^k\sigma_Z^2/(1-\alpha^2) \qquad \text{provided } |\alpha|<1, \\
&= \alpha^k\sigma_X^2.
\end{aligned}
$$

For $k<0$, we find $\gamma(k)=\gamma(-k)$. Since $\gamma(k)$ does not depend on $t$, an AR process of order 1 is second-order stationary provided that $|\alpha|<1$, and the ac.f. is then given by

$$\rho(k)=\alpha^k \qquad k=0,1,2,\ldots.$$

To get an even function defined for all integer $k$ we can use the modulus operator to write

$$\rho(k)=\alpha^{|k|} \qquad k=0,\pm1,\pm2,\ldots.$$

The ac.f. may also be obtained more easily by assuming *a priori* that the process is stationary. Then $E(X_t)$ must be a constant, $\mu$ say, that turns out to be zero using Equation (3.14). If we then multiply through Equation (3.14) by $X_{t-k}$ (not $X_{t+k}$!) and take expectations, we find, for $k>0$, that

$$\gamma(-k)=\alpha\gamma(-k+1),$$

since $E(Z_tX_{t-k})=0$ for $k>0$, by independence of the $Z$'s. Since $\gamma(k)$ is an even function of $k$, we must also have

$$\gamma(k)=\alpha\gamma(k-1) \qquad \text{for } k>0.$$

Now $\gamma(0)=\sigma_X^2$, and so $\gamma(k)=\alpha^k\sigma_X^2$ for $k\geq 0$. Thus $\rho(k)=\alpha^k$ for $k\geq 0$. Now since $|\rho(k)|\leq 1$, we must have $|\alpha|\leq 1$. However, if $|\alpha|=1$, then $|\rho(k)|=1$ for all $k$, which is a degenerate case. Thus we must have $|\alpha|$ strictly less than 1 to have a proper stationary process.

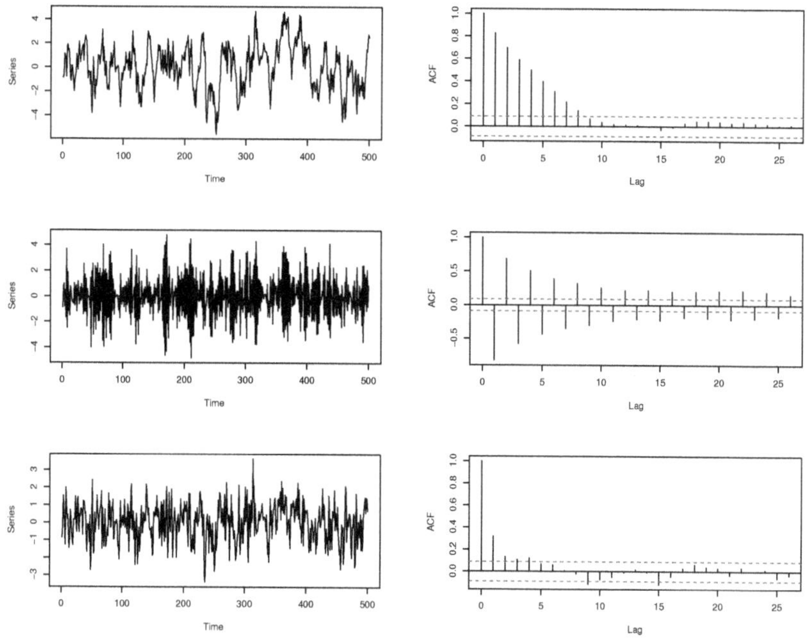


Figure 3.4 *Three simulated AR(1) processes and their correlograms. Top:* $X_t = 0.8X_{t-1} + Z_t$, $Z_t \sim N(0,1)$*; Middle:* $X_t = -0.8X_{t-1} + Z_t$, $Z_t \sim N(0,1)$*; Bottom:* $X_t = 0.3X_{t-1} + Z_t$, $Z_t \sim N(0,1)$.

The above method of obtaining the ac.f. is often used, even though it involves 'cheating' a little by making an initial assumption of stationarity.

Three examples of the ac.f. of a first-order AR process are shown in Figure 3.4 for $\alpha = 0.8, -0.8$, and $0.3$. All three series are constructed from the same noise series by the R script below. Note how quickly the ac.f. decays when $\alpha = 0.3$, and note how the ac.f. alternates when $\alpha$ is negative.

```
> n<-500
> z<-rnorm(n, 0, 1)
> x.ar1.1<-x.ar1.2<-x.ar1.3<-rep(0,n)
> x.ar1.1[1]<-x.ar1.2[1]<-x.ar1.3[1]<-z[1]
> for (i in 2:n){
    x.ar1.1[i]<- 0.8*x.ar1.1[i-1]+z[i]
    x.ar1.2[i]<- -0.8*x.ar1.2[i-1]+z[i]
    x.ar1.3[i]<- 0.3*x.ar1.3[i-1]+z[i]
  }
```

### 3.7.2 *General-order process*

As in the first-order case, we can express an AR process of finite order as an MA process of infinite order. This may be done by successive substitution, or, more easily, by utilizing the backward shift operator. Thus Equation (3.13) may be written as

$$(1 - \alpha_1 B - \cdots - \alpha_p B^p) X_t = Z_t \tag{3.15}$$

or equivalently as

$$X_t = Z_t / (1 - \alpha_1 B - \cdots - \alpha_p B^p) = f(B) Z_t,$$

where

$$f(B) = (1 - \alpha_1 B - \cdots - \alpha_p B^p)^{-1} = (1 + \beta_1 B + \beta_2 B^2 + \cdots).$$

The relationship between the $\alpha$s and the $\beta$s may then be found. Having expressed $X_t$ as an MA process, it follows that $E(X_t) = 0$. The variance is finite provided that $\Sigma \beta_i^2$ converges, and this is a necessary condition for stationarity. The acv.f. is given by

$$\gamma(k) = \sigma_Z^2 \sum_{i=0}^{\infty} \beta_i \beta_{i+k} \qquad \text{where } \beta_0 = 1.$$

A sufficient condition for this to converge, and hence for stationarity, is that $\Sigma |\beta_i|$ converges.

*Yule-Walker equations*

We can in principle find the ac.f. of the general-order AR process using the above procedure, but the $\{\beta_i\}$ may be algebraically hard to find. The alternative simpler way is to *assume* the process is stationary, multiply through Equation (3.13) by $X_{t-k}$, take expectations and divide by $\sigma_X^2$, assuming that the variance of $X_t$ is finite. Then, using the fact that $\rho(k) = \rho(-k)$ for all $k$, we find

$$\rho(k) = \alpha_1 \rho(k-1) + \cdots + \alpha_p \rho(k-p) \qquad \text{for all } k > 0. \tag{3.16}$$

This set of equations is called the Yule–Walker equations after G.U. Yule and Sir Gilbert Walker. It is a set of difference equations and has the general solution

$$\rho(k) = A_1 \pi_1^{|k|} + \cdots + A_p \pi_p^{|k|},$$

where $\{\pi_i\}$ are the roots of the so-called auxiliary equation

$$y^p - \alpha_1 y^{p-1} - \cdots - \alpha_p = 0.$$

The constants $\{A_i\}$ are chosen to satisfy the initial conditions depending on $\rho(0) = 1$, which means that $\Sigma A_i = 1$. The first $(p-1)$ Yule–Walker equations provide $(p-1)$ further restrictions on the $\{A_i\}$ using $\rho(0) = 1$ and $\rho(k) = \rho(-k)$.

*Stationarity conditions*

From the general form of $\rho(k)$, it is clear that $\rho(k)$ tends to zero as $k$ increases provided that $|\pi_i| < 1$ for all $i$, and this is a necessary and sufficient condition for the AR$(p)$ process to be stationary.

It can be shown that an equivalent way of expressing the stationarity condition is to say that the roots of the equation

$$\phi(B) = 1 - \alpha_1 B - \cdots - \alpha_p B^p = 0 \tag{3.17}$$

must lie outside the unit circle (where we again regard $B$ as a complex variable, rather than as an operator, so that the roots, which may be complex, are greater than one in modulus).

Of particular interest is the AR(2) process, when $\pi_1, \pi_2$ are the roots of the quadratic equation

$$y^2 - \alpha_1 y - \alpha_2 = 0.$$

Here $|\pi_i| < 1$ if

$$\left| \frac{\alpha_1 \pm \sqrt{\alpha_1^2 + 4\alpha_2}}{2} \right| < 1$$

from which it can be shown (Exercise 3.6) that the stationarity region is the triangular region satisfying

$$\alpha_1 + \alpha_2 < 1, \qquad \alpha_1 - \alpha_2 > -1, \qquad \alpha_2 > -1.$$

The roots are real if $\alpha_1^2 + 4\alpha_2 > 0$, in which case the ac.f. decreases exponentially with $k$, but the roots are complex if $\alpha_1^2 + 4\alpha_2 < 0$, in which case the ac.f. turns out to be a damped sinusoidal wave. (See Example 3.1 at the end of this section.)

When the roots are real, we have $\rho(k) = A_1 \pi_1^{|k|} + A_2 \pi_2^{|k|}$ where the constants $A_1, A_2$ are also real and may be found as follows. Since $\rho(0) = 1$, we have

$$A_1 + A_2 = 1$$

while the first Yule–Walker equation gives

$$\rho(1) = \alpha_1 \rho(0) + \alpha_2 \rho(-1) = \alpha_1 + \alpha_2 \rho(1).$$

This equation may be solved to give $\rho(1) = \alpha_1/(1 - \alpha_2)$, which in turn must equal

$$A_1 \pi_1 + A_2 \pi_2 = A_1 \pi_1 + (1 - A_1)\pi_2.$$

Hence we find

$$A_1 = [\alpha_1/(1 - \alpha_2) - \pi_2]/(\pi_1 - \pi_2), \qquad A_2 = 1 - A_1$$

and this enables us to write down the general form of the ac.f. of an AR(2) process with real roots. As an example, consider an AR(2) process,

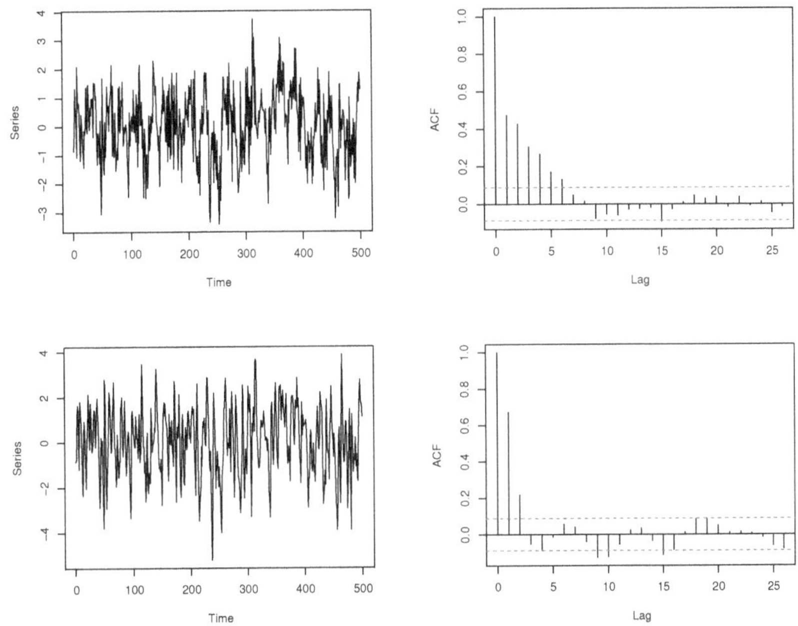


Figure 3.5 *Two simulated AR(2) processes and their correlograms. Top:* $X_t = \frac{1}{3}X_{t-1} + \frac{2}{9}X_{t-2} + Z_t$, $Z_t \sim N(0,1)$*; Bottom:* $X_t = X_{t-1} - \frac{1}{2}X_{t-2} + Z_t$, $Z_t \sim N(0,1)$.

$X_t = 1/3X_{t-1} + 2/9X_{t-2} + Z_t$. The top panels of Figure 3.5 show a realization of this process and its correlogram. Note that the roots of $1 - 1/3B - 2/9B^2 = 0$ are $-3$ and $3/2$; both have absolute values larger than 1. This indicates $X_t$ is stationary and its ac.f. can be calculated with the method above (see Exercise 3.6). For complex roots, the general solution can also be found and an example is given below.

*Example 3.1* Consider the AR(2) process given by

$$X_t = X_{t-1} - \frac{1}{2}X_{t-2} + Z_t.$$

Is this process stationary? If so, what is its ac.f.?

In order to answer the first question we find the roots of Equation (3.17), which, in this case, is

$$\phi(B) = 1 - B + \frac{1}{2}B^2 = 0.$$

The roots of this equation (regarding $B$ as a variable) are complex, namely, $1 \pm \mathrm{i}$. As the modulus of both roots exceeds one, the roots are both outside the unit circle and so the process is stationary.

In order to find the ac.f. of the process, we use the first Yule–Walker equation to give

$$\rho(1) = \rho(0) - \frac{1}{2}\rho(-1) = 1 - \frac{1}{2}\rho(1)$$

giving $\rho(1) = 2/3$.

For $k \geq 2$, the Yule–Walker equations are

$$\rho(k) = \rho(k-1) - \frac{1}{2}\rho(k-2).$$

We could use these equations to find $\rho(2)$, then $\rho(3)$, and so on by successive substitution, but it is easier to find the general solution by solving the set of Yule–Walker equations as a set of difference equations. The general form of the above Yule–Walker equation has the auxiliary equation

$$y^2 - y + \frac{1}{2} = 0$$

with roots $y = (1 \pm \mathrm{i})/2$. They may be rewritten as $[\cos(\pi/4) \pm \mathrm{i}\ \sin(\pi/4)]/\sqrt{2}$ or as $\mathrm{e}^{\pm \mathrm{i}\pi/4}/\sqrt{2}$. Since $\alpha_1^2 + 4\alpha_2 = (1-2)$ is less than zero, and the roots are complex, the ac.f. is a damped sinusoidal wave. Using $\rho(0) = 1$ and $\rho(1) = 2/3$, some messy trigonometry and algebra involving complex numbers gives

$$\rho(k) = \left(\frac{1}{\sqrt{2}}\right)^k \left(\cos\frac{\pi k}{4} + \frac{1}{3}\sin\frac{\pi k}{4}\right)$$

for $k = 0, 1, 2, \ldots$. Note the values of the ac.f. are all real, even though the roots of the auxiliary equation are complex. The bottom two panels of Figure 3.5 show a realization of $X_t$ and its correlogram. □

AR processes have been applied to many situations in which it is reasonable to assume that the present value of a time series depends linearly on the immediate past values together with a random error. For simplicity we have only considered processes with mean zero, but non-zero means may be dealt with by rewriting Equation (3.13) in the form

$$X_t - \mu = \alpha_1(X_{t-1} - \mu) + \cdots + \alpha_p(X_{t-p} - \mu) + Z_t.$$

This does not affect the ac.f. (see Exercise 3.4).

## 3.8 Mixed ARMA Models

A useful class of models for time series is formed by combining MA and AR processes. A mixed autoregressive/moving-average process containing $p$ AR terms and $q$ MA terms is said to be an ARMA process of order $(p,q)$. It is given by

$$X_t = \alpha_1 X_{t-1} + \cdots + \alpha_p X_{t-p} + Z_t + \beta_1 Z_{t-1} + \cdots + \beta_q Z_{t-q}, \qquad (3.18)$$

where $\{Z_t\}$ is a purely random process with mean zero and variance $\sigma_Z^2$. Using the backward shift operator $B$, Equation (3.18) may be written in the form

$$\phi(B)X_t = \theta(B)Z_t, \tag{3.19}$$

where $\phi(B), \theta(B)$ are polynomials of order $p, q$, respectively, such that

$$\phi(B) = 1 - \alpha_1 B - \cdots - \alpha_p B^p$$

and

$$\theta(B) = 1 + \beta_1 B + \cdots + \beta_q B^q.$$

### 3.8.1 *Stationarity and invertibility conditions*

The conditions on the model parameters to make the process stationary and invertible are the same as for a pure AR or pure MA process, namely, that the values of $\{\alpha_i\}$, which make the process stationary, are such that the roots of

$$\phi(B) = 0$$

lie outside the unit circle, while the values of $\{\beta_i\}$, which make the process invertible, are such that the roots of

$$\theta(B) = 0$$

lie outside the unit circle. It is straightforward in principle, though algebraically rather tedious, to calculate the ac.f. of an ARMA process, but this will not be discussed here. (See Exercise 3.11, and Box et al., 1994, Section 3.4.)

The importance of ARMA processes lies in the fact that a stationary time series may often be adequately modelled by an ARMA model involving fewer parameters than a pure MA or AR process by itself. This is an early example of what is often called the **Principle of Parsimony**. This says that we want to find a model with as few parameters as possible, but which gives an adequate representation of the data at hand.

### 3.8.2 *Yule-Walker equations and autocorrelations*

The ac.f. of the general ARMA process can be found using similar procedures as for AR processes. First, multiply through Equation (3.18) by $X_{t-k}$ and take expectations. Note that, for $k \geq q+1$, $Z_t, \ldots, Z_{t-q}$ are independent of $X_{t-k}$. Hence the expected values of $Z_t X_{t-k}, \ldots, Z_{t-q} X_{t-k}$ are all zero. If $k \geq p$, we can further divide both sides by $\gamma(0)$. Then we find the following Yule-Walker equations for general ARMA$(p,q)$ processes

$$\rho(k) = \alpha_1 \rho(k-1) + \cdots + \alpha_p \rho(k-p), \qquad k \geq \max(p, q+1). \tag{3.20}$$

Note that (3.20) has the same form as that of an AR process except that their initial conditions are different. The initial conditions of the Yule-Walker equations are usually computed seperately.

The following example shows how the ac.f. of an ARMA(1,1) process is derived.

*Example 3.2*

Consider the ARMA(1,1) process

$$X_t = \alpha X_{t-1} + Z_t + \beta Z_{t-1}, \tag{3.21}$$

where $|\alpha| < 1$ and $|\beta| < 1$. To derive the ac.f. of the process, the Yule-Walker equations are

$$\rho(k) = \alpha\rho(k-1), \qquad k \geq 2.$$

To obtain the initial conditions, we note that $\gamma(1)$ can be computed as follows.

$$\begin{aligned}\gamma(1) &= \text{Cov}(X_t, X_{t-1}) = \text{Cov}(\alpha X_{t-1} + Z_t + \beta Z_{t-1}, X_{t-1}) \\ &= \alpha\gamma(0) + \beta\text{Cov}(Z_{t-1}, X_{t-1}) \\ &= \alpha\gamma(0) + \beta\text{Cov}(Z_{t-1}, \alpha X_{t-2} + Z_{t-1} + \beta Z_{t-2}) \\ &= \alpha\gamma(0) + \beta\sigma_Z^2. \end{aligned} \tag{3.22}$$

The variance of $X_t$, or $\gamma(0)$, can be calculated as follows

$$\begin{aligned}\gamma(0) &= \text{Var}(\alpha X_{t-1} + Z_t + \beta Z_{t-1}) \\ &= \text{Cov}(\alpha X_{t-1} + Z_t + \beta Z_{t-1}, \alpha X_{t-1} + Z_t + \beta Z_{t-1}) \\ &= \alpha^2\gamma(0) + (1+\beta^2)\sigma_Z^2 + 2\alpha\beta\text{Cov}(X_{t-1}, Z_{t-1}) \\ &= \alpha^2\gamma(0) + (1 + 2\alpha\beta + \beta^2)\sigma_Z^2. \end{aligned}$$

Solving for $\gamma(0)$ yields

$$\gamma(0) = \frac{1 + 2\alpha\beta + \beta^2}{1-\alpha^2}\sigma_Z^2. \tag{3.23}$$

Plug $\gamma(0)$ into Equation (3.22) to obtain an explicit expression for $\gamma(1)$. Then

$$\gamma(1) = \frac{(1+\alpha\beta)(\alpha+\beta)}{1-\alpha^2}\sigma_Z^2.$$

Hence

$$\rho(1) = \frac{\gamma(1)}{\gamma(0)} = \frac{(1+\alpha\beta)(\alpha+\beta)}{1+2\alpha\beta+\beta^2}.$$

Using the Yule-Walker equations, we then have

$$\rho(k) = \frac{(1+\alpha\beta)(\alpha+\beta)}{1+2\alpha\beta+\beta^2}\alpha^{k-1}, \qquad k \geq 1. \tag{3.24}$$

□

In principle, the ac.f. for a general ARMA process can be computed, similarly as in Example 3.2, but are usually more complicated than for AR processes. Figure 3.6 shows two simulated ARMA processes (ARMA(1,1) and ARMA(2,2)) and their correlograms using the following R script.

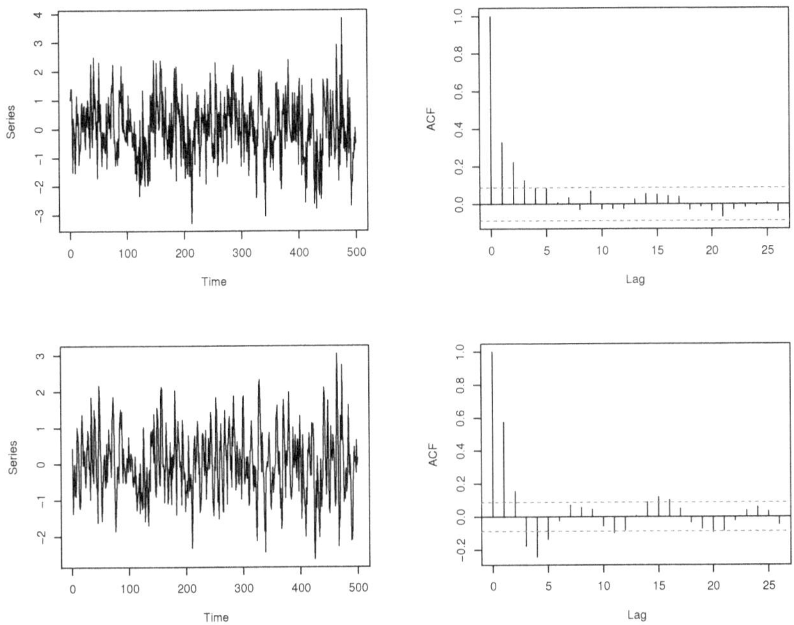


Figure 3.6 *Two simulated ARMA processes and their correlograms. Top:* $X_t = 0.7X_{t-1} + Z_t - 0.4Z_{t-1}$, $Z_t \sim N(0,1)$*; Bottom:* $X_t = 0.9X_{t-1} - 0.5X_{t-2} + Z_t - 0.2Z_{t-1} + 0.25Z_{t-2}$, $Z_t \sim N(0,0.5)$.

```
> x1<-arima.sim(n=500, list(ar=0.7, ma=-0.4))
> x2<-arima.sim(n=500, list(ar=c(0.9,-0.5),
     ma=c(-0.2,0.25)), sd=sqrt(0.5))
```

Note that, though the correlogram of the ARMA(2, 2) process in Figure 3.6 looks similar to the bottom correlogram of the AR(2) process in Figure 3.5, these two processes are completely different.

### 3.8.3 *AR and MA representations*

It is sometimes helpful to express an ARMA model as a pure MA process in the form

$$X_t = \psi(B)Z_t, \tag{3.25}$$

where $\psi(B) = \sum_{i\geq 0} \psi_i B^i$ is the MA operator, which may be of infinite order. The $\psi$ weights, $\{\psi_i\}$, can be useful in calculating forecasts (see Chapter 5) and in assessing the properties of a model. By comparison with Equation (3.19), we see that $\psi(B) = \theta(B)/\phi(B)$. Alternatively, it can be helpful to express an ARMA model as a pure AR process in the form

$$\pi(B)X_t = Z_t, \tag{3.26}$$

where $\pi(B) = \phi(B)/\theta(B)$. By convention we write $\pi(B) = 1 - \sum_{i\geq 1} \pi_i B^i$, since the natural way to write an AR model is in the form

$$X_t = \sum_{i=1}^{\infty} \pi_i X_{t-i} + Z_t.$$

By comparing (3.25) and (3.26), we see that

$$\pi(B)\psi(B) = 1.$$

The $\psi$ weights or $\pi$ weights may be obtained directly by division or by equating powers of $B$ in an equation such as

$$\psi(B)\phi(B) = \theta(B).$$

*Example 3.3* Find the $\psi$ weights and $\pi$ weights for the ARMA$(1,1)$ process given by

$$X_t = 0.5X_{t-1} + Z_t - 0.3Z_{t-1}.$$

Here $\phi(B) = (1 - 0.5B)$ and $\theta(B) = (1 - 0.3B)$. It follows that the process is stationary and invertible, because both equations have roots greater than one (or are outside the unit circle). Then

$$\begin{aligned} \psi(B) = \theta(B)/\phi(B) &= (1-0.3B)(1-0.5B)^{-1} \\ &= (1-0.3B)(1+0.5B+0.5^2B^2+\cdots) \\ &= 1+0.2B+0.1B^2+0.05B^3+\cdots. \end{aligned}$$

Hence

$$\psi_0 = 1, \qquad \psi_i = 0.2 \times 0.5^{i-1} \qquad \text{for } i = 1,2,\ldots.$$

Similarly we find

$$\pi_0 = 1, \qquad \pi_i = 0.2 \times 0.3^{i-1} \qquad \text{for } i = 1,2,\ldots.$$

Note that both the $\psi$ weights and $\pi$ weights die away quickly, and this also indicates a stationary, invertible process. □

## 3.9 Integrated ARMA (or ARIMA) Models

In practice most time series are non-stationary. In order to fit a stationary model, such as those discussed in Sections 3.6–3.8, it is necessary to remove non-stationary sources of variation. If the observed time series is non-stationary in the mean, then we can difference the series, as suggested in Section 2.5.3. Differencing is widely used for econometric data. If $X_t$ is replaced by $\nabla^d X_t$ in Equation (3.18), then we have a model capable of describing certain types of non-stationary series. Such a model is called an 'integrated' model because the stationary model that is fitted to the differenced data has to be summed or 'integrated' to provide a model for the original non-stationary data. Writing

$$W_t = \nabla^d X_t = (1-B)^d X_t, \qquad d = 0,1,2,\ldots$$

the general autoregressive integrated moving average (ARIMA) process is of the form

$$W_t = \alpha_1 W_{t-1} + \cdots + \alpha_p W_{t-p} + Z_t + \cdots + \beta_q Z_{t-q}. \tag{3.27}$$

ARIMA models are mainly used to remove a trend from the data. Note that if $d = 0$, the process is an ARMA$(p,q)$ process. By analogy with Equation (3.19), we may write Equation (3.27) in the form

$$\phi(B)W_t = \theta(B)Z_t \tag{3.28}$$

or

$$\phi(B)(1-B)^d X_t = \theta(B)Z_t. \tag{3.29}$$

Thus we have an ARMA$(p,q)$ model for $W_t$, while the model in Equation (3.29), describing the $d$th differences of $X_t$, is said to be an ARIMA process of order $(p,d,q)$.

The model for $X_t$ is clearly non-stationary, as the AR operator $\phi(B)(1-B)^d$ has $d$ roots on the unit circle (since putting $B = 1$ makes the AR operator equal to zero). In practice, first differencing is often found to be adequate to make a series stationary, and so the value of $d$ is often taken to be one. Note that the random walk in Section 3.5 can be regarded as an ARIMA$(0,1,0)$ process. As we saw in Section 3.5, the random walk is a non-stationary process, and applying the difference operator once makes the process stationary.

Figure 3.7 shows a simulated ARIMA(1,1,1) and a simulated ARIMA(1,1,2) process and their correlograms using the following commands in R:

```
> y1<-arima.sim(list(order=c(1,1,1), ar=-0.5, ma=-0.3),
     n=500)
> y2<-arima.sim(list(order=c(1,1,2), ar=0.3,
     ma=c(-0.3,0.5)), n=500)
```

Note that the ac.f. of the two integrated processes in Figure 3.7 is slowly decaying because of the trend, as already observed in Section 3.5.

ARIMA models can be generalized to include seasonal terms, as will be discussed in Section 4.8.

## 3.10 Fractional Differencing and Long-Memory Models

An interesting variant of ARIMA modelling arises with the use of what is called **fractional differencing**, leading to a fractional integrated ARMA (abbreviated to ARFIMA) model. The ARIMA$(p,d,q)$ model from Section 3.9 is usually written as

$$\phi(B)(1-B)^d X_t = \theta(B)Z_t,$$

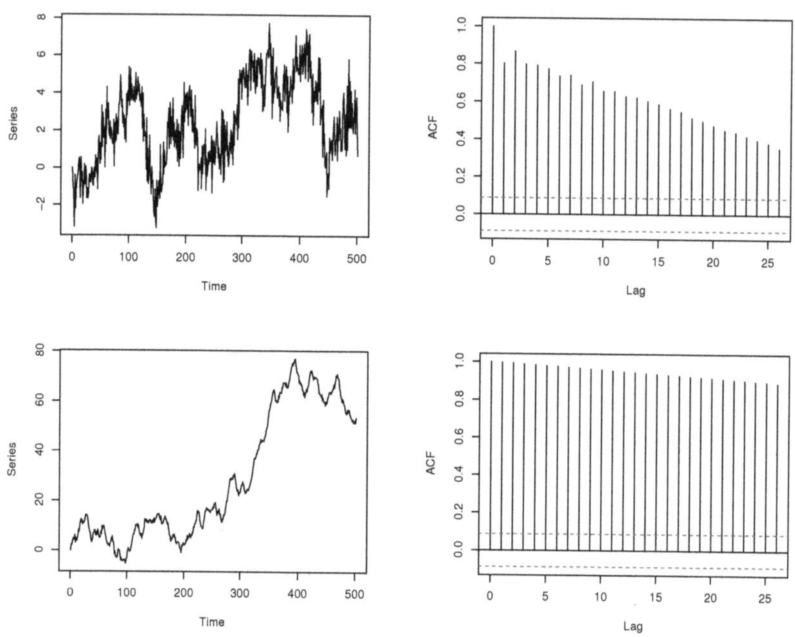


Figure 3.7 *Two simulated ARIMA processes and their correlograms. Top:* $(1+0.5B)(1-B)X_t = (1+0.3B)Z_t$; *Bottom:* $(1-0.6B)(1-B)X_t = (1-0.3B+0.5B^2)Z_t$.

where $\phi(B)$ and $\theta(B)$ are polynomials of order $p, q$, respectively, in the backward shift operator $B$. Here $p, q$ and $d$ are integers and, in particular, the order of differencing $d$ is an integer that is typically one or two. We further assume here that $\phi(B)$ and $\theta(B)$ have all their roots outside the unit circle, so that when $d = 0$ the process is stationary and invertible. Fractional ARIMA models extend the above class of models by allowing $d$ to be a non-integer. In other words, the formula for an ARFIMA model is exactly the same as for an ARIMA$(p,d,q)$ model, except that $d$ is no longer restricted to being an integer.

When $d$ is not an integer, then the $d$th difference $(1-B)^d X_t$ becomes a fractional difference, and may be represented by its binomial expansion, namely

$$(1-B)^d X_t = \left[1 - dB + \frac{d(d-1)}{2!}B^2 - \frac{d(d-1)(d-2)}{3!}B^3 + \ldots\right] X_t.$$

As such, it is an infinite weighted sum of past values. This contrasts with the case where $d$ is an integer when a finite sum is obtained.

It can be shown (e.g. Brockwell and Davis, 1991, Section 13.2) that an ARFIMA process is stationary provided that $-0.5 < d < 0.5$. For $d > \frac{1}{2}$, the process is not stationary in the usual sense, but further integer differencing can be used to give a stationary ARFIMA process. For example, if an observed series is

ARFIMA$(p, d = 1.3, q)$, then the first differences of the series will follow a stationary ARFIMA$(p, d = 0.3, q)$ process.

A drawback to fractional differencing is that it is difficult to give an intuitive interpretation to a non-integer difference. It is also more difficult to calculate them, given that the binomial expansion will need to be truncated, and so the parameter $d$ is often estimated in the frequency domain. Details will not be given here and the reader is referred, for example, to Beran (1994), Crato and Ray (1996) and the earlier references therein, though the reader is warned that the literature is technically demanding.

A stationary ARFIMA model, with $0 < d < 0.5$, is of particular interest as such a process is not only stationary, but is also an example of what is called a **long-memory** model. For most stationary time-series models (including stationary ARMA models), the autocorrelation function (ac.f.) decreases 'fairly fast', as demonstrated, for example, by the exponential decay in the ac.f. of the AR(1) model. However, for some models the correlations decay to zero very slowly, implying that observations far apart are still related to some extent. An intuitive way to describe such behaviour is to say that the process has a long memory, or that there is long-range dependence.

A working definition of a long-memory process is as follows: A stationary process with ac.f. $\rho(k)$ is said to be a long-memory process if $\sum_{k=0}^{\infty} |\rho(k)|$ does not converge. In particular, the latter condition applies when the ac.f. $\rho(k)$ is of the form $\rho(k) \sim Ck^{2d-1}$ as $k \to \infty$, where $C$ is a constant, not equal to zero, and $0 < d < 0.5$. It can be shown that a stationary ARFIMA model, with differencing parameter $d$ in the range $0 < d < 0.5$, has an ac.f. $\rho_k$ whose limiting form as $k \to \infty$ has the required structure. This means that the correlations decay slowly at a hyperbolic rate and are not absolutely summable. Thus an ARFIMA model is a long-memory model for $0 < d < 0.5$.

As an example, Figure 3.8 shows a simulated ARFIMA(1,$d = 0.4$,2) series, $(1 - 0.3B)(1 - B)^d X_t = (1 - 0.3B + 0.5B^2)Z_t$ with $Z_t \sim N(0, 0.2^2)$, and its correlogram using the following command in R. Note that a specific R package, `fracdiff`, needs to be loaded before calling the function `fracdiff.sim`.

```
> library("fracdiff")
> x<-fracdiff.sim(2000, ar=0.3, ma=c(-0.3, 0.5), d=0.4,
     sd=0.2)
> names(x)
[1] "series"  "ar"     "ma"     "d"      "mu"     "n.start"
> par(mfrow=c(2,1), mar=c(4,4,4,4))
> plot(x$series, type="l", xlab="Time", ylab="Series")
> acf(x$series, 50, xlab="Lag", ylab="ACF", main="")
```

In contrast, the ac.f. of a stationary ARMA process satisfies the condition that $|\rho(k)| < C\lambda^k$, where $C$ and $\lambda$ are constants with $0 < \lambda < 1$. Thus these correlations are absolutely summable and such processes may be called **short-memory** models.

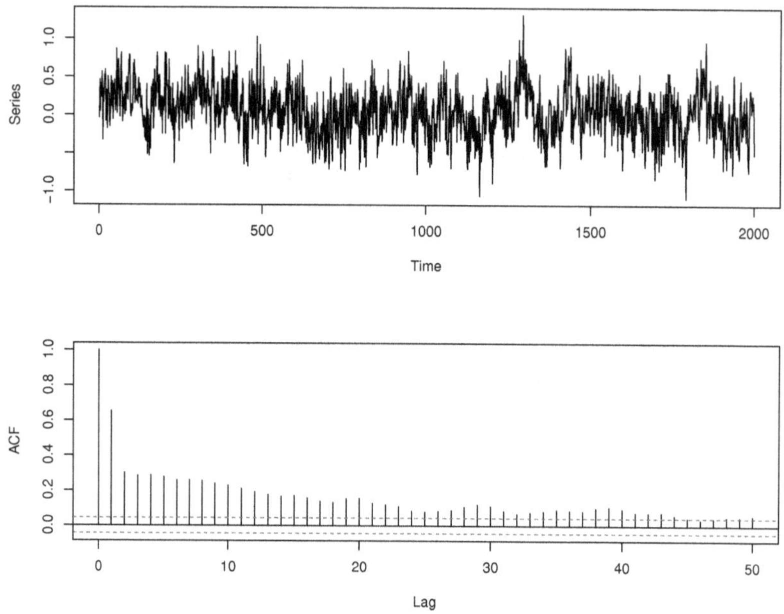


Figure 3.8 *A simulated ARFIMA(1, $d = 0.4$, 2) process,* $(1-0.3B)(1-B)^d X_t = (1-0.3B+0.5B^2)Z_t$ *with* $Z_t \sim N(0, 0.04)$, *and its correlogram.*

Further details on ARFIMA and other long-memory models are given by Beran (1994). It is also possible to find non-linear models that have the long-memory property (e.g. Robinson and Zaffaroni, 1998). For example, while ordinary ARCH models are short memory, some variants for modelling volatility in stock and exchange rate returns are such that the squared returns are long memory (see Section 12.1 for a definition of the term 'returns').

Long-memory models have a number of interesting features. Although it can be difficult to get good estimates of some parameters of a long-memory model, notably the mean, it is usually possible to make better forecasts, at least in theory. As regards estimating the mean, the usual formula for the variance of a sample mean is $\sigma_Z^2/N$, but this applies to the case of $N$ independent observations having constant variance $\sigma^2$. In time-series analysis, successive observations are generally correlated and the variance of a sample mean can be expressed as

$$\frac{\sigma_Z^2}{N}\left[1+2\sum_{k=1}^{N-1}(1-\frac{k}{n})\rho(k)\right];$$

see Section 4.1.2. When the correlations are positive, as they usually are, the latter expression can be much larger than $\sigma^2/N$, especially for long-memory processes where the correlations die out slowly. In contrast to this result, it is intuitively clear

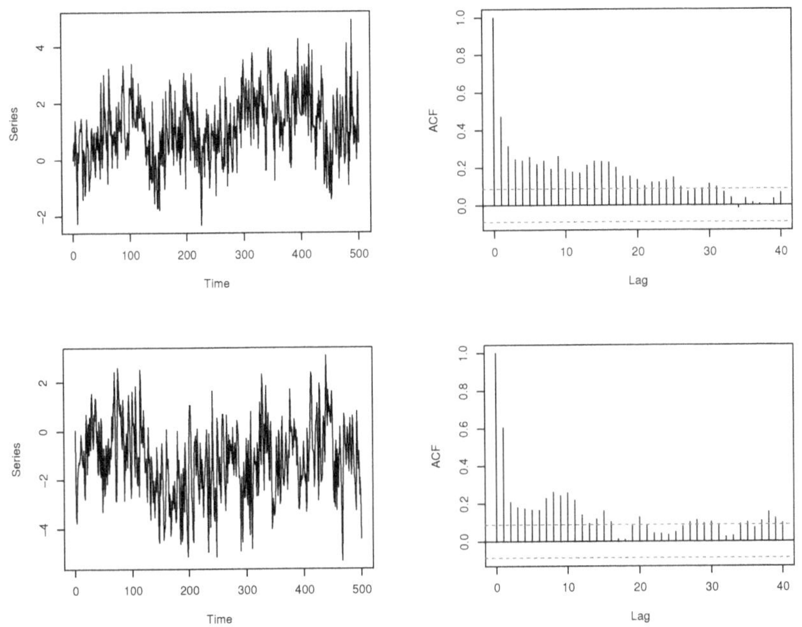


Figure 3.9 *A simulated non-stationary process and a simulated ARFIMA(1, $d = 0.45$, 2) process and their correlograms. Top:* $(1-0.3B)(1-B)X_t = (1+0.9B)Z_t$, $Z_t \sim N(0,1)$. *Bottom:* $(1-0.3B)(1-B)^d X_t = (1-0.3B+0.5B^2)Z_t$, $Z_t \sim N(0,1)$.

that the larger and longer lasting the autocorrelations, the better will be the forecasts from the model. This can readily be demonstrated, both theoretically and practically (Beran, 1994, Section 8.7), but this topic will not be pursued here.

*A long-memory (stationary) process and a non-stationary process*

A major problem in practice is distinguishing between a long-memory (stationary) process and a non-stationary process. A feature of both models is that the empirical ac.f. will die out slowly and the spectrum will be large at zero frequency. Figure 3.9 shows a simulated non-stationary process, a simulated long memory (stationary) process, and their correlograms. The non-stationary process is specified as $(1-0.3B)(1-B)X_t = (1+0.9B)Z_t$, $Z_t \sim N(0,1)$, and the long memory process is given by $(1-0.3B)(1-B)^d X_t = (1-0.3B+0.5B^2)Z_t$ with $d = 0.45$ and $Z_t \sim N(0,1)$. Note that both the series themselves and their correlograms look quite similar in Figure 3.9.

Therefore, given a set of data with these properties, so that it appears to be non-stationary, or at least 'nearly non-stationary', then it may be worth considering a fractional ARIMA model, with $0 < d < 1$, as well as an ordinary ARIMA model with $d = 1$. The question then arises as to whether the resulting forecasts are likely to be better than those from alternative models. While there are some encouraging results (e.g. Sutcliffe, 1994), the results in Smith and Yadav (1994) suggest that little will be lost by taking first, rather than fractional, differences. Research continues on this interesting class of models.

Note that the fractional autoregressive (FAR) model defined by Tong (1990, Section 3.5) is unrelated to ARFIMA models. Also note that a completely different definition of the long-memory property is given by Granger and Teräsvirta (1993, Chapter 5). It concerns the asymptotic properties of the expected value of the forecast for long lead times, and applies to many non-stationary processes.

## 3.11 The General Linear Process

Expanding Equation (3.25), a general class of processes may be written as an MA process, of possibly infinite order, in the form

$$X_t = \sum_{i=0}^{\infty} \psi_i Z_{t-i}. \tag{3.30}$$

A sufficient condition for the sum to converge, and hence for the process to be stationary, is that $\sum_{i=0}^{\infty} |\psi_i| < \infty$, in which case we also have $\sum_{i=0}^{\infty} \psi_i^2 < \infty$ so that $\text{Var}(X_t)$ is finite. A stationary process described by Equation (3.30) is sometimes called a **general linear process**, although some authors use this term when the $Z$'s are merely uncorrelated, rather than independent. Although the model is in the form of an MA process, it is interesting to note that stationary AR and ARMA processes can also be expressed as a general linear process using the duality between AR and MA processes arising, for example, from Equations (3.25) and (3.26).

## 3.12 Continuous Processes

So far, we have only considered stochastic processes in discrete time, because models of this type are nearly always used by the statistician in practice. This subsection[2] gives a brief introduction to processes in continuous time. The latter have been used in some applications, notably in the study of control theory by electrical engineers. Here we are mainly concerned with indicating some of the mathematical problems that arise when time is continuous.

By analogy with a discrete-time purely random process, we might expect to define a continuous-time purely random process as having an ac.f. given by

$$\rho(\tau) = \begin{cases} 1 & \tau = 0, \\ 0 & \tau \neq 0. \end{cases}$$

However, this is a discontinuous function, and it can be shown that such a process would have an infinite variance and hence be a physically unrealizable phenomenon. Nevertheless, some processes that arise in practice do appear to have the properties

[2] This subsection may be omitted at a first reading.

of continuous-time white noise even when sampled at quite small discrete intervals. We may approximate continuous-time white noise by considering a purely random process in discrete time at intervals $\Delta t$, and letting $\Delta t \to 0$, or by considering a process in continuous time with ac.f. $\rho(\tau) = \mathrm{e}^{-\lambda|\tau|}$ and letting $\lambda \to \infty$ so that the ac.f. decays very quickly.

As an example of the difficulties involved with continuous-time processes, we briefly consider a first-order AR process in continuous time. A first-order AR process in discrete time may readily be rewritten[3] in terms of $X_t, \nabla X_t = (1-B)X_t$ and $Z_t$, rather than $X_t, X_{t-1}$ and $Z_t$. Now differencing in discrete time corresponds to differentiation in continuous time, so that an apparently natural way of trying to define a first-order AR process in continuous time is by means of the general equation

$$aX(t) + \frac{\mathrm{d}X(t)}{\mathrm{d}t} = Z(t), \tag{3.31}$$

where $a$ is a constant, and $Z(t)$ denotes continuous white noise. However, as $\{Z(t)\}$ cannot physically exist, it is more legitimate to write Equation (3.31) in a form involving infinitesimal small changes as

$$\mathrm{d}X(t) = -aX(t)\,\mathrm{d}t + \mathrm{d}U(t), \tag{3.32}$$

where $\{U(t)\}$ is a process with orthogonal increments such that the random variables $[U(t_2) - U(t_1)]$ and $[U(t_4) - U(t_3)]$ are uncorrelated for any two non-overlapping intervals $(t_1, t_2)$ and $(t_3, t_4)$. In the theory of Brownian motion, Equation (3.32) arises in the study of the Ornstein–Uhlenbeck model and is sometimes called the Langevin equation. It can be shown that the process $\{X(t)\}$ defined in Equation (3.32) has ac.f.

$$\rho(\tau) = \mathrm{e}^{-a|\tau|},$$

which is similar to the ac.f. of a first-order AR process in discrete time in that both decay exponentially. However, the rigorous study of continuous processes, such as that defined by Equation (3.32), requires considerable mathematical machinery, including a knowledge of stochastic integration. Thus we will not pursue this topic here. The reader is referred to a specialist book on probability theory such as Rogers and Williams (1994).

## 3.13 The Wold Decomposition Theorem

This section[4] gives a brief introduction to a famous result, called the Wold decomposition theorem, which is of mainly theoretical interest. This essentially says that any discrete-time stationary process can be expressed as the sum of two uncorrelated processes, one purely deterministic and one purely indeterministic. The terms 'deterministic' and 'indeterministic' are defined as follows. We can regress $X_t$ on $(X_{t-q}, X_{t-q-1}, \ldots)$ and denote the residual variance from the resulting linear regression model by $\tau_q^2$. As $\tau_q^2 \leq \mathrm{Var}(X_t)$, it is clear that, as $q$ increases, $\tau_q^2$

[3]For example, if $X_t = \alpha X_{t-1} + Z_t$, then $(1-\alpha)X_t + \alpha \nabla X_t = Z_t$.

[4]This section may be omitted at a first reading.

is a non-decreasing bounded sequence and therefore tends to a limit as $q \to \infty$. If $\lim_{q\to\infty} \tau_q^2 = \text{Var}(X_t)$ then linear regression on the remote past is useless for prediction purposes, and we say that $\{X_t\}$ is **purely indeterministic**. Stationary linear stochastic processes, such as AR, MA, and ARMA processes, are of this type. However, if $\lim_{q\to\infty} \tau_q^2$ is zero, then the process can be forecast exactly, and we say that $\{X_t\}$ is purely **deterministic**. Note that this definition of 'deterministic' involves linear models and bears rather little relation to the usual meaning of the word as used in philosophical debate or common parlance.

The Wold decomposition theorem also says that the purely indeterministic component can be written as a linear sum of a sequence of uncorrelated random variables, say $\{Z_t\}$. This has the same form as the general linear process in Equation (3.30) except that the Zs are merely uncorrelated, rather than independent. Of course, if the Zs are normally distributed, then zero correlation implies independence and we really do have a general linear process.

While the concept of a purely indeterministic process may sometimes be helpful, the Wold decomposition itself can be of little assistance. For a linear purely indeterministic process, such as an AR or ARMA model, it is inappropriate to try to model it as an MA($\infty$) process, as there will be too many parameters to estimate (although the MA($\infty$) form may be helpful for computing forecast error variances — see Section 5.3.1). Rather, we normally seek a model that gives an adequate approximation to the given data with as few parameters as possible, and that is where an ARMA representation can help.

For processes generated in a non-linear way, the Wold decomposition is usually of even less interest, as the best predictor may be quite different from the best linear predictor. Consider, for example, a sinusoidal process (see Exercise 3.14), such as

$$X_t = g\cos(\omega t + \theta), \tag{3.33}$$

where $g$ is a constant, $\omega$ is a constant in $(0, \pi)$ called the frequency of the process, and $\theta$ is a random variable, called the phase, which is uniformly distributed on $(0, 2\pi)$ but which is fixed for a single realization. Note that we must include the term $\theta$ so that

$$E(X_t) = 0 \qquad \text{for all } t.$$

If this is not done, Equation (3.33) would not define a stationary process. As $\theta$ is fixed for a single realization, once enough values of $X_t$ have been observed to evaluate $\theta$, all subsequent values of $X_t$ are completely determined. It is then obvious that (3.33) defines a deterministic process. However, it is not 'purely deterministic' as defined above because (3.33) is not linear. In fact the process is 'purely indeterministic' using a linear predictor, even though it *is* deterministic using an appropriate non-linear predictor. These are deep waters!

## Exercises

In all the following questions, $\{Z_t\}$ is a discrete-time, purely random process, such that $E(Z_t) = 0$, $\text{Var}(Z_t) = \sigma_Z^2$, and successive values of $Z_t$ are independent so that

$\text{Cov}(Z_t, Z_{t+k}) = 0, k \neq 0$. Exercise 3.14 is harder than the other exercises and may be omitted.

**3.1** Show that the ac.f. of the second-order MA process

$$X_t = Z_t + 0.7Z_{t-1} - 0.2Z_{t-2}$$

is given by

$$\rho(k) = \begin{cases} 1 & k = 0 \\ 0.37 & k = \pm 1 \\ -0.13 & k = \pm 2 \\ 0 & \text{otherwise} \end{cases}$$

**3.2** Consider the MA($m$) process, with equal weights $1/(m+1)$ at all lags (so it is a *real* moving average), given by

$$X_t = \sum_{k=0}^{m} Z_{t-k}/(m+1)$$

Show that the ac.f. of this process is

$$\rho(k) = \begin{cases} (m+1-k)/(m+1) & k = 0, 1, \ldots, m, \\ 0 & k > m, \\ \rho(-k) & k < 0. \end{cases}$$

**3.3** Consider the infinite-order MA process $\{X_t\}$, defined by

$$X_t = Z_t + C(Z_{t-1} + Z_{t-2} + \cdots)$$

where $C$ is a non-zero constant. Show that the process is non-stationary. Also show that the series of first differences $\{Y_t\}$ defined by

$$Y_t = X_t - X_{t-1}$$

is a first-order MA process and is stationary. Find the ac.f. of $\{Y_t\}$.

**3.4** Find the ac.f. of the first-order AR process with $E(X_t) = \mu$, defined by

$$X_t - \mu = 0.7(X_{t-1} - \mu) + Z_t.$$

Plot $\rho(k)$ for $k = -6, -5, \ldots, -1, 0, +1, \ldots, +6$.

**3.5** If $X_t = \mu + Z_t + \beta Z_{t-1}$, where $\mu$ is a constant, show that the ac.f. does not depend on $\mu$.

**3.6** Find the values of $\lambda_1, \lambda_2$, such that the second-order AR process defined by

$$X_t = \lambda_1 X_{t-1} + \lambda_2 X_{t-2} + Z_t$$

is stationary. If $\lambda_1 = 1/3, \lambda_2 = 2/9$, show that the ac.f. of $X_t$ is given by

$$\rho(k) = \frac{16}{21}\left(\frac{2}{3}\right)^{|k|} + \frac{5}{21}\left(-\frac{1}{3}\right)^{|k|} \qquad k = 0, \pm 1, \pm 2, \ldots.$$

**3.7** Show that the ac.f. of the stationary second-order AR process

$$X_t = \frac{1}{12}X_{t-1} + \frac{1}{12}X_{t-2} + Z_t$$

is given by

$$\rho(k) = \frac{45}{77}\left(\frac{1}{3}\right)^{|k|} + \frac{32}{77}\left(-\frac{1}{4}\right)^{|k|} \qquad k = 0, \pm 1, \pm 2, \ldots.$$

**3.8** Suppose the process $\{X_t\}$ is stationary and has acv.f. $\gamma_X(k)$. A new process $\{Y_t\}$ is defined by $Y_t = X_t - X_{t-1}$. Show that $\{Y_t\}$ is stationary and obtain the acv.f. of $\{Y_t\}$ in terms of $\gamma_X(k)$ and find $\gamma_Y(k)$ when $\gamma_X(k) = \lambda^{|k|}$.

**3.9** For each of the following models:

(a) $X_t = 0.3X_{t-1} + Z_t$

(b) $X_t = Z_t - 1.3Z_{t-1} + 0.4Z_{t-2}$

(c) $X_t = 0.5X_{t-1} + Z_t - 1.3Z_{t-1} + 0.4Z_{t-2}$

express the model using $B$ notation and determine whether the model is stationary and/or invertible. For model (a) find the equivalent MA representation.

**3.10** Suppose that a MA($q$) process, $\{X_t\}$, can be represented in the form $X_t = \psi(B)Z_t$. The autocovariance generating function is defined by

$$\Gamma(s) = \sum_{k=-\infty}^{\infty} \gamma_X(k)s^k,$$

where $\gamma_X(k)$ is the autocovariance coefficient of $X_t$ at lag $k$, and $s$ is a dummy variable. Show that $\Gamma(s) = \sigma_Z^2\psi(s)\psi(1/s)$. (Hint: Equate coefficients of $s^k$.)

**3.11** Compute the ac.f. of the ARMA$(1,1)$ model

$$X_t = 0.7X_{t-1} + Z_t + 0.3Z_{t-1}.$$

**3.12** For the model $(1-0.2B)(1-B)X_t = (1-0.5B)Z_t$:

(a) Classify the model as an ARIMA$(p,d,q)$ process (i.e. find $p,d,q$).

(b) Determine whether the process is stationary and invertible.

(c) Evaluate the first three $\psi$ weights of the model when expressed as a MA$(\infty)$ model.

(d) Evaluate the first four $\pi$ weights of the model when expressed as an AR$(\infty)$ model.

Is the behaviour of the $\psi$ and $\pi$ weights what you would expect, given the type of model?

**3.13** Show that the AR$(2)$ process

$$X_t = X_{t-1} + cX_{t-2} + Z_t$$

is stationary provided $-1 < c < 0$. Find the autocorrelation function when $c = -3/16$.

Show that the AR(3) process

$$X_t = X_{t-1} + cX_{t-2} - cX_{t-3} + Z_t$$

is non-stationary for all values of $c$.

**3.14** For a complex-valued process $X(t)$, with (possibly complex) mean $\mu$, the acv.f. is defined by

$$\gamma(\tau) = E\{[X(t) - \mu][\bar{X}(t+\tau) - \bar{\mu}]\}$$

where the overbar denotes the complex conjugate. If $Y$ is a complex random variable which does not depend on $t$ with mean zero, show that the process $X(t) = Y e^{i\omega t}$ is second-order stationary, where $\omega$ is a real constant. One useful form for the random variable $Y$ occurs when it takes the form $g\, e^{i\theta}$, where $g$ is a constant and $\theta$ is a random variable that is uniformly distributed on $(0, 2\pi)$. Show that $E(Y) = 0$ in this case.

**3.15** Consider the stationary process $X_t = Z_t - \theta Z_{t-1} - 6\theta^2 Z_{t-2}$, where $\theta$ is a real non-zero number.

(a) Under which condition does $X_t$ become invertible?

(b) Find the ac.f. of $\{X_t\}$.

**3.16** Consider the AR(2) process ($a$ is a real number):

$$X_t - \frac{2}{a(a+2)} X_{t-1} - \frac{1}{a(a+2)} X_{t-2} = Z_t.$$

(a) Under which conditions, is the process $X_t$ stationary?

(b) Find the ac.f. of $\{X_t\}$.

**3.17** Consider the ARMA(2,2) process $X_t - \alpha X_{t-2} = Z_t - \beta Z_{t-2}$, where $\alpha, \beta$ are different real non-zero numbers.

(a) Under which conditions, is process $X_t$ stationary and invertible?

(b) Compute $\gamma(0), \gamma(1)$, and $\gamma(2)$, where $\gamma(\cdot)$ is the acv.f. of $X_t$. Find the values of $\rho(1)$ and $\rho(2)$, where $\rho(\cdot)$ is the ac.f. of $X_t$.

Chapter 4

# Fitting Time Series Models in the Time Domain

Chapter 3 introduced several different types of probability models that may be used to describe time series. This chapter discusses the problem of fitting a suitable model to an observed time series. We restrict attention to the discrete-time case and the major diagnostic tool used in this chapter is the sample autocorrelation function (ac.f.). Inference based on this function is often called an **analysis in the time domain**.

## 4.1 Estimating Autocovariance and Autocorrelation Functions

We have already noted in Section 3.3 that the theoretical ac.f. is an important tool for describing the properties of a stationary stochastic process. In Section 2.7 we heuristically introduced the sample ac.f. of an observed time series, and this is an intuitively reasonable estimate of the theoretical ac.f., provided the series is stationary. This section investigates the properties of the sample ac.f. more closely.

Let us look first at the autocovariance function (acv.f.). Suppose we have $N$ observations on a stationary process, say $x_1, x_2, \ldots, x_N$. Then the sample autocovariance coefficient at lag $k$ (see Equation (2.8)) given by

$$c_k = \sum_{t=1}^{N-k} (x_t - \bar{x})(x_{t+k} - \bar{x})/N \tag{4.1}$$

is the usual estimator for the theoretical autocovariance coefficient $\gamma(k)$ at lag $k$. It can be shown (e.g. Priestley, 1981, Chapter 5) that this estimator is biased, but that the bias is of order $1/N$. Moreover

$$\lim_{N \to \infty} E(c_k) = \gamma(k),$$

so that the estimator is asymptotically unbiased.

It can also be shown that

$$\mathrm{Cov}(c_k, c_m) \simeq \sum_{r=-\infty}^{\infty} \{\gamma(r)\gamma(r+m-k) + \gamma(r+m)\gamma(r-k)\}/N. \tag{4.2}$$


DOI: 10.1201/9781003646112-4

When $m = k$, Equation (4.2) gives us the variance of $c_k$. Equation (4.2) also highlights the fact that successive values of $c_k$ may be (highly) correlated and this increases the difficulty of interpreting the correlogram.

The estimator (4.1) may be compared with the alternative estimator

$$c'_k = \sum_{t=1}^{N-k} (x_t - \bar{x})(x_{t+k} - \bar{x})/(N-k), \tag{4.3}$$

where the divisor is $(N-k)$, rather than $N$. This is used by some authors because it is claimed to have a smaller bias, though this is, in fact, not always the case. Indeed some authors call $c'_k$ the 'unbiased estimator' even though it is biased when, as here, the population mean, $\mu$, is unknown and replaced by the sample mean, $\bar{x}$, as is usually the case in practice. It turns out that the earlier estimator in Equation (4.1) leads to a sample acv.f. having a useful property called positive semi-definiteness, which means that its finite Fourier transform is non-negative, among other consequences. This property is useful in estimating the spectrum (see Chapter 7) and so we generally prefer to use Equation (4.1), rather than (4.3). Note that when $k = 0$, we get the same estimate of variance using both formulae, but that this estimate, involving $\bar{x}$, will generally still be biased (Percival, 1993).

Having estimated the acv.f., we then take

$$r_k = c_k/c_0 \tag{4.4}$$

as an estimator for $\rho(k)$. The properties of $r_k$ are rather more difficult to find than those of $c_k$ because it is the ratio of two random variables. It can be shown that $r_k$ is generally biased. A general formula for the variance of $r_k$ is given by Kendall et al. (1983, Section 48.1) and depends on *all* the autocorrelation coefficients of the process. We will only consider the properties of $r_k$ when sampling from a purely random process, when all the theoretical autocorrelation coefficients are zero except at lag zero. These results help us to decide if the observed values of $r_k$ from a given time series are significantly different from zero.

Suppose that $x_1, \ldots, x_N$ are observations on independent and identically distributed random variables with arbitrary mean. Then it can be shown (Kendall et al., 1983, Chapter 48) that

$$E(r_k) \simeq -1/N, \qquad Var(r_k) \simeq 1/N$$

and that $r_k$ is asymptotically normally distributed under weak conditions. Thus having plotted the correlogram, as described in Section 2.7, we can check for randomness by plotting approximate 95% confidence limits at

$$-1/N \pm 1.96/\sqrt{N} = \frac{-1 \pm 1.96\sqrt{N}}{N} \approx \frac{\pm 1.96\sqrt{N}}{N} = \pm\frac{1.96}{N}.$$

Observed values of $r_k$ which fall outside these limits are 'significantly' different from zero at the 5% level. However, when interpreting a correlogram, it must be remembered that the overall probability of getting at least one coefficient outside these limits, given that the data really are random, increases with the number of coefficients

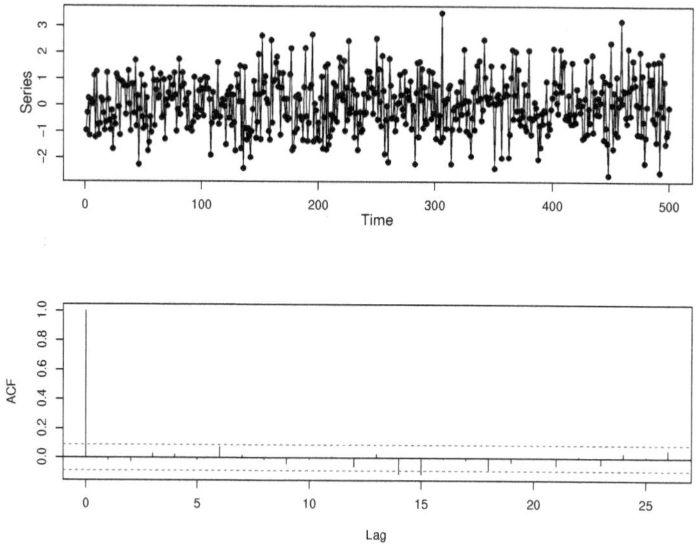


Figure 4.1 *500 'independent' normally distributed observations and their correlogram. The dashed lines in the bottom panel are at* $\pm 1.96/\sqrt{500}$.

plotted. For example, if the first 20 values of $r_k$ are plotted, then one expects one 'significant' value (at the 5% level) on average even if the data really are random. Thus, if only one or two coefficients are 'significant', the size and lag of these coefficients must be taken into account when deciding if a set of data is random. A single coefficient just outside the 'null' 95% confidence limits may be ignored, but two or three values well outside the 'null' limits will be taken to indicate non-randomness. A single 'significant' coefficient at a lag which has some physical interpretation, such as lag 1 or a lag corresponding to seasonal variation, will also provide plausible evidence of non-randomness.

Figure 4.1 shows the correlogram for 400 observations, generated on a computer, which are independent normally distributed variables. The 'null' 95% confidence limits are approximately $\pm\, 1.96/\sqrt{500} \approx \pm 0.088$. We see that 2 of the first 20 values of $r_k$ are just 'significant'. However, they occur at apparently arbitrary lags (namely, 14 and 15). Thus we conclude that there is no firm evidence to reject the hypothesis that the observations are independent and identically distributed. This in turn means that the way that the computer generates 'random' numbers appears to be satisfactory.

### *4.1.1 Using the correlogram in modelling*

We have already given some general advice on interpreting correlograms in Section 2.7.2, while their use in assessing randomness was considered above. The correlogram is also helpful in trying to identify a suitable class of models for a given time series, and, in particular, for selecting the most appropriate type of autoregressive integrated moving average (ARIMA) model. A correlogram like that in Figure 2.7, where the values of $r_k$ do not come down to zero reasonably quickly, indicates non-stationarity and so the series needs to be differenced. For stationary series, the correlogram is compared with the theoretical ac.f.s of different ARMA processes in order to choose the one which seems to be the 'best' representation. In Section 3.6, we saw that the ac.f. of a MA($q$) process is easy to recognize as it 'cuts off' at lag $q$. However, the ac.f. of an AR($p$) process is rather more difficult to categorize, as it is a mixture of damped exponentials and sinusoids and dies out slowly (or attenuates). The ac.f. of a mixed ARMA model will also generally attenuate rather than 'cut off'.

Suppose, for example, that we find that $r_1$ is significantly different from zero but that subsequent values of $r_k$ are all close to zero. Then an MA(1) model is indicated because its theoretical ac.f. is of this form. Alternatively, if $r_1, r_2, r_3, \ldots$ appear to be decreasing exponentially, then an AR(1) model may be appropriate.

The interpretation of correlograms is one of the hardest aspects of time-series analysis and practical experience is a 'must'. Inspection of the partial autocorrelation function (see Section 4.2.2) can provide additional help.

### *4.1.2 Estimating the mean*

The first inferential problem considered in most statistics texts is the estimation of a population mean using a sample mean. However, in time-series analysis, the topic is often overlooked and omitted, because of the special problems that relate to using a time-series sample mean as an estimate of some underlying population mean. Although we have used the sample mean in computing the sample acv.f. and ac.f. — see Equations (4.1) and (4.4) — we have already noted, at the start of Chapter 2, that the sample mean is a potentially misleading summary statistic unless all systematic components have been removed. Thus the sample mean should only be considered as a summary statistic for data thought to have come from a stationary process. Even when this is so, it is important to realize that the statistical properties of the sample mean are quite different from those that usually apply.

Suppose we have time-series data $\{x_i \text{ for } i = 1, \ldots, N\}$ from a stationary process having mean $\mu$, variance $\sigma^2$ and theoretical ac.f. $\rho(k)$. Let $\overline{X} = \sum_{i=1}^{N} X_i/N$ denote the sample mean value expressed as a random variable. Then the usual result for independent observations is that $\text{Var}(\overline{X}) = \sigma^2/N$. However, for correlated observations, it can be shown (e.g. Priestley, 1981, p. 319) that

$$\text{Var}(\overline{X}) = \frac{\sigma^2}{N}\left[1 + 2\sum_{r=1}^{N-1}(1 - \frac{r}{N})\rho(r)\right]$$

and this quantity can differ considerably from $\sigma^2/N$ when autocorrelations are substantial. In particular, for an AR(1) process with parameter $\alpha$, the formula reduces to $\text{Var}(\overline{X}) = \frac{\sigma^2}{N}(\frac{1+\alpha}{1-\alpha})$ for large $N$ (see Exercise 4.1). Put another way, the equivalent number of independent observations is $N(1-\alpha)/(1+\alpha)$. When $\alpha > 0$ (the usual case), the positive autocorrelation means that there is less information in the data than might be expected in regard to estimating the mean. On the other hand, when $\alpha < 0$, there is *more* information. Any tests on the sample mean should remember to adjust for this effect.

Another way that estimating a time-series mean differs from the rest of Statistics is that any results depend upon the underlying process having a property called ergodicity. The difficult ideas involved are introduced briefly in the next optional subsection.

### *4.1.3 Ergodicity*

This subsection[1] gives a brief introduction to the idea of ergodicity. It is not immediately obvious that one can obtain consistent estimates of the properties of a stationary process from a single finite realization. This requires that an average over time for a single time series, like $\bar{x} = \sum_{t=1}^{N} x_t/N$, can be used to estimate the ensemble properties of the underlying process at a particular time, like $E(X_t)$. This means that, as we only ever get one observation actually at time $t$, the properties of the series at time $t$ are estimated using data collected at other time points. Fortunately, some theorems, called **ergodic theorems**, have been proved, showing that for most stationary processes, which are likely to be met in practice, the sample moments of an observed time series do indeed converge to the corresponding population moments. In other words, a time average like $\sum_{t=1}^{N} x_t/N$ converges to a population quantity like $E(X_t)$ as $N \to \infty$. A sufficient condition for this to happen is that $\rho_k \to 0$ as $k \to \infty$ and the process is then called 'ergodic in the mean'.

We will not pursue the topic here but rather simply assume that appropriate ergodic properties are satisfied when estimating the properties of stationary processes. More details may be found, for example, in Hamilton (1994, p. 46).

## 4.2 Fitting an Autoregressive Process

Having estimated the ac.f. of a given time series, we should have some idea as to which stochastic process will provide a suitable model. If an autoregressive (AR) process is thought to be appropriate, there are two related questions:

1. What is the order of the process?
2. How can we estimate the parameters of the process?

It is convenient to consider the second question first.

### *4.2.1 Estimating parameters of an AR process*

Suppose we have an AR process of order $p$, with mean $\mu$, given by

$$X_t - \mu = \alpha_1(X_{t-1} - \mu) + \cdots + \alpha_p(X_{t-p} - \mu) + Z_t. \qquad (4.5)$$

[1] This subsection may be omitted at a first reading.

Given $N$ observations $x_1, \ldots, x_N$, the parameters $\mu$, $\alpha_1, \ldots, \alpha_p$ may be estimated by least squares by minimizing

$$S = \sum_{t=p+1}^{N} \left[ x_t - \mu - \alpha_1(x_{t-1} - \mu) - \cdots - \alpha_p(x_{t-p} - \mu) \right]^2$$

with respect to $\mu, \alpha_1, \ldots, \alpha_p$. If the $Z_t$ process is normal, then the least squares estimates are also maximum likelihood estimates (Jenkins and Watts, 1968, Section 5.4) conditional on the first $p$ values in the time series being fixed.

In the first-order case, with $p = 1$, we find (see Exercise 4.1)

$$\widehat{\mu} = \frac{\bar{x}_{(2)} - \widehat{\alpha}_1 \bar{x}_{(1)}}{1 - \widehat{\alpha}_1} \tag{4.6}$$

and

$$\widehat{\alpha}_1 = \frac{\sum_{t=1}^{N-1}(x_t - \widehat{\mu})(x_{t+1} - \widehat{\mu})}{\sum_{t=1}^{N-1}(x_t - \widehat{\mu})^2} \tag{4.7}$$

where $\bar{x}_{(1)}, \bar{x}_{(2)}$ are the means of the first and last $(N-1)$ observations. Now

$$\bar{x}_{(1)} \simeq \bar{x}_{(2)} \simeq \bar{x}$$

and so we have approximately that

$$\widehat{\mu} = \bar{x}. \tag{4.8}$$

This approximate estimator is intuitively appealing and is nearly always preferred to Equation (4.6). Substituting this estimator into Equation (4.7) gives

$$\widehat{\alpha}_1 \approx \frac{\sum_{t=1}^{N-1}(x_t - \bar{x})(x_{t+1} - \bar{x})}{\sum_{t=1}^{N-1}(x_t - \bar{x})^2}. \tag{4.9}$$

It is interesting to note that this is exactly the same estimator that would arise if we were to treat the autoregressive equation

$$X_t - \bar{x} = \alpha_1(x_{t-1} - \bar{x}) + Z_t$$

as an ordinary regression with $(x_{t-1} - \bar{x})$ as the 'independent' variable (which of course it isn't). In fact Mann and Wald (1943) showed that, asymptotically, much classical regression theory can be applied to AR models.

A further approximation, which is often used, is obtained by changing the denominator of (4.9) slightly to

$$\sum_{t=1}^{N}(x_t - \bar{x})^2 \qquad \text{so that} \quad \widehat{\alpha}_1 \simeq c_1/c_0 = r_1.$$

This approximate estimator for $\widehat{\alpha}_1$ is also intuitively appealing since $r_1$ is an estimator for $\rho(1)$ and $\rho(1) = \alpha_1$ for a first-order AR process. A confidence interval for $\alpha_1$ may be obtained from the fact that the asymptotic standard deviation of $\widehat{\alpha}_1$ is $\sqrt{(1-\alpha_1^2)/N}$, although the confidence interval will not be symmetric for $\widehat{\alpha}_1$ away from zero. When $\alpha_1 = 0$, the standard deviation of $\widehat{\alpha}_1$ is $1/\sqrt{N}$, and so a test for $\alpha_1 = 0$ is given by seeing whether $\widehat{\alpha}_1 \approx 1$ lies within the range $\pm\ 1.96/\sqrt{N}$. This is equivalent to the test for $\rho(1) = 0$ already noted in Section 4.1.

For a second-order AR process, where $p = 2$, similar approximations may be made to give

$$\widehat{\mu} \simeq \bar{x}$$

$$\widehat{\alpha}_1 \simeq r_1(1-r_2)/(1-r_1^2) \tag{4.10}$$

$$\widehat{\alpha}_2 \simeq (r_2-r_1^2)/(1-r_1^2). \tag{4.11}$$

These results are also intuitively reasonable in that if we fit a second-order model to what is really a first-order process, then as $\alpha_2 = 0$ we have $\rho(2) = \rho(1)^2 = \alpha_1^2$ and so $r_2 \simeq r_1^2$. Thus Equations (4.10) and (4.11) become $\widehat{\alpha}_1 \simeq r_1$ and $\widehat{\alpha}_2 \simeq 0$, as we would hope. The coefficient $\widehat{\alpha}_2$ is called the (sample) **partial autocorrelation coefficient** of order two, as it measures the excess correlation between observations two steps apart (e.g. between $X_t$ and $X_{t+2}$) not accounted for by the autocorrelation at lag 1, namely, $r_1$ – see Section 4.2.2.

In addition to point estimates of $\alpha_1$ and $\alpha_2$ it is also possible to find a confidence region in the $(\alpha_1, \alpha_2)$ plane (Jenkins and Watts, 1968, p. 192).

Higher-order AR processes may also be fitted by least squares in a straightforward way. Two alternative approximate methods are commonly used, which both involve taking $\widehat{\mu} = \bar{x}$. The first method fits the data to the model

$$X_t - \bar{x} = \alpha_1(x_{t-1}-\bar{x}) + \cdots + \alpha_p(x_{t-p}-\bar{x}) + Z_t$$

treating it as if it were an ordinary regression model. A standard multiple regression computer program may be used with appropriate modification.

The second method involves substituting the sample autocorrelation coefficients into the first $p$ Yule–Walker equations (see Section 3.7) and solving for $(\widehat{\alpha}_1, \ldots, \widehat{\alpha}_p)$. In matrix form these equations are

$$R\widehat{\alpha} = \mathbf{r} \tag{4.12}$$

where $\widehat{\alpha}^{\mathrm{T}} = (\widehat{\alpha}_1, \ldots, \widehat{\alpha}_p)$, $\mathbf{r}^{\mathrm{T}} = (r_1, \ldots, r_p)$ and

$$\mathrm{R} = \begin{pmatrix} 1 & r_1 & r_2 & \cdots & r_{p-1} \\ r_1 & 1 & r_1 & \cdots & r_{p-2} \\ r_2 & r_1 & 1 & \cdots & r_{p-3} \\ \cdots & \cdots & \cdots & \cdots & \cdots \\ r_{p-1} & r_{p-2} & r_{p-3} & \cdots & 1 \end{pmatrix}$$

is a $(p \times p)$ matrix. For $N$ reasonably large, both methods will give estimated values 'very close' to the exact least squares estimates for which $\widehat{\mu}$ is close to, but not necessarily equal to, $\bar{x}$. However, for smaller $N$ (e.g. less than about 50), the Yule–Walker estimates are not so good, especially when the parameter values are such that the model is 'close' to being non-stationary.

### 4.2.2 *Determining the order of an AR process*

It is often difficult to assess the order of an AR process from the sample ac.f. alone. For a first-order process the theoretical ac.f. decreases exponentially and the sample function should have a similar shape. However, for higher-order processes, the ac.f. may be a mixture of damped exponential or sinusoidal functions and is difficult to identify. One approach is to fit AR processes of progressively higher order, to calculate the residual sum of squares for each value of $p$ and to plot this against $p$. It may then be possible to see the value of $p$ where the curve 'flattens out' and the addition of extra parameters gives little improvement in fit.

Another aid to determining the order of an AR process is the **partial autocorrelation function**, which is defined as follows. When fitting an AR($p$) model, the last coefficient $\alpha_p$ will be denoted by $\pi_p$ and measures the excess correlation at lag $p$ which is not accounted for by an AR($p-1$) model. It is called the $p$th partial autocorrelation coefficient and, when plotted against $p$, gives the partial ac.f. The first partial autocorrelation coefficient $\pi_1$ is simply equal to $\rho(1)$, and this is equal to $\alpha_1$ for an AR(1) process. It can be shown (see Exercise 4.3 and the discussion following Equation (4.11) above) that the second partial correlation coefficient is $[\rho(2)-\rho(1)^2]/[1-\rho(1)^2]$, and we note that this is zero for an AR(1) process where $\rho(2)=\rho(1)^2$.

The sample partial ac.f. is usually estimated by fitting AR processes of successively higher order and taking $\widehat{\pi}_1 = \widehat{\alpha}_1$ when an AR(1) process is fitted, taking $\widehat{\pi}_2 = \widehat{\alpha}_2$ when an AR(2) process is fitted, and so on. Values of $\widehat{\pi}_j$, which are outside the range $\pm\, 1.96/\sqrt{N}$, are significantly different from zero at the 5% level. It can be shown that the partial ac.f. of an AR($p$) process 'cuts off' at lag $p$ so that the 'correct' order is assessed as that value of $p$ beyond which the sample values of $\{\pi_j\}$ are not significantly different from zero.

In contrast, the partial ac.f. of an MA process will generally attenuate (or die out slowly). Thus the partial ac.f. has reverse properties to those of the ac.f. in regard to identifying AR and MA models. Note that McCullough (1998) recommends that Yule–Walker estimates should *not* be used to estimate partial autocorrelations.

Some additional tools to aid model identification are discussed in Section 4.5.

*Example 4.1*

We now show an example of using the above procedure to fit an AR process to two simulated series. The first series follows an AR(2) process

$$X_t = -0.4X_{t-1} + 0.3X_{t-2} + Z_t, \qquad Z_t \sim N(0,1),$$

and the second follows an AR(3) process

$$X_t = \frac{1}{4}X_{t-1} + \frac{1}{2}X_{t-2} - \frac{1}{8}X_{t-3} + Z_t, \qquad Z_t \sim N(0,1).$$

Both series have $N = 500$ observations and are shown in the top left and top right panels of Figure 4.2, respectively. To see the dependence of time series observations

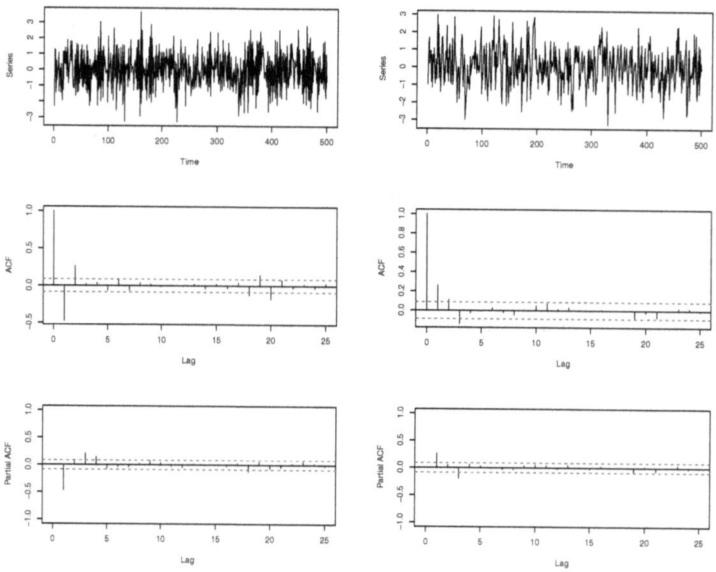

Figure 4.2 *Two simulated AR processes (Left:* $(1+0.4B-0.3B^2)X_t = Z_t$*; Right:* $(1-1/4B-1/2B^2+1/8B^3)X_t = Z_t$*) and their sample ac.f.'s and sample partial ac.f.'s. (Top: Simulated series; Middle: Sample ac.f.'s; Bottom: Sample partial ac.f.'s.)*

at different lags, their sample ac.f.'s and sample partial ac.f.'s are calculated and shown in the middle and bottom panels of Figure 4.2, respectively. Note that both sample ac.f.'s decay rapidly, and the sample partial ac.f.'s cut off at lags 2 and 3, respectively. Figure 4.2 can be reproduced using the following R command.

```
> set.seed(4)
> n<-500
> x1<-arima.sim(list(order=c(2,0,0), ar=c(-0.4, 0.3)), n)
> x2<-arima.sim(list(order=c(3,0,0), ar=c(1/4,1/2,-1/8)),n)

> par(mfrow=c(3,2), mar=c(4,4,4,4))
> plot(x1, type="l", xlab="Time", ylab="Series")
> plot(x2, type="l", xlab="Time", ylab="Series")
> acf(x1, 25, xlab="Lag", ylab="ACF", main="")
> acf(x2, 25, xlab="Lag", ylab="ACF", main="")
> acf(x1,25,type="partial",xlab="Lag",ylab="Partial ACF",
    main="", ylim=c(-1,1),xlim=c(0,25))
> acf(x2,25,type="partial",xlab="Lag",ylab="Partial ACF",
    main="", ylim=c(-1,1),xlim=c(0,25))
```

If AR models are thought to be appropriate for these two series, one may consider fitting an AR(2) process and an AR(3) process, respectively, to the data because the sample partial acf's cut off at lags 2 and 3, respectively. Depending on whether one assumes the mean of the series is zero or not, parameters of an AR process with zero or nonzero mean should be estimated. This can be easily done by using the `arima` function in `R`, which estimates model parameters by maximum likelihood or minimizing conditional sum-of-squares. The following `R` commands demonstrate the usage of this function.

```
> x1.fit1<-arima(x1, order=c(2,0,0))
> x1.fit1
Call:
arima(x = x1, order = c(2, 0, 0))

Coefficients:
          ar1     ar2  intercept
      -0.4450  0.2959    -0.0600
s.e.   0.0426  0.0426     0.0374

sigma^2 estimated as 0.9254: log likelihood=-690.42,aic=1388.84

> x2.fit1<-arima(x2, order=c(3,0,0))
> x2.fit1
Call:
arima(x = x2, order = c(3, 0, 0))

Coefficients:
         ar1     ar2      ar3  intercept
      0.1737  0.5043  -0.1175    -0.0946
s.e.  0.0445  0.0393   0.0448     0.0988

sigma^2 estimated as 0.9495: log likelihood=-696.82,aic=1403.65
```

Hence the fitted models are

$$X_t = -0.0600_{(.0426)} - 0.4450_{(.0426)}X_{t-1} + 0.2959_{(.0426)}X_{t-2} + Z_t,$$

with $Z_t \sim N(0, 0.9254)$, for the first series, and

$$\begin{aligned} X_t = -0.0946_{(.0988)} &+ 0.1737_{(.0445)}X_{t-1} + 0.5043_{(.0393)}X_{t-2} \\ &-0.1175_{(.0448)}X_{t-3} + Z_t, \end{aligned}$$

with $Z_t \sim N(0, 0.9495)$, for the second series. Note that the 95% confidence intervals of the intercepts in the two fitted models, $-0.0600 \pm 1.96 \cdot 0.0426$ and $-0.0946 \pm 1.96 \cdot 0.0988$, cover zero, indicating that the means of two AR processes might be zero. An AR model with zero mean can be estimated by adding an argument `include.mean=FALSE` to the `arima` command. In particular, fitting AR models with zero means to the two series can be done by the following `R` commands.

```
> x1.fit2<-arima(x1, order=c(2,0,0), include.mean=FALSE)
> x1.fit2

Call:
arima(x = x1, order = c(2, 0, 0), include.mean = FALSE)

Coefficients:
          ar1     ar2
      -0.4384  0.3025
s.e.   0.0425  0.0425

sigma^2 estimated as 0.93: log likelihood=-691.69, aic=1389.37

> x2.fit2<-arima(x2, order=c(3,0,0), include.mean=FALSE)
> x2.fit2

Call:
arima(x = x2, order = c(3, 0, 0), include.mean = FALSE)

Coefficients:
         ar1     ar2      ar3
      0.1755  0.5069  -0.1156
s.e.  0.0445  0.0392   0.0448

sigma^2 estimated as 0.9511: log likelihood=-697.28, aic=1402.55
```

Therefore, the fitted AR models with zero means can be summarized as

$$X_t = -0.4384_{(.0425)}X_{t-1} + 0.3025_{(.0425)}X_{t-2} + Z_t,$$

with $Z_t \sim N(0, 0.93)$, for the first series, and

$$X_t = 0.1755_{(.0445)}X_{t-1} + 0.5069_{(.0392)}X_{t-2} - 0.1156_{(.0448)}X_{t-3} + Z_t,$$

with $Z_t \sim N(0, 0.9511)$, for the second series. We further note that, for the first series, the 95% confidence intervals of coefficients of $X_{t-1}$ and $X_{t-2}$ in the fitted model are $-0.4384 \pm 1.96 \cdot 0.0425$ and $0.3025 \pm 1.96 \cdot 0.0425$, respectively, and hence cover the true values of the coefficients $-0.4$ and $0.3$, respectively. For the second series, the 95% confidence intervals of coefficients of $X_{t-1}$, $X_{t-2}$ and $X_{t-3}$ in the fitted model are $0.1755 \pm 1.96 \cdot 0.0445$, $0.5069 \pm 1.96 \cdot 0.0392$, and $-0.1156 \pm 1.96 \cdot 0.0448$, respectively, and hence cover the true values of the coefficients $0.25$, $0.5$, and $-0.125$, respectively.

## 4.3 Fitting a Moving Average Process

Suppose now that a moving average (MA) process is thought to be an appropriate model for a given time series. As for an AR process, we have two problems:

(1) Finding the order of the process

(2) Estimating the parameters of the process.

As for an AR process, it is convenient to consider the second problem first.

### 4.3.1 *Estimating parameters of an MA process*

Estimation is more difficult for an MA process than an AR process, because efficient explicit estimators cannot be found. Instead some form of numerical iteration must be performed, albeit greatly facilitated by modern computers.

Let us begin by considering the first-order MA process, with mean $\mu$, given by

$$X_t = \mu + Z_t + \beta_1 Z_{t-1} \tag{4.13}$$

where $\mu$, $\beta_1$ are constants and $Z_t$ denotes a purely random process. We would like to be able to write the residual sum of squares $\Sigma Z_t^2$ solely in terms of the observed $x$s and the parameters $\mu$, $\beta_1$, as we did for the AR process. Then we could differentiate with respect to $\mu$ and $\beta_1$, and hence find the least squares estimates. Unfortunately this cannot be done and so explicit least squares estimates cannot be found; nor is it wise to simply equate sample and theoretical first-order autocorrelation coefficients by

$$r_1 = \widehat{\beta}_1 / (1 + \widehat{\beta_1^2}) \tag{4.14}$$

and choose the solution $\widehat{\beta}_1$ such that $|\widehat{\beta}_1| < 1$, because it can be shown that this gives rise to an inefficient estimator.

One possible alternative approach is as follows: (i) Select suitable starting values for $\mu$ and $\beta_1$, such as $\mu = \bar{x}$ and the value of $\beta_1$ given by the solution of Equation (4.14) (see Box et al., 1994, Part 5, Table A). (ii) Calculate the corresponding residual sum of squares using Equation (4.13) recursively in the form

$$Z_t = X_t - \mu - \beta_1 Z_{t-1}. \tag{4.15}$$

Taking $z_0 = 0$, we calculate $z_1 = x_1 - \widehat{\mu}$, and then $z_2 = x_2 - \widehat{\mu} - \widehat{\beta}_1 z_1$, and so on until $z_N = x_N - \widehat{\mu} - \widehat{\beta}_1 z_{N-1}$. Then the residual sum of squares $\sum_{t=1}^{N} z_t^2$ is calculated conditional on the given values of the parameters $\mu$ and $\beta_1$ and $z_0 = 0$. (iii) Repeat this procedure for other neighbouring values of $\mu$ and $\beta_1$ so that the residual sum of squares $\Sigma z_t^2$ is computed on a grid of points in the $(\mu, \beta_1)$ plane. (iv) Determine by inspection the values of $\mu$ and $\beta_1$ that minimize $\Sigma z_t^2$.

The above procedure gives least squares estimates, which are also maximum likelihood estimates conditional on $z_0 = 0$ provided that $Z_t$ is normally distributed. The procedure can be further refined by **back-forecasting** the value of $z_0$ (see Box et al., 1994, Section 6.4.3), but this is unnecessary except when $N$ is small or when $\beta_1$ is 'close' to plus or minus one. Nowadays the values of $\mu$ and $\beta$, which minimize $\Sigma z_t^2$, would normally be found by some iterative optimization procedure, such as hill-climbing, although a grid search can still sometimes be useful to see what the sum of squares surface looks like.

For higher-order processes a similar type of iterative procedure to that described above could be used. For example, with a second-order MA process one would guess starting values for $\mu, \beta_1, \beta_2$, compute the residuals recursively using

$$z_t = x_t - \widehat{\mu} - \widehat{\beta}_1 z_{t-1} - \widehat{\beta}_2 z_{t-2}$$

and compute $\Sigma z_t^2$. Then other values of $\widehat{\mu}, \widehat{\beta}_1, \widehat{\beta}_2$ could be tried, perhaps over a grid of points, until the minimum value of $\Sigma z_t^2$ is found. Nowadays computers can take advantage of an appropriate numerically efficient optimization procedure to minimize the residual sum of squares. Box et al. (1994, Section 7.2) describe such a procedure, which they call 'non-linear estimation', because the residuals are non-linear functions of the parameters. However, this term could give rise to confusion.

For a completely new set of data, it may be a good idea to use the method based on evaluating the residual sum of squares at a grid of points. A visual examination of the sum of squares surface will sometimes provide useful information. In particular, it is interesting to see how 'flat' the surface is; whether the surface is approximately quadratic; and whether the parameter estimates are approximately uncorrelated (because the major axes of the 'hill' are parallel to the coordinate axes).

In addition to point estimates, an approximate confidence region for the model parameters may be found as described by Box et al. (1994, Chapter 7) by assuming that the $Z_t$ are normally distributed. However, there is some doubt as to whether the asymptotic normality of maximum likelihood estimators will apply even for moderately large sample sizes (e.g. $N = 200$).

It should now be clear that it is much harder to estimate the parameters of an MA model than those of an AR model, because the 'errors' in an MA model are non-linear functions of the parameters and iterative methods are required to minimize the residual sum of squares. Because of this, many analysts prefer to fit an AR model to a given time series even though the resulting model may contain more parameters than the 'best' MA model.

### 4.3.2 *Determining the order of an MA process*

Although parameter estimation is harder for a given MA process than for a given AR process, the reverse is true as regards determining the order of the process. If an MA process is thought to be appropriate for a given set of data, the order of the process is usually evident from the sample ac.f. The theoretical ac.f. of an MA($q$) process has a very simple form in that it 'cuts off' at lag $q$ (see Section 3.6.1), and so the analyst should look for the lag beyond which the values of $r_k$ are close to zero. The partial ac.f. is generally of little help in identifying pure MA models because of its attenuated form.

*Example 4.2*

We now show an example of using the above procedure to fit a MA process to two simulated series. The first series follows an MA(2) process

$$X_t = Z_t - 0.5Z_{t-1} + 0.4Z_{t-2}, \qquad Z_t \sim N(0,1),$$

and the second follows an MA(3) process

$$X_t = Z_t + \frac{1}{3}Z_{t-1} + \frac{1}{5}Z_{t-2} - \frac{1}{9}Z_{t-3}, \qquad Z_t \sim N(0,1).$$

Both series have $N = 500$ observations and are shown in the top left and top right panels of Figure 4.3, respectively. To see the dependence of time series observations at different lags, their sample ac.f.'s and sample partial ac.f.'s are calculated and shown in the middle and bottom panels of Figure 4.3, respectively. Note that the sample ac.f.'s cut off at lags 2 and 3, respectively, while it is difficult to see patterns in their sample partial ac.f.'s. Figure 4.3 can be reproduced using the following R command.

```
> set.seed(5)
> n<-500
> x3<-arima.sim(list(order=c(0,0,2), ma=c(-0.5, 0.4)), n)
> x4<-arima.sim(list(order=c(0,0,3), ma=c(1/3,1/5,-1/9)),n)

> par(mfrow=c(3,2), mar=c(2,2,2,2))
> plot(x3, type="l", xlab="", ylab="")
> plot(x4, type="l", xlab="", ylab="")
> acf(x3, 25, ylab="", main="")
> acf(x4, 25, ylab="", main="")
> acf(x3, 25, type="partial", ylab="", ylim=c(-1,1),
      xlim=c(0,25))
> acf(x4, 25, type="partial", ylab="", ylim=c(-1,1),
      xlim=c(0,25))
```

If MA models are thought to be appropriate for these two series, the correlograms in Figure 4.3 suggest an MA(2) process and an MA(3) process respectively, to the data since the sample acf's cut off at lags 2 and 3, respectively. Depending on whether one assumes the mean of the series is zero or not, parameters of an MA process with zero or nonzero mean should be estimated. This can be easily done by using the `arima` function in R, which estimates model parameters by maximum likelihood or minimizing conditional sum-of-squares. The following R commands demonstrate the usage of this function.

```
> x3.fit1<-arima(x3, order=c(0,0,2))
> x3.fit1
Call:
arima(x = x3, order = c(0, 0, 2))

Coefficients:
          ma1     ma2  intercept
      -0.5165  0.4227    -0.0046
s.e.   0.0422  0.0434     0.0407
```

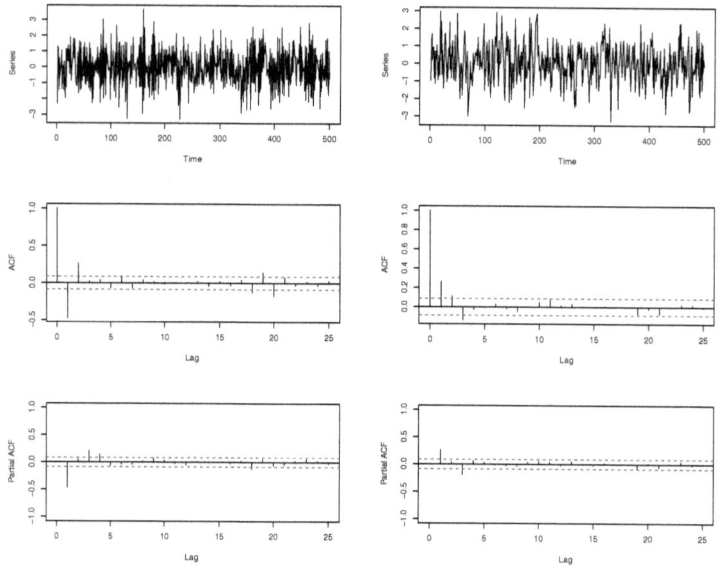

Figure 4.3 *Two simulated MA processes (Left: $X_t = Z_t - 0.5Z_{t-1} + 0.4Z_{t-2}$; Right: $X_t = Z_t + \frac{1}{3}Z_{t-1} + \frac{1}{5}Z_{t-2} - \frac{1}{9}Z_{t-3}$) and their sample ac.f.'s and partial ac.f.'s. (Top: Simulated series; Middle: Sample ac.f.'s; Bottom: Sample partial ac.f.'s.)*

```
sigma^2 estimated as 1.009:  log likelihood = -712.09,
aic = 1432.17

> x4.fit1<-arima(x4, order=c(0,0,3))
> x4.fit1
Call:
arima(x = x4, order = c(0, 0, 3))

Coefficients:
         ma1     ma2      ma3  intercept
      0.2644  0.1607  -0.1245     0.0528
s.e.  0.0448  0.0470   0.0446     0.0591

sigma^2 estimated as 1.033:  log likelihood = -717.77,
aic = 1445.54
```

Hence the fitted models are

$$X_t = -0.0046_{(.0407)} + Z_t - 0.5165_{(.0422)}Z_{t-1} + 0.4227_{(.0434)}Z_{t-2},$$

with $Z_t \sim N(0, 1.009)$, for the first series, and

$$X_t = -0.0528_{(.0591)} + Z_t + 0.2644_{(.0448)}Z_{t-1} \\ +0.1607_{(.0470)}Z_{t-2} - 0.1245_{(.0446)}Z_{t-3}$$

with $Z_t \sim N(0, 1.033)$, for the second series. Note that the 95% confidence intervals of the intercepts in the two fitted models, $-0.0046 \pm 1.96 \cdot 0.0407$ and $0.0528 \pm 1.96 \cdot 0.0591$, cover zero, indicating that the means of the two MA processes might be zero. MA models with zero means can be fitted to the series using the following `R` commands.

```
> x3.fit2<-arima(x3, order=c(0,0,2), include.mean=FALSE)
> x3.fit2
Call:
arima(x = x3, order = c(0, 0, 2), include.mean = FALSE)

Coefficients:
          ma1     ma2
      -0.5165  0.4228
s.e.   0.0422  0.0434

sigma^2 estimated as 1.009:  log likelihood = -712.09,
aic = 1430.18

> x4.fit2<-arima(x4, order=c(0,0,3), include.mean=FALSE)
> x4.fit2

Call:
arima(x = x4, order = c(0, 0, 3), include.mean = FALSE)

Coefficients:
         ma1     ma2      ma3
      0.2664  0.1630  -0.1226
s.e.  0.0448  0.0469   0.0446

sigma^2 estimated as 1.035:  log likelihood = -718.17,
aic = 1444.34
```

Therefore, the fitted MA models with zero means can be summarized as

$$X_t = Z_t - 0.5165_{(.0422)} Z_{t-1} + 0.4228_{(.0434)} Z_{t-2},$$

with $Z_t \sim N(0, 1.009)$, for the first series, and

$$X_t = Z_t + 0.2664_{(.0448)} Z_{t-1} + 0.1630_{(.0469)} Z_{t-2} - 0.1226_{(.0446)} Z_{t-3},$$

with $Z_t \sim N(0, 1.035)$, for the second series. For the first series, the 95% confidence intervals of coefficients of $Z_{t-1}$ and $Z_{t-2}$ are $-0.5165 \pm 1.96 \cdot 0.0422$ and $0.4228 \pm 1.96 \cdot 0.0434$, respectively, and hence cover the true values of the coefficients $-0.5$ and $0.4$, respectively. For the second series, the 95% confidence intervals of coefficients of $Z_{t-1}$, $Z_{t-2}$, and $Z_{t-3}$ are $0.2664 \pm 1.96 \cdot 0.0448$, $0.1630 \pm 1.96 \cdot 0.0469$, and $-0.1226 \pm 1.96 \cdot 0.0446$, respectively, or equivalently, $(0.1786, 0.3542)$,

$(0.0710, 0.2549)$, and $(-0.2099, -0.0353)$, respectively. These intervals cover the true values of the corresponding coefficients, $1/3$, $1/5$, and $-1/9$, indicating that the fitted MA(3) model is close to the true MA(3) model.

## 4.4 Estimating Parameters of an ARMA Model

Suppose now that a model with both AR and MA terms is thought to be appropriate for a given time series. The estimation problems for an ARMA model are similar to those for an MA model in that an iterative procedure has to be used. The residual sum of squares can be calculated at every point on a suitable grid of the parameter values, and the values, which give the minimum sum of squares may then be assessed. Alternatively some sort of optimization procedure may be used.

As an example, consider the ARMA$(1,1)$ process whose ac.f. decreases exponentially after lag 1 (see Example 3.2). This model may be recognized as appropriate if the sample ac.f. has a similar form. The model is given by

$$X_t - \mu = \alpha_1(X_{t-1} - \mu) + Z_t + \beta_1 Z_{t-1}.$$

Given $N$ observations, $x_1, \ldots, x_N$, we guess values for $\widehat{\mu}, \widehat{\alpha}_1, \widehat{\beta}_1$, set $z_0 = 0$ and $x_0 = \widehat{\mu}$, and then calculate the residuals recursively by

$$\begin{aligned} z_1 &= x_1 - \widehat{\mu} \\ z_2 &= x_2 - \widehat{\mu} - \widehat{\alpha}_1(x_1 - \mu) - \widehat{\beta}_1 z_1 \\ &\text{up to} \\ z_N &= x_N - \widehat{\mu} - \widehat{\alpha}_1(x_{N-1} - \mu) - \widehat{\beta}_1 z_{N-1}. \end{aligned}$$

The residual sum of squares $\sum_{t=1}^{N} z_t^2$ may then be calculated. Then other values of $\widehat{\mu}, \widehat{\alpha}_1, \widehat{\beta}_1$ may be tried until the minimum residual sum of squares is found.

Many variants of the above estimation procedure have been studied – see, for example, the reviews by Priestley (1981, Chapter 5) and Kendall et al. (1983, Chapter 50). Nowadays exact maximum likelihood estimates are often preferred, since the extra computation involved is no longer a limiting factor. The conditional least squares estimates introduced above are conceptually easier to understand and can also be used as starting values for exact maximum likelihood estimation. The Hannan–Rissanen recursive regression procedure (e.g. Granger and Newbold, 1986) is primarily intended for model identification but can additionally be used to provide starting values as well. The Kalman filter (see Section 10.1.4) may be used to calculate exact maximum likelihood estimates to any desired degree of approximation. We will say no more about this important, but rather advanced, topic here. Modern computer software for time-series modelling should incorporate sound estimation procedures for fitting MA and ARMA models, though it is worth finding out exactly what method is used, provided this information is given in the software documentation (as it should be, but often isn't!). For short series, and models with roots near the unit circle, it is worth stressing that different estimation procedures can give noticeably different results. It is also worth noting that least squares estimates can be biased for short series.

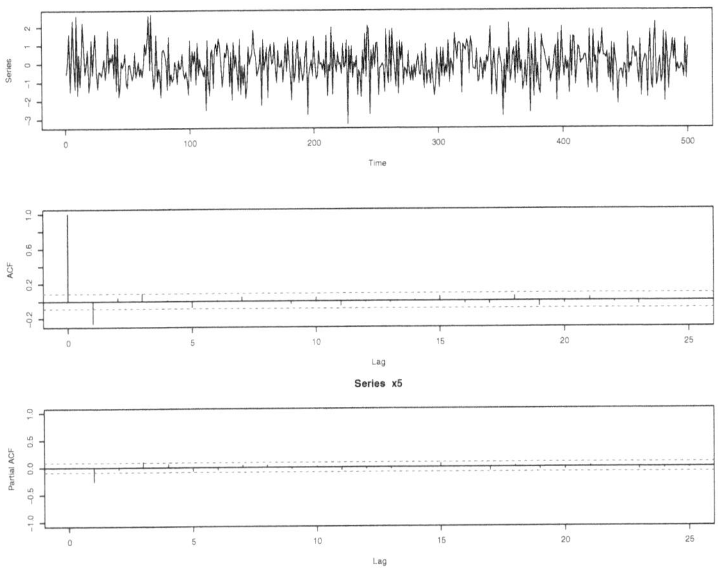


Figure 4.4 *A simulated ARMA(2,2) process (Top),* $X_t = 0.4X_{t-1} - 0.3X_{t-2} + Z_t - 0.6Z_{t-1} + 0.5Z_{t-2}$*, and its sample ac.f.'s (Middle) and partial ac.f.'s (Bottom).*

*Example 4.3*

We simulate an ARMA(2,2) process with $N = 500$ observations,

$$X_t = 0.4X_{t-1} - 0.3X_{t-2} + Z_t - 0.6Z_{t-1} + 0.5Z_{t-2}, \quad Z_t \sim N(0,1).$$

We then use the simulated series to compute its sample ac.f. and the sample partial ac.f. Figure 4.4 shows the simulated series and its sample ac.f. and sample partial ac.f.

```
> set.seed(6)
> x5<-arima.sim(list(order=c(2,0,2), ar=c(0.4,-0.3),
     ma=c(-0.6,0.5)), 500)
```

To fit an ARMA model to the series, we need to specify the orders of AR and MA components. However, it is difficult to determine the orders of the AR and MA components from Figure 4.4. As model identification tools are introduced in the next section, we consider the issue of estimating model parameters when the orders of AR and MA components are given. The following commands in R show how to fit an ARMA(2,2) model to the series.

```
> x5.fit1<-arima(x5, order=c(2,0,2))
> x5.fit1
```

```
Call:
arima(x = x5, order = c(2, 0, 2))

Coefficients:
          ar1      ar2      ma1     ma2  intercept
      -0.0404  -0.4311  -0.2188  0.4872    -0.0445
s.e.   0.2843   0.1498   0.2838  0.1162     0.0376

sigma^2 estimated as 0.9503: log likelihood=-696.8, aic=1405.6
```

Then we may conclude the fitted ARMA(2,2) model is

$$X_t = -0.0445_{(.0376)} - 0.0404_{(.2843)} X_{t-1} - 0.4311_{(.1498)} X_{t-2} + Z_t \\ -0.2188_{(.2838)} Z_{t-1} + 0.4872_{(.1162)} Z_{t-2}$$

with $Z_t \sim N(0, 0.9503)$. Note that the 95% confidence intervals of the intercept in the fitted model, $-0.0445 \pm 1.96 \cdot 0.0376$, covers zero, suggesting that the means of the ARMA(2,2) process might be zero. ARMA models with zero means can be fitted to the series using the following R commands.

```
> x5.fit2<-arima(x5, order=c(2,0,2), include.mean=FALSE)
> x5.fit2

Call:
arima(x = x5, order = c(2, 0, 2), include.mean = FALSE)

Coefficients:
          ar1      ar2      ma1     ma2
      -0.0267  -0.4300  -0.2305  0.4915
s.e.   0.2928   0.1514   0.2919  0.1147

sigma^2 estimated as 0.953: log likelihood=-697.5, aic=1405
```

Hence the fitted ARMA(2,2) model with zero means is

$$X_t = -0.0267_{(.2928)} X_{t-1} - 0.4300_{(.1514)} X_{t-2} + Z_t \\ -0.2305_{(.2919)} Z_{t-1} + 0.4915_{(.1147)} Z_{t-2},$$

with $Z_t \sim N(0, 0.953)$. The 95% confidence intervals of the coefficients of $X_{t-1}$, $X_{t-2}$, $Z_{t-1}$, and $Z_{t-2}$ are $-0.0267 \pm 1.96 \cdot 0.2928$, $-0.4300 \pm 1.96 \cdot 0.1514$, $-0.2305 \pm 1.96 \cdot 0.2919$, and $0.4915 \pm 1.96 \cdot 0.1147$, respectively, which are equivalents to $(-0.6006, 0.5472)$, $(-0.7267, -0.1332)$, $(-0.8026, 0.3416)$, and $(0.2667, 0.7163)$, respectively. These intervals cover the true values of the coefficients of $X_{t-1}$, $X_{t-2}$, $Z_{t-1}$, and $Z_{t-2}$, which are 0.4, $-0.3$, $-0.6$, and 0.5, respectively, suggesting the fitted model is close to the true model.

## 4.5 Model Identification Tools

Model building is a key element of most statistical work. Some general remarks on how to formulate an appropriate model for a given time series are given in Section 4.8. As well as finding out about any background knowledge, the analyst will typically look at the time plot of the data and at various diagnostic tools. The two standard tools, which are used to identify an appropriate autoregressive moving average (ARMA) model for a given stationary time series (see Section 4.4), are the sample autocorrelation function (ac.f.) and the sample partial ac.f. This choice is often made subjectively, using the analyst's experience to match an appropriate model to the observed characteristics. However, various additional diagnostic tools are also available, and this section gives a brief introduction to some of them. Detailed reviews of methods for determining the order of an ARMA process are given by Gooijer et al. (1985) and Choi (1992); see also Newbold (1988).

An alternative to the partial ac.f. is the **inverse ac.f.**, whose use in identifying ARMA models is described by Chatfield (1979). The inverse ac.f. of the general ARMA model, as written in Equation (3.19), namely

$$\phi(B)X_t = \theta(B)Z_t$$

is exactly the same as the ordinary ac.f. of the corresponding inverse ARMA model given by

$$\theta(B)X_t = \phi(B)Z_t,$$

where $\theta$ and $\phi$ are interchanged. It turns out that the inverse ac.f. has similar properties as the partial ac.f. of the ARMA process in that it 'cuts off' at lag $p$ for an AR($p$) process but generally dies out slowly for MA and ARMA processes. The inverse ac.f. often contains more information than the partial ac.f., extends easily to seasonal ARMA models and is a viable competitor to the partial ac.f.

Instead of (or in addition to) subjectively examining functions like the ac.f., two alternative approaches to model selection are to carry out a series of **hypothesis tests** or to use a **model-selection criterion**. Econometricians tend to favour the former approach and test null hypotheses such as normality, constant variance and non-linearity. Statisticians tend to favour the latter approach and several such model-selection criteria have been proposed. Given a class of models, such as the ARIMA class, the idea is to choose a model from that class so as to optimize a suitably chosen function of the data.

What criterion should we use to select a model in the 'best' way? It is not sensible to simply choose a model to give the best fit by minimizing the residual sum of squares, as the latter will generally decrease as the number of parameters is increased regardless of whether the additional complexity is really worthwhile. There is an alternative fit statistic, called adjusted-$R^2$, which makes some attempt to take account of the number of parameters fitted, but more sophisticated model-selection statistics are generally preferred. **Akaike's Information Criterion** (AIC) is the most commonly used and is given (approximately) by:

$$\text{AIC} = -2\ln(\text{max. likelihood}) + 2r$$

where $r$ denotes the number of independent parameters that are fitted for the model being assessed. Thus the AIC essentially chooses the model with the best fit, as measured by the likelihood function, subject to a penalty term, to prevent over-fitting, that increases with the number of parameters in the model. For an ARMA$(p,q)$ model, note that $r = p+q+1$ as the residual variance is included as a parameter. Ignoring arbitrary constants, the first (likelihood) term is usually approximated by $N\ln(S/N)$, where $S$ denotes the residual sum of squares, and $N$ is the number of observations. It turns out that the AIC is biased for small samples, and a bias-corrected version, denoted by $\text{AIC}_\text{C}$, is increasingly used. The latter is given (approximately) by replacing the quantity $2r$ in the ordinary AIC with the expression $2rN/(N-r-1)$. The $\text{AIC}_\text{C}$ is recommended, for example, by Brockwell and Davis (1991, Section 9.3) and Burnham and Anderson (2002).

An alternative, widely used criterion is the **Bayesian Information Criterion** (BIC) that essentially replaces the term $2r$ in the AIC with the expression $(r+r\ln N)$. Thus it penalizes the addition of extra parameters more severely than the AIC. **Schwartz's Bayesian criterion** is yet another alternative that is similar to BIC in its dependence on $\ln N$, but replaces $(r+r\ln N)$ with $r\ln N$.

Several other possible criteria have been proposed including Parzen's **autoregressive transfer function criterion (CAT)** and Akaike's **final prediction error (FPE)** criterion, which are both primarily intended for choosing the order of an AR process. Note that all the above criteria may not have a unique minimum and depend on assuming that the data are (approximately) normally distributed. (Priestley, 1981, Chapter 5) and Burnham and Anderson (2002) give a general review of these criteria. Following the results in Faraway and Chatfield (1998), we recommend to use the $\text{AIC}_\text{C}$ or BIC. Computer packages routinely produce numerical values for several such criteria so that analysts can pick the one they like best. The guiding principle throughout is to apply the Principle of Parsimony, introduced in Section 3.8. In brief this says, '*Adopt the simplest acceptable model*'. Of course, there is a real danger that the analyst will try many different models, pick the one that appears to fit best according to one of these criteria, but then make predictions as if certain that the best-fit model is the true model. Further remarks on this problem are made in Section 14.2.

As noted above, an alternative approach to model selection relies on carrying out a series of hypothesis tests. However, little will be said here about this approach, because the authors prefer to rely on the subjective interpretation of diagnostic tools, such as the ac.f., allied to the model-selection criteria given above. There are three reasons for this preference:

1. A model-selection criterion gives a numerical-valued ranking of all models, so that the analyst can see whether there is a clear winner or, alternatively, several close competing models.
2. Some model-selection criteria can be used to compare non-nested[2] models, as

[2] Two models are nested when one is a subset of the other. For example, a first-order AR model is a special case of a second-order AR model when the coefficient at lag two is zero. This makes it relatively easy to test the null hypothesis that a first-order model is adequate.

would arise, for example, when trying to decide whether to compute forecasts using an ARIMA, neural network or econometric model. It is very difficult to use hypothesis tests to compare non-nested models.

3. A hypothesis test requires the specification of an appropriate null hypothesis, and effectively assumes the existence of a true model that is contained in the set of candidate models.

*Example 4.4*

We take the ARMA(2,2) series simulated in Example 4.3 as an example. The following R commands show how to select a best model using AIC. We fit a set of ARMA models and compute the AIC for each fitted model.

```
> x5.aic<-matrix(0,4,4)
> for (i in 0:3) for (j in 0:3) {
    x5.fit<-arima(x5, order=c(i,0,j))
    x5.aic[i+1,j+1]<-x5.fit$aic
  }
> x5.aic
         [,1]     [,2]     [,3]     [,4]
[1,] 1439.570 1409.792 1408.332 1406.792
[2,] 1408.238 1409.955 1408.699 1408.510
[3,] 1409.748 1410.947 1405.603 1407.509
[4,] 1407.752 1409.147 1407.489 1409.022
```

The model with the minimum AIC (1405.603) is ARMA$(p,q)$ with $p=2$ and $q=2$. Given the values of $p$ and $q$, the inference on the ARMA(2, 2) model in the next step is same as that in Section 4.4. We refer to the reader to Example 4.3 for further analysis of the series 'x5'. Model selection procedures with criteria other than AIC can be similarly carried out as this example.

## 4.6 Testing for Unit Roots

In the previous subsection, we referred to the difficulty in distinguishing between a long-memory and a non-stationary process. More generally, a major problem in practice, especially with relatively short series, is distinguishing between a non-stationary process and one that is 'nearly non-stationary' in some general sense. There is no simple mathematical definition of the latter description, but we have in mind processes whose ac.f. decays slowly. With finite samples, the sample ac.f. of non-stationary and nearly non-stationary processes will both decay (very) slowly towards zero. This makes it hard to distinguish between these different types of processes on the basis of sample data.

Long-memory stationary processes are 'nearly non-stationary' in the above sense, as are ordinary AR stationary processes with roots near the unit circle. For example, the (stationary) AR(1) process with parameter 0.95, namely, $X_t = 0.95X_{t-1} + Z_t$, will yield data that, for short series, will have properties that look much like those of data generated by a (non-stationary) random walk, namely, $X_t = X_{t-1} + Z_t$. The time plots and sample ac.f.s will have similar characteristics.

For some purposes, it won't matter too much if we identify the wrong model. For example, the short-term forecasts from an AR(1) model with parameter 0.95 and from a random walk are quite close to each other. However, the long-term forecasts will be substantially different. For the AR(1) model the long-run forecasts revert to the overall mean, whereas those for the random walk are all equal to the most recent observed value. Moreover, the forecast error variance remains finite for a stationary model but increases without bound for a non-stationary model. Thus it can be very important to distinguish between the two different types of processes.

One way of tackling this problem in regard to ARIMA models, is to ask whether the differencing parameter $d$ is exactly equal to one. If it is, then that would mean that a *unit root* is present. Put another way, if the original data are non-stationary but the first differences are stationary, then a unit root is said to be present. One way of trying to answer this question is to carry out an appropriate significance test, and there is a large literature in econometrics on **testing for a unit root**, e.g. DeJong and Whiteman (1993), Hamilton (1994, Chapter 15), Diebold and Kilian (2000), and Diebold (2001, Chapter 12).

Several types of tests designed for assessing different null and alternative hypotheses are available. Perhaps the most important example is the so-called **augmented Dickey–Fuller test**, details of which may be found, for example, in Enders (1995, Chapter 5) or Harvey (1993, Section 5.4). The tests generally take the *null* hypothesis to be that there *is* a unit root (so that $d = 1$), partly on the grounds that many economic series are known to be close to a random walk or ARIMA(0,1,0) process. The following commands in `R` show how to use the Augmented Dickey-Fuller test, `adf.test` (in `tseries` library), to test the null hypothesis that a series has a unit root.

```
> set.seed(8)
> y1<-cumsum(rnorm(500))
> library("tseries")
> y1.adf<-adf.test(y1, alternative="stationary")
> y1.adf

 Augmented Dickey-Fuller Test

data:  y1
Dickey-Fuller = -2.6149, Lag order = 7, p-value = 0.318
alternative hypothesis: stationary
```

However, it is not obvious that this is always a sensible choice of null hypothesis, and the statistician is likely to be interested in assessing an appropriate value for $d$, as well as for $p$ and $q$, with no prior assumptions other than assuming that some member of the ARIMA class of models is appropriate. Whereas the econometrician tends to rely on a series of tests, not only for the presence of a unit root but also for other features such as constant variance and uncorrelated residuals, the statistician is more likely to choose a model from a general class of models by looking at appropriate diagnostic tools and perhaps minimizing a criterion such as AIC — see Section 14.1.

Unfortunately, tests for unit roots generally have poor power, even for moderate size samples, quite apart from the question as to whether and when it is sensible to take $d = 1$ as the null hypothesis. The alternative hypothesis is typically 'close' to the null hypothesis and the testing procedure can be sensitive to the way that lag structure is modelled. Newbold et al. (1993) go so far as to say that 'testing for unit autoregressive roots is misguided', although there is some simulation evidence (Diebold and Kilian, 2000) that unit root tests can help to select models that give superior forecasts. The topic is too complex and unresolved to make a definitive assessment, but a formal test for a unit root can only ever be a small contribution to the important task of modelling (near) non-stationary behaviour. The fact that we cannot reject a unit root, does not mean that we should necessarily impose one, as, for example, if we want an explicit estimate of any trend. Conversely, there could still be practical reasons why we might wish to difference our data, even when a unit root is rejected, as, for example, when a model for the differenced data appears to be more robust to unexpected changes. The key question is not whether a unit-root test helps select the 'true model' (that probably doesn't exist), but whether the chosen model (that we fully expect to be misspecified in some respects) helps to solve the given problem better than alternatives.

It may be helpful at this point to make some further remarks on different types of trend. Section 2.5 distinguished between a global and a local linear trend. Econometricians make a similar distinction but use a different vocabulary, by referring to what they call **difference-stationary** series, where stationarity can be induced by first differencing, and **trend-stationary** series where the deviations from a deterministic trend are stationary. The random walk is a simple example of a difference-stationary series, whereas a global trend model with stationary residuals is trend-stationary. In the latter case, the trend is deterministic, while econometricians generally say that there is a **stochastic trend** for difference-stationary series. Most economic series are difference-stationary rather than trend-stationary, and there is empirical evidence that difference-stationary models tend to give better out-of-sample forecasts for non-stationary data.

Spurious autocorrelations can readily be induced in a series showing trend, either by mistakenly removing a deterministic trend from difference-stationary data, or by differencing trend-stationary data. This illustrates the importance of identifying the appropriate form of trend so that an appropriate level of differencing may be applied. However, from what is said above, it is clear that it is generally rather difficult to distinguish between the cases (1) $d = 1$, (2) $0 < d < 1$, (3) $d = 0$ and (4) trend-stationarity. This is perhaps not too surprising given that it is possible to construct examples where models with different orders of differencing or with different trend structures can be made in some sense arbitrarily close. We illustrate this problem by considering the AR(1) model as the coefficient $\phi$ approaches one. For $\phi = 0.99$, we have a stationary model with $d = 0$, but, when $\phi = 1$, then the model reduces to a difference-stationary random walk model with $d = 1$.

Alternatively, consider the model

$$X_t = \alpha X_{t-1} + Z_t + \beta Z_{t-1}$$

When $\alpha = 0.95$ and $\beta = -0.9$, the process is a stationary ARMA$(1,1)$ model so that $d = 0$ when expressed as an ARIMA$(1,0,1)$ model. When the value of $\alpha$ is changed slightly to $\alpha = 1$, the operator $(1-B)$ appears on the left-hand side of the equation so that the process becomes a non-stationary ARIMA$(0,1,1)$ model with $d = 1$. However, if in addition we now change the value of $\beta$ slightly to $-1$, then the operator $(1-B)$ will appear on both sides of the equation and can be cancelled. Then $X_t$ is white noise and we are back to the stationary case $(d = 0)$. With a sample size of say 100, it would be nearly impossible to distinguish between these three cases.

The implications of the above discussion are that identification of an appropriate trend model is a key element of time-series modelling. However, it can be very difficult and a unit root test may not help much. Rather than agonize about the 'correct' order of differencing to use, there is much to be said for fitting a model that makes few assumptions about the form of the trend, but rather is designed to be adaptive in form and to be *robust* to changes in the underlying model. For this reason, state-space models may be preferred to ARIMA models, and a local trend model may be preferred to a global trend model.

## 4.7 Estimating Parameters of an ARIMA Model

In practice, many time series are clearly non-stationary, and so the stationary models we have studied so far cannot be applied directly. In Box–Jenkins ARIMA modelling, the general approach is to *difference* an observed time series until it appears to come from a stationary process. An AR, MA or ARMA model may then be fitted to the differenced series as described in Sections 4.2–4.4. It is often found that first-order differencing of non-seasonal data is adequate — see Section 3.9 and Example 14.2 — although second-order differencing is occasionally required. The resulting model for the undifferenced series is the fitted ARIMA model. For seasonal data, seasonal differencing may also be required (see Section 4.8 and Example 14.3).

It is unnecessary to say anything further about fitting ARIMA models because the possible use of differencing is the only feature that changes from fitting ARMA models.

## 4.8 Box–Jenkins Seasonal ARIMA Models

In practice, many time series contain a seasonal periodic component, which repeats every $s$ observations. For example, with monthly observations, where $s = 12$, we may typically expect $X_t$ to depend on values at annual lags, such as $X_{t-12}$, and perhaps $X_{t-24}$, as well as on more recent non-seasonal values such as $X_{t-1}$ and $X_{t-2}$. Box and Jenkins (1970) generalized the ARIMA model to deal with seasonality, and defined a general multiplicative seasonal ARIMA (SARIMA) model as

$$\phi_p(B)\Phi_P(B^s)W_t = \theta_q(B)\Theta_Q(B^s)Z_t, \tag{4.16}$$

where $B$ denotes the backward shift operator, $\phi_p, \Phi_P, \theta_q, \Theta_Q$ are polynomials of order $p, P, q, Q$, respectively, $Z_t$ denotes a purely random process and

$$W_t = \nabla^d \nabla_s^D X_t \tag{4.17}$$

denotes the differenced series. If the integer $D$ is not zero, then seasonal differencing is involved (see Section 2.6). The above model is called a SARIMA model of order $(p,d,q) \times (P,D,Q)_s$.

The SARIMA model looks rather complicated at first sight, so let us look at the different components of the model in turn. First, the differenced series $\{W_t\}$ is formed from the original series $\{X_t\}$ by appropriate differencing to remove non-stationary terms. If $d$ is non-zero, then there is simple differencing to remove trend, while seasonal differencing $\nabla_s$ may be used to remove seasonality. For example, if $d = D = 1$ and $s = 12$, then

$$W_t = \nabla\nabla_{12} X_t = \nabla_{12} X_t - \nabla_{12} X_{t-1} = (X_t - X_{t-12}) - (X_{t-1} - X_{t-13}).$$

In practice, the values of $d$ and $D$ are usually zero or one, and rarely two.

Now let us look at the seasonal AR term, $\Phi_P(B^s)$. Suppose for simplicity that $P = 1$. Then $\Phi_1(B^s)$ will be of the form $(1 - C \times B^s)$, where $C$ denotes a constant, which simply means that $W_t$ will depend on $W_{t-s}$, since $B^s W_t = W_{t-s}$. Similarly, a seasonal MA term of order one means that $W_t$ will depend on $Z_{t-s}$ as well as on $Z_t$.

As an example, consider a SARIMA model of order $(1,0,0) \times (0,1,1)_{12}$, where we note $s = 12$. Here we have one non-seasonal AR term, one seasonal MA term and one seasonal difference. Then Equations (4.16) and (4.17) can be written as

$$(1 - \alpha B) W_t = (1 + \Theta B^{12}) Z_t,$$

where $W_t = \nabla_{12} X_t$ and $\Theta$ is a constant parameter (rather than a function as in Equation (4.16)). It may help to write this out in terms of the original observed variable $X_t$ as

$$X_t = X_{t-12} + \alpha(X_{t-1} - X_{t-13}) + Z_t + \Theta Z_{t-12},$$

so that $X_t$ depends on $X_{t-1}, X_{t-12}$ and $X_{t-13}$ as well as the innovations at times $t$ and $(t-12)$. You may be surprised to see that $X_{t-13}$ is involved, but this is a consequence of mixing first-order autocorrelation with seasonal differencing.

When fitting a seasonal model to data, the first task is to assess values of $d$ and $D$, which remove most of the trend and seasonality and make the series appear to come from a stationary process. Then the values of $p, P, q$ and $Q$ need to be assessed by looking at the ac.f. and partial ac.f. of the differenced series and choosing a SARIMA model whose ac.f. and partial ac.f. are of similar form. Finally, the model parameters may be estimated by some suitable iterative procedure. Full details are given by Box et al. (1994, Chapter 9), but the wide availability of suitable computer software means that the average analyst does not need to worry about the practical details of fitting a particular model. Instead, the analyst can concentrate on choosing a sensible model to fit and ensuring that appropriate diagnostic checks are carried out.

*Example 4.5*

Figure 1.4 shows the domestic sales of Australian fortified wine by winemakers in successive quarters over a 28-year period. For convenience, we show the time series again in the top panel of Figure 4.5. Also shown in the middle and bottom panels of Figure 4.5 are the sample ac.f. and sample partial ac.f. of the sales series. Note that the series consists of quarterly observations, so we denote $X_t$ the sales at quarter $t$. The sale series demonstrates both a trend and seasonal variation, and the analysis below shows how to use Box-Jenkins models to fit a time series model to the data.

```
> library("tseries")
> wine<-read.csv("mydata/aus_wine_sales.csv", header=F)
> wine.ts<-ts(wine[,2], start=c(1985,1), frequency=4)

> par(mfrow=c(3,1), mar=c(4,4,4,4))  ### Figure 4.5
> plot(wine.ts, type="l", xlab="Year",
    ylab="Sales in thousand liters")
> acf(wine[,2], 25, xlab="Lag", ylab="ACF", main="")
> acf(wine[,2], 25, type="partial", xlab="Lag",
    ylab="Partial ACF", main="")
```

First, regardless of the trend effect, we find that the sale series shows cyclic behaviour every year. As the series consists of quarterly observations, it is reasonable to conclude the period of the seasonal effect is $d = 4$. This motivates us to remove the seasonal effect by taking seasonal difference. Consider the seasonal differencing operator $\nabla_d = 1 - B^d$ with $d = 4$, and we use it to obtain the seasonal differenced series $Y_t$,

$$Y_t = \nabla_d X_t = (1 - B^d) X_t, \qquad d = 4.$$

```
> dwine<-diff(wine[,2], lag=4)
> par(mfrow=c(3,1), mar=c(4,4,4,4))  ### Figure 4.6
> plot(diff(wine.ts, lag=4), type="l", xlab="Year",
    ylab="Differenced Series")
> acf(dwine, 25, xlab="Lag", ylab="ACF", main="")
> acf(dwine, 25, type="partial", xlab="Lag",
    ylab="Partial ACF", main="")
```

Figures 4.6 show the time series plot, the sample ac.f., and the sample partial ac.f. of the differenced series $Y_t$. Note that the differenced series does not show any seasonal effect in the time series plot; furthermore, the sample ac.f. and sample partial ac.f. of $Y_t$ are significantly less than those of the original series $X_t$. We then try to fit an ARMA model to $Y_t$. Since the order of the ARMA model is not given, we fit a set of ARMA models and compute their AICs.

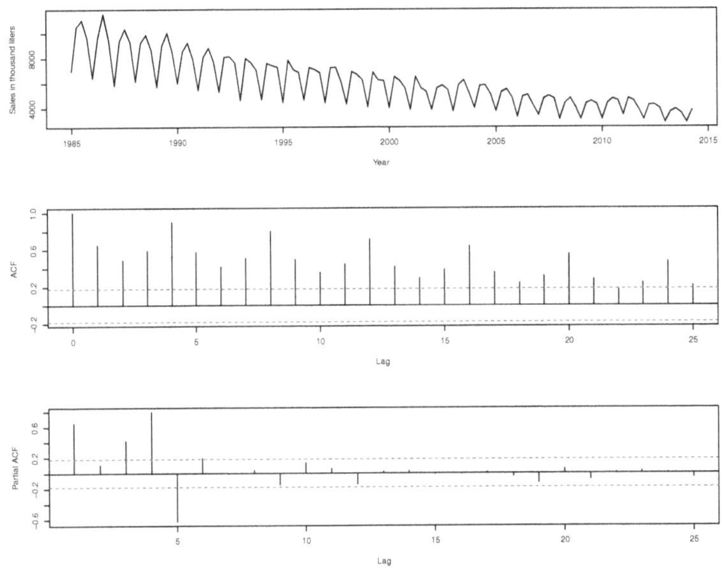


Figure 4.5 *The series of domestic sales of Australian fortified wine (Top) and its sample ac.f. (Middle) and sample partial ac.f. (Bottom).*

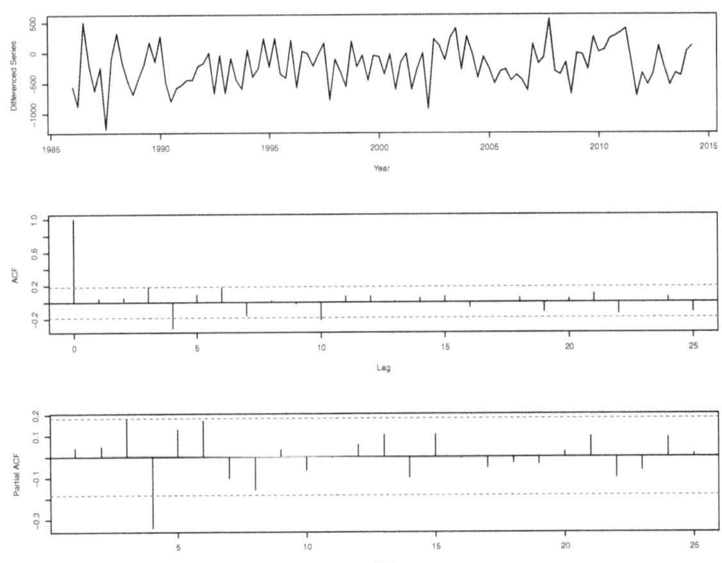


Figure 4.6 *The seasonal differenced series of domestic sales of Australian fortified wine (Top) and its sample ac.f. (Middle) and sample partial ac.f. (Bottom).*

```
> dwine.aic<-matrix(0,5,5)
> for (i in 0:4) for (j in 0:4) {
    dwine.fit<-arima(dwine, order=c(i,0,j))
    dwine.aic[i+1,j+1]<-dwine.fit$aic
  }
> dwine.aic
         [,1]     [,2]     [,3]     [,4]     [,5]
[1,] 1650.682 1652.504 1653.521 1650.854 1637.755
[2,] 1652.489 1654.140 1655.125 1644.301 1638.960
[3,] 1654.209 1651.112 1643.001 1630.410 1632.405
[4,] 1651.993 1645.663 1632.891 1632.401 1634.360
[5,] 1638.412 1639.506 1631.802 1633.801 1633.935
```

The model with minimum AIC (1630.410) is ARMA$(p,q)$ with $p=2$ and $q=3$. Hence we fit an ARMA(2,3) model to the differenced series $Y_t$.

```
> dwine.fit<-arima(dwine, order=c(2,0,3))
> dwine.fit

Call:
arima(x = dwine, order = c(2, 0, 3))

Coefficients:
          ar1      ar2     ma1     ma2     ma3  intercept
      -1.0226  -0.8662  1.1775  1.1854  0.4210  -213.3700
s.e.   0.0691   0.0717  0.1011  0.1348  0.0915    35.2152

sigma^2 estimated as 82830: log likelihood=-808.21, aic=1630.41
```

The fitted result suggests the following ARMA(2,3) model for $Y_t$.

$$
\begin{aligned}
Y_t = & -213.3700_{(35.2152)} - 1.0226_{(.0691)}Y_{t-1} - 0.8662_{(0.0717)}Y_{t-2} + Z_t \\
& +1.1775_{(0.1011)}Z_{t-1} + 1.1854_{(0.1348)}Z_{t-2} + 0.4210_{(0.0915)}Z_{t-3},
\end{aligned}
$$

where $Z_t$ are i.i.d. series with $Z_t \sim N(0, 82830)$. Note that the next step one should perform is residual diagnostics, which will be discussed in the next section.

## 4.9 Residual Analysis

When a model has been fitted to a time series, it is advisable to check that the model really does provide an adequate description of the data. As with most statistical models, this is usually done by looking at the **residuals**, which are generally defined by

$$
\text{residual} = \text{observation} - \text{fitted value}.
$$

For a univariate time-series model, the fitted value is the one-step-ahead forecast so that the residual is the one-step-ahead forecast error. For example, for the AR(1) model, $X_t = \alpha X_{t-1} + Z_t$ where $\alpha$ is estimated by least squares, the fitted value at time $t$ is $\widehat{\alpha}x_{t-1}$ so that the residual corresponding to the observed value, $x_t$, is

$$\widehat{z}_t = x_t - \widehat{\alpha}x_{t-1}.$$

Note the 'hat' on $z_t$ as the residual is the estimated error term. Of course if $\alpha$ were known exactly, then the exact error $z_t = x_t - \alpha x_{t-1}$ could be calculated, but this situation never arises in practice (except in simulated exercises).

If we have a 'good' model, then we expect the residuals to be 'random' and 'close to zero', and model validation usually consists of plotting residuals in various ways to see whether this is the case. With time-series models we have the added feature that the residuals are ordered in time and it is natural to treat them as a time series.

Two obvious steps are to plot the residuals as a time plot, and to calculate the correlogram of the residuals. The time plot will reveal any outliers and any obvious autocorrelation or cyclic effects. The correlogram of the residuals will enable autocorrelation effects to be examined more closely. Let $r_{z,k}$ denote the sample autocorrelation coefficient at lag $k$ of the residuals $\{\widehat{z}_t\}$. If we could fit the true model[3], with the correct model parameter values, then the true errors $\{z_t\}$ form a purely random process and, from Section 4.1, their correlogram is such that each autocorrelation coefficient is approximately normally distributed, with mean 0 and variance $1/N$, for reasonably large values of $N$. Of course, in practice, the true model is unknown and the correlogram of the residuals from the fitted model has somewhat different properties. For example, suppose we know the form of the model is an AR(1) process, but have to estimate the parameter. If the true value of the parameter is $\alpha = 0.7$, it can be shown that approximate 95% confidence limits for the true residual autocorrelations are at $\pm\ 1.3/\sqrt{N}$ for $r_{z,1}$, at $\pm\ 1.7/\sqrt{N}$ for $r_{z,2}$ and at $\pm\ 2/\sqrt{N}$ for values of $r_{z,k}$ at higher lags. Thus for lags greater than 2, the confidence limits *are* the same as for the correlogram of the true errors. On the other hand, if we fit the wrong form of the model, then the distribution of residual autocorrelations will be quite different, and we hope to get some 'significant' values so that the wrong model is rejected. The analysis of residuals from ARMA processes is discussed more generally by Box et al. (1994, Chapter 8). As in the AR(1) example above, it turns out that $1/\sqrt{N}$ supplies an *upper bound* for the standard error of the residual autocorrelations, so that values, which lie outside the range $\pm\ 2\sqrt{N}$, are significantly different from zero at the 5% level and give evidence that the wrong form of model has been fitted.

Instead of looking at the residual autocorrelations one at a time, it is possible to carry out what is called a **portmanteau lack-of-fit test**. This looks at the first $K$ values of the residual correlogram all at once. The test statistic is

$$Q = N \sum_{k=1}^{K} r_{z,k}^2, \tag{4.18}$$

where $N$ is the number of terms in the differenced series and $K$ is typically chosen in the range 15 to 30. If the fitted model is appropriate, then $Q$ should be

[3]Whether, or not, a 'true model' really exists, is the subject of debate — see Section 14.2.

approximately distributed as $\chi^2$ with $(K-p-q)$ degrees of freedom, where $p$, $q$ are the number of AR and MA terms, respectively, in the model. Unfortunately the $\chi^2$ approximation can be rather poor for $N < 100$, and various alternative statistics have been proposed. For example, the modified Ljung–Box–Pierce statistic, given by $N(N+2)\sum_{k=1}^{K} r_{z,k}^2/(N-k)$, is often used — see Box et al. (1994, Section 8.2.2). However, these tests have rather poor power properties (e.g. Davies and Newbold, 1979) and rarely give significant results, except when the model is obviously inadequate. Several other procedures for looking at residuals have also been proposed (e.g. Newbold, 1988, Section 4), but we can just 'look' at the first few values of $r_{z,k}$, particularly at lags 1, 2 and the first seasonal lag (if any), and see if any are significantly different from zero using the crude limits of $\pm\ 1.96/\sqrt{N}$. If they are, then the model can be modified in an appropriate way by putting in extra terms to account for the significant autocorrelation(s). However, if only one (or two) values of $r_{z,k}$ are just significant at lags having no obvious physical interpretation, (e.g. $k = 5$), then this should not be regarded as compelling evidence to reject the model.

Another statistic used for testing residuals is the Durbin–Watson statistic (e.g. Granger and Newbold, 1986, Section 6.2). This often appears in computer output. The statistic is defined by

$$d = \sum_{t=2}^{N} (\widehat{z}_t - \widehat{z}_{t-1})^2 / \sum_{t=1}^{N} \widehat{z}_t^2. \tag{4.19}$$

Now since

$$\sum_{t=2}^{N} (\widehat{z}_t - \widehat{z}_{t-1})^2 \simeq 2\sum_{t=1}^{N} \widehat{z}_t^2 - 2\sum_{t=2}^{N} \widehat{z}_t \widehat{z}_{t-1},$$

we find $d \simeq 2(1-r_{z,1})$, where $r_{z,1} = \Sigma \widehat{z}_t \widehat{z}_{t-1} / \Sigma \widehat{z}_t^2$ is the sample autocorrelation coefficient of the residuals at lag 1 (since the mean residual should be virtually zero). Thus, in this sort of application, the Durbin–Watson statistic is really the value $r_{z,1}$ in a slightly different guise. If the true model has been fitted, then we expect $r_{z,1} \simeq 0$, so that $d \simeq 2$. Thus a 'typical' value for $d$ is around two and not zero. Furthermore, a test on $d$ is asymptotically equivalent to a test on the value of $r_{z,1}$ for the residuals.

The Durbin–Watson statistic was originally proposed for use with multiple regression models as applied to time-series data. Suppose we have $N$ observations on a dependent variable $y$, and $m$ explanatory variables, say $x_1, \ldots, x_m$, and we fit the model

$$Y_t = \beta_1 x_{1t} + \cdots + \beta_m x_{mt} + Z_t \qquad t = 1, \ldots, N.$$

Having estimated the parameters $\{\beta_i\}$ by least squares, we want to see whether the error terms are really independent. The residuals are therefore calculated by

$$\widehat{z}_t = y_t - \widehat{\beta}_1 x_{1t} - \cdots - \widehat{\beta}_m x_{mt} \qquad t = 1, \ldots, N.$$

The statistic $d$ may now be calculated, and the distribution of $d$ under the null hypothesis that the $z_t$ are independent has been investigated. Tables of critical values are available (e.g. Kendall et al., 1983) and they depend on the number of explanatory variables. Since $d$ corresponds to the value of $r_1$ for the residuals, this test implies

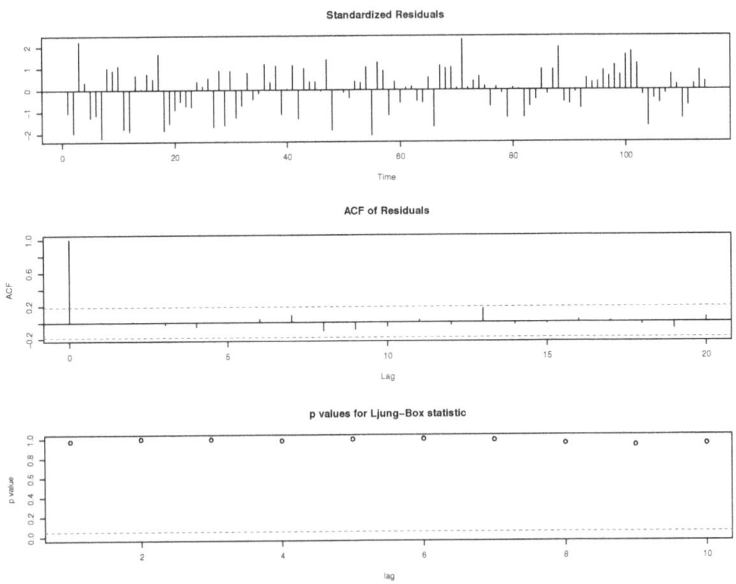


Figure 4.7 *The standardized residual series (Top), the sample ac.f. of residuals (Middle) and p-values for Ljung-Box statistic (Bottom).*

that we are only considering an AR(1) process as an alternative to a purely random process for $z_t$. Although it may be possible to modify the use of the Durbin–Watson statistic for models other than multiple regression models, it is usually better to look at the correlogram of the residuals as described earlier.

If the residual analysis indicates that the fitted model is inadequate in some way, then alternative models may need to be tried, and there are various tools for comparing the fit of several competing models (see Section 4.5). An iterative strategy for building time-series models, which is an integral part of the Box–Jenkins approach, is discussed more fully in Sections 4.10 and 5.3.

*Example 4.5 (Continued)*

For the estimated ARMA(2,2) model in Example 4.5, we show the standardized residual series, the sample ac.f. of residuals, and $p$-values for Ljung-Box statistic. in Figure 4.7. We also carry out a portmanteau test for residuals. In particular, the $p$-value of the following test is much larger than 0.05, indicating the fitted model is appropriate.

```
> tsdiag(dwine.fit)
> Box.test(dwine.fit$resid)

        Box-Pierce test
```

```
data:  dwine.fit$resid
X-squared = 9e-04, df = 1, p-value = 0.976
```

## 4.10 General Remarks on Model Building

How do we find a suitable model for a given time series? The answer depends on various considerations, including the properties of the series as assessed by a visual examination of the data, the number of observations available, the context and the way the model is to be used.

It is important to understand that model building has three main stages, which can be described as:

(1) Model formulation (or model specification)

(2) Model estimation (or model fitting)

(3) Model checking (or model verification).

Textbooks often concentrate on estimation, but say rather little about the more important topic of formulating the model. This is unfortunate because modern computer software makes model fitting straightforward for many types of models, so that the real problem is knowing which model to fit in the first place. For example, it is relatively easy nowadays to fit an ARIMA model, but it can still be difficult to know *when* ARIMA modelling is appropriate and *which* ARIMA model to fit.

Model checking is also very important, and the assessment of residuals is an essential step in the analysis. Modern software makes this relatively painless and may result in an initial model being discredited. Then an alternative model or models will be tried. Sometimes there are several cycles of model fitting as a model is modified and improved in response to residual checks or in response to additional data. Thus model building is an iterative, interactive process.

This section concentrates on model formulation. The analyst should consult appropriate 'experts' about the given problem, ask questions to get relevant background knowledge, look at a time plot of the data to assess their more important features, and make sure that a proposed model is consistent with empirical and/or theoretical knowledge and with the objectives of the investigation.

There are many classes of time-series models to choose from. Chapter 3 introduced a general class of (univariate) models called ARIMA models, which includes AR, MA and ARMA models as special cases. This useful class of processes provides a good fit to many different types of time series and should generally be considered when more than about 50 observations are available. Another general class of models is the trend and seasonal type of model introduced in Chapter 2. Later in this book several more classes of models will be introduced, including multivariate models of various types, and structural models.

In areas such as oceanography and electrical engineering, long stationary series may occur. If a parametric model is required, an ARMA model should be considered. As well as the time plot, the correlogram and the partial ac.f. should be examined in order to identify an appropriate ARMA model. The model can then be fitted, checked and possibly modified in the usual way. However, as we will see later in Chapters 6 and 7, we may be more interested in the frequency properties of the time series, in which case an ARMA model may not be very helpful.

In many other areas, such as economics and marketing, non-stationary series often arise and in addition may be fairly short. It is possible to treat non-stationarity by differencing the observed time series until it becomes stationary and then fitting an ARMA model to the differenced series. For seasonal series, the seasonal ARIMA model may be used. However, it should be clearly recognized that when the variation of the systematic part of the time series (i.e. the trend and seasonality) is dominant, the effectiveness of the ARIMA model is mainly determined by the initial differencing operations and not by the subsequent fitting of an ARMA model to the differenced series, even though the latter operation is much more time-consuming. Thus the simple models discussed in Chapter 2 may be preferable for time series with a pronounced trend and/or large seasonal effect. Models of this type have the advantage of being simple, easy to interpret and fairly robust. In addition they can be used for short series where it is impossible to fit an ARIMA model. An alternative approach is to model the non-stationary effects explicitly, rather than to difference them away, and this suggests the use of a state-space or structural model as will be described in Chapter 10.

Sometimes the analyst may have several competing models in mind and then it may help to look at a *model-selection statistic* such as *Akaike's Information Criterion* (AIC). These statistics try to strike a balance between the need for a 'parsimonious' model, which uses as few parameters as possible, and a model that is too simple and overlooks important effects. A useful reference on model building, in general, and model-selection statistics and the Principle of Parsimony, in particular, is Burnham and Anderson (2002).

Whatever model is fitted, it is important to realise that it is only an approximation to the 'truth', and the analyst should always be prepared to modify a model in the light of new evidence. The effects of model uncertainty are discussed later in Section 14.2.

## Exercises

**4.1** Suppose that time series data $\{x_i\ ;\ i = 1,\ldots,N\}$ are generated from an AR(1) process with parameter $\alpha$. Show that the variance of their sample mean can be approximated as follows for large $N$,

$$\mathrm{Var}(\overline{X}) \approx \frac{\sigma^2}{N}\left(\frac{1+\alpha}{1-\alpha}\right).$$

**4.2** Derive the least squares estimates for an AR(1) process having mean $\mu$ (i.e. derive Equations (4.6) and (4.7), and check the approximations in Equations (4.8) and (4.9)).

**4.3** Derive the least squares normal equations for an AR($p$) process, taking $\widehat{\mu} = \bar{x}$, and compare with the Yule–Walker equations (Equation (4.12)).

**4.4** Show that the (theoretical) partial autocorrelation coefficient of order 2, $\pi_2$, is given by

$$[\rho(2)-\rho(1)^2]/[1-\rho(1)^2].$$

Compare with Equation (4.11).

**4.5** Find the partial ac.f. of the AR(2) process given by

$$X_t = \frac{1}{3}X_{t-1} + \frac{2}{9}X_{t-2} + Z_t$$

(see Exercise 3.6).

**4.6** Suppose that the correlogram of a time series consisting of 100 observations has $r_1 = 0.31, r_2 = 0.37, r_3 = -0.05, r_4 = 0.06, r_5 = -0.21, r_6 = 0.11, r_7 = 0.08, r_8 = 0.05, r_9 = 0.12, r_{10} = -0.01$. Suggest an ARMA model, which may be appropriate.

**4.6** Sixty observations are taken on a quarterly economic index, $x_t$. The first eight values of the sample ac.f., $r_k$, and the sample partial ac.f., $\widehat{\pi}_k$, of $x_t$, and of the first differences, $\nabla x_t$, are shown below:

| | Lag | 1 | 2 | 3 | 4 | 5 | 6 | 7 | 8 |
|---|---|---|---|---|---|---|---|---|---|
| $x_t$ | $r_k$ | 0.95 | 0.91 | 0.87 | 0.82 | 0.79 | 0.74 | 0.70 | 0.67 |
| | $\widehat{\pi}_k$ | 0.95 | 0.04 | −0.05 | 0.07 | 0.00 | 0.07 | −0.04 | −0.02 |
| $\nabla x_t$ | $r_k$ | 0.02 | 0.08 | 0.12 | 0.05 | −0.02 | −0.05 | −0.01 | 0.03 |
| | $\widehat{\pi}_k$ | 0.02 | 0.08 | 0.06 | 0.03 | −0.05 | −0.06 | −0.04 | −0.02 |

Identify a model for the series. What else would you like to know about the data in order to make a better job of formulating a 'good' model?

**4.7** Use the Box-Jenkins seasonal ARIMA models to analyse the Beverage price index series shown in Figure 1.1.

# Chapter 5

# Forecasting

Forecasting is the art of saying what will happen, and then explaining why it didn't! – Anonymous

## 5.1 Introduction

Forecasting the future values of an observed time series is an important problem in many areas, including economics, production planning, sales forecasting and stock control.

Suppose we have an observed time series $x_1, x_2, \ldots, x_N$. Then the basic problem is to estimate future values such as $x_{N+h}$, where the integer $h$ is called the **lead time** or **forecasting horizon** – $h$ for horizon. The forecast of $x_{N+h}$ made at time $N$ for $h$ steps ahead is typically denoted by $\widehat{x}_N(h)$. Note that some authors still use the imprecise notation $\widehat{x}_{N+h}$, which assumes implicitly that the forecast is made at time $N$.

A wide variety of different forecasting procedures is available and it is important to realize that no single method is universally applicable. Rather, the analyst must choose the procedure that is most appropriate for a given set of conditions. It is also worth bearing in mind that forecasting is a form of extrapolation, with all the dangers that it entails. Forecasts are conditional statements about the future based on specific assumptions. Thus forecasts are not sacred and the analyst should always be prepared to modify them as necessary in the light of any external information. For long-term forecasting, it can be helpful to produce a range of forecasts based on different sets of assumptions so that alternative 'scenarios' can be explored.

Forecasting methods may be broadly classified into three groups as follows:

(1) *Subjective*

Forecasts can be made on a subjective basis using judgement, intuition, commercial knowledge and any other relevant information. Methods range widely from bold freehand extrapolation to the Delphi technique, in which a group of forecasters tries to obtain a consensus forecast with controlled feedback of other analysts' predictions and opinions as well as other relevant information. These methods will not be described here, as most statisticians will want their forecasts to be at least partly objective. The interested reader is referred, for example, to Webby and O'Connor (1996), Rowe and Wright (1999) and the relevant sections of Armstrong (2001).

DOI: 10.1201/9781003646112-5

However, note that some subjective judgement is often used in a more statistical approach, for example, to choose an appropriate model and perhaps make adjustments to the resulting forecasts.

(2) *Univariate*

Forecasts of a given variable are based on a model fitted only to present and past observations of a given time series, so that $\widehat{x}_N(h)$ depends only on the values of $x_N, x_{N-1}, \ldots$, possibly augmented by a simple function of time, such as a global linear trend. This would mean, for example, that univariate forecasts of the future sales of a given product would be based entirely on past sales, and would not take account of other economic factors. Methods of this type are sometimes called naive or projection methods and will be discussed in Sections 5.2 and 5.3.

(3) *Multivariate*

Forecasts of a given variable depend at least partly on values of one or more additional series, called **predictor** or **explanatory** variables[1]. For example, sales forecasts may depend on stocks and/or on economic indices. Models of this type are sometimes called causal models and will be introduced in Section 5.4.

In practice, a forecasting procedure may involve a *combination* of the above approaches. For example, marketing forecasts are often made by combining statistical predictions with the subjective knowledge and insight of people involved in the market. A more formal type of combination is to compute a weighted average of two or more objective forecasts, as this often proves superior on average to the individual forecasts. Unfortunately, an informative model may not result.

An alternative way of classifying forecasting methods is between an **automatic** approach requiring no human intervention, and a **non-automatic** approach requiring some subjective input from the forecaster. The latter applies to subjective methods and most multivariate methods. Most univariate methods can be made fully automatic but can also be used in a non-automatic form, and there can be a surprising difference between the results.

The choice of method depends on a variety of considerations, including:

- How the forecast is to be used.
- The type of time series (e.g. macroeconomic series or sales figures) and its properties (e.g. are trend and seasonality present?). Some series are very regular and hence 'very predictable', but others are not. As always, a time plot of the data is very helpful.
- How many past observations are available.
- The length of the forecasting horizon. This book is mainly concerned with short-term forecasting. For example, in stock control the lead time for which forecasts are required is the time between ordering an item and its delivery.

[1]They are also sometimes called *independent variables* but this terminology is misleading, as they are typically *not* independent of each other.

- The number of series to be forecast and the cost allowed per series.
- The skill and experience of the analyst. Analysts should select a method with which they feel 'happy' and for which relevant computer software is available. They should also consider the possibility of trying more than one method.

It is particularly important to *clarify the objectives* (as in any statistical investigation). This means finding out how a forecast will actually be used, and whether it may even influence the future. In the latter case, some forecasts turn out to be self-fulfilling. In a commercial environment, forecasting should be an integral part of the management process leading to what is sometimes called a *systems approach*.

This chapter concentrates on calculating **point forecasts**, where the forecast for a particular future time period consists of a single number. Point forecasts are adequate for many purposes, but a **prediction interval** is often helpful to give a better indication of future uncertainty. Instead of a single value, a prediction interval consists of upper and lower limits between which a future value is expected to lie with a prescribed probability. Some methods for calculating prediction intervals are considered in Section 5.3.3. Taking one more step away from a point forecast, it may be desirable to calculate the entire probability distribution of a future value of interest. This is called **density forecasting**. The reader is referred to Tay and Wallis (2000). A practical halfway house between prediction intervals and density forecasting is the use of *fan charts*. The latter essentially plot prediction intervals at several different probability levels, by using darker shades for central values, and lighter shades for outer bands, which cover less likely values. These graphs can be very effective for presenting the future range of uncertainty in a simple, visually effective way. The reader is referred to Wallis (1999).

Whatever forecasting method is used, some sort of *forecast monitoring scheme* is often advisable, particularly with large numbers of series, to ensure that forecast errors are not systematically positive or negative. A variety of *tracking signals* for detecting 'trouble' are discussed, for example, by Gardner (1983) and McLain (1988).

## 5.2 Extrapolation and Exponential Smoothing

This section introduces the many projection methods that are now available. Further details may be found in Abraham and Ledolter (1983), Chan and Tong (2001, Chapters 3–4), Diebold (2001), Granger and Newbold (1986), and Montgomery et al. (1990).

### *5.2.1 Extrapolation of trend curves*

For long-term forecasting of non-seasonal data, it is often useful to fit a **trend curve** (or **growth curve**) to successive values and then extrapolate. This approach is most often used when the data are yearly totals, and hence clearly non-seasonal. A variety of curves may be tried including polynomial, exponential, logistic and Gompertz curves (see also Section 2.5.1). When the data are annual totals, at least 7 to 10 years of historical data are required to fit such curves. The method is worth considering for short annual series where fitting a complicated model to past data is unlikely to be worthwhile. Although primarily intended for long-term forecasting, it is inadvisable to make forecasts ahead for a longer period than about half the number of past years for which data are available.

A drawback to the use of trend curves is that there is no logical basis for choosing among the different curves except by goodness-of-fit. Unfortunately it is often the case that one can find several curves that fit a given set of data almost equally well but which, when projected forward, give widely different forecasts. Further details about trend curves are given by Meade (1984).

### 5.2.2 *Simple exponential smoothing*

**Exponential smoothing** (ES) is the name given to a general class of forecasting procedures that rely on simple updating equations to calculate forecasts. The most basic form, introduced in this subsection, is called **simple exponential smoothing** (SES), but this should only be used for non-seasonal time series showing no systematic trend. Of course many time series that arise in practice *do* contain a trend or seasonal pattern, but these effects can be measured and removed to produce a stationary series for which simple ES is appropriate. Alternatively, more complicated versions of ES are available to cope with trend and seasonality – see Section 5.2.3. Thus adaptations of exponential smoothing are useful for many types of time series – see Gardner (1985) for a detailed general review of these popular procedures.

Given a non-seasonal time series, say $x_1, x_2, \ldots, x_N$, with no systematic trend, it is natural to forecast $x_{N+1}$ by means of a weighted sum of the past observations:

$$\widehat{x}_N(1) = c_0 x_N + c_1 x_{N-1} + c_2 x_{N-2} + \cdots, + c_{N-1} x_1, \tag{5.1}$$

where the $\{c_i\}$ are weights. It seems sensible to give more weight to recent observations and less weight to observations further in the past. An intuitively appealing set of weights are *geometric* weights, which decrease by a constant ratio for every unit increase in the lag. In order that the weights sum to one, we take

$$c_i = \alpha(1-\alpha)^i \qquad i = 0, 1, \ldots, N-1,$$

where $\alpha$ is a constant such that $0 < \alpha < 1$. Then Equation (5.1) becomes

$$\widehat{x}_N(1) = \alpha x_N + \alpha(1-\alpha)x_{N-1} + \alpha(1-\alpha)^2 x_{N-2} + \cdots + \alpha(1-\alpha)^{N-1} x_1. \tag{5.2}$$

Strictly speaking, Equation (5.2) implies an infinite number of past observations, but in practice there will only be a finite number. Thus Equation (5.2) is customarily rewritten in the **recurrence** form as

$$\begin{aligned} \widehat{x}_N(1) &= \alpha x_N + (1-\alpha)[\alpha x_{N-1} + \alpha(1-\alpha)x_{N-2} + \cdots] \\ &= \alpha x_N + (1-\alpha)\widehat{x}_{N-1}(1). \end{aligned} \tag{5.3}$$

If we set $\widehat{x}_1(1) = x_1$, then Equation (5.3) can be used recursively to compute forecasts. Equation (5.3) also reduces the amount of arithmetic involved since forecasts can easily be updated using only the latest observation and the previous forecast of that latest observation.

The procedure defined by Equation (5.3) is called simple exponential smoothing. The adjective 'exponential' arises from the fact that the geometric weights lie on an exponential curve, but the procedure could equally well have been called geometric smoothing.

Equation (5.3) is sometimes rewritten in the equivalent **error-correction** form

$$\widehat{x}_N(1) = \alpha[x_N - \widehat{x}_{N-1}(1)] + \widehat{x}_{N-1}(1) = \alpha e_N + \widehat{x}_{N-1}(1), \tag{5.4}$$

where $e_N = x_N - \widehat{x}_{N-1}(1)$ is the prediction error at time $N$. Equations (5.3) and (5.4) look different at first sight, but give identical forecasts, and it is a matter of practical convenience as to which one should be used.

Although intuitively appealing, it is natural to ask when SES is a 'good' method to use. It can be shown (see Exercise 5.6) that SES is optimal if the underlying model for the time series is given by

$$X_t = \mu + \alpha \sum_{j<t} Z_j + Z_t,$$

where $\{Z_t\}$ denotes a purely random process. This infinite-order moving average (MA) process is non-stationary, but the first differences $(X_{t+1} - X_t)$ form a stationary first-order MA process (see Exercise 3.3). Thus $X_t$ is an ARIMA process of order $(0,1,1)$. In fact it can be shown (Chatfield et al., 2001) that there are many other models for which SES is optimal. This helps to explain why SES appears to be such a robust method.

The value of the smoothing constant $\alpha$ depends on the properties of the given time series. Values between 0.1 and 0.3 are commonly used and produce a forecast that depends on a large number of past observations. Values close to one are used rather less often and give forecasts that depend much more on recent observations. When $\alpha = 1$, the forecast is equal to the most recent observation.

The value of $\alpha$ may be estimated from past data by a similar procedure to that used for estimating the parameters of an MA process. Given a particular value of $\alpha$, one-step-ahead forecasts are produced iteratively through the series, and then the sum of squares of the one-step-ahead prediction errors is computed. This can be repeated for different values of $\alpha$ so that the value, which minimizes the sum of squares, can be found. In more detail, for a given value of $\alpha$, calculate

$$\begin{aligned}
\widehat{x}_1(1) &= x_1 \\
e_2 &= x_2 - \widehat{x}_1(1) \\
\widehat{x}_2(1) &= \alpha e_2 + \widehat{x}_1(1) \\
e_3 &= x_3 - \widehat{x}_2(1)
\end{aligned}$$

and so on until

$$e_N = x_N - \widehat{x}_{N-1}(1)$$

and then compute $\sum_{i=2}^{N} e_i^2$. Repeat this procedure for other values of $\alpha$ between 0 and 1, say in steps of 0.1, and select the value that minimizes $\sum e_i^2$, either by inspection or using an algorithmic numerical procedure. Modern computers make this all easy to do. Usually the sum of squares surface is quite flat near the minimum and so the choice of $\alpha$ is not critical.

### *5.2.3 The Holt and Holt–Winters forecasting procedures*

Exponential smoothing may readily be generalized to deal with time series containing trend and seasonal variation. The version for handling a trend with non-seasonal data is usually called Holt's (two-parameter) exponential smoothing, while the version that also copes with seasonal variation is usually referred to as the Holt–Winters (three-parameter) procedure. These names honour the pioneering work of C.C. Holt and P.R. Winters around 1960. The general idea is to generalize the equations for SES by introducing trend and seasonal terms, which are also updated by exponential smoothing.

We first consider Holt's ES. In the absence of trend and seasonality, the one-step-ahead forecast from simple ES can be thought of as an estimate of the local mean level of the series, so that simple ES can be regarded as a way of updating the local **level** of the series, say $L_t$. This suggests rewriting Equation (5.3) in the form

$$L_t = \alpha x_t + (1-\alpha)L_{t-1}.$$

Suppose we now wish to include a trend term, $T_t$ say, which is the expected increase or decrease per unit time period in the current level. Then a plausible pair of equations for updating the values of $L_t$ and $T_t$ in recurrence form are the following

$$\begin{aligned} L_t &= \alpha x_t + (1-\alpha)(L_{t-1} + T_{t-1}) \\ T_t &= \gamma(L_t - L_{t-1}) + (1-\gamma)T_{t-1}. \end{aligned}$$

Then the $h$-step-ahead forecast at time $t$ will be of the form

$$\hat{x}_t(h) = (L_t + hT_t)$$

for $h = 1, 2, 3, \ldots$. There are now two updating equations, involving two smoothing parameters, $\alpha$ and $\gamma$, which are generally chosen to lie in the range $(0, 1)$. It is natural to call this the two-parameter version of ES.

The above procedure may readily be generalized again to cope with seasonality. Let $L_t, T_t, I_t$ denote the local level, trend and seasonal index, respectively, at time $t$. The interpretation of $I_t$ depends on whether seasonality is thought to be additive or multiplicative – see Section 2.6. In the former case, $x_t - I_t$ is the deseasonalized value, while in the multiplicative case, it is $x_t/I_t$. The values of the three quantities, $L_t, T_t$ and $I_t$, all need to be estimated and so we need three updating equations with three smoothing parameters, say $\alpha, \gamma$ and $\delta$. As before, the smoothing parameters are usually chosen in the range (0,1). The form of the updating equations is again intuitively plausible. Suppose the observations are monthly, and that the seasonal variation is multiplicative. Then the (recurrence form) equations for updating $L_t, T_t, I_t$, when a new observation $x_t$ becomes available, are

$$\begin{aligned} L_t &= \alpha(x_t/I_{t-12}) + (1-\alpha)(L_{t-1}+T_{t-1}) \\ T_t &= \gamma(L_t - L_{t-1}) + (1-\gamma)T_{t-1} \\ I_t &= \delta(x_t/L_t) + (1-\delta)I_{t-12} \end{aligned}$$

and the forecasts from time $t$ are then

$$\widehat{x}_t(h) = (L_t + hT_t)I_{t-12+h}$$

for $h = 1, 2, \ldots, 12$. There are analogous formulae for the additive seasonal case. There are also analogous formulae for the case where the seasonality is of length $s$ say, rather than 12 as for monthly observations. In particular, $s = 4$ for quarterly data, when we would, for example, compare $I_t$ with $I_{t-4}$. Unfortunately, the literature is confused by many different notations and by the fact that the updating equations may be presented in an equivalent error-correction form, which can look quite different. For example, the above formula for updating the trend in the monthly multiplicative case can be rewritten (after some algebra) in the form

$$T_t = T_{t-1} + \alpha\gamma e_t/I_{t-12}$$

where $e_t = x_t - \widehat{x}_{t-1}(1)$ denotes the one-step-ahead forecast error as before. This formula looks quite different and, in particular, it looks as though $\alpha\gamma$ is the smoothing parameter. Clearly, great care needs to be taken when comparing formulae from different sources expressed in different ways.

In order to apply Holt–Winters smoothing to seasonal data, the analyst should carry out the following steps:

(1) Examine a graph of the data to see whether an additive or a multiplicative seasonal effect is the more appropriate.

(2) Provide starting values for $L_1$, and $T_1$ as well as seasonal values for the first year, say $I_1, I_2, \ldots, I_s$, using the first few observations in the series in a fairly simple way; for example, the analyst could choose $L_1 = \sum_1^s x_i/s$.

(3) Estimate values for $\alpha, \gamma, \delta$ by minimizing $\sum e_t^2$ over a suitable fitting period for which historical data are available.

(4) Decide whether to normalize the seasonal indices at regular intervals by making them sum to zero in the additive case or have an average of one in the multiplicative case.

(5) Choose between a fully automatic approach (for a large number of series) and a non-automatic approach. The latter allows subjective adjustments for particular series, for example, by allowing the removal of outliers and a careful selection of the appropriate form of seasonality.

Further details on the Holt–Winters method are given by Chatfield and Yar (1988). The method is straightforward and is widely used in practice.

Another variation of ES that deserves mention here is the use of a **damped trend** (Gardner, 1985). This procedure can be used with the Holt and Holt–Winters methods and introduces another smoothing parameter, say $\phi$, where $0 < \phi < 1$, such that the estimate of the trend or growth rate at time $t$, namely, $T_t$, is damped to $\phi T_t$ in the subsequent time period. The intuitive justification for this is that most trends do not in fact go on forever, but rather damp down towards zero over time. For Holt's method, the $h$-steps-ahead forecast at time $t$ then becomes

$$\widehat{x}_t(h) = [L_t + (\sum_{i=1}^{h} \phi^i)T_t)]$$

rather than $(L_t + hT_t)$. There are analogous formulae for damped Holt–Winters. The method involves estimating one extra smoothing parameter, but can give more accurate forecasts (Gardner and McKenzie, 1985) and so may justify the extra effort. A successful empirical example is given by Grubb and Mason (2001) using U.K. airline passenger data from 1949 to 1998, but note that they damp towards the average trend rather than towards zero.

We have still not exhausted the many variants of ES. Brown (1963) has suggested a technique called **general exponential smoothing**, which consists of fitting polynomial, sinusoidal or exponential functions to the data and finding appropriate updating formulae. This method is more complicated than earlier ES methods, even though there is rather little emphasis on identifying the correct functional forms. It should be noted that sinusoids are used to describe any seasonality rather than a set of seasonal indices as in the Holt–Winters method. One special case of this approach is **double exponential smoothing**, which is applicable to series containing a linear trend. Unlike Holt's method, double ES uses a single smoothing parameter. This makes the method simpler, but may lead to poorer forecast performance. Note that Brown suggests fitting by *discounted* least squares, in which more weight is given to recent observations, rather than using (global) least squares.

Further details on the many variants of ES are given by Gardner (1985).

## 5.3 The Box–Jenkins Methodology

This section gives a brief outline of the forecasting procedure, based on autoregressive integrated moving average (ARIMA) models, which is usually known as the Box–Jenkins approach. The beginner may find it easier to get further information from books such as Vandaele (1983), Granger and Newbold (1986) or Jenkins (1979), rather than the original 1970 Box–Jenkins book now revised as Box et al. (1994), although the latter is still an essential reference source.

### *5.3.1 The Box-Jenkins procedures*

Now AR, MA and ARMA models have been around for many years and are associated, in particular, with early work by G.U. Yule and H.O. Wold. A major contribution of Box and Jenkins has been to provide a general *strategy* for time-series forecasting, which emphasizes the importance of identifying an appropriate model in an iterative way as outlined briefly in Section 4.8. Indeed the iterative approach to model building that they suggested has since become standard in many areas of statistics. Furthermore, Box and Jenkins showed how the use of differencing can extend ARMA models to ARIMA models and hence cope with non-stationary series. In addition, Box and Jenkins show how to incorporate seasonal terms into seasonal ARIMA (SARIMA) models. Because of all these fundamental contributions, ARIMA models are often referred to as Box–Jenkins models.

In brief, the main stages in setting up a Box–Jenkins forecasting model are as follows:

(1) *Model identification*

Examine the data to see which member of the class of ARIMA processes appears to be most appropriate.

(2) *Estimation*

Estimate the parameters of the chosen model as described in Chapter 4.

(3) *Diagnostic checking*

Examine the residuals from the fitted model to see if it is adequate.

(d) *Consideration of alternative models if necessary*

If the first model appears to be inadequate for some reason, then alternative ARIMA models may be tried until a satisfactory model is found. When such a model has been found, it is usually relatively straightforward to calculate forecasts as conditional expectations.

We now consider these stages in more detail. In order to identify an appropriate ARIMA model, the first step in the Box–Jenkins procedure is to *difference* the data until they are stationary. This is achieved by examining the correlograms of various differenced series until one is found that comes down to zero 'fairly quickly' and from which any seasonal cyclic effect has been largely removed, although there could still be some 'spikes' at the seasonal lags $s, 2s$, and so on, where $s$ is the number of observations per year. For non-seasonal data, first-order differencing is usually sufficient to attain stationarity. For monthly data (of period 12), the operator $\nabla\nabla_{12}$ is often used if the seasonal effect is additive, while the operator $\nabla_{12}^2$ may be used if the seasonal effect is multiplicative. Sometimes the operator $\nabla_{12}$ by itself will be sufficient. Over-differencing should be avoided. For a seasonal period of length $s$, the operator $\nabla_s$ may be used, and, in particular, for quarterly data we may use $\nabla_4$.

The differenced series will be denoted by $\{w_t; t = 1, \ldots, N - c\}$, where $c$ terms are 'lost' by differencing. For example, if the operator $\nabla\nabla_{12}$ is used, then $c = 13$.

For non-seasonal data, an ARMA model can now be fitted to $\{w_t\}$ as described in Chapter 4. If the data are seasonal, then the SARIMA model defined in Equation (4.16) may be fitted as follows. Plausible values of $p, P, q, Q$ are selected by examining the correlogram and the partial autocorrelation function (ac.f.) of the differenced series $\{w_t\}$. Values of $p$ and $q$ are selected by examining the first few values of $r_k$, as outlined in Chapter 4. Values of $P$ and $Q$ are selected primarily by examining the values of $r_k$ at $k = 12, 24, \ldots$, when the seasonal period is given by $s = 12$. If, for example, $r_{12}$ is 'large' but $r_{24}$ is 'small', this suggests one seasonal moving average term, so we would take $P = 0, Q = 1$, as this SARIMA model has an ac.f. of similar form. Box et al. (1994, Table A9.1) list the autocovariance functions of various SARIMA models.

Having tentatively identified what appears to be a reasonable SARIMA model, least squares estimates of the model parameters may be obtained by minimizing the residual sum of squares in a similar way to that proposed for ordinary ARMA models. In the case of seasonal series, it is advisable to estimate initial values of $a_t$ and $w_t$ by backforecasting (or backcasting) rather than setting them equal to zero. This procedure is described by Box et al. (1994, Section 9.2.4). In fact, if the model contains a seasonal MA parameter that is close to one, several cycles of forward and backward iteration may be needed. Nowadays several alternative estimation procedures are available, based on, for example, the exact likelihood function, on conditional or unconditional least squares, or on a Kalman filter approach (see references in Section 4.4).

For both seasonal and non-seasonal data, the adequacy of the fitted model should be checked by what Box and Jenkins call 'diagnostic checking'. This essentially consists of examining the residuals from the fitted model to see whether there is any evidence of non-randomness. The correlogram of the residuals is calculated and we can then see how many coefficients are significantly different from zero and whether any further terms are indicated for the ARIMA model. If the fitted model appears to be inadequate, then alternative ARIMA models may be tried until a satisfactory one is found.

When a satisfactory model is found, forecasts may readily be computed. Given data up to time $N$, these forecasts will involve the observations and the fitted residuals (i.e. the one-step-ahead forecast errors) up to and including time $N$. The minimum mean square error forecast of $X_{N+h}$ at time $N$ is the conditional expectation of $X_{N+h}$ at time $N$, namely,

$$\widehat{x}_N(h) = E(X_{N+h} | X_N, X_{N-1}, \ldots).$$

In evaluating this conditional expectation, we use the fact that the 'best' forecast of all future $Z$s is simply zero (or more formally that the conditional expectation of $Z_{N+h}$, given data up to time $N$, is zero for all $h > 0$). Box et al. (1994) describe three general approaches to computing forecasts.

(1) *Using the model equation directly*

Point forecasts are usually computed most easily directly from the ARIMA model equation, which Box et al. (1994) call the difference equation form. Assuming that the model equation is known exactly, then $\widehat{x}_N(h)$ is obtained from the model equation by replacing (i) future values of $Z$ by zero, (ii) future values of $X$ by their conditional expectation and (iii) present and past values of $X$ and $Z$ by their observed values.

As an example, consider the SARIMA$(1,0,0) \times (0,1,1)_{12}$ model used as an example in Section 4.8, where

$$X_t = X_{t-12} + \alpha(X_{t-1} - X_{t-13}) + Z_t + \theta Z_{t-12}.$$

Then we find

$$\begin{aligned} \widehat{x}_N(1) &= x_{N-11} + \alpha(x_N - x_{N-12}) + \theta z_{N-11} \\ \widehat{x}_N(2) &= x_{N-10} + \alpha[\widehat{x}_N(1) - x_{N-11}] + \theta z_{N-10}. \end{aligned}$$

Forecasts further into the future can be calculated recursively in an obvious way. It is also possible to find ways of updating the forecasts as new observations become available. For example, when $x_{N+1}$ becomes known, we have

$$\begin{aligned} \widehat{x}_{N+1}(1) &= x_{N-10} + \alpha(x_{N+1} - x_{N-11}) + \theta z_{N-10} \\ &= \widehat{x}_N(2) + \alpha[x_{N+1} - \widehat{x}_N(1)] \\ &= \widehat{x}_N(2) + \alpha z_{N+1}. \end{aligned}$$

(2) *Using the $\psi$ weights*

An ARMA model can be rewritten as an infinite-order MA process and the resulting $\psi$ weights, as defined in Equation (3.25), could also be used to compute forecasts, but are primarily helpful for calculating forecast error variances. Since

$$X_{N+h} = Z_{N+h} + \psi_1 Z_{N+h-1} + \psi_2 Z_{N+h-2} + \cdots \tag{5.5}$$

and future Zs are unknown at time $N$, it is clear that $\widehat{x}_N(h)$ is equal to $\sum_{j=0}^{\infty} \psi_{h+j} z_{N-j}$ (since future $z$s cannot be included). Thus the $h$-steps-ahead forecast error is $(Z_{N+h} + \psi_1 Z_{N+h-1} + \cdots + \psi_{h-1} Z_{N+1})$. Hence the variance of the $h$-steps-ahead forecast error is $(1 + \psi_1^2 + \cdots + \psi_{h-1}^2)\sigma_z^2$, by independence of the $Z$'s.

(3) *Using the $\pi$ weights*

An ARMA model can also be rewritten as an infinite-order AR process and the resulting $\pi$ weights, as defined in Equation (3.26), can also be used for compute point forecasts. Since

$$X_{N+h} = \pi_1 X_{N+h-1} + \cdots + \pi_h X_N + \cdots + Z_{N+h}$$

it is intuitively clear that $\widehat{x}_N(h)$ is given by

$$\begin{aligned} \widehat{x}_N(h) &= \pi_1 \widehat{x}_N(h-1) + \pi_2 \widehat{x}_N(h-2) + \cdots + \pi_{h-1} \widehat{X}_N(1) \\ &\quad + \pi_h X_N + \pi_{h+1} X_{N-1} + \cdots . \end{aligned}$$

These forecasts can be computed recursively, replacing future values of $X$ with predicted values as necessary.

In general, methods (1) or (3) are used for point forecasts while method (2) is used for forecast error variances.

In practice, the model will not be known exactly, and we have to estimate the model parameters (and hence the $\psi$s and $\pi$s if they are needed); we also have to estimate the past observed values of $Z$, namely, $z_t$, by the observed residuals or one-step-ahead forecasts errors, namely, $\widehat{z}_t$. Thus for the SARIMA$(1,0,0) \times (0,1,1)_{12}$ model given above, we would have, for example, that

$$\widehat{x}_N(1) = x_{N-11} + \widehat{\alpha}(x_N - x_{N-12}) + \widehat{\theta}\widehat{z}_{N-11}.$$

Except for short series, this generally makes little difference to forecast error variances.

Although some packages have been written to carry out ARIMA modelling and forecasting in an automatic way, the Box–Jenkins procedure is primarily intended for a non-automatic approach where the analyst uses subjective judgement to select an appropriate model from the large family of ARIMA models according to the properties of the individual series being analyzed. Thus, although the procedure is more versatile than many competitors, it is also more complicated and considerable experience is required to identify an appropriate ARIMA model. Unfortunately, the analyst may find several different models, which fit the data equally well but give rather different forecasts, while sometimes it is difficult to find any sensible model. Of course, an inexperienced analyst will sometimes choose a 'silly' model. Another drawback is that the method requires several years of data (e.g. at least 50 observations for monthly seasonal data).

### 5.3.2 *Other methods*

Many other univariate forecasting procedures have been proposed, and we briefly mention a few of them.

Granger and Newbold (1986, Section 5.4) describe a procedure called **stepwise autoregression**, which can be regarded as a subset of the Box–Jenkins procedure. It has the advantage of being fully automatic and relies on the fact that AR models are much easier to fit than MA or ARMA models even though an AR model may require extra parameters to give as good a representation of the data. The first step is to take first differences of the data to allow for non-stationarity in the mean. Then a maximum possible lag, say $p$, is chosen and the best AR model with just one lagged variable at a lag between 1 and $p$, is found, namely

$$W_t = \mu + \alpha_k^{(1)} W_{t-k} + e_t^{(1)},$$

where $W_t = X_t - X_{t-1}$, $1 \leq k \leq p$, $\alpha_k^{(1)}$ is the autoregression coefficient at lag $k$ when fitting one lagged variable only, and $e_t^{(1)}$ is the corresponding error term. Then the best AR model with $2, 3, \ldots$ lagged variables is found. The procedure is terminated when the reduction in the sum of squared residuals at the $j$th stage is less than some preassigned quantity. Thus an integrated AR model is fitted, which is a special case of the Box–Jenkins ARIMA class. Granger and Newbold suggest choosing $p = 13$ for quarterly data and $p = 25$ for monthly data.

Harrison (1965) has proposed a modification of seasonal exponential smoothing, which consists essentially of performing a Fourier analysis of the seasonal factors and replacing them by smoothed factors. Parzen's **ARARMA** approach (Parzen, 1982; Meade and Smith, 1985) relies on fitting an AR model to remove the trend (rather than just differencing the trend away) before fitting an ARMA model. An apparently new method, called the **theta** method, gave promising results (Makridakis and Hibon, 2000), but subsequent research has shown that it is actually equivalent to a form of exponential smoothing.

There are two general forecasting methods, called **Bayesian forecasting** (West and Harrison, 1997) and **structural modelling** (Harvey, 1989), which rely on updating model parameters by a technique called **Kalman filtering**. The latter is introduced in Chapter 10, and so we defer consideration of these methods until then.

### 5.3.3 *Prediction intervals*

Thus far, we have concentrated on calculating point forecasts, but it is sometimes better to calculate an *interval forecast* to give a clearer indication of future uncertainty. A **prediction interval** (P.I.) consists of upper and lower limits between which a future value is expected to lie with a prescribed probability. This interval can be calculated in several ways, depending on the forecasting method used, the properties of the data, and so on.

Most P.I.s used in practice are essentially of the following general form. A $100(1-\alpha)\%$ P.I. for $X_{N+h}$ is given by :

$$\widehat{x}_N(h) \pm z_{\alpha/2}\sqrt{\mathrm{Var}[e_N(h)]}, \tag{5.6}$$

where $e_N(h) = X_{N+h} - \widehat{x}_N(h)$ denotes the forecast error made at time $N$ when forecasting $h$ steps ahead. Here $z_{\alpha/2}$ denotes the percentage point of a standard normal distribution with a proportion $\alpha/2$ above it. Equation (5.6) assumes that an appropriate expression for $\mathrm{Var}[e_N(h)]$ can be found for the method or model being used. As the P.I. in Equation (5.6) is symmetric about $\widehat{x}_N(h)$, it effectively assumes that the forecast is unbiased. The formula also assumes that the forecast errors are normally distributed.

In practice, the forecast errors are unlikely to be exactly normal, because the estimation of model parameters produces small departures from normality, while the assumption of a known, invariant model with normal errors is also unlikely to be exactly true. Nevertheless, because of its simplicity, Equation (5.6) is the formula that is generally used to compute P.I.s, though preferably after checking that the underlying assumptions (e.g. forecast errors are normally distributed) are at least approximately satisfied. For any given forecasting method, the main problem will then lie with evaluating $\mathrm{Var}[e_N(h)]$. Fortunately, formulae for this variance are available for many classes of model. Perhaps the best-known formula is for Box–Jenkins ARIMA forecasting, where the variance may be evaluated by writing an ARIMA model in infinite-moving-average form as

$$X_t = Z_t + \psi_1 Z_{t-1} + \psi_2 Z_{t-2} + \cdots \tag{5.7}$$

— see Equation (5.5). Then the best forecast at time $N$ of $X_{N+h}$ can only involve the values of $Z_t$ up to time $t = N$, so that $e_N(h) = [X_{N+h} - \widehat{X}_N(h)] = Z_{N+h} + \sum_{j=1}^{h-1} \psi_j Z_{N+h-j}$. Thus

$$\text{Var}[e_N(h)] = [1 + \psi_1^2 + \cdots + \psi_{h-1}^2]\sigma_z^2.$$

Formulae can also be found for most variations of exponential smoothing, by assuming that the given method is optimal. For example, for simple exponential smoothing, with smoothing parameter $\alpha$, it can be shown that

$$\text{Var}[e_N(h)] = [1 + (h-1)\alpha^2]\sigma_e^2,$$

where $\sigma_e^2$ denotes the variance of the one-step-ahead forecast errors. Analogous formulae are available in the literature for some other methods and models — see Chatfield (1993) and Chatfield (2001, Chapter 7).

A completely different alternative approach to calculating P.I.s is to work empirically by using the 'forecast' errors obtained by fitting a model to past data, finding the observed variance of the within-sample 'forecast' errors at different lead times and using these values in Equation (5.6). A detailed review of different methods of calculating P.I.s, including simulation, resampling and Bayesian methods, is given by Chatfield (2001, Chapter 7).

Unfortunately, whichever way P.I.s are calculated, they tend to be too narrow in practice. The empirical evidence for this is discussed by Chatfield (2001, Section 7.7), who also suggests various reasons why this phenomenon occurs. The main reason seems to be that the underlying model may change in the future. The forecaster should bear this in mind and should not think that a narrow P.I. is necessarily 'good'.

*Example 5.1*

We now use the Box-Henkins procedure to analyse the monthly unemployment rates in the U.S. from January 1948 to October 2018. The series can be downloaded from the U.S. Federal Reserve at St. Louis. Figure 5.1 plots the series and its sample ac.f. and sample partial ac.f. We note that all unemployment rates vary from 2.1% to 10.4%, and the sample ac.f. of the series decays slowly with the increase of lags.

```
> data<-read.csv("unemrate_LNU04000024.csv", header=T)
> unem<-data[,2]
> unem.ts<-ts(unem, start=c(1948,1), frequency=12)

> ### Figure 5.1
> par(mfrow=c(3,1), mar=c(4,4,4,4))
> plot(unem, type="l", xlab="Month", ylab="Rates", xaxt="n")
> x.pos<-c(1,145,289,433,577,721,850)
> x.label<-c("1948/01", "1960/01", "1972/01", "1984/01",
   "1996/01", "2008/01", "2018/10")
> axis(1, x.pos, x.label)
> acf(unem, 25, xlab="Lag", ylab="ACF", main="")
```

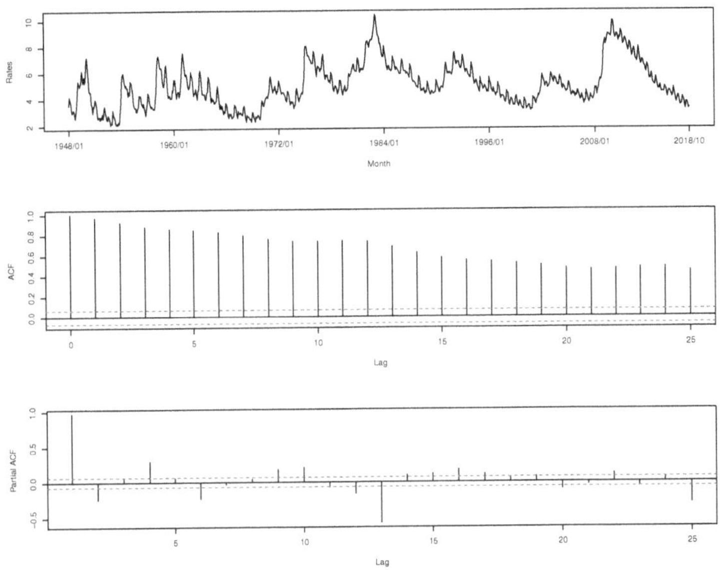


Figure 5.1 *The unemployment rate series (Top) and its sample ac.f. (Middle) and sample partial ac.f. (Bottom).*

```
> acf(unem, 25, type="partial", xlab="Lag",
   ylab="Partial ACF",main="")
```

To find a time series model with good out-of-sample performance, we split the time series into a training sample of historical data from January 1948 to December 2017 and a second sample of 'test data' from January 2018 to October 2018. Denote $\{X_t^{\text{in}}\}$ the unemployment rates in the training period, and $\{X_t^{\text{out}}\}$ the rates in the test period. We will use $\{X_t^{\text{out}}\}$ to measure the performance of the out-of-sample forecasts developed from the training sample $\{X_t^{\text{in}}\}$.

Since there are obvious seasonal effects on unemployment, we use the `R` function `stl` to decompose $X_t^{\text{in}}$ in the training period into a trend, a seasonal component, and remainder series, with a period of 12 months for the seasonal component. Let $S_t$ be the seasonal component in $X_t$, and denote the deseasonalized series as $\widetilde{X}_t := X_t^{\text{in}} - S_t$. Figure 5.2 shows the deseasonalized series $\widetilde{X}_t$ in the training period and its sample ac.f. and sample partial ac.f.

```
> ### Partition the series into training and test data
> unem.ts<-ts(unem, start=c(1948,1), frequency=12)
> unem.train<-ts(unem[1:840], start=c(1948,1),
    end=c(2017,12), frequency=12)
> unem.test<-ts(unem[841:850], start=c(2018,1),
    end=c(2018,10), frequency=12)
```

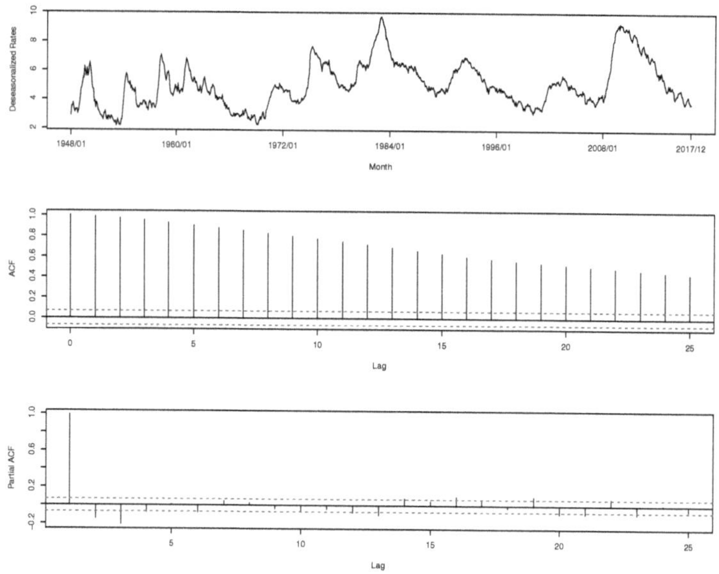


Figure 5.2 *The deseasonalized series $\widetilde{X}_t$ and its sample ac.f. and sample partial ac.f.*

```
> unem.stl<-stl(unem.train,"periodic")
> unem.sea<-unem.stl$time[,1]
> unem.desea<- unem.train-unem.sea
> unem.desea.series<-as.numeric(unem.desea)

> ### Figure 5.2
> par(mfrow=c(3,1), mar=c(4,4,4,4))
> plot(unem.desea.series, type="l", xlab="Month",
 ylab="Deseasonalized Rates", xaxt="n")
> x.pos<-c(1,145,289,433,577,721,840)
> x.label<-c("1948/01", "1960/01", "1972/01", "1984/01",
   "1996/01", "2008/01", "2017/12")
> axis(1, x.pos, x.label)
> acf(unem.desea.series, 25, xlab="Lag",
   ylab="ACF", main="")
> acf(unem.desea.series, 25, type="partial",
   xlab="Lag", ylab="Partial ACF", main="")
```

The sample ac.f. of the deseasonalized series $\widetilde{X}_t$ in Figure 5.2 shows that the deseasonalized series is very persistent. As this might suggest the series is not stationary, we consider a unit-root test for the deseasonalized series. There are several unit-root tests developed in the literature; we consider the augmented Dickey-Fuller test introduced in Section 4.6. The following R implementation shows that the $p$-value of the test is 0.03078, smaller than 5%; hence, the null hypothesis that the series $\widetilde{X}_t$ has a unit root should be rejected and an ARIMA$(p,1,q)$ model might be appropriate for series $\widetilde{X}_t$.

```
> # ADF test indicates that there is a unit root
> library(tseries)
> adf.test(unem.desea)

 Augmented Dickey-Fuller Test

data:  unem.desea
Dickey-Fuller = -3.6198, Lag order = 9, p-value = 0.03078
alternative hypothesis: stationary
```

Since the orders $p$ and $q$ are unknown, we fit a set of ARIMA$(p,1,q)$ models to series $\widetilde{X}_t$ and compute their AICs.

```
> # Select ARIMA models via AICs
> unem.desea.aic<-matrix(0, 4, 4);
> for (i in 0:3) for (j in 0:3) {
   fit.arima <- arima(unem.desea, order=c(i,1,j))
   unem.desea.aic[i+1,j+1]<-fit.arima$aic
> }
> unem.desea.aic
          [,1]      [,2]      [,3]      [,4]
[1,] -102.6192 -121.3318 -164.0700 -168.9881
[2,] -130.6713 -159.3643 -172.0880 -170.2128
[3,] -171.1796 -171.1427 -170.1949 -169.0363
[4,] -172.0392 -174.2756 -173.2259 -171.7914
```

The model with minimum AIC (-174.2756) is ARIMA$(p,1,q)$ with $p=3$ and $q=1$. Hence we fit an ARIMA$(3,1,1)$ model to the series $\widetilde{X}_t$.

```
> fit.arima311<-arima(unem.desea.series, order=c(3,1,1))
> fit.arima311

Call:
arima(x = unem.desea.series, order = c(3, 1, 1))

Coefficients:
          ar1     ar2     ar3     ma1
      -0.4935  0.3039  0.2109  0.6296
s.e.   0.1517  0.0434  0.0406  0.1516

sigma^2 estimated as 0.04699:  log likelihood=92.14, aic=-174.28
```

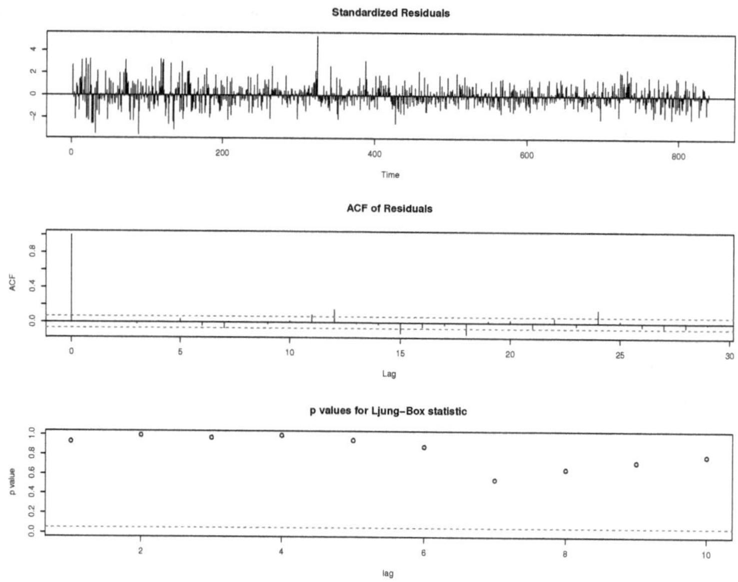


Figure 5.3 *The standardized residual series (Top), the sample ac.f. of residuals (Middle) and p-values for Ljung-Box statistic (Bottom).*

The fitted result suggests the following ARIMA(3,1,1) model for $\widetilde{X}_t$,

$$Y_t = \widetilde{X}_t - \widetilde{X}_{t-1}$$
$$Y_t = -0.4935_{(.1517)}Y_{t-1} + 0.3039_{(.0434)}Y_{t-2} + 0.2109_{(.0406)}Y_{t-3} + Z_t + 0.6296_{(.1516)}Z_{t-1},$$

where $Z_t \sim N(0, 0.0470)$. Note that the fitted model can also be expressed as

$$(1-B)(1+0.4935_{(.1517)}B - 0.3039_{(.0434)}B^2 - 0.2109_{(.0406)}B^3)\widetilde{X}_t = Z_t + 0.6296_{(.1516)}Z_{t-1}.$$

To perform a residual diagnostic, we show in Figure 5.3 the time series plot of standardized $\widehat{Z}_t$, the sample ac.f. of $\widehat{Z}_t$, and $p$-values for the Ljung-Box statistic (see Section 4.9). We also carry out a portmanteau test for the residuals. Note that the $p$-value of the test is much larger than 0.05, indicating the fitted model is appropriate.

```
> # Figure 5.3
> tsdiag(fit.arima311)
> Box.test(fit.arima311$resid)

  Box-Pierce test
```

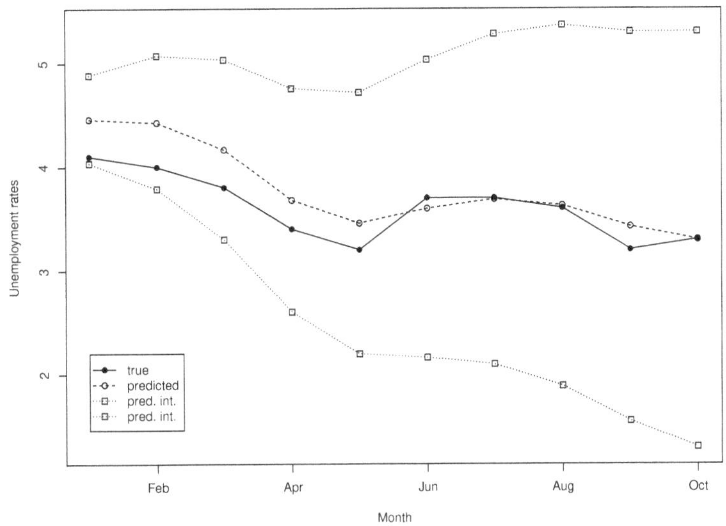


Figure 5.4 *The true and predicted unemployment rates with 95% prediction intervals at January, February, ..., October in 2018.*

```
data:  fit.arima311$resid
X-squared = 0.0072558, df = 1, p-value = 0.9321
```

We then use the fitted model to forecast the rates at January, February, ..., October in 2018. We compute the $k$-step-ahead forecast and its 95% prediction intervals for $\widetilde{X}_t$ at $k = 1, \ldots, 10$. Note that to obtain predicted values for $\{X_t^{(\mathrm{out}}\}$, seasonal effects need to be added back. Figure 5.4 shows the true and predicted values of $\{X_t^{(\mathrm{out}}\}$, and prediction intervals. We can see the following. First, all values of $\{X_t^{(\mathrm{out}}\}$ lie in the corresponding 95% prediction intervals. Second, the predicted values for $X_t^{(\mathrm{out}}$ are quite close to the values of $X_t^{(\mathrm{out}}$. This indicates that the estimated model also achieves a good performance in the out-of-sample analysis.

```
> # Prediction
> fit.arima311<-arima(unem.desea, order=c(3,1,1))
> unem.pred<-predict(fit.arima311, n.ahead=10)
> unem.pred.summary<-cbind(unem.test,
    unem.pred$pred+unem.sea[1:10],
    unem.pred$pred-1.96*unem.pred$se+unem.sea[1:10],
    unem.pred$pred+1.96*unem.pred$se+unem.sea[1:10])
> colnames(unem.pred.summary)<-c("true", "predict",
```

```
      "ci.lower", "ci.upper")

> ### Figure 5.4
> plot(c(1,10), range(unem.pred.summary), type="n",
    xlab="Month", ylab="Unemployment rates", xaxt="n")
> lines(seq(1,10),unem.pred.summary[,1], lty=1)
> points(seq(1,10),unem.pred.summary[,1], pch=16)
> lines(seq(1,10),unem.pred.summary[,2], lty=2)
> points(seq(1,10),unem.pred.summary[,2], pch=1)
> lines(seq(1,10),unem.pred.summary[,3], lty=3)
> points(seq(1,10),unem.pred.summary[,3], pch=0)
> lines(seq(1,10),unem.pred.summary[,4], lty=3)
> points(seq(1,10),unem.pred.summary[,4], pch=0)
> legend(1,2.2,c("true", "predicted", "pred. int.",
    "pred. int."), lty=c(1,2,3,3), pch=c(16,1,0,0))
> x.pos<-c(2,4,6,8,10)
> x.label<-c("Feb", "Apr", "Jun", "Aug", "Oct")
> axis(1, x.pos, x.label)
```

## 5.4 Multivariate Procedures

This section provides a brief introduction to some multivariate forecasting procedures. Further details on multivariate modelling are given later in this book, especially in Section 9.4.2 and in Chapter 12. More detailed coverage is given by Chatfield (2001, Chapter 5). The concepts are much more advanced than for univariate modelling and more sophisticated tools, such as the cross-correlation function, need to be developed in later chapters before we can make much progress.

### *5.4.1 Multiple regression*

One common forecasting method makes use of the multiple regression model, which will be familiar to many readers. This model assumes that the response variable of interest, say $y$, is linearly related to $p$ explanatory variables, say $x_1, x_2, \dots, x_p$. The usual multiple regression model can be written as

$$y = \beta_1 x_1 + \beta_2 x_2 + \cdots + \beta_p x_p + u, \tag{5.8}$$

where $\{\beta_i\}$ are constants, and $u$ denotes the 'error' term. This equation is linear in terms of the parameters $\{\beta_i\}$, but could involve non-linear functions of observed variables. When building a regression model, it is helpful to distinguish between explanatory variables that can be controlled (like stock levels) and those that cannot (like air temperature). Now Equation (5.8) does not specifically involve time. Of course, we could regard time as a predetermined variable and introduce it as one of the explanatory variables, but regression on time alone would normally be regarded as a univariate procedure. More generally, we need to specify when each of the variables in Equation (5.8) is measured and so each variable really needs a subscript indicating when the variable is measured. When lagged values of the explanatory variables are included, they may be called **leading indicators**. Such variables are much more useful for forecasting. If lagged values of the response variable $y$ are included, then they are of an autoregressive nature and change the character of the model.

Multiple regression is covered in numerous statistics texts and the details need not be repeated here. The models are widely used and sometimes work well. However, there are many dangers in applying such models to time-series data. Modern computer software makes it easy (perhaps too easy!) to fit regression models, and the ease of computation makes it tempting to include lots of explanatory variables, even though including too many may yield dubious results.

In fact applying regression models to time-series data is really *not* straightforward, especially as standard results assume that successive values of the 'errors' $\{u\}$ are independent, which is unlikely to be the case in practice. Although a high multiple correlation coefficient $R^2$ may result from fitting a regression model to time-series data, this apparent good fit may be spurious and does not mean that good forecasts necessarily result. This may be demonstrated both theoretically, e.g. Phillips (1986) and Hamilton (1994, Section 18.3), and empirically, e.g. Granger and Newbold (1974) and Granger and Newbold (1986, Section 6.4), for non-stationary data. It is advisable to restrict the value of $p$ to perhaps 3 or 4, and keep back part of the data to check forecasts from the fitted model. When doing this, it is important to distinguish between *ex ante* forecasts of $y$, which replace future values of explanatory variables by their forecasts (and so are true out-of-sample forecasts), and *ex post* forecasts, which use the true values of explanatory variables. The latter can look misleadingly good.

Problems can arise when the explanatory variables are themselves correlated, as often happens with time-series data. It is advisable to begin by looking at the correlations between explanatory variables so that, if necessary, selected explanatory variables can be removed to avoid possible singularity problems. The quality and characteristics of the data also need to be checked. For example, if a crucial explanatory variable has been held more or less constant in the past, then it is impossible to assess its effect using past data. Another type of problem arises when the response variable can, in turn, affect values of the explanatory variables to give what is called a closed-loop system. This is discussed later in Sections 9.4.3 and 13.1.

Perhaps the most important danger arises from mistakenly assuming that the 'error' terms form an independent sequence. This assumption is often inappropriate and can lead to a badly misspecified model and poor forecasts (Box and Newbold, 1971). The residuals from a regression model should always be checked for possible autocorrelation – see Section 4.7. A standard regression model, with independent errors, is usually fitted by Ordinary Least Squares (OLS), but this is seldom applicable directly to time-series data without suitable modification. Several alternative estimation procedures, such as Generalized Least Squares (GLS), have been developed over the years to cope with autocorrelated errors, but in such a way as still to be able to use OLS software. It can be shown that GLS and OLS are sometimes equivalent asymptotically, but such results may have little relevance for short series. Moreover, it is disturbing that autocorrelated errors may arise because certain lagged variables have been omitted from the model so that efforts to overcome such problems (e.g. by using GLS) are likely to lead to failure in the presence of a mis-specified model (e.g. Mizon, 1995). Mis-specifying the error structure also causes problems. Further details about methods, such as GLS, may be found, for example, in an econometrics text such as Hamilton (1994). Nowadays, full maximum likelihood is likely to be used once an appropriate model for the errors (e.g. AR, MA or ARMA) has been identified (Choudhury et al., 1999).

In summary, the use of multiple regression can be dangerous except when there are clear contextual reasons why one or more series should explain variation in another. There are various precautions that should be taken, and various alternative strategies that should be considered. They include: (1) Using the context to choose the explanatory variables with care, and limiting their total number to perhaps 3 or 4; (2) Including appropriate lagged values of variables as variables in their own right; (3) Removing obvious sources of non-stationarity before fitting a regression model; (4) Carrying out careful diagnostic checks on any fitted model; (5) Allowing for correlated errors in the fitting procedure and (6) Considering alternative families of models such as transfer function models, vector AR models or a model allowing for co-integration — see Sections 9.4.2, 13.3 and 13.6, respectively.

### *5.4.2 Econometric models*

Econometric models (e.g. Harvey, 1990) often assume that an economic system can be described, not by a single equation, but by a set of simultaneous equations. For example, not only do wage rates depend on prices but also prices depend on wage rates. Economists distinguish between *exogenous* variables, which affect the system but are not themselves affected, and *endogenous* variables, which interact with each other. The simultaneous equation system involving $k$ dependent (endogenous) variables $\{Y_i\}$ and $g$ predetermined (exogenous) variables $\{X_i\}$ may be written

$$Y_i = f_i(Y_1, \ldots, Y_{i-1}, Y_{i+1}, \ldots, Y_k, X_1, \ldots, X_g) + \text{error}$$

for $i = 1, 2, \ldots, k$. Some of the exogenous variables may be lagged values of the $Y_i$. The above set of equations, often called the *structural form* of the system, can be solved to give what is called the *reduced form* of the system, namely

$$Y_i = F_i(X_1, \ldots, X_g) + \text{error} \qquad \text{for } i = 1, 2, \ldots, k.$$

The principles and problems involved in constructing econometric models are too broad to be discussed in detail here (see, e.g. Mizon, 1995, Section 6.3). A key issue is the extent to which the form of the model should be based on judgement, on economic theory and/or on empirical data. While some econometricians have been scornful[2] of univariate time-series models, which do not 'explain' what is going on, statisticians have been generally sceptical of some econometric model building in

[2]This occurs especially when the univariate forecasts are 'better' than alternatives!

which the structure of the model is determined *a priori* by economic theory and little attention is paid to identifying an appropriate 'error' structure or to using empirical data. Fortunately, mutual understanding has improved in recent years as developments in multivariate time-series modelling have brought statisticians and econometricians closer together to the benefit of both. In fact, the uncontrolled nature of much economic data makes it difficult to identify econometric models solely on an empirical statistical basis, while overreliance on economic theory should also be avoided. It is now widely recognized that a balanced middle way is sensible, and that econometric model building should be an iterative process involving both theory *and* data. See Chapter 13 for more detailed discussions on multivariate time-series modelling in recent years.

### 5.4.3 *Other multivariate models*

There are many other types of multivariate models that may be used to produce forecasts. The multivariate generalization of ARIMA models is considered in Chapter 13. One special case is the class of *vector autoregressive* models, while another useful class of models is that called *transfer function models* (see Section 9.4.2). The latter concentrates on describing the relationship between one 'output' variable and one or more 'input' or explanatory variables. It is helpful to understand the interrelationships between all these classes of multivariate models, e.g. see Granger and Newbold (1986, Chapters 6–8), Priestley (1981, Chapter 9), and Chatfield (2001, Chapter 5).

Of course, more specialized multivariate models may occasionally be required. For example, forecasts of births must take account of the number and age of women of child-bearing age. Common sense and background knowledge of the problem context should indicate what is required.

## 5.5 Comparative Review of Forecasting Procedures

We noted in Section 5.1 that there is no such thing as a 'best' forecasting procedure, but rather that the choice of method depends on a variety of factors such as the objective in producing forecasts, the degree of accuracy required and the properties of the given time series. This section gives a brief review of relevant research and makes recommendations on which method to use and when.

The context and the reason for making a forecast are, of course, of prime importance. Forecasts may be used, for example, for production planning in industry. They can also be used to assess the effects of different strategies by producing several forecasts. They may also be used as a 'norm' (or yardstick) against which the actual outcome may be assessed to see whether anything is changing. Univariate forecasts are usually used to provide such a 'norm' and are also particularly suitable when there are large numbers of series to be forecast (e.g. in stock control) so that a relatively simple method has to be used. They are also suitable when the analyst's skill is limited or when they are otherwise judged appropriate for the client's needs and level of understanding. Multivariate models are particularly appropriate for assessing the effects of explanatory variables, for understanding the economy, and for evaluating alternative economic policy proposals by constructing a range of 'what if' forecasts. The latter strategy of computing more than one forecast is often very helpful in assessing the effects of different assumptions or strategies.

### *5.5.1 Forecasting competitions*

First, we review the empirical evidence as to which method is 'best'. In order to clarify the choice between different univariate methods, there have been several 'competitions' to compare the forecasting accuracy of different methods on a given collection of time series. Some of the more important competitions are described by Newbold and Granger (1974); Makridakis and Hibon (1979); Makridakis et al. (1984, 1993) and Makridakis and Hibon (2000). The last three studies are commonly known as the M-competition, the M2-competition and the M3-competition, respectively. The M-competition was designed to be more wide ranging than earlier studies, and compared 24 methods on 1001 series, while the M3-competition compared 24 methods on no fewer than 3003 series – a mammoth task! The M2 competition was much smaller, just 29 series, but this allowed non-automatic procedures to be investigated.

Given different analysts and data sets, it is perhaps not too surprising that the results from different competitions have not always been consistent. For example, Newbold and Granger (1974) found that Box–Jenkins tended to give more accurate forecasts than other univariate methods, but this was not the case in the M- and M3-competitions. A detailed assessment of the strengths and weaknesses of forecasting competitions is given by Chatfield (2001, Section 6.4). It is essential that results be replicable and that appropriate criteria are used. Moreover, accuracy is only one aspect of forecasting, and practitioners think that cost, ease of use and ease of interpretation are of almost equal importance. Furthermore, competitions mainly analyze large numbers of series in a completely automatic way. Thus although they tell us something, competitions only tell part of the story and are mainly concerned with comparing automatic forecasts. In practice, large gains in accuracy can often be made by applying a carefully chosen method that is tailored to a particular context for a particular type of series – see, for example, Tashman and Kruk (1996) and Section 5.4.2. This will not be evident in the results from automatic forecasting competitions, where, for example, simple exponential smoothing (SES) has sometimes been applied to all series in a group regardless of whether a particular series exhibits trend or not. This is unfair to SES, which does not pretend to be able to cope with trend.

One important question is how forecasting accuracy should be assessed – see, for example, Armstrong and Collopy (1992), Fildes (1992), and Chatfield (2001, Section 6.3). While average prediction mean square error is the obvious measure, care needs to be taken when averaging across series with widely differing variances. Because of this, the mean absolute percentage error is often used, rather than mean square error. Of course, a model that is 'best' under one criterion need not be best under some other criterion. In particular, it can be useful to use a robust measure that is not inflated too much by the occasional large forecast error.

When comparing results from different competitions, it is also essential to ensure that genuine out-of-sample forecasts have been used throughout. When one method appears to give much better forecasts than all alternative methods, there must be a suspicion that it has an unfair advantage in some way. Usually, a method that seems to be a clear winner may not be using genuine out-of-sample forecasts, but rather may be 'cheating' in some way, perhaps unwittingly. For example, multivariate forecasts sometimes outperform univariate methods only by dubiously using future values of explanatory variables.

It is also worth noting that unexciting results tend to be suppressed. For example, forecasters seem eager to publish results that show a new method to be better, but not to publish results that show the reverse – see Chatfield (1995) for examples. This sort of publication bias is endemic in other areas of statistical investigation, as people like to publish 'significant' results, but not the reverse. As a result, new methods tend to have an unfair advantage. Finally, it is worth saying that empirical results tend to be ignored (Fildes and Makridakis, 1995), especially when they run counter to orthodox thinking by showing, for example, that simple forecasting methods do as well as more complicated ones.

*Choosing an automatic method.* If an automatic approach is desirable or unavoidable, perhaps because a large number of series is involved, then my interpretation of the competition results is as follows. While there could be significant gains in being selective, most users will want to apply the same method to all series for obvious practical reasons. Some methods should be discarded, but there are several automatic methods for which average differences in accuracy are small. Thus the choice between them may depend on other practical considerations such as availability of computer programs. The methods include Holt's exponential smoothing (applied to seasonally adjusted data where appropriate), Holt–Winters and Bayesian forecasting. In particular, the Holt–Winters method provides a generally reliable, easy to understand method for seasonal data. The results in Chen (1997) have shown that Holt–Winters is robust to departures from the model for which the method is optimal. Holt's method can be used for non-seasonal data or data that have already been deseasonalized. The use of a damped trend term is often advisable in both the Holt and Holt–Winters procedures.

### *5.5.2 Choosing a non-automatic method*

Suppose instead that a non-automatic approach is indicated because the number of series is small and/or because external information is available that cannot be ignored. Then sensible forecasters will use their skill and knowledge to interact with their clients, incorporate background knowledge, plot the data and generally use all relevant information to build an appropriate model and compute sensible forecasts. The choice then lies between some form of multivariate method and a non-automatic univariate procedure. Here forecasting competitions are of limited value and it is easy to cite case studies where subjective adjustment of automatic forecasts leads to improvements (e.g. Chatfield, 1978). Moreover, the differences in accuracy for different methods when averaged over many series are relatively small compared with the large differences in accuracy that can arise when different methods are applied to individual series. The potential rewards in selecting an appropriate, perhaps non-automatic, method indicate that the distinction between an automatic and a non-automatic *approach* may be more fundamental than the differences between different forecasting *methods*.

We look first at *multivariate methods*. It is possible to cite case studies (e.g. Jenkins and McLeod, 1982) where statistician and client collaborate to develop a successful multivariate model. However, it is difficult to make general statements about the relative accuracy of multivariate methods. Many people expect multivariate forecasts to be at least as good as univariate forecasts, but this is not true either in theory or in practice, and univariate methods outperform multivariate ones in many studies. One reason for this is that the computation of multivariate forecasts of a response variable may require the prior computation of forecasts of explanatory variables, and the latter must be sufficiently accurate to make this viable (Ashley, 1988). Multivariate models work best when one or more variables are leading indicators for the response variables of interest and so do not need to be forecast, at least for shorter lead times. Two other general points are that multivariate models are more difficult to identify than univariate models, and that they are generally less robust to departures from model assumptions. For all these reasons, simple methods and models are often better.

The empirical evidence is reviewed by Chatfield (1988, 2001). Regression models do rather better on average than univariate methods, though not by any means in every case (Fildes, 1985). Econometric simultaneous equation models have a patchy record and it is easy to cite cases where univariate forecasts are more accurate (e.g. Makridakis and Hibon, 1979, Section 2). There have been some encouraging case studies using transfer function models, e.g. Jenkins (1979) and Jenkins and McLeod (1982), but such models assume there is no feedback, and this state of affairs will not apply to much multivariate economic data. Vector AR models, introduced later in Chapter 13, have a mixed record with some successes, but some failures as well. Moreover, many researchers fail to compare their multivariate forecasts with those from simpler alternatives, perhaps because they do not wish to be embarrassed by the possibility that univariate forecasts turn out better. It is arguable that multivariate models are more useful for understanding relationships than for forecasting. Of course multivariate models can usually be made to give a better *fit* to given data than univariate models, but this superiority does not necessarily translate into better forecasts, perhaps because multivariate models are more sensitive to changes in structure.

It has to be realized that the nature of economic time-series data is such as to make it difficult to fit reliable multivariate time-series models. Most economic variables are simply observed, rather than controlled, and there are usually high autocorrelations within each series. In addition there may be high correlations between series, not necessarily because of a real relationship but simply because of a mutual correlation with time. Feedback between 'output' and 'input' variables is another problem. There are special difficulties in fitting regression models to time-series data anyway, as already noted in Section 5.3.1, and an apparent good fit may be spurious. Simultaneous equation and vector AR models are also difficult to construct, and their use seems likely to be limited to the analyst who is as interested in the modelling process as in forecasting. Thus although the much greater effort required to construct multivariate models will sometimes prove fruitful, there are many situations where a univariate method will be preferred.

With a non-automatic *univariate* approach, the main choice is between the Box–Jenkins approach and the non-automatic use of a simple method, such as Holt–Winters, which is more often used in automatic mode. The Box–Jenkins approach has been one of the most influential developments in time-series analysis. However, the accuracy of the resulting forecasts has been rather mixed in practice, particularly when one realizes that forecasting competitions are biased in favour of Box–Jenkins by implementing other methods in a completely automatic way. The advantage of being able to choose from the broad class of ARIMA models is clear, but, as noted in Section 5.3, there are also dangers in that considerable experience is needed to interpret correlograms and other indicators. Moreover, when the variation in a series is dominated by trend and seasonality, the effectiveness of the fitted ARIMA model is mainly determined by the differencing procedure rather than by the identification of the autocorrelation structure of the differenced (stationary) series. Yet the latter is what is emphasized in the Box–Jenkins approach. Nevertheless, some writers have suggested that all exponential smoothing models should be regarded as special cases of Box–Jenkins, the implication being that one might as well use Box–Jenkins. However, this view is now discredited Chatfield (1988) because exponential smoothing methods are actually applied in a different way to Box–Jenkins.

In some situations, a large expenditure of time and effort can be justified and then Box–Jenkins is worth considering. However, for routine sales forecasting, simple methods are more likely to be understood by managers and workers who have to utilize or implement the results. Thus that Box–Jenkins is only recommended when the following conditions are satisfied: (1) the analyst is competent to implement the method and has appropriate software; (2) the objectives justify the additional complexity; and (3) the variation in the series is *not* dominated by trend and seasonality. If these conditions are not satisfied, then a non-automatic version of a simple method, such as Holt–Winters, may be 'best'.

Of course, another alternative to the Box–Jenkins approach is to use a *more* complicated method, such as one of the multivariate methods discussed earlier in this subsection. This can be especially rewarding when there is an obvious leading indicator to include. However, there are many situations when it is advisable to restrict attention to univariate forecasts.

### *5.5.3 A strategy for non-automatic univariate forecasting*

If circumstances suggest that a non-automatic univariate approach is appropriate, then the following eight steps could generally provide a sensible strategy that covers many forecasting situations:

1. Get appropriate background information and carefully define the objectives.
2. Plot the data and look for trend, seasonal variation, outliers and any changes in structure such as slow changes in variance, sudden discontinuities and anything else that looks unusual.

3. 'Clean' the data if necessary, for example, by adjusting any suspect observations, preferably after taking account of external information. Consider the possibility of transforming the data.
4. Decide whether the seasonal variation is non-existent, multiplicative, additive or something else.
5. Decide whether the trend is non-existent, global linear, local linear or non-linear.
6. Fit an appropriate model where possible. It can be helpful to distinguish the following four types of series (although the reader may, of course, come across series that do not fit into any of these categories. Then common sense has to be applied):
   (i) *Discontinuities present.* For a series having a major discontinuity, it is generally unwise to produce any univariate forecasts. There is further discussion of this series in Section 14.1.
   (ii) *Trend and seasonality present.* Figures 1.3 shows series whose variation is dominated by trend and seasonality. Here the Holt–Winters exponential smoothing method is a suitable candidate. The correct seasonal form must be chosen and the smoothing parameters can then be estimated by optimizing one-step-ahead forecasts over the period of fit. Full details are given by Chatfield and Yar (1988). For data showing trend, but no seasonality, Holt's method may be used.
   (iii) *Short-term correlation present.* Many economic indicator series are of this form and it is essential to try to understand the autocorrelation structure. Here, the Box–Jenkins approach is recommended.
   (iv) *Exponential growth present.* Series of this type are difficult to handle because exponential forecasts are inherently unstable. No one really believes that economic growth or population size, for example, can continue forever to increase exponentially. Two alternative strategies are to fit a model that explicitly includes exponential (or perhaps quadratic) growth terms, or to fit a model to the logarithms of the data (or to some other suitable transformation of the data). It may help to damp any trend that is fitted to the transformed data as in Gardner and McKenzie (1985).
7. Check the adequacy of the fitted model. In particular, study the one-step-ahead forecast errors over the period of fit to see whether they have any undesirable properties such as significant autocorrelation. Modify the model if necessary.
8. Compute forecasts. Decide whether the forecasts need to be adjusted subjectively because of anticipated changes in other variables, or because of any other reason.

Of course, much of the above procedure could be automated, making the dividing line between automatic and non-automatic procedures rather blurred. We should also mention **rule-based forecasting** (Collopy and Armstrong, 1992), which is a form of **expert system** that integrates skilled judgement with domain knowledge and also takes account of the features of the data. The rule base, consisting of as many as 99 rules, entertains a variety of forecasting procedures, which may be combined in an appropriate way. This rule base may be modified (e.g. Adya et al., 2000) depending on the particular situation.

### 5.5.4 *Summary*

It is difficult to summarize the many empirical findings and other general principles that have been established in regard to forecasting, but the following general observations and recommendations can be made:

- As in all statistical work, it is essential to formulate the forecasting problem carefully, clarify objectives and be clear exactly how a forecast will be used.
- The more frequent and the greater the number of forecasts required, the more desirable it is to use a simple approach.
- Fitting the 'best' model to historical data does not necessarily lead to the most accurate out-of-sample forecast errors in the future. In particular, complex models usually give a better fit than simpler models but the resulting forecasts need not be more accurate. It is essential that all comparisons between different forecasting models and methods should be made using genuine out-of-sample forecasts.
- Prediction intervals, calculated on the assumption that the model fitted to past data will also be true in the future, are generally too narrow.
- If an automatic univariate method is required, then the Holt and Holt–Winters versions of exponential smoothing are suitable candidates, but there are several close competitors.
- When a non-automatic approach is appropriate, there is a wide choice from judgemental and multivariate methods through to (univariate) Box–Jenkins and the 'thoughtful' use of univariate methods that are usually regarded as being automatic. A general strategy for non-automatic univariate forecasting has been proposed that involves looking carefully at the data and selecting an appropriate method from a set of candidates including the Box–Jenkins, Holt and Holt–Winters methods. Whatever approach is used, the analyst should be prepared to improvise and modify 'objective' forecasts using subjective judgement.
- It is often possible to get more accurate results by combining forecasts from different methods, perhaps by using a weighted average, but this does not lead to an informative model.

Comprehensive coverage of the general principles involved in time-series forecasting, together with further guidance on empirical results, is given by Armstrong (2001) and Chatfield (2001).

## 5.6 Prediction Theory

This section[3] gives a brief introduction to the general theory of linear prediction, which has been developed by Wiener (1949), Yaglom (1962) and Whittle (1983) among others. All these authors avoid the use of the word 'forecasting', although most of the univariate methods considered in Section 5.2 are in the general class of linear predictors. The theory of linear prediction has applications in control and communications engineering and is of considerable theoretical interest, but readers who

[3]This section may be omitted at first reading.

wish to tackle the sort of forecasting problem we have been considering earlier in this chapter will find this literature less accessible, and less relevant to real-life problems, than earlier references, such as those describing the Box–Jenkins approach.

Two types of problems are often distinguished. In the first type of problem we have data up to time $N$, say $\{x_N, x_{N-1}, \ldots\}$, and wish to predict the value of $x_{N+h}$. One approach is to use the predictor

$$\widehat{x}_N(h) = \sum_{j \geq 0} c_j x_{N-j}$$

which is a linear function of the available data. The weights $\{c_j\}$ are chosen so as to minimize the expected mean square prediction error $E(X_{N+h} - \widehat{x}_N(h))^2$. This is often called the *prediction* problem (e.g. Cox and Miller, 1968), while Yaglom (1962) refers to it as the *extrapolation* problem and Whittle (1983) calls it *pure prediction.* As an example of the sort of result that has been obtained, Wiener (1949) has considered the problem of evaluating the weights $\{c_j\}$, so as to find the best linear predictor, when the underlying process is assumed to be stationary with a known autocorrelation function, and when the entire past sequence of observations, namely, $\{x_t\}$ for $t \leq N$, is assumed known. It is interesting to compare this sort of approach with the forecasting techniques proposed earlier in this chapter. Wiener, for example, says little or nothing about identifying an appropriate model and then estimating the model parameters from a finite sample, and yet these are the sort of problems that have to be faced in real-life applications. The Box–Jenkins approach does tackle this sort of problem, and aims to employ a linear predictor that is optimal for a particular ARIMA process, while recognizing that the form of the model and the model parameters are usually *not* known beforehand and have to be inferred from the data.

The second type of problem tackled in the more theoretical literature arises when the process of interest, called the *signal*, is contaminated by *noise*, and we actually observe the process

$$y(t) = s(t) + n(t),$$

where $s(t)$ and $n(t)$ denote the signal and noise, respectively. In some situations the noise is simply measurement error; in engineering applications the noise could be an interference process of some kind. The problem now is to separate the signal from the noise. Given measurements on $y(t)$ up to time $T$ we may want to reconstruct the signal up to time $T$ or alternatively make a prediction of $s(T+\tau)$. The problem of reconstructing the signal is often called *smoothing* or *filtering*. The problem of predicting the signal is also sometimes called *filtering* (Yaglom, 1962; Cox and Miller, 1968), but is sometimes called *prediction* (Astrom, 1970). To make progress, it is often assumed that the signal and noise processes are uncorrelated and that $s(t)$ and $n(t)$ have known autocorrelation functions. These assumptions, which are unlikely to hold in practice, make the results of limited practical value.

It is clear that both the above types of problems are closely related to the control problem because, if we can predict how a process will behave, then we can adjust the process so that the achieved values are, in some sense, as close as possible to the target value. Control theory is generally outside the scope of this book, although we do make some further brief remarks on the topic later in Section 14.3 after we have studied the theory of linear systems.

## Exercises

**5.1** For the MA(1) model given by

$$X_t = Z_t + \theta Z_{t-1},$$

show that $\widehat{x}_N(1) = \theta z_N$ and that $\widehat{x}_N(h) = 0$ for $h = 2, 3, \ldots$.

Show that the variance of the $h$-steps-ahead forecast error is given by $\sigma_Z^2$ for $h = 1$, and by $(1+\theta^2)\sigma_Z^2$ for $h \geq 2$, provided the true model is known. (In practice we would take $\widehat{x}_N(1) = \widehat{\theta}\widehat{z}_N$, where $\widehat{\theta}$ is the least squares estimate of $\theta$ and $\widehat{z}_N$ is the observed residual at time $N$.)

**5.2** For the AR(1) model given by

$$X_t = \alpha X_{t-1} + Z_t,$$

show that $\widehat{x}_N(h) = \alpha^h x_N$ for $h = 1, 2, \ldots$. Show that the variance of the $h$-steps-ahead forecast error is given by $(1-\alpha^{2h})\sigma_Z^2/(1-\alpha^2)$.

For the AR(1) model, with non-zero mean $\mu$, given by

$$X_t - \mu = \alpha(X_{t-1} - \mu) + Z_t$$

show that $\widehat{x}_N(h) = \mu + \alpha^h(x_N - \mu)$ for $h = 1, 2, \ldots$. (In practice the least squares estimates of $\alpha$ and $\mu$ would need to be substituted into the above formulae.)

**5.3** Consider the SARIMA$(1,0,0) \times (0,1,1)_{12}$ model used as an example in Section 5.2.4. Show that

$$\widehat{x}_N(2) = x_{N-10} + \alpha^2(x_N - x_{N-12}) + \theta\alpha z_{N-11} + \theta z_{N-10}.$$

**5.4** For the SARIMA$(0,0,1) \times (1,1,0)_{12}$ model, find forecasts at time $N$ for up to 12 steps ahead in terms of observations and estimated residuals up to time $N$.

**5.5** For the model $(1-B)(1-0.2B)X_t = (1-0.5B)Z_t$ in Exercise 3.12, find forecasts for one and two steps ahead, and show that a recursive expression for forecasts three or more steps ahead is given by

$$\widehat{x}_N(h) = 1.2\widehat{x}_N(h-1) - 0.2\widehat{x}_N(h-2).$$

Find the variance of the one-, two- and three-steps-ahead forecast errors. If $z_N = 1, x_N = 4, x_{N-1} = 3$ and $\sigma_Z^2 = 2$, show that $\widehat{x}_N(2) = 3.64$ and that the standard error of the corresponding forecast error is 1.72.

**5.6** Consider the ARIMA$(0,1,1)$ process

$$(1-B)X_t = (1-\theta B)Z_t.$$

Show that $\widehat{x}_N(1) = x_N - \theta z_N$, and $\widehat{x}_N(h) = \widehat{x}_N(h-1)$ for $h \geq 2$. Express $\widehat{x}_N(1)$ in terms of $x_N$ and $\widehat{x}_{N-1}(1)$ and show that this is equivalent to exponential smoothing. By considering the $\psi$ weights of the process, show that the variance of the $h$-steps-ahead prediction error is

$$[1+(h-1)(1-\theta)^2]\sigma_Z^2.$$

**5.7** Consider the AR(2) process ($a$ is real)

$$X_t + \frac{1}{a(a+1)}X_{t-1} - \frac{1}{a(a+1)}X_{t-2} = Z_t.$$

(a) For what values of real number $a$, is the process stationary?

(b) What is the autocorrelation of the process at lag $k$, i.e. $\rho(k)$? $(k \geq 0)$.

(c) What are the 1-step and 2-step ahead forecasts of the process at forecast origin $X_n$?

**5.8** Consider the ARMA(2,1) process $X_t - \alpha X_{t-2} = Z_t - \beta Z_{t-1}$, where $\alpha, \beta$ are real and $\alpha, \beta \neq 0, \alpha \neq \beta$.

(a) Under which conditions is the process $X_t$ stationary and invertible?

(b) Given the observations $X_1, \ldots, X_n$, what is your best prediction for $X_{n+1}$ and $X_{n+2}$ at the forecast origin $X_n$? What is the corresponding forecasting error?

Chapter 6

# Stationary Processes in the Frequency Domain

## 6.1 Introduction

Chapter 3 described several types of stationary stochastic processes, such as autoregressive processes. When discussing their properties, we emphasized the autocovariance (or autocorrelation) function, as being the natural tool for considering the evolution of a process through time. This chapter introduces a complementary function, called the **spectral density function**, which is the natural tool for considering the frequency properties of a time series. Inference regarding the spectral density function is called an analysis in the **frequency domain**.

Some statisticians initially have difficulty in understanding the frequency approach, but the advantages of frequency methods are widely appreciated in such fields as electrical engineering, geophysics and meteorology. These advantages will become apparent in the next few chapters.

We confine attention to real-valued processes. Some authors consider the more general problem of complex-valued processes, and this has some mathematical advantages. However, in our view, the reader is more likely to understand an approach restricted to real-valued processes. The vast majority of practical problems are covered by this approach.

## 6.2 The Spectral Distribution Function

In order to introduce the idea of a spectral density function, we first consider a function called the **spectral distribution function**. The approach adopted is heuristic and not mathematically rigorous, but will hopefully give the reader a better understanding of the subject than a more theoretical approach.

Suppose we suspect that a time series contains a periodic sinusoidal component with a known wavelength. Then a natural model is

$$X_t = R\cos(\omega t + \phi) + Z_t \tag{6.1}$$

where $\omega$ is called the **frequency** of the sinusoidal variation, $R$ is called the **amplitude** of the variation, $\phi$ is called the **phase** and $\{Z_t\}$ denotes a stationary random series.

DOI: 10.1201/9781003646112-6

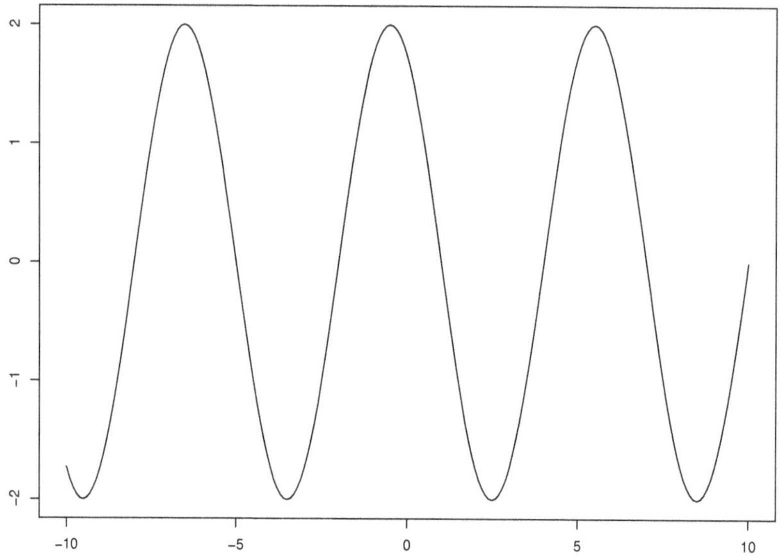


Figure 6.1 *A graph of* $R\cos(\omega t+\phi)$ *with* $R=2, \omega=\pi/3$ *and* $\phi=\pi/6$.

Note that the angle $(\omega t+\phi)$ is usually measured in units called *radians*, where $\pi$ radians $= 180^\circ$. As $\omega$ is the number of radians per unit time, it is sometimes called the **angular** frequency, but in keeping with most authors we simply call $\omega$ the frequency. However, some authors, notably Jenkins and Watts (1968), use the term 'frequency' to refer to the number of cycles per unit time, namely, $f=\omega/2\pi$, and this form of frequency is easier to interpret from a physical point of view. We usually use the angular frequency $\omega$ in mathematical formulae, because it makes them more concise, but we will often use the frequency $f=\omega/2\pi$ to interpret the results of a data analysis. The **period** of a sinusoidal cycle, sometimes called the **wavelength**, is clearly $1/f$ or $2\pi/\omega$. Figure 6.1 shows an example of a sinusoidal function with angular frequency $\omega=\pi/3$, so that $f=\omega/2\pi=1/6$. The wavelength is the reciprocal of $f$, namely, 6, and inspection of Figure 6.1 shows that this is indeed the number of time units between successive peaks or successive troughs of the sinusoid.

Figure 6.1 is a very simple model, but in practice the variation in a time series may be caused by variation at several different frequencies. For example, sales figures may contain weekly, monthly, yearly and other cyclical variation. In other words the data show variation at high, medium and low frequencies. It is natural therefore to generalize Equation (6.1) to

$$X_t=\sum_{j=1}^{k} R_j\cos(\omega_j t+\phi_j)+Z_t, \tag{6.2}$$

where $R_j$, $\phi_j$ denote the amplitude and phase, respectively, at frequency $\omega_j$. Figure 6.2 shows an example of the mixture (6.2) with $k=3$ constructed in the following way. First, for $t = 1,\dots,100$), we generate four series

$$X_{t1} = \cos\left(\frac{10\pi t}{150} + \frac{\pi}{8}\right), \qquad X_{t2} = 3\cos\left(\frac{30\pi t}{150} + \frac{3\pi}{8}\right),$$
$$X_{t3} = 5\cos\left(\frac{60\pi t}{150} + \frac{5\pi}{8}\right), \qquad Z_t \sim N(0,1).$$

Then the series $X_t$ is constructed as

$$X_t = X_{t1} + X_{t2} + X_{t3} + Z_t.$$

The series of $X_{t1}, X_{t2}, X_{t3}$ and $X_t$ are displayed in Figure 6.2, which can be reproduced by the following R code.

```
> y<-seq(0, 100, 0.3)
> x1<- 1*cos(pi*y*10/150 + pi/8)
> x2<- 3*cos(pi*y*30/150 + 3*pi/8)
> x3<- 5*cos(pi*y*60/150 + 5*pi/8)
> set.seed(1)
> z<- rnorm(length(y), 0, 1)
> x<-x1+x2+x3+z
> par(mfrow=c(2,2), mar=c(2,2,2,2))
> plot(x1, type="l", xlab="", ylab="x1")
> plot(x2, type="l", xlab="", ylab="x1")
> plot(x3, type="l", xlab="", ylab="x1")
> plot(x, type="l", xlab="", ylab="x1")
```

The reader will notice that the models in Equations (6.1) and (6.2) are *not* stationary if parameters such as $\{R_j\}$, $\{\omega_j\}$ and $\{\phi_j\}$ are all fixed constants, because $E(X_t)$ will change with time. In order to apply the theory of stationary processes to models like Equation (6.2), it is customary to assume that $\{R_j\}$ are (uncorrelated) random variables with mean zero, or that $\{\phi_j\}$ are random variables with a uniform distribution on $(0, 2\pi)$, which are fixed for a single realization of the process (see Section 3.13 and Exercise 3.14). This is something of a 'mathematical trick', but it does enable us to treat time series containing one or more deterministic sinusoidal components as stationary series.

Since $\cos(\omega t + \phi) = \cos(\omega t)\cos\phi - \sin(\omega t)\sin\phi$, the model in Equation (6.2) can alternatively be expressed as a sum of sine and cosine terms in the form

$$X_t = \sum_{j=1}^{k} (a_j \cos(\omega_j t) + b_j \sin(\omega_j t)) + Z_t, \tag{6.3}$$

where $a_j = R_j \cos\phi_j$ and $b_j = -R_j \sin\phi_j$.

However, we may now ask why there should only be a finite number of frequencies involved in Equations (6.2) or (6.3). In fact, letting $k \to \infty$, the work of Wiener

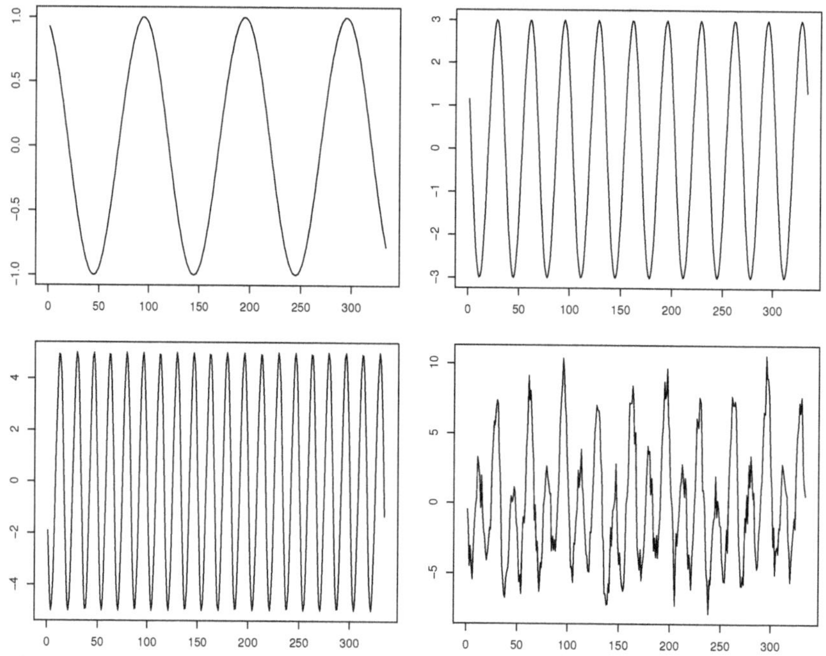


Figure 6.2 *Periodic components and their sum (Top left: $X_{t1}$; Top right: $X_{t2}$; Bottom left: $X_{t3}$; Bottom right: $X_t$).*

and others has shown that any discrete-time stationary process measured at unit intervals may be represented in the form

$$X_t = \int_0^{\pi} \cos \omega t \, \mathrm{d}u(\omega) + \int_0^{\pi} \sin \omega t \, \mathrm{d}v(\omega), \tag{6.4}$$

where $u(\omega), v(\omega)$ are uncorrelated continuous processes, with orthogonal increments (see Section 3.12), which are defined for all $\omega$ in the range $(0, \pi)$. Equation (6.4) is called the **spectral representation** of the process; it involves stochastic integrals, which require considerable mathematical skill to handle properly. It is intuitively more helpful to ignore these mathematical problems and simply regard $X_t$ as a linear combination of orthogonal sinusoidal terms. Thus the derivation of the spectral representation will not be considered here (see, e.g. Cox and Miller, 1968, Chapter 8).

The reader may wonder why the upper limits of the integrals in Equation (6.4) are $\pi$ rather than $\infty$. For a continuous process the upper limits would indeed be $\infty$, but for a discrete-time process measured at unit intervals of time there is no loss of generality in restricting $\omega$ to the range $(0, \pi)$, since

$$\cos[(\omega + k\pi)t] = \begin{cases} \cos \omega t & \text{for } k, t \text{ integers with } k \text{ even} \\ \cos(\pi - \omega)t & \text{for } k, t \text{ integers with } k \text{ odd} \end{cases}$$

and so variation at frequencies higher than $\pi$ cannot be distinguished from variation at a corresponding frequency in $(0, \pi)$. The frequency $\omega = \pi$ is called the **Nyquist frequency**. We will say more about this in Section 7.2.1. For a discrete-time process measured at equal intervals of time of length $\Delta t$, the Nyquist frequency is $\pi/\Delta t$. In the next two sections we consider discrete-time processes measured at unit intervals of time, but the arguments carry over to any discrete-time process, measured at an arbitrary time interval $\Delta t$ if we replace $\pi$ by $\pi/\Delta t$.

The main point of introducing the spectral representation in Equation (6.4) is to show that every frequency in the range $(0, \pi)$ may contribute to the variation of the process. However, the processes $u(\omega)$ and $v(\omega)$ in Equation (6.4) are of little direct practical interest. Instead we introduce a single function, $F(\omega)$, called the **(power) spectral distribution function**. This function is related to the autocovariance function and provides an intuitively understandable description of the frequency properties of a stationary process. It arises from a theorem, called the Wiener–Khintchine theorem, named after N. Wiener and A.Y. Khintchine, which says that for any real-valued stationary stochastic process with autocovariance function $\gamma(k)$, there exists a monotonically[1] increasing function $F(\omega)$ such that

$$\gamma(k) = \int_0^{\pi} \cos \omega k \, \mathrm{d}F(\omega). \tag{6.5}$$

Equation (6.5) is called the spectral representation of the autocovariance function, and involves a type of integral (called Stieltjes) that may be unfamiliar to some readers. However, it can be shown that the function $F(\omega)$ has a direct physical interpretation: it is the contribution to the variance of the series, which is accounted for by frequencies in the range $(0, \omega)$. If variation at negative frequencies is not allowed, then

$$F(\omega) = 0 \quad \text{for } \omega < 0.$$

For a discrete-time process measured at unit intervals of time, the highest possible frequency is the Nyquist frequency $\pi$ and so all the variation is accounted for by frequencies less than $\pi$. Thus

$$F(\pi) = \mathrm{Var}(X_t) = \sigma_X^2.$$

This last result also comes directly from Equation (6.5) by putting $k = 0$, when we have

$$\gamma(0) = \sigma_X^2 = \int_0^{\pi} \mathrm{d}F(\omega) = F(\pi).$$

In between $\omega = 0$ and $\omega = \pi, F(\omega)$ is monotonically increasing.

If the process contains a deterministic sinusoidal component at frequency $\omega_0$, say $R\cos(\omega_0 t + \phi)$ where $R$ is a constant and $\phi$ is uniformly distributed on $(0, 2\pi)$, then there will be a step increase in $F(\omega)$ at $\omega_0$ equal to the contribution to variance of this particular component. As the component has mean zero, the contribution to variance is just the average squared value, namely, $E[R^2 \cos^2(\omega_0 t + \phi)] = \frac{1}{2}R^2$.

---

[1] A function, say $f(x)$, is said to be monotonically increasing with $x$ if $f(x)$ never decreases as $x$ increases. Thus $f(x)$ will either increase or stay constant.

As $F(\omega)$ is monotonically increasing, it can be decomposed into two functions, $F_1(\omega)$ and $F_2(\omega)$, such that

$$F(\omega) = F_1(\omega) + F_2(\omega), \tag{6.6}$$

where $F_1(\omega)$ is a non-decreasing continuous function and $F_2(\omega)$ is a non-decreasing step function. This decomposition usually corresponds to the Wold decomposition, with $F_1(\omega)$ relating to the purely indeterministic component of the process and $F_2(\omega)$ relating to the deterministic component. We are mainly concerned with purely indeterministic processes, where $F_2(\omega) \equiv 0$, in which case $F(\omega)$ is a continuous function on $(0, \pi)$.

The adjective 'power', which is sometimes prefixed to 'spectral distribution function', derives from the engineer's use of the word in connection with the passage of an electric current through a resistance. For a sinusoidal input, the power is directly proportional to the squared amplitude of the oscillation. For a more general input, the power spectral distribution function describes how the power is distributed with respect to frequency. In the case of a time series, the variance may be regarded as the total power.

Note that some authors use a normalized form of $F(\omega)$ given by

$$F^*(\omega) = F(\omega)/\sigma_X^2. \tag{6.7}$$

Thus $F^*(\omega)$ is the *proportion* of variance accounted for by frequencies in the range $(0, \omega)$. Since $F^*(0) = 0$, $F^*(\pi) = 1$, and $F^*(\omega)$ is monotonically increasing in $(0, \pi)$, $F^*(\omega)$ has similar properties to the cumulative distribution function of a random variable.

## 6.3 The Spectral Density Function

For a purely indeterministic discrete-time stationary process, the spectral distribution function is a continuous (monotone bounded) function in $[0, \pi]$, and may therefore be differentiated[2] with respect to $\omega$ in $(0, \pi)$. We will denote the derivative by $f(\omega)$, so that

$$f(\omega) = \frac{\mathrm{d}F(\omega)}{\mathrm{d}\omega}. \tag{6.8}$$

This is the **power spectral density function**. The term 'spectral density function' is often shortened to **spectrum**, and the adjective 'power' is sometimes omitted.

When $f(\omega)$ exists, Equation (6.5) can be expressed in the form

$$\gamma(k) = \int_0^{\pi} \cos \omega k \, f(\omega) \, \mathrm{d}\omega. \tag{6.9}$$

[2]Strictly speaking, $F(\omega)$ may not be differentiable on a set of measure zero, but this is usually of no practical importance. In particular, $F(\omega)$ may not be differentiable at the endpoints 0 and $\pi$ and so it is more accurate to define the derivative $f(\omega)$ over the open interval $(0, \pi)$ that excludes the endpoints. This does not affect any integrals involving $f(\omega)$. In fact the formula in Equation (6.12) for $f(\omega)$, which is given later, usually can be evaluated at the end-points.

This is an ordinary (Riemann) integral and therefore much easier to handle than (6.5). Putting $k = 0$, we have

$$\gamma(0) = \sigma_X^2 = \int_0^\pi f(\omega)\,\mathrm{d}\omega = F(\pi). \tag{6.10}$$

The physical interpretation of the spectrum is that $f(\omega)\,\mathrm{d}\omega$ represents the contribution to variance of components with frequencies in the range $(\omega, \omega + \mathrm{d}\omega)$. When the spectrum is drawn, Equation (6.10) indicates that the total area underneath the curve is equal to the variance of the process. A peak in the spectrum indicates an important contribution to variance at frequencies near the value that corresponds to the peak. An example of a spectrum is shown in Figure 6.3, together with the corresponding normalized spectral distribution function.

It is important to realise that the autocovariance function (acv.f.) and the power spectral density function are equivalent ways of describing a stationary stochastic process. From a practical point of view, they are complementary to each other. Both functions contain the same information but express it in different ways. In some situations a time-domain approach based on the acv.f. is more useful, while in other situations a frequency-domain approach is preferable.

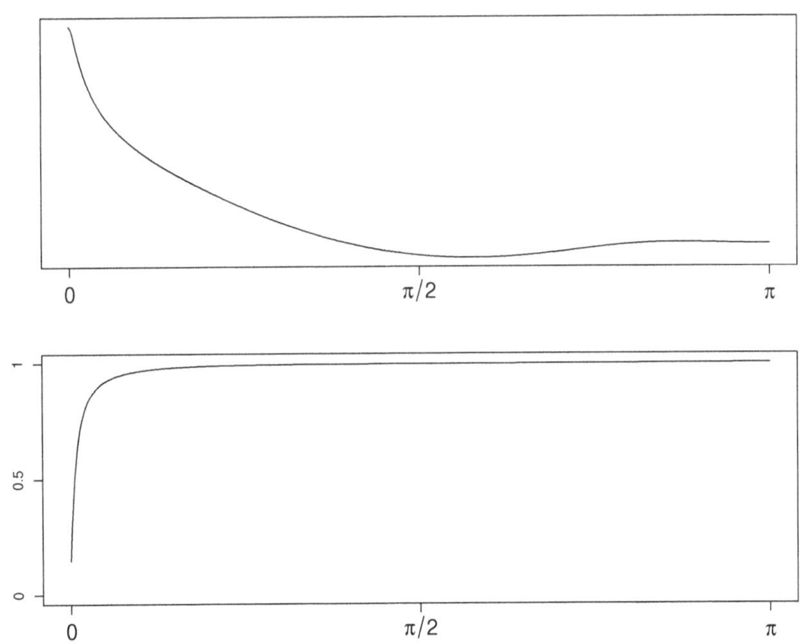


Figure 6.3 *An example of a spectrum, together with the corresponding normalized spectral distribution function.*

Equation (6.9) expresses $\gamma(k)$ in terms of $f(\omega)$ as a cosine transform. It can be shown that the corresponding inverse relationship is given by

$$f(\omega) = \frac{1}{\pi} \sum_{k=-\infty}^{\infty} \gamma(k) \mathrm{e}^{-\mathrm{i}\omega k}, \tag{6.11}$$

so that the spectrum is the **Fourier transform**[3] of the acv.f. Since $\gamma(k)$ is an even function of $k$, Equation (6.11) is often written in the equivalent form

$$f(\omega) = \frac{1}{\pi} \left[ \gamma(0) + 2 \sum_{k=1}^{\infty} \gamma(k) \cos \omega k \right]. \tag{6.12}$$

Note that if we try to apply Equation (6.12) to a process containing a deterministic component at a particular frequency $\omega_0$, then $\sum \gamma(k) \cos \omega k$ will not converge when $\omega = \omega_0$. This arises because $F(\omega)$ has a step change at $\omega_0$ and so will not be differentiable at $\omega_0$. Thus its derivative $f(\omega)$ will not be defined at $\omega = \omega_0$.

The reader should note that several other definitions of the spectrum are given in the literature, most of which differ from Equation (6.12) by a constant multiple and by the range of definition of $f(\omega)$. The most popular approach is to define the spectrum in the range $(-\pi, \pi)$ by

$$f(\omega) = \frac{1}{2\pi} \sum_{k=-\infty}^{\infty} \gamma(k) \mathrm{e}^{-\mathrm{i}\omega k}, \tag{6.13}$$

whose inverse relationship (see Appendix A) is

$$\gamma(k) = \int_{-\pi}^{\pi} \mathrm{e}^{\mathrm{i}\omega k} f(\omega) \mathrm{d}\omega. \tag{6.14}$$

Jenkins and Watts (1968) use these equations, except that they take $f = \omega/2\pi$ as the frequency variable (see Equations (A.3) and (A.4)). Equations (6.13) and (6.14), which form a Fourier transform pair, are the more usual form of the Wiener–Khintchine relations. The formulation is slightly more general in that it can be applied to complex-valued time series. However, for real time series we find that $f(\omega)$ is an even function of $\omega$, and then we need only consider $f(\omega)$ for $\omega > 0$. As we are concerned only with real-valued processes, we prefer Equation (6.11) defined on $(0, \pi)$.

It is sometimes useful to use a normalized form of the spectral density function, given by

$$f^*(\omega) = f(\omega)/\sigma_X^2 = \frac{\mathrm{d}F^*(\omega)}{\mathrm{d}\omega}. \tag{6.15}$$

This is the derivative of the normalized spectral distribution function (see Equation (6.7)). Then we find that $f^*(\omega)$ is the Fourier transform of the *autocorrelation* function (ac.f.), namely,

$$f^*(\omega) = \frac{1}{\pi} \left[ 1 + 2 \sum_{k=1}^{\infty} \rho(k) \cos \omega k \right]. \tag{6.16}$$

[3]See Appendix A for details on the Fourier transform.

This means that $f^*(\omega)\,\mathrm{d}\omega$ is the *proportion*[4] of variance in the interval $(\omega, \omega+\mathrm{d}\omega)$.

## 6.4 The Spectrum of a Continuous Process

For a continuous purely indeterministic stationary process, $X(t)$, the autocovariance function, $\gamma(\tau)$, is defined for all $\tau$ and the (power) spectral density function, $f(\omega)$, is defined for all positive $\omega$. The relationship between these functions is very similar to that in the discrete-time case except that there is no upper bound to the frequency. We have

$$f(\omega) = \frac{1}{\pi}\int_{-\infty}^{\infty} \gamma(\tau)\mathrm{e}^{-\mathrm{i}\omega\tau}\,\mathrm{d}\tau = \frac{2}{\pi}\int_{0}^{\infty} \gamma(\tau)\cos\,\omega\tau\,\mathrm{d}\tau \tag{6.17}$$

for $0 < \omega < \infty$, with the inverse relationship

$$\gamma(\tau) = \int_{0}^{\infty} f(\omega)\cos\,\omega\tau\,\mathrm{d}\omega. \tag{6.18}$$

## 6.5 Derivation of Selected Spectra

This section derives the spectral density functions of six simple, but important, types of discrete-time stationary processes.

(1) *Purely random processes*

A purely random process $\{Z_t\}$ as defined in Section 3.4, is a sequence of independent random variables. Thus, if $\mathrm{Var}(Z_t) = \sigma_Z^2$, then the acv.f. is given by

$$\gamma(k) = \begin{cases} \sigma_Z^2 & k = 0 \\ 0 & \text{otherwise} \end{cases}$$

so that the power spectral density function (or spectrum) is given by

$$f(\omega) = \sigma_Z^2/\pi \tag{6.19}$$

using Equation (6.12). In other words the spectrum is constant in the range $(0, \pi)$.

In continuous time, we have already pointed out that a continuous white noise process is physically unrealizable. A process is regarded as a practical approximation to continuous white noise if its spectrum is substantially constant over the frequency band of interest, even if it then approaches zero at high frequency.

(2) *First-order moving average processes*

The first-order moving average (MA) process (see Section 3.6)

$$X_t = Z_t + \beta Z_{t-1}$$

---

[4]As an example of the difficulties in comparing formulae from different sources, Kendall et al. (1983, Equation 47.20) define the spectral density function in terms of the autocorrelation function over the same range, $(0, \pi)$, as we do, but omit the constant $1/\pi$ from Equation (6.16). This makes it more difficult to give the function a physical interpretation. Instead they introduce an intensity function that corresponds to our power spectrum.

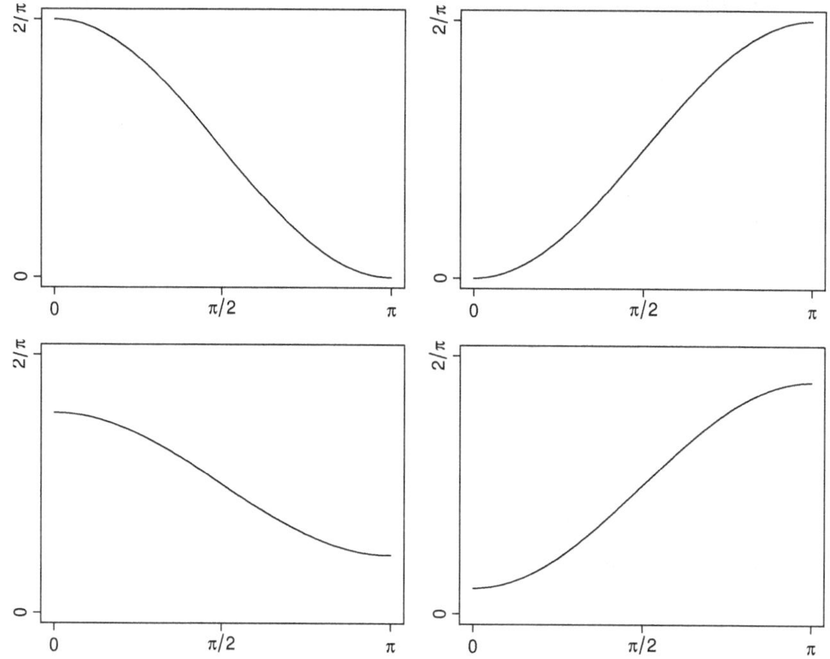


Figure 6.4 *Examples of normalized spectra, $f^*(\omega)$, of first-order moving average processes with $\beta = 1$ (Top left), $\beta = -1$ (Top right), $\beta = 0.3$ (Bottom left), and $\beta = -0.5$ (Bottom right).*

has an ac.f. given by

$$\rho(k) = \begin{cases} 1 & k = 0 \\ \beta/(1+\beta^2) & k = \pm 1 \\ 0 & \text{otherwise.} \end{cases}$$

Thus, using Equation (6.16), the normalized spectral density function is given by

$$f^*(\omega) = \frac{1}{\pi}\left[1 + (2\beta \cos \omega)/(1+\beta^2)\right] \tag{6.20}$$

for $0 < \omega < \pi$. The power spectral density function is then

$$f(\omega) = \sigma_X^2 f^*(\omega),$$

where $\sigma_X^2 = (1+\beta^2)\sigma_Z^2$.

The shape of the spectrum depends on the value of $\beta$. When $\beta > 0$ the power is concentrated at low frequencies, giving what is called a **low-frequency spectrum**; if $\beta < 0$ the power is concentrated at high frequencies, giving a **high-frequency spectrum**. Examples are shown in Figure 6.4.

(3) *First-order autoregressive processes*

The first-order autoregressive (AR) process (see Section 3.7)

$$X_t = \alpha X_{t-1} + Z_t \tag{6.21}$$

has an acv.f. given by

$$\gamma(k) = \sigma_X^2 \alpha^{|k|} \qquad k = 0, \pm 1, \pm 2, \ldots.$$

Then, using Equation (6.11), the power spectral density function is given by

$$\begin{aligned} f(\omega) &= \frac{\sigma_X^2}{\pi}\left(1 + \sum_{k=1}^{\infty} \alpha^k e^{-ik\omega} + \sum_{k=1}^{\infty} \alpha^k e^{ik\omega}\right) \\ &= \frac{\sigma_X^2}{\pi}\left(1 + \frac{\alpha e^{-i\omega}}{1-\alpha e^{-i\omega}} + \frac{\alpha e^{i\omega}}{1-\alpha e^{i\omega}}\right) \end{aligned}$$

which, after some algebra, gives

$$\begin{aligned} f(\omega) &= \sigma_X^2(1-\alpha^2)/[\pi(1-2\alpha\cos\omega+\alpha^2)] & (6.22) \\ &= \sigma_Z^2/[\pi(1-2\alpha\cos\omega+\alpha^2)] & (6.23) \end{aligned}$$

since $\sigma_Z^2 = \sigma_X^2(1-\alpha^2)$.

The shape of the spectrum depends on the value of $\alpha$. When $\alpha > 0$, the spectral density function is 'large' when $\omega$ is 'small', so that power is concentrated at low frequencies – a low-frequency spectrum. On the other hand, when $\alpha < 0$, the power is concentrated at high frequencies – a high-frequency spectrum. Examples are shown in Figure 6.5.

It is hoped that the reader finds the shapes of the spectra in Figure 6.4 intuitively reasonable. For example, if $\alpha$ is negative, then it is clear from Equation (6.21) that values of $X_t$ will tend to change sign at every time point, and rapid oscillations like this correspond to high-frequency variation.

(4) *ARMA(1, 1) processes*

The ARMA(1, 1) process (see Section 3.8)

$$X_t = \alpha X_{t-1} + Z_t + \beta Z_{t-1}$$

has an acv.f. given by

$$\begin{aligned} &\sigma_X^2 = \gamma(0) = \frac{1+2\alpha\beta+\beta^2}{1-\alpha^2}\sigma_Z^2, \\ &\gamma(k) = \gamma(0)\rho(k) = \frac{(1+\alpha\beta)(\alpha+\beta)}{1+2\alpha\beta+\beta^2}\sigma_X^2\alpha^{k-1}, \qquad k \geq 1. \end{aligned}$$

Then, using Equation (6.11), the power spectral density function is given by

$$\begin{aligned} f(\omega) &= \frac{\sigma_X^2}{\pi}\left[1 + \frac{(1+\alpha\beta)(\alpha+\beta)}{\alpha(1+2\alpha 2\beta+\beta^2)}\left(\sum_{k=1}^{\infty} \alpha^k e^{-ik\omega} + \sum_{k=1}^{\infty} \alpha^k e^{ik\omega}\right)\right] \\ &= \frac{\sigma_X^2}{\pi}\left[1 + \frac{(1+\alpha\beta)(\alpha+\beta)}{\alpha(1+2\alpha 2\beta+\beta^2)}\left(\frac{\alpha e^{-i\omega}}{1-\alpha e^{-i\omega}} + \frac{\alpha e^{i\omega}}{1-\alpha e^{i\omega}}\right)\right]. \end{aligned}$$

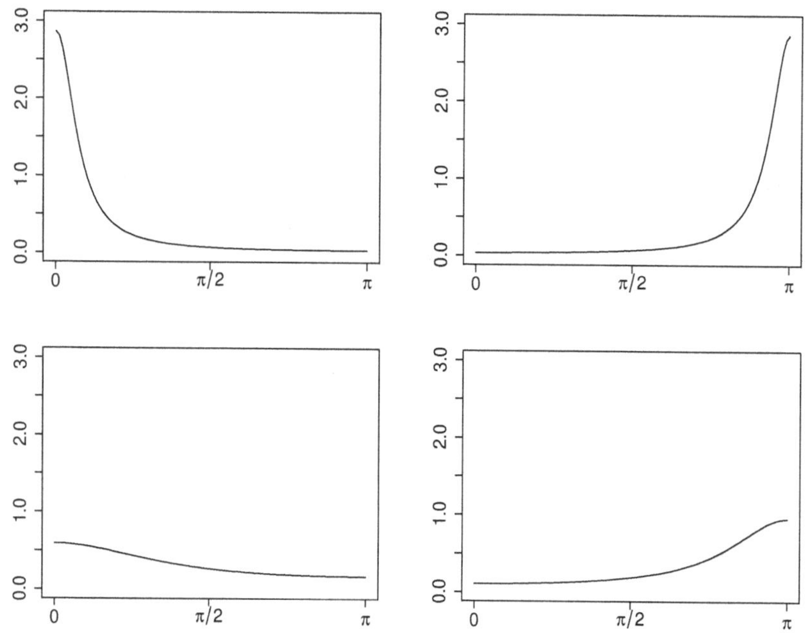


Figure 6.5 *Examples of normalized spectra, $f^*(\omega)$, of first-order autoregressive processes with $\alpha = 0.8$ (Top left), $\alpha = -0.8$ (Top right), $\alpha = 0.3$ (Bottom left), and $\alpha = -0.5$ (Bottom right).*

After some algebra, we have

$$f(\omega) = \frac{\sigma_X^2}{\pi} \cdot \frac{1-\alpha^2}{1+2\alpha\beta+\beta^2} \cdot \frac{1+2\beta\cos\omega+\beta^2}{1-2\alpha\cos\omega+\alpha^2} \tag{6.24}$$

$$= \frac{\sigma_Z^2}{\pi} \cdot \frac{1+2\beta\cos\omega+\beta^2}{1-2\alpha\cos\omega+\alpha^2}. \tag{6.25}$$

The shape of the spectrum depends on the values of $\alpha$ and $\beta$. When $\alpha > 0$ and $\beta > 0$, the power is concentrated at low frequencies. When $\alpha < 0$ and $\beta < 0$, the power is concentrated at high frequencies. For other values of $\alpha$ and $\beta$, the shape of the spectrum becomes complicated. Examples are shown in Figure 6.6.

(5) *Higher order AR, MA and ARMA processes*

For a second-order AR process

$$X_t = \alpha_1 X_{t-1} + \alpha_2 X_{t-2} + Z_t,$$

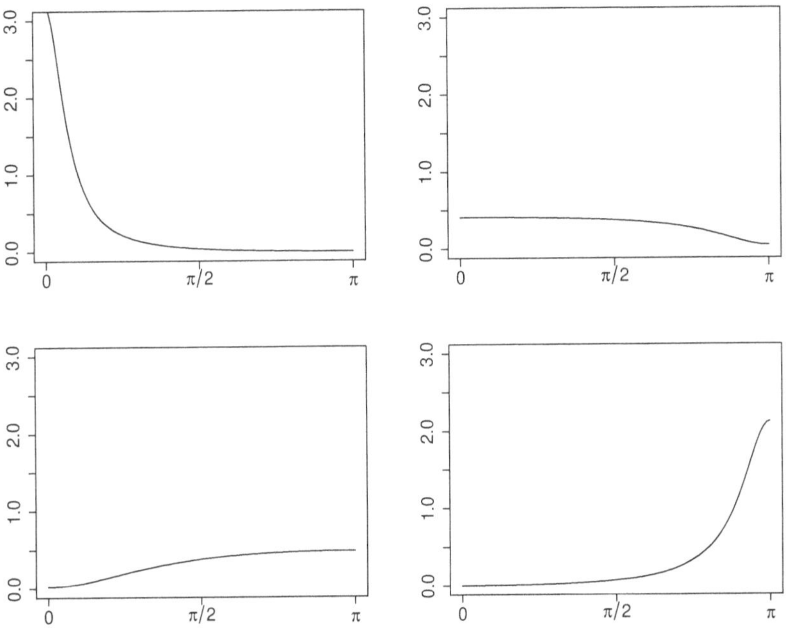


Figure 6.6 *Examples of normalized spectra, $f^*(\omega)$, of ARMA(1, 1) processes (Top left: $\alpha = 0.8, \beta = 0.8$; Top right: $\alpha = -0.5, \beta = 0.8$; Bottom left: $\alpha = 0.3, \beta = -0.8$; Bottom right: $\alpha = -0.7, \beta = -0.8$).*

after some algebra, it can be shown that its spectrum is given by

$$\begin{aligned} f(\omega) &= \sigma_Z^2/\pi[1+\alpha_1^2+\alpha_2^2-2\alpha_1(1-\alpha_2)\cos\omega-2\alpha_2\cos 2\omega] \\ &= \frac{\sigma_Z^2}{\pi}\cdot\frac{1}{(1-\alpha_1 e^{-i\omega}-\alpha_2 e^{-i2\omega})(1-\alpha_1 e^{i\omega}-\alpha_2 e^{i2\omega})} \\ &= \frac{\sigma_Z^2}{\pi}\cdot\frac{1}{|1-\alpha_1 e^{i\omega}-\alpha_2 e^{i2\omega}|^2} \end{aligned}$$

for $0 < \omega < \pi$, where $|\cdot|$ represents the modulus of a complex number. The shape of the spectrum depends on the values of $\alpha_1$ and $\alpha_2$. It is possible to find values of $\alpha_1$ and $\alpha_2$ that give a spectrum with a peak at zero frequency (a low-frequency spectrum), a peak at frequency $\pi$ (a high-frequency spectrum), a peak *between* 0 and $\pi$, or a *minimum* between 0 and $\pi$.

For AR processes of higher order than first order, their spectrum can be similarly computed. For an AR process of order $p$

$$\phi(B)X_t = Z_t, \qquad \phi(B) = 1-\alpha_1 B - \cdots - \alpha_p B^p,$$

it can be shown that its spectrum is given by

$$f(\omega) = \frac{\sigma_Z^2}{\pi} \cdot \frac{1}{|\phi(e^{i\omega})|^2}.$$

For a MA process of order $q$

$$X_t = \theta(B)Z_t, \qquad \theta(B) = 1 + \beta_1 B + \cdots + \beta_q B^q,$$

it can be shown that its spectrum is given by

$$f(\omega) = \frac{\sigma_Z^2}{\pi} \cdot |\theta(e^{i\omega})|^2.$$

Similarly, for an ARMA process of order $(p,q)$,

$$\phi(B)X_t = \theta(B)Z_t,$$

its spectrum can be expressed as

$$f(\omega) = \frac{\sigma_Z^2}{\pi} \cdot \frac{|\theta(e^{i\omega})|^2}{|\phi(e^{i\omega})|^2}.$$

(6) *Deterministic sinusoidal perturbations*

Suppose that

$$X_t = \cos(\omega_0 t + \phi) \tag{6.26}$$

where $\omega_0$ is a constant in $(0,\pi)$ and $\phi$ is a random variable having a uniform distribution on $(0,2\pi)$. As explained in Section 3.13, $\phi$ is fixed for a single realization of the process and Equation (6.26) defines a purely deterministic process.

The acv.f. of the process is given by

$$\gamma(k) = \frac{1}{2}\cos\,\omega_0 k,$$

which we note does *not* tend to zero as $k$ increases. This contrasts with the behaviour of the ac.f. of a stationary linearly indeterministic process, which does tend to zero as $k$ increases. Put loosely, we can say that the above deterministic process has a long memory.

From the model Equation (6.24), it is obvious that all the 'power' of the process is concentrated at the frequency $\omega_0$. Since $E(X_t) = 0$, we find that $\text{Var}(X_t) = E(X_t^2) = \frac{1}{2}$, so that the power spectral distribution function is given by

$$F(\omega) = \begin{cases} 0 & \omega < \omega_0, \\ \frac{1}{2} & \omega \geq \omega_0 \end{cases}$$

Since this is a step function, it has no derivative at $\omega_0$ and so the spectrum is not defined at $\omega_0$. If we nevertheless try to use Equation (6.12) to obtain the spectrum as the Fourier transform of the acv.f., then we find that

$$f(\omega) = 0 \qquad \omega \neq \omega_0$$

as expected, but that $\sum \gamma(k) \cos \omega k$ does not converge at $\omega = \omega_0$. This confirms that the spectrum is not defined at $\omega_0$.

(7) *A mixture of deterministic and stochastic components*

Our final example contains a mixture of deterministic and stochastic components, namely,

$$X_t = \cos(\omega_0 t + \phi) + Z_t,$$

where $\omega_0, \phi$ are as defined in Example 5 above, and $\{Z_t\}$ is a purely random process with mean zero and variance $\sigma_Z^2$. Then we find that the acv.f. is given by

$$\gamma(k) = \begin{cases} \frac{1}{2} + \sigma_Z^2 & k = 0 \\ \frac{1}{2} \cos \omega_0 k & k = \pm 1, \pm 2, \ldots. \end{cases}$$

As in Example 5 above, note that $\gamma(k)$ does *not* tend to zero, as $k$ increases, because $X_t$ contains a periodic deterministic component.

We can obtain the power spectral distribution function of $X_t$ by using Equation (6.6), since the deterministic component $\cos(\omega_0 t + \phi)$ has power spectral distribution function

$$F_1(\omega) = \begin{cases} 0 & \omega < \omega_0, \\ \frac{1}{2} & \omega \geq \omega_0. \end{cases}$$

while the stochastic component $Z_t$ has power spectral distribution function

$$F_2(\omega) = \sigma_Z^2 \, \omega / \pi \qquad 0 < \omega < \pi$$

on integrating Equation (6.19). Thus the combined power spectral distribution function is given by

$$F(\omega) = \begin{cases} \sigma_Z^2 \omega / \pi & 0 < \omega < \omega_0 \\ \frac{1}{2} + \sigma_Z^2 \omega / \pi & \omega_0 \leq \omega < \pi \end{cases}$$

As in Example 5, the distribution function has a step at $\omega_0$ and so the power spectrum is not defined at $\omega = \omega_0$.

## Exercises

In the following questions $\{Z_t\}$ denotes a purely random process, mean zero and variance $\sigma_Z^2$.

**6.1** Find (a) the power spectral density function, (b) the normalized spectral density function of the first-order AR process

$$X_t = \lambda X_{t-1} + Z_t$$

with $|\lambda| < 1$. (Note: (a) is covered in the text, but see if you can do it without looking it up.)

**6.2** Find the power spectral density functions of the following MA processes:
(a) $X_t = Z_t + Z_{t-1} + Z_{t-2}$
(b) $X_t = Z_t + 0.5Z_{t-1} - 0.3Z_{t-2}$

**6.3** If $\mu$ denotes a constant, show that the second-order MA process

$$X_t = \mu + Z_t + 0.8Z_{t-1} + 0.5Z_{t-2}$$

is second-order stationary. Find the acv.f. and ac.f. of $\{X_t\}$ and show that its normalized spectral density function is given by

$$f^*(\omega) = (1 + 1.27\cos\omega + 0.53\cos 2\omega)/\pi \qquad 0 < \omega < \pi.$$

**6.4** A stationary time series $(X_t; t = \ldots, -1, 0, +1, \ldots)$ has normalized spectral density function

$$f^*(\omega) = 2(\pi - \omega)/\pi^2 \qquad 0 < \omega < \pi.$$

Show that its ac.f. is given by

$$\rho(k) = \begin{cases} 1 & k = 0, \\ (2/\pi k)^2 & k \text{ odd}, \\ 0 & k \text{ even } (\neq 0). \end{cases}$$

**6.5** A two-state Markov process may be set up as follows. Alpha particles from a radioactive source are used to trigger a flip-flop device that takes the states $+1$ and $-1$ alternately. The times $t_i$ at which changes occur constitute a Poisson process, with mean event rate $\lambda$. Let $X(t)$ denote the state variable at time $t$. If the process is started at time $t = 0$ with the two possible states having equal probability (so that $P[X(0) = 1] = P[X(0) = -1] = \frac{1}{2}$, show that the process is second-order stationary, with autocorrelation function

$$\rho(u) = e^{-2\lambda|u|} \qquad -\infty < u < \infty$$

and spectral density function

$$f(\omega) = 4\lambda/[\pi(4\lambda^2 + \omega^2)] \qquad 0 < \omega < \infty.$$

**6.6** Show that if $\{X_t\}$ and $\{Y_t\}$ are independent, stationary processes with power spectral density functions $f_x(\omega)$ and $f_y(\omega)$, then $\{V_t\} = \{X_t + Y_t\}$ is also stationary with power spectral density function $f_v(\omega) = f_x(\omega) + f_y(\omega)$. If $X_t$ is a first-order AR process

$$X_t = \alpha X_{t-1} + W_t \qquad -1 < \alpha < +1$$

and $\{Y_t\}, \{W_t\}$ are independent purely random processes with zero mean and common variance $\sigma^2$, show that the power spectral density function of $\{V_t\}$ is given by

$$f_v(\omega) = \sigma^2(2 - 2\alpha \cos \omega + \alpha^2)/\pi(1 - 2\alpha \cos \omega + \alpha^2) \qquad 0 < \omega < \pi.$$

**6.7** Show that the normalized spectral density function of the ARMA$(1,1)$ process

$$X_t = \alpha X_{t-1} + Z_t + \beta Z_{t-1}$$

(with $|\alpha| < 1$ and $|\beta| < 1$) is given by

$$f^*(\omega) = \frac{1}{\pi}[1 + 2\rho(1)(\cos \omega - \alpha)/(1 - 2\alpha \cos \omega + \alpha^2)] \qquad 0 < \omega < \pi.$$

Chapter 7

# Spectral Analysis

**Spectral analysis** is the name given to methods of estimating the spectral density function, or spectrum, of a given time series.

Before about 1900, research workers such as A. Schuster were essentially concerned with looking for 'hidden periodicities' in data at one or two specific frequencies. Spectral analysis as we know it today is concerned with estimating the spectrum over the whole range of frequencies. The techniques are widely used by many scientists, particularly in electrical engineering, physics, meteorology and marine science.

We are mainly concerned with purely indeterministic processes, which have a continuous spectrum, but the techniques can also be used for deterministic processes to pick out periodic components in the presence of noise.

## 7.1 Fourier Analysis

Traditional spectral analysis is essentially a modification of Fourier analysis so as to make it suitable for stochastic rather than deterministic functions of time. **Fourier analysis** (e.g. Priestley, 1981) is essentially concerned with approximating a function by a sum of sine and cosine terms, called the Fourier series representation. Suppose that a function $f(t)$ is defined on $(-\pi, \pi]$[1] and satisfies the so-called Dirichlet conditions. These conditions ensure that $f(t)$ is reasonably 'well behaved', meaning that, over the range $(-\pi, \pi]$, $f(t)$ is absolutely integrable, has a finite number of discontinuities, and has a finite number of maxima and minima. Then $f(t)$ may be approximated by the Fourier series

$$\frac{a_0}{2} + \sum_{r=1}^{k} (a_r \cos rt + b_r \sin rt),$$

[1]The different-shaped brackets indicate that the lower limit $-\pi$ is *not* included in the interval, while the square bracket indicates that the upper limit $+\pi$ *is* included.


DOI: 10.1201/9781003646112-7

where

$$\begin{aligned} a_0 &= \frac{1}{\pi}\int_{-\pi}^{\pi} f(t)\,\mathrm{d}t, \\ a_r &= \frac{1}{\pi}\int_{-\pi}^{\pi} f(t)\cos rt\,\mathrm{d}t \qquad r=1,2,\ldots, \\ b_r &= \frac{1}{\pi}\int_{-\pi}^{\pi} f(t)\sin rt\,\mathrm{d}t \qquad r=1,2,\ldots. \end{aligned}$$

It can be shown that this Fourier series converges to $f(t)$ as $k \to \infty$ except at points of discontinuity, where it converges to halfway[2] up the step change.

In order to apply Fourier analysis to discrete time series, we need to consider the Fourier series representation of $f(t)$ when $f(t)$ is defined only on the integers $1,2,\ldots,N$. Rather than write down the formula, we demonstrate that the required Fourier series emerges naturally by considering a simple sinusoidal model.

## 7.2 A Simple Sinusoidal Model

Suppose we suspect that a given time series, with observations made at unit time intervals, contains a deterministic sinusoidal component at a known frequency $\omega$, together with a random error term. Then we will consider the model

$$X_t = \mu + \alpha\cos\omega t + \beta\sin\omega t + Z_t, \tag{7.1}$$

where $Z_t$ denotes a purely random process, and $\mu,\alpha,\beta$ are parameters to be estimated from the data.

The observations will be denoted by $(x_1,x_2,\ldots,x_N)$. The algebra in the next few sections is somewhat simplified if we confine ourselves to the case where $N$ is *even*. There is no real difficulty in extending the results to the case where $N$ is odd (e.g. Anderson, 1971), and indeed many of the later estimation formulae apply for both odd and even $N$, but some results require one to consider odd $N$ and even $N$ separately. Thus, if $N$ happens to be odd and a spectral analysis is required, computation can be made somewhat simpler by removing the first observation so as to make $N$ even. If $N$ is reasonably large, little information is lost.

Expected values for the model in Equation (7.1) can be represented in matrix notation by

$$E(X) = A\,\theta$$

where

$$\begin{aligned} X^T &= (X_1,\ldots,X_N) \\ \theta^T &= (\mu,\alpha,\beta) \\ A &= \begin{pmatrix} 1 & \cos\omega & \sin\omega \\ 1 & \cos 2\omega & \sin 2\omega \\ \cdots & \cdots & \cdots \\ 1 & \cos N\omega & \sin N\omega \end{pmatrix}. \end{aligned}$$

[2] Mathematicians say that this is the average of the limit from below and the limit from above, sometimes written as $\frac{1}{2}[f(t-0)+f(t+0)]$.

As this model is linear in the parameters $\mu, \alpha$ and $\beta$, it is an example of a general linear model. In that case the least squares estimate of $\theta$, which minimizes $\sum_{t=1}^{N}(x_t - \mu - \alpha\cos\omega t - \beta\sin\omega t)^2$, is 'well known' to be

$$\widehat{\theta} = (A^T A)^{-1} A^T x,$$

where

$$x^T = (x_1, \ldots, x_N).$$

The above formulae hold for any value of the frequency $\omega$, but they only make practical sense for values of $\omega$ that are not too high or too low. As noted in Section 6.2, the highest frequency we can uniquely fit to the data is the Nyquist frequency, given by $\omega = \pi$, which completes one cycle every two observations. At the other end of the spectrum, the lowest frequency we can reasonably fit completes one cycle in the whole length of the time series. These upper and lower limits will be explained further in Section 7.2.1. By equating the cycle length $2\pi/\omega$ to $N$, we find that the lowest frequency is given by $2\pi/N$. The formulae for the least squares estimates of $\widehat{\theta}$ turn out to be particularly simple if $\omega$ is restricted to one of the values

$$\omega_p = 2\pi p/N \quad p = 1, \ldots, N/2,$$

which lie in equal steps from the lowest frequency $2\pi/N$ to the Nyquist frequency $\pi$. In this case, it turns out that $(A^T A)$ is a diagonal matrix in view of the following 'well-known' trigonometric results (all summations are for $t = 1$ to N):

$$\sum \cos\omega_p t = \sum \sin\omega_p t = 0, \tag{7.2}$$

$$\sum \cos\omega_p t \cos\omega_q t = \begin{cases} 0 & p \neq q, \\ N & p = q = N/2, \\ N/2 & p = q \neq N/2, \end{cases} \tag{7.3}$$

$$\sum \sin\omega_p t \sin\omega_q t = \begin{cases} 0 & p \neq q, \\ 0 & p = q = N/2, \\ N/2 & p = q \neq N/2, \end{cases} \tag{7.4}$$

$$\sum \cos\omega_p t \sin\omega_q t = 0 \qquad \text{for all } p, q. \tag{7.5}$$

With $(A^T A)$ diagonal, it is easy to evaluate the least squares estimate of $\theta$, as the inverse $(A^T A)^{-1}$ will also be diagonal. For $\omega_p$ such that $p \neq N/2$, we find (Exercise 7.2)

$$\begin{aligned} \widehat{\mu} &= \sum x_t/N = \bar{x}, \\ \widehat{\alpha} &= 2\left[\sum x_t \cos\omega_p t\right]/N, \\ \widehat{\beta} &= 2\left[\sum x_t \sin\omega_p t\right]/N. \end{aligned} \tag{7.6}$$

If $p = N/2$, we ignore the term in $\beta \sin \omega t$, which is zero for all $t$, and find

$$\widehat{\mu} = \bar{x},$$
$$\widehat{\alpha} = \sum(-1)^t x_t / N.$$

The model in Equation (7.1) is essentially the one used before about 1900 to search for hidden periodicities, but this model has now gone out of fashion. However, it can still be useful if there is reason to suspect that a time series does contain a deterministic periodic component at a known frequency and it is desired to isolate this component (e.g. Bloomfield, 2000, Chapters 2, 3).

Readers who are familiar with the analysis of variance (ANOVA) technique will be able to work out that the total corrected sum of squared deviations, namely,

$$\sum_{t=1}^{N}(x_t - \bar{x})^2$$

can be partitioned into two components, namely, the residual sum of squares and the sum of squares 'explained' by the periodic component at frequency $\omega_p$. This latter component is given by

$$\sum_{t=1}^{N}(\widehat{\alpha}\cos\omega_p t + \widehat{\beta}\sin\omega_p t)^2$$

which, after some algebra (Exercise 7.2), can be shown to be

$$\begin{array}{ll} (\widehat{\alpha}^2 + \widehat{\beta}^2)N/2 & p \neq N/2 \\ \widehat{\alpha}^2 N & p = N/2 \end{array} \tag{7.7}$$

using Equations (7.2)–(7.5).

When fitting the simple sinusoidal model in Equation (7.1), we restricted the frequency $\omega$ to one of the values $(2\pi/N, 4\pi/N, \ldots, \pi)$, assuming that $N$ is even. Here we examine the practical rationale for the upper and lower limits, namely, $\pi$ and $2\pi/N$.

In Section 6.2, we pointed out that, for a discrete-time process measured at unit intervals of time, there is no loss of generality in restricting the spectral distribution function to the range $(0, \pi)$. We now demonstrate that the upper bound $\pi$, called the **Nyquist frequency**, is indeed the highest frequency about which we can get meaningful information from a set of data.

First, we give a more general form for the Nyquist frequency. If observations are taken at equal intervals of time of length $\Delta t$, then the Nyquist (angular) frequency is given by $\omega_N = \pi/\Delta t$. The equivalent frequency expressed in cycles per unit time is $f_N = \omega_N/2\pi = 1/2\Delta t$.

Consider the following example. Suppose that temperature readings are taken every day in a certain town at noon. It is clear that these observations will tell us nothing about temperature variation *within* a day. In particular, they will not tell us whether nights are hotter or cooler than days. With only one observation per day, the Nyquist frequency is $\omega_N = \pi$ radians per day or $f_N = \frac{1}{2}$ cycle per day (or 1 cycle per

2 days). This is lower than the frequencies, which correspond to variation within a day. For example, variation with a period of 1 day has (angular) frequency $\omega = 2\pi$ radians per day or $f = 1$ cycle per day. In order to get information about variation within a day at these higher frequencies, we must increase the sampling rate and take two or more observations per day.

A similar example is provided by yearly sales figures. These will obviously give no information about any seasonal effects, whereas monthly or quarterly observations *will* give information about seasonality.

At the other end of the spectrum, we will now explain why there is a *lowest* frequency below which it is not sensible to try to fit to a set of data. If we had just 6 months of temperature readings from winter to summer, the analyst would not be able to decide, from the data alone, whether there is an upward trend in the observations or whether winters are colder than summers. However, with 1 year's data, it *would* become clear that winters are colder than summers. Thus if we are interested in variation at the low frequency of 1 cycle per year, then we should have at least 1 year's data, in which case the lowest frequency we can fit is at 1 cycle per year. With weekly observations, for example, 1 year's data have $N = 52$, $\Delta t = 1$ week, and the lowest angular frequency of $2\pi/N\Delta t$ corresponds to a frequency of $1/N\Delta t$ cycles per week. (Note that all time units must be expressed in terms of the same period, here a week.) The lowest frequency is therefore 1/52 cycles per week, which can now be converted to 1 cycle per year.

The lowest frequency, namely, $2\pi/N\Delta$, is sometimes called the **fundamental Fourier frequency**, because the Fourier series representation of the data is normally evaluated at the frequencies $\omega_p = 2\pi p/N\Delta$ for $p = 1,\ldots,N/2$, which are all integer multiples of the fundamental frequency. These integer multiples are often called **harmonics**. The phrase *fundamental frequency* is perhaps more typically, and more helpfully, used when a function, $f(t)$ say, is **periodic** with period $T$ so that $f(t+nT) = f(t)$ for all integer values of $n$. Then $f = 1/T$, or $\omega = 2\pi/T$, is called the fundamental frequency and the Fourier series representation of $f(t)$ is a sum over integer multiples, or harmonics, of the fundamental frequency. When $T = N\Delta =$ (the length of the observed time series), the fundamental frequencies coincide. This raises a practical point, in regard to choosing the length of a time series. Suppose, for example, that you are collecting weekly data and are particularly interested in annual variation. As noted above, you should collect at least 1 year's data. If you collect exactly 52 weeks of data[3], then the fundamental frequency will be at exactly 1 cycle per year. We will see that this makes it much easier to interpret the results of a spectral analysis. The fundamental frequency is at 1 cycle per year and the harmonics are at 2 cycles per year, 3 cycles per year, and so on. However, if you have, say, an extra 12 weeks of data making 64 weeks, then it will be much harder to interpret the results at frequencies $\omega_p = 2\pi p/N\Delta$. Wherever possible, you should choose the length of the time series so that the harmonics cover the frequencies of particular interest. The easiest option is to collect observations covering an integer multiple of the lowest wavelength of particular interest. This ensures that this frequency is an

[3]For simplicity, ignore day 365, and day 366 if a leap year.

integer multiple of the fundamental frequency. Thus, to investigate annual variation, 2 years of data is good and 3 or 4 years of data even better.

The reader will notice that the Nyquist frequency does *not* depend on $N$, but rather only on the sampling frequency, whereas the lowest frequency *does* depend on $N$. Put another way, the lower the frequency we are interested in, the longer the time period over which we need to take measurements, whereas the higher the frequency we are interested in, the more frequently must we take observations.

## 7.3 Periodogram Analysis

Early attempts at discovering hidden periodicities in a given time series basically consisted of repeating the analysis of Section 7.2 at all the frequencies $2\pi/N, 4\pi/N, \ldots, \pi$. In view of Equations (7.3)–(7.5), the different terms are orthogonal and we end up with the finite Fourier series representation of the $\{x_t\}$, namely

$$x_t = a_0 + \sum_{p=1}^{(N/2)-1} [a_p \cos(2\pi pt/N) + b_p \sin(2\pi pt/N)] + a_{N/2} \cos \pi t \tag{7.8}$$

for $t = 1, 2, \ldots, N$, where the coefficients $\{a_p, b_p\}$ are of the same form as Equations (7.6) and (7.7), namely

$$\begin{aligned} a_0 &= \bar{x} \\ a_{N/2} &= \sum (-1)^t x_t / N \\ \left.\begin{aligned} a_p &= 2\left[\sum x_t \cos(2\pi pt/N)\right]/N \\ b_p &= 2\left[\sum x_t \sin(2\pi pt/N)\right]/N \end{aligned}\right\} & \quad p = 1, \ldots, (N/2) - 1. \end{aligned} \tag{7.9}$$

An analysis along these lines is sometimes called a **Fourier analysis** or a **harmonic analysis**. The Fourier series representation in Equation (7.8) has $N$ parameters to describe $N$ observations and so can be made to fit the data exactly (just as a polynomial of degree $N-1$ involving $N$ parameters can be found that goes exactly through $N$ observations in polynomial regression). This explains why there is no error term in Equation (7.8) in contrast to Equation (7.1). Also note that there is no term in $\sin \pi t$ in Equation (7.8) as $\sin \pi t$ is zero for all integer $t$.

It is worth stressing that the Fourier series coefficients in Equation (7.9) at a given frequency $\omega$ are exactly the same as the least squares estimates for the coefficients of the model in Equation (7.1).

The overall effect of the Fourier analysis of the data is to partition the variability of the series into components at frequencies $2\pi/N, 4\pi/N, \ldots, \pi$. The component at frequency $\omega_p = 2\pi p/N$ is called the $p$th harmonic. For $p \neq N/2$, it can be useful to write the $p$th harmonic in the equivalent form

$$a_p \cos \omega_p t + b_p \sin \omega_p t = R_p \cos(\omega_p t + \phi_p) \tag{7.10}$$

where

$$R_p = \sqrt{a_p^2 + b_p^2} \tag{7.11}$$

is the **amplitude** of the $p$th harmonic, and

$$\phi_p = \tan^{-1}(-b_p/a_p) \tag{7.12}$$

is the **phase** of the $p$th harmonic.

We have already noted in Section 7.2 that, for $p \neq N/2$, the contribution of the $p$th harmonic to the total sum of squares is given by $N(a_p^2 + b_p^2)/2$. Using Equation (7.11), this is equal to $NR_p^2/2$. Extending this result using Equations (7.2)–(7.5) and (7.8), we have, after some algebra (Exercise 7.3), that

$$\sum_{t=1}^{N}(x_t - \bar{x})^2 = N \sum_{p=1}^{(N/2)-1} R_p^2/2 + Na_{N/2}^2.$$

Dividing through by $N$ we have

$$\sum(x_t - \bar{x})^2/N = \sum_{p=1}^{(N/2)-1} R_p^2/2 + a_{N/2}^2 \tag{7.13}$$

which is known as **Parseval's theorem**. The left-hand side of Equation (7.13) is effectively the variance[4] of the observations. Thus $R_p^2/2$ is the contribution of the $p$th harmonic to the variance, and Equation (7.13) shows how the total variance is partitioned.

If we plot $R_p^2/2$ against $\omega_p = 2\pi p/N$, we obtain a line spectrum. A different type of line spectrum occurs in the physical sciences when light from molecules in a gas discharge tube is viewed through a spectroscope. The light has energy at discrete frequencies and this energy can be seen as bright lines. However, most time series have continuous spectra, and then it is inappropriate to plot a line spectrum. If we regard $R_p^2/2$ as the contribution to variance in the range $\omega_p \pm \pi/N$, we can plot a histogram whose height in the range $\omega_p \pm \pi/N$ is such that

$$\begin{aligned} R_p^2/2 &= \text{area of histogram rectangle} \\ &= \text{height of histogram} \times 2\pi/N. \end{aligned}$$

Thus the height of the histogram at $\omega_p$, denoted by $I(\omega_p)$, is given by

$$I(\omega_p) = NR_p^2/4\pi. \tag{7.14}$$

As usual, Equation (7.14) does not apply for $p = N/2$; we may regard $a_{N/2}^2$ as the contribution to variance in the range $[(N-1)\pi/N, \pi]$ so that

$$I(\pi) = Na_{N/2}^2/\pi.$$

The plot of $I(\omega)$ against $\omega$ is usually called the **periodogram**, even though $I(\omega)$ is a function of frequency rather than period. It follows from Parseval's theorem in

[4]The divisor is $N$ rather than the more usual $(N-1)$, but this makes little difference for large $N$.

Equation (7.13) that the total area under the periodogram is equal to the variance of the time series.

Note that the formula for $R_p^2$, and hence for $I(\omega_p)$, can be written in several equivalent ways that look quite different. For example, after some algebra, it can be shown that

$$I(\omega_p) = \frac{1}{\pi N}\left|\sum_{t=1}^{N} x_t e^{it\omega_p}\right|^2 \tag{7.15}$$

or we can replace $e^{it\omega}$ with $e^{-it\omega}$ in Equation (7.15). The usual way to actually calculate the periodogram directly from the data uses the expression

$$I(\omega_p) = \left[\left(\sum x_t \cos 2\pi pt/N\right)^2 + \left(\sum x_t \sin 2\pi pt/N\right)^2\right] \Big/ N\pi. \tag{7.16}$$

Equation (7.16) also applies for $p = N/2$.

Other authors define the periodogram in what appear to be slightly different ways, but the differences usually arise from allowing negative frequencies or using the cyclic frequency $f = \omega/2\pi$, rather than $\omega$. The expressions generally turn out to be some other multiple of $I(\omega_p)$ or $R_p^2$. For example, Hannan (1970, Equation (3.8)) and Koopmans (1995, Equation (8.7)) give expressions that correspond to $\frac{1}{2}\times$ expression (7.15). As to terminology, Anderson (1971, Section 4.3.2) describes the graph of $R_p^2$ against the *period* $N/p$, as the periodogram, and suggests the term **spectrogram** to describe the graph of $R_p^2$ against frequency. Jenkins and Watts (1968) define a similar expression to Equation (7.16) in terms of the variable $f = \omega/2\pi$, but call it the 'sample spectrum'. As always, when comparing terms and formulae from different sources, the reader needs to take great care.

The periodogram appears to be a natural way of estimating the power spectral density function, but Section 7.3.2 shows that, for a process with a **continuous** spectrum, it provides a poor estimate and needs to be modified. First, we derive the relationship between the periodogram of a given time series and the corresponding autocovariance function (acv.f.).

### *7.3.1 The relationship between the periodogram and the autocovariance function*

The periodogram ordinate $I(\omega_p)$ and the autocovariance coefficient $c_k$ are both quadratic forms of the data $\{x_t\}$. It is therefore natural to enquire how they are related. In fact, we will show that the periodogram is the finite Fourier transform of $\{c_k\}$.

Using Equation (7.2), we may rewrite Equation (7.16) for $p \neq N/2$ as

$$\begin{aligned} I(\omega_p) &= \left\{\left[\sum (x_t - \bar{x})\cos\omega_p t\right]^2 + \left[\sum (x_t - \bar{x})\sin\omega_p t\right]^2\right\} \Big/ N\pi \\ &= \sum_{s,t=1}^{N} (x_t - \bar{x})(x_s - \bar{x})(\cos\omega_p t \cos\omega_p s + \sin\omega_p t \sin\omega_p s) \Big/ N\pi. \end{aligned}$$

However, (see Equation (4.1))

$$\sum_{t=1}^{N-k} (x_t - \bar{x})(x_{t+k} - \bar{x})/N = c_k$$

and $\cos \omega_p t \cos \omega_p(t+k) + \sin \omega_p t \sin \omega_p(t+k) = \cos \omega_p(t+k-t) = \cos \omega_p k$ so that

$$I(\omega_p) = \left( c_0 + 2\sum_{k=1}^{N-1} c_k \cos \omega_p k \right) \Big/ \pi \tag{7.17}$$

$$= \sum_{k=-(N-1)}^{N-1} c_k e^{-i\omega_p k} \Big/ \pi. \tag{7.18}$$

The formula in Equation (7.18) is an expression called a **discrete finite Fourier transform** (assuming that $c_k = 0$ for $|k| \geq N$). Any reader not familiar with the Fourier transform, is recommended to read Appendix A – see especially Equation (A.5).

### *7.3.2 Properties of the periodogram*

When the periodogram is expressed in the form of Equation (7.17), it appears to be the 'obvious' estimate of the power spectrum

$$f(\omega) = \left( \gamma_0 + 2\sum_{k=1}^{\infty} \gamma_k \cos \omega k \right) \Big/ \pi$$

simply replacing $\gamma_k$ by its estimate $c_k$ for values of $k$ up to $(N-1)$, and putting subsequent estimates of $\gamma_k$ equal to zero. However, although we find

$$\underset{N\to\infty}{E}[I(\omega)] \to f(\omega) \tag{7.19}$$

so that the periodogram is asymptotically unbiased, we see below that the variance of $I(\omega)$ does not decrease as $N$ increases. Thus $I(\omega)$ is *not a consistent estimator* for $f(\omega)$. The lack of consistency is perhaps not too surprising when one realizes that the Fourier series representation in Equation (7.8) requires one to evaluate $N$ parameters from $N$ observations, however long the series may be. Thus in Section 7.4 we will consider alternative ways of estimating a power spectrum that are essentially ways of *smoothing* the periodogram.

We complete this section by proving that $I(\omega)$ is not a consistent estimator for $f(\omega)$ in the case where the observations are assumed to be independent $N(\mu, \sigma^2)$ variates, so that they form a discrete-time purely random process with a uniform spectrum. This result can be extended to other stationary processes with continuous spectra, but this does not need to be demonstrated here. If the periodogram estimator does not 'work' for a uniform spectrum, it cannot be expected to 'work' for more complicated spectra. Given the above assumptions, Equation (7.9) shows that $a_p$ and $b_p$ are linear combinations of normally distributed random variables and so will themselves be normally distributed. Using Equations (7.2)–(7.4), it can be shown (Exercise 7.4) that $a_p$ and $b_p$ each have mean zero and variance $2\sigma^2/N$ for $p \neq N/2$. Furthermore we have

$$\begin{aligned}\text{Cov}(a_p, b_p) &= 4\text{Cov}\left[\left(\sum X_t \cos\omega_p t\right), \left(\sum X_t \sin\omega_p t\right)\right]/N^2 \\ &= 4\sigma^2\left(\sum \cos\omega_p t \sin\omega_p t\right)/N^2\end{aligned}$$

since the observations are assumed to be independent. Thus, using Equation (7.5), we see that $a_p$ and $b_p$ are uncorrelated. Since $(a_p, b_p)$ are bivariate normal, zero correlation implies that $a_p$ and $b_p$ are independent. The variables $a_p$ and $b_p$ can be standardized by dividing by $\sqrt{2}\sigma/\sqrt{N}$ to give standard $N(0,1)$ variables. Now a result from distribution theory says that if $Y_1, Y_2$ are independent N(0, 1) variables, then $(Y_1^2 + Y_2^2)$ has a $\chi^2$ distribution with two degrees of freedom, which is written $\chi_2^2$. Thus

$$\frac{N(a_p^2 + b_p^2)}{2\sigma^2} = \frac{I(\omega_p)2\pi}{\sigma^2}$$

is $\chi_2^2$. Put another way, this means that $2I(\omega)/f(\omega)$ is $\chi_2^2$ when $f(\omega) = \sigma^2/\pi$ as in this case, although this result does, in fact, generalize to spectra that are not constant. Now the variance of a $\chi^2$ distribution with $\nu$ degrees of freedom is $2\nu$, so that

$$\text{Var}[I(\omega_p)2\pi/\sigma^2] = 4$$

and

$$\text{Var}[I(\omega_p)] = \sigma^4/\pi^2.$$

As this variance is a constant, it does *not* tend to zero as $N \to \infty$, and hence $I(\omega_p)$ is not a consistent estimator for $f(\omega_p)$. Furthermore it can be shown that neighbouring periodogram ordinates are asymptotically independent, which further explains the very irregular form of an observed periodogram. This all means that the periodogram needs to be modified in order to obtain a good estimate of a continuous spectrum.

## 7.4 Some Consistent Estimation Procedures

This section describes several alternative ways of estimating a spectrum. The different methods will be compared in Section 7.6. Each method provides a *consistent* estimator for the (power) spectral density function, in contrast to the (raw) periodogram. However, although the periodogram is itself an inconsistent estimator, the procedures described in this section are essentially based on smoothing the periodogram in some way.

Throughout the section we will assume that any obvious trend and seasonal variation have been removed from the data. If this is not done, the results of the spectral analysis are likely to be dominated by these effects, making any other effects difficult or impossible to see. Trend produces a peak at zero frequency, while seasonal variation produces peaks at the seasonal frequency and at integer multiples of the seasonal frequency – the seasonal **harmonics** (see Section 7.2.1). For a nonstationary series, the estimated spectrum of the detrended, deseasonalized data will depend to some extent on the method chosen to remove trend and seasonality. We assume throughout that a 'good' method is used to do this.

The methods described in this chapter are essentially *non-parametric* in that no model fitting is involved. It is possible to use a model-based approach and an alternative, parametric approach, called **autoregressive spectrum estimation**, will be introduced later in Section 13.6.1.

### 7.4.1 *Transforming the truncated autocovariance function*

One type of estimation procedure consists of taking a Fourier transform of the truncated weighted sample acv.f. From Equation (7.17), we know that the periodogram is the discrete finite Fourier transform of the complete sample acv.f. However, it is clear that the precision of the values of $c_k$ decreases as $k$ increases, because the coefficients are based on fewer and fewer terms. Thus, it would seem intuitively reasonable to give less weight to the values of $c_k$ as $k$ increases. An estimator, which has this property is

$$\widehat{f}(\omega) = \frac{1}{\pi}\left\{\lambda_0 c_0 + 2\sum_{k=1}^{M} \lambda_k c_k \cos \omega k\right\} \tag{7.20}$$

where $\{\lambda_k\}$ are a set of weights called the **lag window**, and $M(< N)$ is called the **truncation point**. Comparing Equation (7.20) with (7.17) we see that values of $c_k$ for $M < k < N$ are no longer used, while values of $c_k$ for $k \leq M$ are weighted by a factor $\lambda_k$. The latter are chosen so as to get smaller as $k$ approaches $M$.

In order to use the above estimator, the analyst must choose a suitable lag window and a suitable truncation point. The two best-known lag windows are as follows.

*Tukey window*

$$\lambda_k = \frac{1}{2}\left(1 + \cos\frac{\pi k}{M}\right) \qquad k = 0, 1, \ldots, M.$$

This window is sometimes called the Tukey–Hanning or Blackman–Tukey window.

*Parzen window*

$$\lambda_k = \begin{cases} 1 - 6\left(\frac{k}{M}\right)^2 + 6\left(\frac{k}{M}\right)^3 & 0 \leq k \leq M/2 \\ 2(1 - k/M)^3 & M/2 \leq k \leq M. \end{cases}$$

These two windows are illustrated in Figure 7.1 with $M = 20$.

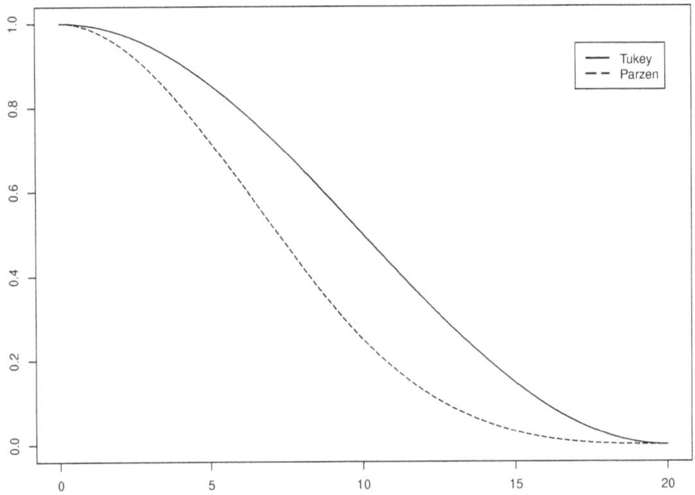


Figure 7.1 *The Tukey and Parzen lag windows with $M = 20$.*

The Tukey and Parzen windows give very similar estimated spectra for a given time series, although the Parzen window has a slight advantage in that it cannot give negative estimates. Many other lag windows have been suggested and 'window carpentry' was a popular research topic in the 1950s. Ways of comparing different windows will be discussed in Section 7.6. The well-known **Bartlett window**, with $\lambda_k = 1 - k/M$ for $k = 0, 1, \ldots, M$, is very simple but is now rarely used as its properties are inferior to those of the Tukey and Parzen windows.

The choice of the truncation point $M$ is more difficult and it is not easy to give clear-cut advice. It has to be chosen subjectively so as to *balance 'resolution' against 'variance'*. The smaller the value of $M$, the smaller will be the variance of $\hat{f}(\omega)$ but the larger will be the bias. If $M$ is too small, important features of $f(\omega)$ may be smoothed out, but if $M$ is too large the behaviour of $\hat{f}(\omega)$ becomes more like that of the periodogram with erratic variation. Thus a compromise value must be chosen. A useful rough guide is to choose $M$ to be about $2\sqrt{N}$, so that if, for example, $N$ is 200, then $M$ will be round about the value 28. This choice of $M$ ensures the asymptotic situation that as $N \to \infty$, so also does $M \to \infty$ but in such a way that $M/N \to 0$. A somewhat larger value of $M$ is required for the Parzen window than for the Tukey window. Other writers have suggested $\sqrt{N}$, rather than $2\sqrt{N}$, while results from density estimation suggest that a different power of $N$ may be appropriate. Percival (1993, Chapter 6) point out that an appropriate value of $M$ depends on the properties of the underlying process and give more detailed guidance. However, our advice is to try three or four different values of $M$. A low value will give an idea where the large peaks in $f(\omega)$ are, but the curve is likely to be too smooth. A high value is likely to produce a curve showing a large number of peaks, some of which may be spurious. A compromise can then be achieved with an in-between value of $M$.

In principle, Equation (7.20) may be evaluated at any value of $\omega$ in $(0, \pi)$, but it is usually evaluated at equal intervals at $\omega = \pi j/Q$ for $j = 0, 1 \ldots, Q$, where $Q$ is chosen sufficiently large to show up all features of $\widehat{f}(\omega)$. Often $Q$ is chosen to be equal to $M$. The graph of $\widehat{f}(\omega)$ against $\omega$ can then be plotted and examined. An example is given later in Figure 7.5, for the data plotted in Figure 1.2, using the Tukey window with $M = 24$.

### 7.4.2 *Hanning*

This procedure, named after Julius Von Hann, is equivalent to the use of the Tukey window as described in Section 7.4.1, but adopts a different computational procedure. The estimated spectrum is calculated in two stages. First, a truncated unweighted cosine transform of the acv.f. of the data is taken to give

$$\widehat{f}_1(\omega) = \frac{1}{\pi}\left(c_0 + 2\sum_{k=1}^{M} c_k \cos \omega k\right). \tag{7.21}$$

This is the same as Equation (7.20) except that the lag window is taken to be unity (i.e. $\lambda_k = 1$). The estimates given by Equation (7.21) are calculated at $\omega = \pi j/M$ for $j = 0, 1, \ldots, M$. These estimates are then smoothed using the weights $(\frac{1}{4}, \frac{1}{2}, \frac{1}{4})$ to give the Hanning estimates

$$\widehat{f}(\omega) = \frac{1}{4}\widehat{f}_1(\omega - \pi/M) + \frac{1}{2}\widehat{f}_1(\omega) + \frac{1}{4}\widehat{f}_1(\omega + \pi/M) \tag{7.22}$$

at $\omega = \pi j/M$ for $j = 1, 2, \ldots, (M-1)$. At zero frequency, and at the Nyquist frequency $\pi$, we take

$$\begin{aligned} \widehat{f}(0) &= \frac{1}{2}[\widehat{f}_1(0) + \widehat{f}_1(\pi/M)] \\ \widehat{f}(\pi) &= \frac{1}{2}[\widehat{f}_1(\pi) + \widehat{f}_1(\pi(M-1)/M)]. \end{aligned}$$

It can easily be shown algebraically that this procedure is equivalent to the use of the Tukey window. Substituting Equation (7.21) into (7.22) we find

$$\widehat{f}(\omega) = \frac{1}{\pi}\left\{c_o + 2\sum_{k=1}^{M} c_k\left[\frac{1}{4}\cos(\omega - \pi/M)k + \frac{1}{2}\cos \omega k + \frac{1}{4}\cos(\omega + \pi/M)k\right]\right\}$$

and, using $\cos(\omega - \pi/M)k + \cos(\omega + \pi/M)k = 2\cos \omega k \cos(\pi k/M)$, a comparison with Equation (7.20) shows that the lag window is indeed the Tukey window.

There is relatively little difference in the computational efficiency of Hanning and the straightforward use of the Tukey window. Both methods should yield the same estimates and so it does not matter which of the two procedures is used in practice.

### 7.4.3 *Hamming*

This technique is very similar to Hanning and has a very similar title, which sometimes leads to confusion. In fact Hamming is named after a quite different person, namely R.W. Hamming. The technique is nearly identical to Hanning except that the weights $(\frac{1}{4}, \frac{1}{2}, \frac{1}{4})$ in Equation (7.22) are changed to (0.23, 0.54, 0.23). At the frequencies $\omega = 0$ and $\omega = \pi$, the weights are 0.54 (at the 'end' frequency) and 0.46. The procedure gives similar estimates to those produced by Hanning.

### 7.4.4 *Smoothing the periodogram*

The methods of Sections 7.4.1–7.4.3 are based on transforming the truncated sample acv.f. An alternative type of approach is to smooth the periodogram ordinates in some way, the simplest approach being to group the periodogram ordinates in sets of size $m$ and find their average value. The latter approach is based on a suggestion made by P.J. Daniell as long ago as 1946. However, the use of lag window estimators was standard for many years because less computation was involved. Nowadays, some form of smoothed periodogram is used much more widely, particularly with the advent of the fast Fourier transform — see Section 7.4.5.

The basic idea of the simple smoothed periodogram can be expressed in the following formula:

$$\widehat{f}(\omega) = \frac{1}{m}\sum_j I(\omega_j) \tag{7.23}$$

where $\omega_j = 2\pi j/N$ and $j$ varies over $m$ consecutive integers so that the $\omega_j$ are symmetric about the frequency of interest, namely, $\omega$. In order to estimate $f(\omega)$ at the end-points $\omega = 0$ and $\omega = \pi$, Equation (7.23) has to be modified in an obvious way, treating the periodogram as being symmetric about 0 and $\pi$. Then, taking $m$ to be odd with $m^* = (m-1)/2$, we have

$$\begin{aligned}
\widehat{f}(0) &= I(0) + 2\sum_{j=1}^{m^*} I(2\pi j/N)/m \\
\widehat{f}(\pi) &= \left[I(\pi) + 2\sum_{j=1}^{m^*} I(\pi - 2\pi j/N)\right]\Big/ m.
\end{aligned}$$

The expression for $\widehat{f}(0)$ can be simplified as the first term $I(0)$ is zero.

Now we know that the periodogram is asymptotically unbiased but inconsistent for the true spectrum. Since neighbouring periodogram ordinates are asymptotically uncorrelated, it is clear that the variance of Equation (7.23) will be of order $1/m$. It is also clear that the estimator in Equation (7.23) may be biased since

$$E[\widehat{f}(\omega)] \simeq \frac{1}{m}\sum_j f(\omega_j),$$

which is only equal to $f(\omega)$ if the spectrum is linear over the relevant interval. However, the bias will be 'small' provided that $f(\omega)$ is a reasonably smooth function and $m$ is not too large compared with $N$.

The consequence of the above remarks is that the choice of group size $m$ is rather like the choice of the truncation point $M$ in Section 7.4.1 in that it has to be chosen so as to balance resolution against variance. However, the choice is different in that changes in $m$ and in $M$ act in opposite directions. An increase in $m$ has a similar effect to a reduction in $M$. The larger the value of $m$ the smaller will be the variance of the resulting estimate but the larger will be the bias. If $m$ is too large, then interesting features of $f(\omega)$, such as peaks, may be smoothed out. Of course, as $N$ increases, we can in turn allow $m$ to increase, just as we allowed $M$ to increase with $N$ in Section 7.4.1.

There is relatively little advice in the literature on the choice of $m$. As in Section 7.4.1, it seems advisable to try several values for $m$. A 'high' value should give some idea as to whether large peaks in $f(\omega)$ exist, but the curve is likely to be too smooth and some real peaks may be hidden. A 'low' value is likely to produce a much more uneven curve showing many peaks, some of which will be spurious. A compromise between the effects of bias and variance can then be made. A rule of thumb is to choose values of $m$ near $2\sqrt{N}$.

Although the procedure described in this section is computationally quite different from that of Section 7.4.1, there are in fact close theoretical links between the two procedures. In Section 7.3.1 we derived the relationship between the periodogram and the sample acv.f., and, if we substitute Equation (7.17) into (7.23), we can express the smoothed periodogram estimate of the spectrum in terms of the sample acv.f. in a similar form to Equation (7.20). After some algebra (Exercise 7.5), it can be shown that the truncation point is $(N-1)$ and the lag window is given by

$$\lambda_k = \begin{cases} 1 & k=0 \\ \sin(m\pi k/N)/[m\sin(\pi k/N)] & k=1,2,\ldots,N-1. \end{cases}$$

Thus, the formula uses values of $c_k$ right up to $k=(N-1)$, rather than having a truncation point much lower than $N$. Moreover, the lag window has the undesirable property that it does not tend to zero as $k$ tends to $N$. The smoothed periodogram effectively uses a rectangular window in the frequency domain and the resulting lag window shows that a sudden cut-off in the frequency domain can give rise to 'nasty' effects in the time domain (and vice versa). The smoothed periodogram often works reasonably well, but its window properties suggest that it may be possible to find a way of smoothing the periodogram, using a non-uniform averaging procedure, that has better time-domain properties. In fact, the simple smoothed periodogram is rarely used today, but rather a windowed form of averaging is used instead. Various alternative smoothing procedures have been suggested, with the idea of giving more weight to the periodogram ordinate at the particular frequency of interest and progressively less weight to periodogram ordinates further away. The analyst can think of this as applying a window in the frequency domain rather than in the time domain, but in a way that corresponds to the use of a lag window as in Section 7.4.1. It is possible to use a triangular (Bartlett window) or aim for a bell-shaped curve, perhaps by applying a simple smoother, such as Hanning, more than once. These approaches will not be considered here and the reader is referred, for example, to Hayes (1996, Chapter 8) or Bloomfield (2000, Chapter 8).

Historically, the smoothed periodogram was not much used until the 1990s because it apparently requires much more computational effort than transforming the truncated acv.f. Calculating the periodogram using Equation (7.16) at $\omega_p$ for $p = 1,2,\ldots,N/2$ would require about $N^2$ arithmetic operations (each one a multiplication and an addition), whereas using Equation (7.20) fewer than $MN$ operations are required to calculate the $\{c_k\}$ so that the total number of operations is only of order $M(N+M)$ if the spectrum is evaluated at $M$ frequencies. Two factors have led to the increasing use of the smoothed periodogram. First, the advent of high-speed computers means that it is unnecessary to restrict attention to the method requiring fewest calculations. The second factor has been the widespread use of an algorithm called the **fast Fourier transform**, which makes it much quicker to compute the periodogram. This procedure will now be described.

### *7.4.5 The fast Fourier transform (FFT)*

The computational procedure described in this section is usually abbreviated to FFT[5] and we adopt this abbreviation. For long series, the technique can substantially reduce the time required to perform a Fourier analysis of a set of data on a computer, and can also give more accurate results.

The history of the FFT dates back to the early 1900s. However, it was the work of J.W. Cooley, J.W. Tukey and G. Sande in about 1965 coupled with the arrival of faster computers that stimulated the application of the technique to time-series analysis. Much of the early work was published in the various Transactions of the IEEE, but more recent coverage is given, for example, by Bendat and Piersol (2000), Bloomfield (2000) and Priestley (1981). We will only give a broad outline of the technique here.

The FFT requires that the value of $N$ should be composite, meaning that $N$ is not a prime number and so can be factorized. The basic idea of the FFT will be illustrated for the case when $N$ can be factorized in the form $N = rs$, where $r$ and $s$ are integers. If we assume that $N$ is even, then at least one of the factors, say $r$, will be even. Using complex numbers for mathematical simplicity, the Fourier coefficients from Equation (7.9) can be expressed in the form

$$a_p + \mathrm{i}b_p = 2\left[\sum x_t\, \mathrm{e}^{2\pi \mathrm{i}pt/N}\right]/N \tag{7.24}$$

for $p = 0,1,2,\ldots,(N/2)-1$. For mathematical convenience, we denote the observations by $x_0, x_1, \ldots, x_{N-1}$, so that the summation in Equation (7.24) is from $t = 0$ to $N-1$. Now we can write $t$ in the form

$$t = rt_1 + t_0$$

where $t_1 = 0,1\ldots,s-1$, and $t_0 = 0,1,\ldots,r-1$, as $t$ goes from 0 to $N-1$, in view of the fact that $N = rs$. Similarly we can decompose $p$ in the form

$$p = sp_1 + p_0$$

[5] Some authors have used this abbreviation to denote the *finite Fourier transform*.

where $p_1 = 0, 1, \ldots, (r/2) - 1$, and $p_0 = 0, 1, \ldots, s-1$, as $p$ goes from 0 to $(N/2) - 1$. Then the summation in Equation (7.24) may be written

$$\sum_{t_0=0}^{r-1} e^{2\pi i p t_0/N} \sum_{t_1=0}^{s-1} x_t \, e^{2\pi i p r t_1/N}.$$

However,

$$e^{2\pi i p r t_1/N} = e^{2\pi i (s p_1 + p_0) r t_1/N} = e^{2\pi i p_0 r t_1/N}$$

since $e^{2\pi i s p_1 r t_1/N} = e^{2\pi i p_1 t_1} = 1$ for all $p_1, t_1$. Thus $\sum_{t_1=0}^{s-1} x_t \, e^{2\pi i p r t_1/N}$ does not depend on $p_1$ and is therefore a function of $t_0$ and $p_0$ only, say $A(p_0, t_0)$. Then Equation (7.24) may be written

$$a_p + ib_p = 2\left[\sum_{t_0=0}^{r-1} A(p_0, t_0) \, e^{2\pi i p t_0/N}\right] \Big/ N$$

Now there are $N = rs$ functions of type $A(p_0, t_0)$ to be calculated, each requiring $s$ complex multiplications and additions. There are $N/2$ values of $(a_p + ib_p)$ to be calculated, each requiring $r$ further complex multiplications and additions. This gives a grand total of $Ns + \frac{N}{2}r = N(s + r/2)$ calculations instead of the $N \times N/2 = N^2/2$ calculations required to use Equation (7.24) directly. By a suitable choice of $s$ and $r$, we can usually arrange for $(s + r/2)$ to be (much) less than $N/2$.

Much bigger reductions in computing can be made by an extension of the above procedure when $N$ is highly composite (i.e. has many small factors). In particular, if $N$ is of the form $2^k$, then we find that the number of operations is of order $Nk$ (or $N\log_2 N$) instead of $N^2/2$. Substantial gains can also be made when $N$ has several factors (e.g. $N = 2^p 3^q 5^r \ldots$).

In practice it is unlikely that $N$ will naturally be of a simple form such as $2^k$, unless the value of $N$ can be chosen before measurement starts. However, there are other things we can do. It may be possible to make $N$ highly composite by the simple expedient of omitting a few observations from the beginning or end of the series. For example, with 270 observations, we can omit the last 14 to make $N = 256 = 2^8$. More generally we can *increase* the length of the series by adding zeros to the (mean-corrected) sample record until the value of the revised $N$ becomes a suitable integer. Then a procedure called **tapering** or **data windowing** (e.g. Priestley, 1981) is often recommended[6] to avoid a discontinuity at the end of the data. Suppose, for example, that we happen to have 382 observations. This value of $N$ is *not* highly composite and we might proceed as follows:

- Remove any linear trend from the data, and keep the residuals (which should have mean zero) for subsequent analysis. If there is no trend, simply subtract the overall mean from each observation.

[6] Note that some researchers view tapering with suspicion as the data are modified — see, for example, the discussion in (Percival, 1993, p. 215).

- Apply a linear taper to about 5% of the data at each end. In this example, if we denote the detrended mean-corrected data by $x_0, x_1, \ldots, x_{381}$, then the tapered series is given by

$$x_t^* = \begin{cases} (t+1)x_t/20 & t = 0, 1, \ldots, 18, \\ (382-t)x_t/20 & t = 363, \ldots, 381, \\ x_t & t = 19, 20, \ldots, 362. \end{cases}$$

- Add $512 - 382 = 130$ zeros at one end of the tapered series, so that $N = 512 = 2^9$.
- Carry out an FFT on the data, calculate the Fourier coefficients $a_p + ib_p$ and average the values of $(a_p^2 + b_p^2)$ in groups of about 10.

In fact with $N$ as low as 382, the computational advantage of the FFT is limited and we could equally well calculate the periodogram directly, which avoids the need for tapering and adding zeros. The FFT really comes into its own when there are several thousand observations.

It is also worth explaining that the FFT is still useful when the analyst prefers to look at the autocorrelation function (ac.f.) *before* carrying out a spectral analysis, either because inspecting the ac.f. is thought to be an invaluable preliminary exercise or because the analyst prefers to transform the truncated weighted acv.f. rather than smooth the periodogram. It can be quicker to calculate the sample acv.f. by performing *two* FFTs (e.g. Priestley, 1981, Section 7.6), rather than directly as a sum of lagged products. The procedure is as follows. Compute the Fourier coefficients $(a_p, b_p)$ with an FFT of the mean-corrected data at $\omega_p = 2\pi p/N$ for $p = 0, 1, \ldots, N-1$ rather than for $p = 0, 1, \ldots, N/2$ as we usually do. The extra coefficients are normally redundant for real-valued processes since $a_{N-k} = a_k$ and $b_{N-k} = -b_k$. However, for calculating the autocovariances, we can compute $R_p^2 = a_p^2 + b_p^2$ at these values of $p$ and then fast Fourier *re*transform the sequence $(R_p^2)$ to get the mean lagged products. We will not give the algebra here. For several thousand observations, this can be much faster than calculating them directly. However, when using the FFT in this way, the analyst should take care to add enough zeros to the data (*without* tapering) to make sure that **non-circular** sums of lagged products are calculated, as defined by Equation (4.1) and used throughout this book. **Circular** coefficients result if zeros are not added where, for example, the circular autocovariance coefficient at lag 1 is

$$c_1^* = \left[ \sum_{t=1}^{N} (x_t - \bar{x})(x_{t+1} - \bar{x}) \right] \Big/ N.$$

where $x_{N+1}$ is taken to be equal to $x_1$ to make the series 'circular'. Note that, if $x_1 = \bar{x}$, then the circular and non-circular coefficients at lag 1 are the same. If we use mean-corrected data, which will have mean zero, then adding zeros will make circular and non-circular coefficients be the same. In order to calculate all the non-circular autocovariance coefficients of a set of $N$ mean-corrected observations, the analyst should add $N$ zeros, to make $2N$ 'observations' in all.

## 7.5 Confidence Intervals for the Spectrum

The methods of Section 7.4 all produce *point* estimates of the spectral density function, and hence give no indication of their likely accuracy. This section shows how to find appropriate confidence intervals.

In Section 7.3.2, we showed that data from a white noise process, with constant spectrum $f(\omega) = \sigma^2/\pi$, yields a periodogram ordinate $I(\omega)$ at frequency $\omega$, which is such that $2I(\omega)/f(\omega)$ is distributed as $\chi_2^2$. Note that this distribution does not depend on $N$, which explains why $I(\omega)$ is not a consistent estimator for $f(\omega)$. Wide confidence intervals would result if $I(\omega)$ was used as an estimator. Suppose instead that we use the estimator of Section 7.4.1, namely

$$\hat{f}(\omega) = \left[ \sum_{k=-M}^{M} \lambda_k c_k \cos \omega k \right] \bigg/ \pi.$$

Then, it can be shown (Jenkins and Watts, 1968, Section 6.4.2) that $\nu \hat{f}(\omega)/f(\omega)$ is asymptotically distributed as an approximate $\chi_\nu^2$ random variable, where

$$\nu = 2N \bigg/ \sum_{k=-M}^{M} \lambda_k^2 \tag{7.25}$$

is called the number of *degrees of freedom* of the lag window. It follows that

$$P(\chi^2_{\nu,1-\alpha/2} < \nu \hat{f}(\omega)/f(\omega) < \chi^2_{\nu,\alpha/2}) = 1 - \alpha,$$

so that a $100(1-\alpha)\%$ confidence interval for $f(\omega)$ is given by

$$\frac{\nu \hat{f}(\omega)}{\chi^2_{\nu,\alpha/2}} \quad \text{to} \quad \frac{\nu \hat{f}(\omega)}{\chi^2_{\nu,1-\alpha/2}}.$$

Some simple algebra shows that the degrees of freedom for the Tukey and Parzen windows turn out to be $2.67N/M$ and $3.71N/M$, respectively. Although relying on asymptotic results, Neave (1972a) has shown that the above formulae are also quite accurate for short series.

For the smoothed periodogram estimator of Section 7.4.4, there is no need to apply Equation (7.25), because smoothing the periodogram in groups of size $m$ is effectively the same as averaging independent $\chi_2^2$ random variables. Thus, it is clear that the smoothed periodogram will have $\nu = 2m$ degrees of freedom, and we can then apply the same formula for the confidence interval as given above.

## 7.6 Comparison of Different Estimation Procedures

Several factors need to be considered when comparing the different estimation procedures that were introduced in Section 7.4. Although we concentrate on the theoretical properties of the different procedures, the analyst will also need to consider practical questions such as computing time and the availability of suitable computer software.

Alternative comparative discussions are given by Jenkins and Watts (1968), Neave (1972), Priestley (1981, Section 7.5) and Bloomfield (2000).

It is helpful to introduce a function called the **spectral window** or **kernel**, which is defined to be the Fourier transform of the lag window $\{\lambda_k\}$ introduced in Equation (7.20). Assuming that $\lambda_k$ is zero for $k > M$, and symmetric, so that $\lambda_{-k} = \lambda_k$, then the spectral window is defined by

$$K(\omega) = \frac{1}{2\pi} \sum_{k=-\infty}^{\infty} \lambda_k \mathrm{e}^{-ik\omega} \tag{7.26}$$

for $(-\pi < \omega < \pi)$[7]. The corresponding inverse Fourier transform is given by

$$\lambda_k = \int_{-\pi}^{\pi} K(\omega) \mathrm{e}^{\mathrm{i}\omega k} \mathrm{d}\omega. \tag{7.27}$$

All the estimation procedures for the spectrum that we have studied so far can be put in the general form

$$\begin{aligned}
\widehat{f}(\omega_0) &= \frac{1}{\pi} \sum_{k=-N+1}^{N-1} \lambda_k c_k \mathrm{e}^{-\mathrm{i}\omega_0 k} \\
&= \frac{1}{\pi} \sum \left[ \int_{-\pi}^{\pi} K(\omega) \mathrm{e}^{\mathrm{i}\omega k} \mathrm{d}\omega \right] c_k \mathrm{e}^{-\mathrm{i}\omega_0 k} \\
&= \frac{1}{\pi} \int_{-\pi}^{\pi} K(\omega) \left[ \sum c_k \mathrm{e}^{\mathrm{i}k(\omega-\omega_0)} \right] \mathrm{d}\omega \\
&= \int_{-\pi}^{\pi} K(\omega) I(\omega_0 - \omega) \mathrm{d}\omega
\end{aligned} \tag{7.28}$$

using Equation (7.18). Equation (7.28) shows that all the estimation procedures are essentially smoothing the periodogram using the weight function $K(\omega)$. The value of the lag window at lag zero is usually specified to be one, so that from Equation (7.27) we have

$$\lambda_0 = 1 = \int_{-\pi}^{\pi} K(\omega) \mathrm{d}\omega$$

which is a desirable property for a smoothing function.

Taking expectations in Equation (7.28) we have asymptotically that

$$E[\widehat{f}(\omega_0)] = \int_{-\pi}^{\pi} K(\omega) f(\omega_0 - \omega) \mathrm{d}\omega. \tag{7.29}$$

Thus the spectral window is a weight function expressing the contribution of the spectral density function at each frequency to the expectation of $\widehat{f}(\omega_0)$. The name 'window' arises from the fact that $K(\omega)$ determines the part of the periodogram that is 'seen' by the estimator.

[7]Note that we cannot avoid negative frequencies here, as the spectral window looks at differences in frequency from some specified frequency. If the lag window is symmetric about zero, then so is $K(\omega)$.

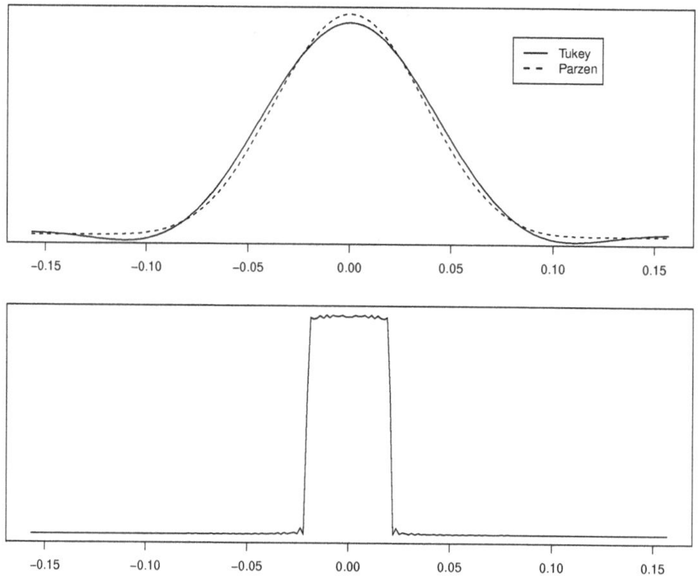


Figure 7.2 *The spectral windows for three common methods of spectral analysis. Top: Parzen ($M = 93$), Tukey ($M = 67$); Bottom: Smoothed periodogram $N = 1000, m = 20$.*

Examples of the spectral windows for three common methods of spectral analysis are shown in Figure 7.2. Taking $N = 1000$, the spectral window for the smoothed periodogram with $m = 20$ is shown in the bottom panel of Figure 7.2. The other two windows are the Parzen and Tukey windows, shown as dashed and solid lines in the top panel. The values of the truncation point $M$ were chosen to be 93 for the Parzen window and 67 for the Tukey window. These values of $M$ were chosen so that the Parzen and Tukey windows gave estimators with equal variance. Formulae for variances will be given later in this section.

Inspecting Figure 7.2, we see that the Parzen and Tukey windows look very similar, although the Parzen window has the advantage of being non-negative and of having smaller side lobes. The shape of the periodogram window is quite different. It is approximately rectangular with a sharp cut-off and is close to the 'ideal' band-pass filter, which would be exactly rectangular but which is unattainable in practice. The periodogram window also has the advantage of being non-negative.

In comparing different windows, we should consider both the bias and the variance of the estimator. This is sometimes called the *variance-bias trade-off* question, as well as *balancing resolution against variance*. By taking a wider window, we generally get a lower variance but a larger bias and some sort of compromise has to be made in practice. This is often achieved by using trial and error, as, for example, in the choice of the truncation point for a lag window as discussed earlier in Section 7.4.1. It is not easy to get general formulae for the bias produced by the different procedures. However, it is intuitively clear from Equation (7.29) and from earlier remarks that the wider the window, the larger will be the bias. In particular, it is clear that all the smoothing procedures will tend to lower peaks and raise troughs.

As regards variance, we noted in Section 7.5 that $\nu\widehat{f}(\omega)/f(\omega)$ is approximately distributed as $\chi^2_\nu$, where $\nu = 2m$, for the smoothed periodogram, and, using Equation (7.25), $3.71N/M$ and $8N/3M$ for the Parzen and Tukey windows, respectively. Since

$$\text{Var}(\chi^2_\nu) = 2\nu$$

and

$$\text{Var}[\nu\widehat{f}(\omega)/f(\omega)] = \nu^2\text{Var}[\widehat{f}(\omega)/f(\omega)]$$

we find $\text{Var}[\widehat{f}(\omega)/f(\omega)]$ turns out to be $1/m, 2M/3.71N$, and $3M/4N$, respectively, for the three windows. Equating these expressions gives the values of $M$ chosen for Figure 7.2.

When comparing the different estimators, the notion of a **bandwidth** may be helpful. Roughly speaking, the bandwidth is the width of the spectral window, as might be expected. Various formal definitions are given in the literature, but we adopt the one given by Jenkins and Watts (1968), namely, the width of the 'ideal' rectangular window that would give an estimator with the same variance. The window of the smoothed periodogram is so close to being rectangular for $m$ 'large' that it is clear from Figure 7.2 that the bandwidth will be approximately $2m\pi/N$ (as area must be unity and height is $N/2m\pi$). The bandwidths for the Bartlett, Parzen and Tukey windows turn out to be $3/2M$, $2\pi(1.86/M)$ and $8\pi/3M$, respectively. When plotting a graph of an estimated spectrum, it is a good idea to indicate the bandwidth that has been used.

The choice of bandwidth is equivalent to the choice of $m$ or $M$, depending on the method used. This choice is an important step in spectral analysis, though it is important to remember that the effects of changing $m$ and $M$ act in opposite directions. For the Bartlett, Parzen and Tukey windows, the bandwidth is inversely proportional to $M$. Figure 7.3 shows how the window changes as $M$ varies, using the Bartlett window as a representative example. As $M$ gets larger, the window gets narrower, the bias gets smaller but the variance of the resulting estimator gets larger. For the smoothed periodogram, the reverse happens. The bandwidth is directly proportional to $m$, and as $m$ gets larger, the window gets wider, the bias increases but the variance reduces. For the unsmoothed periodogram, with $m = 1$, the window is very tall and narrow giving an estimator with large variance as we have already shown. All in all, the choice of bandwidth is rather like the choice of class interval when constructing a histogram.

We are now in a position to give guidance on the relative merits of the different estimation procedures. As regards theoretical properties, it is arguable that the smoothed periodogram has the better-shaped spectral window in that it is approximately rectangular, although there are some side lobes. For the transformed acv.f., the Parzen and Tukey windows are preferred to the Bartlett window. Computationally, the smoothed periodogram can be much slower for large $N$ unless the FFT is used. However, if the FFT *is* used, then the smoothed periodogram can be faster. Moreover it is possible to calculate the ac.f. quickly using *two* FFTs. One drawback

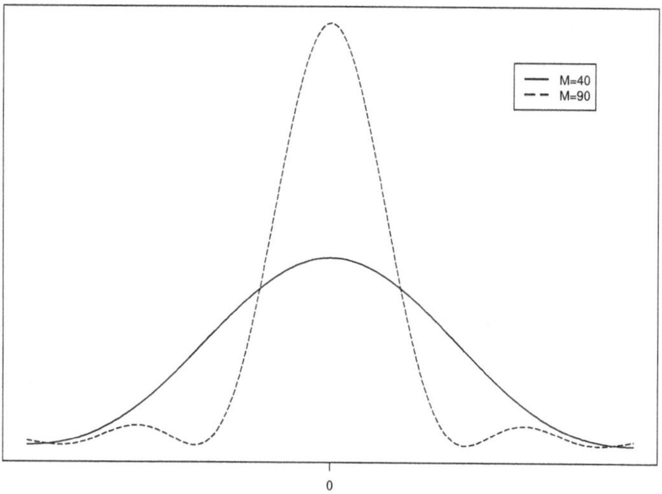


Figure 7.3 *The Bartlett spectral window for different values of M.*

to the use of the FFT is that it may require data-tapering, whose use is still somewhat controversial. Of course, for small $N$, computing time is a relatively unimportant consideration. As regards computer software, it is much easier to write a program for the Parzen or Tukey windows, but programs and algorithms for the FFT are becoming readily available. Thus the use of the smoothed periodogram has become more general, either with equal weights as discussed in Section 7.4.4, or with a more 'bell-shaped' set of weights.

All the above assumes that a *non-parametric* approach is used in that no model fitting is attempted prior to carrying out a spectral analysis. As noted earlier, an alternative approach gaining ground is to use a *parametric* approach, fitting an autoregressive (AR) or ARMA model to the data. The spectrum of the fitted model is then used to estimate the spectrum. This approach will be described later in Section 13.7.1.

## 7.7 Analysing a Continuous Time Series

Up to now, we have been concerned with the spectral analysis of time series recorded at *discrete* time intervals. However, time series are sometimes recorded as a *continuous trace*. For example, variables such as air temperature, humidity and the moisture content of tobacco emerging from a processing plant are often recorded by machines that give continuous-time readings. For series that contain components at very high frequencies, such as those arising in acoustics and speech processing, it may be possible to analyze such records mechanically using tuned filters, but the more usual procedure is to **digitize** the series by reading off the values of the trace at discrete intervals. If values are taken at equal time intervals of length $\Delta t$, we have converted a continuous time series into a standard discrete-time time series and can use the methods already described.

In sampling a continuous time series, the main question is how to choose the sampling interval $\Delta t$. It is clear that sampling leads to some loss of information and that this loss gets worse as $\Delta t$ increases. However, sampling costs increase as $\Delta t$ gets smaller and so a compromise value must be sought.

For the sampled series, the Nyquist frequency is $\pi/\Delta t$ radians per unit time, and we can get no information about variation at higher frequencies. Thus we clearly want to choose $\Delta t$ so that variation in the continuous series is negligible at frequencies higher than $\pi/\Delta t$. In fact most measuring instruments are **band-limited** in that they do not respond to frequencies higher than a certain maximum frequency. If this maximum frequency, say $\omega_{max}$, is known or can be guessed, then the choice of $\Delta t$ is straightforward in that it should be less than $\pi/\omega_{max}$. However, if $\Delta t$ is chosen to be too large, then a phenomenon called **aliasing** may occur. This can be illustrated by the following theorem.

*Theorem 7.1* Suppose that a continuous time series, with spectrum $f_c(\omega)$ for $0 < \omega < \infty$, is sampled at equal time intervals of length $\Delta t$. The resulting discrete time series will have a somewhat different spectrum, say $f_d(\omega)$ defined over $0 < \omega < \pi/\Delta t$. We will see that the two spectra will only be equal if $f_c(\omega)$ is zero for $\omega > \pi/\Delta t$. More generally, it can be shown that $f_d(\omega)$ and $f_c(\omega)$ are related by

$$f_d(\omega) = \sum_{s=0}^{\infty} f_c(\omega + 2\pi s/\Delta t) + \sum_{s=1}^{\infty} f_c(-\omega + 2\pi s/\Delta t). \tag{7.30}$$

*Proof* The proof will be given for the case $\Delta t = 1$. The extension to other values of $\Delta t$ is straightforward. Suppose that the acv.f.s of the continuous and sampled series are given by $\gamma(\tau)$ and $\gamma_k$, respectively. Here $\gamma(\tau)$ is defined for all $\tau$, while $\gamma_k$ is only defined for integer $k$. Of course if $\tau$ takes an integer value, say $k$, then the two functions are equal as in

$$\gamma(k) = \gamma_k. \tag{7.31}$$

Now from Equation (6.18) we have

$$\gamma(\tau) = \int_0^{\infty} f_c(\omega) \cos \omega\tau \, d\omega,$$

while, from Equation (6.9), we have

$$\gamma_k = \int_0^{\pi} f_d(\omega) \cos \omega k \, d\omega.$$

Thus, using Equation (7.31), we have

$$\int_0^{\pi} f_d(\omega) \cos \omega k \, d\omega = \int_0^{\infty} f_c(\omega) \cos \omega k \, d\omega$$

for $k = 0, \pm 1, \pm 2, \ldots$. The next step is to split the infinite integral into sections of length $2\pi$, and then of $\pi$, using $\cos \omega k = \cos(\omega + 2\pi s)k = \cos(2\pi s - \omega)k$ for all integers $s$. We get

$$\begin{aligned}
\int_0^\infty f_c(\omega) \cos \omega k \, d\omega &= \sum_{s=0}^{\infty} \int_{2\pi s}^{2\pi(s+1)} f_c(\omega) \cos \omega k \, d\omega \\
&= \sum_{s=0}^{\infty} \int_0^{2\pi} f_c(\omega + 2\pi s) \cos \omega k \, d\omega \\
&= \sum_{s=0}^{\infty} \int_0^{\pi} \{ f_c(\omega + 2\pi s) + f_c[2\pi(s+1) - \omega] \} \cos \omega k \, d\omega \\
&= \int_0^{\pi} \left\{ \sum_{s=0}^{\infty} f_c(\omega + 2\pi s) + \sum_{s=1}^{\infty} f_c(2\pi s - \omega) \right\} \cos \omega k \, d\omega
\end{aligned}$$

and the result follows. □

The implications of this theorem may now be considered. First, as noted earlier, if the continuous series contains no variation at frequencies above the Nyquist frequency, so that $f_c(\omega) = 0$ for $\omega > \pi/\Delta t$, then $f_d(\omega) = f_c(\omega)$. In this case no information is lost by sampling. However, the more general result is that sampling *will* have an effect in that variation at frequencies above the Nyquist frequency in the continuous series will be 'folded back' to produce apparent variation in the sampled series at a frequency lower than the Nyquist frequency. If we denote the Nyquist frequency $\pi/\Delta t$ by $\omega_N$, then the frequencies $\omega, 2\omega_N - \omega, 2\omega_N + \omega, 4\omega_N - \omega, \ldots$ are called **aliases** of one another. Variation at all these frequencies in the continuous series will appear as variation at frequency $\omega$ in the sampled series.

From a practical point of view, aliasing will cause trouble unless $\Delta t$ is chosen to be sufficiently small so that $f_c(\omega) \simeq 0$ for $\omega > \pi/\Delta t$. If we have no advance knowledge about $f_c(\omega)$, then we have to guesstimate a value for $\Delta t$. If the resulting estimate of $f_d(\omega)$ approaches zero near the Nyquist frequency $\pi/\Delta t$, then our choice of $\Delta t$ is almost certainly sufficiently small. However, if $f_d(\omega)$ does not approach zero near the Nyquist frequency, then it is probably wise to try a smaller value of $\Delta t$. Alternatively, if the analyst is only interested in the low-frequency components, then it may be easier to filter the continuous series so as to remove the high-frequency components and remove the need for selecting a small value of $\Delta t$.

## 7.8 Examples and Discussion

Spectral analysis can be a useful exploratory diagnostic tool in the analysis of many types of time series. With the aid of examples, this section discusses how to interpret an estimated spectrum, and tries to indicate when spectral analysis is likely to be most useful and when it is likely to be unhelpful. We also discuss some of the practical problems arising in spectral analysis.

We begin with an example to give the reader some 'feel' for the sorts of spectrum shapes that may arise. Figure 7.4 shows three series and their corresponding spectra. In the top panel, the series shows cyclic behaviours over different periods. For

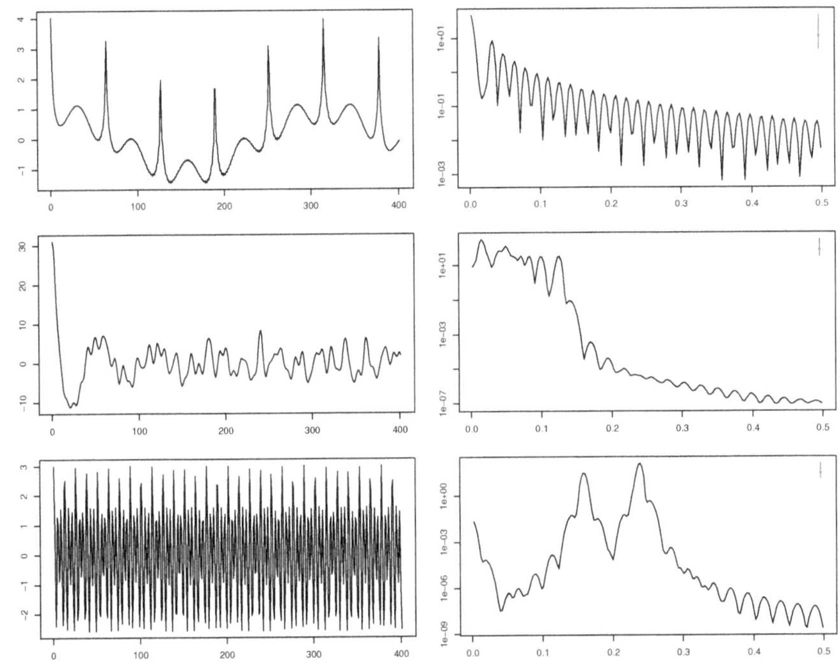

Figure 7.4 *Three time series (Left) and their spectra (Right).*

example, the series shows peaks for a period around 60, and it also has fluctuations over the whole sample period, indicating the spectrum is concentrated on different frequencies. In the middle panel, the series has more oscillations than the first series. The largest variation is in the period $(1,50)$, which never occurs again afterwards. As this feature dominates other oscillations of the series, it indicates the spectrum has more concentration on the low frequency. In the bottom panel, the series shows strong seasonal effects, and the seasonal effect in one period seems been aggregated from oscillations over a different frequency. Actually, its spectrum on the right shows two modes, indicating the third series is essentially an aggregation of signals over two or several frequencies.

Figure 7.4 can be reproduced using the following R commands.

```
> x<-seq(0,400)
> for (i in 1:30)  y1<-y1+cos(x*(i+1)/10)/(i+1)
> for (i in 1:30)  y2<-y2+cos(pi*x/(i+3))
> y3<-cos(x)+2*cos(1.5*x)

> par(mfrow=c(3,2), mar=c(2,2,2,2))
> plot(y1, type="l");
> spec.pgram(y1,spans=c(3,3),main="")
> plot(y2, type="l");
> spec.pgram(y2,spans=c(3,3),main="")
```

```
> plot(y3, type="l");
> spec.pgram(y3,spans=c(3,3),main="")
```

*An example using air temperature*

The next example considers the air temperature series at Alaska in the United States, plotted in Figure 1.3. There the regular seasonal variation is quite obvious from a visual inspection of the time plot, but in this case the deterministic component accounts for about 94% of the total variation. If we nevertheless carry out a spectral analysis of the air temperature series, we get the spectrum shown in the top panel of Figure 7.5 with a large peak at a frequency of one cycle per year. However, it is arguable that the spectral analysis is not really necessary here, as the seasonal effect is so obvious anyway. In fact, if the analyst has a series containing an obvious trend or seasonal component, then it is advisable to remove such variation from the data **before** carrying out a spectral analysis, as any other effects will be relatively small and may not be visible in the spectrum of the raw data.

The middle panel of Figure 7.5 shows the spectrum of the Alaska air temperature data when the seasonal variation has been removed. The variance is concentrated at low frequencies, indicating either a trend, which is not apparent in Figure 1.3, or short-term correlation as in a first-order AR process with a positive coefficient (cf. Figure 6.4). The latter seems the more likely explanation here, given that there is no contextual reason to expect a trend in temperature (other than global warming, which is relatively small compared with other effects). As noted earlier, the corresponding periodogram in the bottom panel of Figure 7.5 shows a graph that oscillates up and down very quickly and is not helpful for interpreting the properties of the data. This demonstrates again that the periodogram has to be smoothed to get a consistent estimate of the underlying spectrum.

Removing trend and seasonality is a simple form of the general procedure usually called **prewhitening**. As the name suggests, this aims to construct a series having properties which are closer to those of white noise. In spectral analysis, this is useful because it is easier to estimate the spectrum of a series having a relatively flat spectrum, than one with sharp peaks and troughs. Prewhitening is often carried out by making a linear transformation of the raw data. Then the spectrum of the transformed series can be found, after which the spectrum of the original series can be found, if desired, by using the properties of the linear transformation[8] used to carry out the prewhitening. In spectral analysis, this procedure is often limited to removing trend and seasonality, though in other applications more sophisticated model fitting is often used[9].

Having estimated the spectrum of a given time series, how do we interpret the results? There are various features to look for. First, are there any peaks in the spectrum and, if so, at what frequency? Can we find a contextual reason for a peak at

[8]The frequency response function of the linear transformation is defined later in Chapter 9 and leads directly to the required spectrum.

[9]For example, with two time series it is advisable to prewhiten the series by removing as much autocorrelation as possible before calculating quantities called cross-correlations – see Chapter 8.

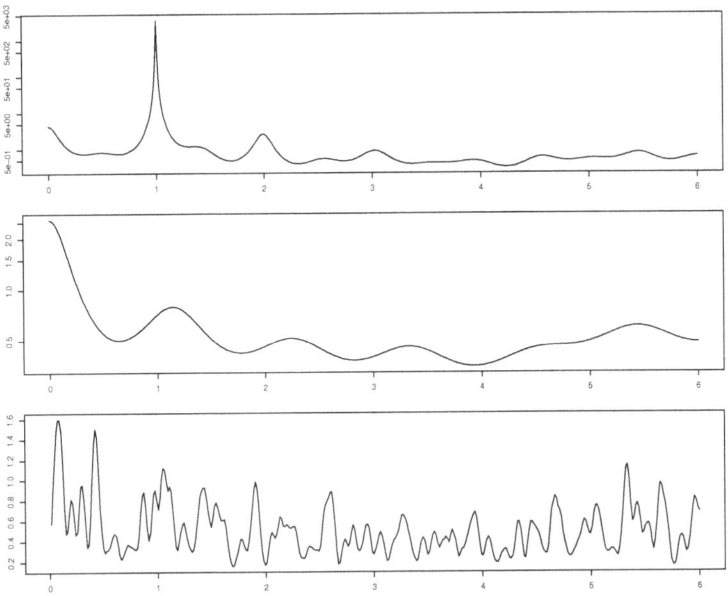

Figure 7.5 *Spectra for average monthly air temperature readings in Alaska. Top: The raw data; Middle: Spectrum for the seasonally adjusted data; Bottom: The periodogram of the seasonally adjusted data is shown for comparison.*

this frequency? Second, what is the general shape of the spectrum? In particular, does the spectrum get larger as the frequency tends to zero? This often happens with economic variables and indicates a business cycle with a very long period or an underlying long-term non-stationarity in the mean that has not been removed by prior filtering. Economists, who expect to find a clear peak at low frequency, will usually be disappointed, especially if looking for business cycles with a period around 5–7 years. There is usually little evidence of anything so clear-cut. Rather the low-frequency variation is typically spread over a range of frequencies.

The general shape of the spectrum could in principle be helpful in indicating an appropriate parametric model. For example, the shape of the spectrum of various ARMA models could be found and listed in a similar way to that used for specifying ac.f.s for different ARMA models. The use of the correlogram is a standard diagnostic tool in the Box–Jenkins procedure for identifying an appropriate ARIMA process, but the observed spectrum has rarely been used in this way. Why is this? Spectral analysis, as described in this chapter, is essentially a non-parametric procedure in which a finite set of observations is used to estimate a function defined over the whole range from $(0, \pi)$. The function is not constrained to any particular functional form and so one is effectively trying to estimate more items than in a correlogram analysis, where the analyst may only look at values for a few low lags. Being non-parametric, spectral analysis is in one sense more general than inference based on a particular parametric class of models, but the downside is that it is likely to be less accurate if a parametric model really is appropriate. In our experience, spectral analysis is typically used when there is a suspicion that cyclic variation may be present at some unknown frequency, and the spectrum shape is rarely used for diagnosing a parametric model.

Spectral analysis is arguably at its most useful for series that has no obvious trend or 'seasonal' variation. Such series arise mostly in the physical sciences. In economics, spectral techniques have perhaps not proved as useful as was first hoped, although there have been a few successes. Attempts have also been made to apply spectral analysis to marketing data, but it can be argued that marketing series are usually too short and the seasonal variation too large for spectral analysis to give useful results. In meteorology and oceanography, spectral analysis can be very useful (e.g. Craddock, 1965; Snodgrass et al., 1966) but, even in these sciences, spectral analysis may produce no worthwhile results, other than those that are obvious anyway. It is often the case that, once obvious cyclic effects have been removed (e.g. annual variation from monthly rainfall data; daily variation from hourly temperature data), the spectrum will show no clear peaks, but rather a tendency to get larger as the frequency tends to zero. The spectrum in Figure 7.5(b) is a case in point. The two examples in Percival (1993, Chapter 6), featuring ocean wave data and ice profile data, yield similar results. Sometimes a small peak is observed but tests usually show that this has dubious significance.

We conclude this section by commenting on some practical aspects of spectral analysis.

As most aspects, such as the choice of truncation point, have already been discussed, one problem that has not been discussed, is whether to plot the estimated spectrum on a linear or logarithm scale. An advantage of using a logarithmic scale is that the asymptotic variance of the estimated spectrum is then independent of the level of the spectrum, and so confidence intervals for the spectrum are of constant width on a logarithmic scale. For spectra showing large variations in power, a logarithmic scale also makes it possible to show more detail over a wide range. A similar idea is used by engineers when measuring sound in decibels, as the latter take values on a logarithmic scale. Jenkins and Watts (1968, p. 266) suggest that spectrum estimates should always be plotted on a logarithmic scale. However, Anderson (1971, p. 547) points out that this exaggerates the visual effects of variations where the spectrum is small. It may be easier to interpret a spectrum plotted on an arithmetic scale, as the area under the graph corresponds to power and this makes it easier to assess the relative importance of different peaks. Thus, while it is often useful to plot $\widehat{f}(\omega)$ on a logarithmic scale in the initial stages of a spectral analysis, especially when trying different truncation points and testing the significance of peaks, it is often better to plot the final version of the estimated spectrum on a linear scale in order to get a clearer interpretation of the final result.

It is also generally easier to interpret a spectrum if the frequency scale is measured in cycles per unit time ($f$) rather than radians per unit time ($\omega$). This has been done in Figures 7.4 and 7.5. A linear transformation of frequency does not affect the *relative* heights of the spectrum at different frequencies, though it does change the absolute heights by a constant multiple.

Another point worth mentioning is the possible presence in estimated spectra of **harmonics**. As noted earlier, when a series has a strong cyclic component at some frequency $\omega$, then the estimated spectrum may additionally show related peaks at $2\omega, 3\omega, \ldots$. These multiples of the fundamental frequency are called harmonics and generally speaking simply indicate that the main cyclical component is not exactly sinusoidal in character.

Finally, a question that is often asked is how large a value of $N$ is required to get a reasonable estimate of the spectrum. It is often recommended that between 100 and 200 observations is the minimum. With smaller values of $N$, only very large peaks can be found. However, if the data are prewhitened to make the spectrum fairly flat, then reasonable estimates may be obtained even with values of $N$ around 100, as we have shown in the middle panel of Figure 7.5. However, much longer series are to be preferred and are the norm when spectral analysis is contemplated.

## Exercises

**7.1** *Revision of Fourier series*. Show that the Fourier series, which represents the function

$$f(x) = x^2 \qquad \text{for } -\pi \le x \le \pi$$

is given by

$$f(x) = \frac{\pi^2}{3} - 4\left(\frac{\cos x}{1} - \frac{\cos 2x}{2^2} + \frac{\cos 3x}{3^2} - \cdots\right).$$

**7.2** Derive Equations (7.6) and (7.7).

**7.3** Derive Parseval's theorem, given by Equation (7.13).

**7.4** If $X_1, \ldots, X_N$ are independent $N(\mu, \sigma^2)$ variates show that

$$a_p = 2\left[\sum X_t \cos(2\pi pt/N)\right]/N$$

is $N(0, 2\sigma^2/N)$ for $p = 1, 2, \ldots, (N/2) - 1$.

**7.5** Derive the lag window for smoothing the periodogram in sets of size $m$. For algebraic simplicity take $m$ odd, with $m = 2m^* + 1$, so that

$$\widehat{f}(\omega_p) = \frac{1}{m} \sum_{j=-m^*}^{m^*} I\left(\omega_p + \frac{2\pi j}{N}\right).$$

(Hint: The answer is given in Section 7.4.4. The algebra is rather messy. Use Equation (7.17) and the following two trigonometric results:

$$\cos[2\pi k(p+j)/N] + \cos[2\pi k(p-j)/N] = 2\cos[2\pi kp/N]\cos[2\pi kj/N]$$

$$\sin A - \sin B = 2\sin[(A-B)/2]\cos[(A+B)/2]\,).$$

Chapter 8

# Bivariate Processes

Thus far, we have been concerned with analysing a single time series. We now turn our attention to the situation where we have observations on *two* time series and we are interested in the relationship between them.

We may distinguish two types of situations. First, we may have two series that arise 'on an equal footing'. For example, it is often of interest to analyze seismic signals received at two recording sites. Here, we are not usually interested in trying to predict one variable from the other, but rather are primarily interested in measuring the correlations between the two series. In the second type of situation, the two series are thought to be 'causally related', in that one series is regarded as being the **input** to some sort of processor or system, while the second series is regarded as the **output**; we are then interested in finding the properties of the system that converts the input into the output. The two types of situations are roughly speaking the time-series analogues of **correlation** and **regression**.

The first type of situation is considered in this chapter, where the cross-correlation function and the cross-spectrum are introduced, and again in Chapter 13, where vector ARMA models are introduced. The second type of situation is discussed in Chapter 9 where it is assumed that the system can be described as a **linear system**.

## 8.1 Cross-Covariance and Cross-Correlation

Suppose we make $N$ observations on two variables at unit time intervals over the same period and denote the observations by $(x_1, y_1), \ldots, (x_N, y_N)$. We assume that these observations may be regarded as a finite realization of a discrete-time bivariate stochastic process $(X_t, Y_t)$.

In order to describe a bivariate process it is useful to know the first- and second-order moments. For a univariate process, the first-order moment is the mean while the second-order moment is the autocovariance function (acv.f.), which includes the variance as a special case at lag zero. For a bivariate process, the moments up to second order consist of the mean and acv.f.s for each of the two components plus a new function, called the cross-covariance function. We will only consider bivariate processes that are *second-order stationary*, meaning that all moments up to second order do not change with time (as in the univariate case).


DOI: 10.1201/9781003646112-8

We use the following notation:

*Means.* $E(X_t) = \mu_X$; $E(Y_t) = \mu_Y$.
*Autocovariances.* $\text{Cov}(X_t, X_{t+k}) = \gamma_X(k)$; $\text{Cov}(Y_t, Y_{t+k}) = \gamma_Y(k)$.

Then the **cross-covariance function** is defined by

$$\text{Cov}(X_t, Y_{t+k}) = E[(X_t - \mu_X)(Y_{t+k} - \mu_Y) = \gamma_{XY}(k) \tag{8.1}$$

and is a function of the lag only, because the processes are assumed to be stationary.

Note that some authors define the cross-covariance function in the 'opposite direction' by

$$\text{Cov}(X_t, Y_{t-k}) = \gamma^*_{XY}(k).$$

Comparing with Equation (8.1) we see that

$$\gamma_{XY}(k) = \gamma^*_{XY}(-k).$$

It doesn't matter which definition is used as long as it is stated clearly and used consistently.

The cross-covariance function differs from the acv.f. in that it is *not* an even function, since in general

$$\gamma_{XY}(k) \neq \gamma_{XY}(-k).$$

Instead we have the relationship

$$\gamma_{XY}(k) = \gamma_{YX}(-k),$$

where the subscripts are reversed.

The size of the cross-covariance coefficients depends on the units in which $X_t$ and $Y_t$ are measured. Thus for interpretative purposes, it is useful to standardize the cross-covariance function to produce a function called the **cross-correlation function**, $\rho_{XY}(k)$, which is defined by

$$\rho_{XY}(k) = \gamma_{XY}(k)/\sqrt{\gamma_X(0)\gamma_Y(0)} = \gamma_{XY}(k)/\sigma_X\sigma_Y, \tag{8.2}$$

where $\sigma_X = \sqrt{\gamma_X(0)}$ denotes the standard deviation of the $X$-process, and similarly for $\sigma_Y$. This function measures the correlation between $X_t$ and $Y_{t+k}$ and has these two properties:

(1) $\rho_{XY}(k) = \rho_{YX}(-k)$ (Note the subscripts reverse in order.)

(2) $|\rho_{XY}(k)| \leq 1$ (See Exercise 8.2.)

Whereas $\rho_X(0)$, $\rho_Y(0)$ are both equal to one, the value of $\rho_{XY}(0)$ is usually *not* equal to one, a fact that is sometimes overlooked.

### 8.1.1 Examples

Before discussing the estimation of cross-covariance and cross-correlation functions, we will derive the theoretical functions for two examples of bivariate processes. The first example is rather 'artificial', but the model in Example 8.2 can be useful in practice.

*Example 8.1* Suppose that $\{X_t\}, \{Y_t\}$ are both formed from the same purely random process $\{Z_t\}$, which has mean zero, variance $\sigma_Z^2$, by

$$\begin{aligned} X_t &= Z_t, \\ Y_t &= 0.5Z_{t-1} + 0.5Z_{t-2}. \end{aligned}$$

Then, using (8.1), we have

$$\gamma_{XY}(k) = \begin{cases} 0.5\sigma_Z^2 & k = 1, 2 \\ 0 & \text{otherwise.} \end{cases}$$

Now the variances of the two components are given by

$$\begin{aligned} \gamma_X(0) &= \sigma_Z^2, \\ \gamma_Y(0) &= \sigma_Z^2/2, \end{aligned}$$

so that, using (8.2), we have

$$\rho_{XY}(k) = \begin{cases} 0.5\sqrt{2} & k = 1, 2 \\ 0 & \text{otherwise.} \end{cases}$$

*Example 8.2* Suppose that

$$\begin{aligned} X_t &= Z_{1,t}, \\ Y_t &= X_{t-d} + Z_{2,t}, \end{aligned} \tag{8.3}$$

where $\{Z_{1,t}\}, \{Z_{2,t}\}$ are uncorrelated purely random processes with mean zero and variance $\sigma_Z^2$, and where $d$ is a positive integer. Then we find

$$\begin{aligned} \gamma_{XY}(k) &= \begin{cases} \sigma_Z^2 & k = d, \\ 0 & \text{otherwise,} \end{cases} \\ \rho_{XY}(k) &= \begin{cases} 1/\sqrt{2} & k = d, \\ 0 & \text{otherwise,} \end{cases} \end{aligned}$$

since $\sigma_X = \sigma_Z$ and $\sigma_Y = \sqrt{2}\sigma_Z$.

In Chapter 9 we will see that Equation (8.3) corresponds to putting noise into a linear system, which consists of a simple delay of lag $d$ and then adding more noise. The cross-correlation function has a peak at lag $d$ corresponding to the delay in the system, a result that the reader should find intuitively reasonable.

### 8.1.2 *Estimation*

The 'obvious' way of estimating the cross-covariance and cross-correlation functions is by means of the corresponding sample functions. Suppose we have $N$ pairs of observations $\{(x_i, y_i); i = 1 \text{ to } N\}$, on two series labelled $x$ and $y$. Then the **sample cross-covariance function** is

$$c_{XY}(k) = \begin{cases} \sum_{t=1}^{N-k} (x_t - \bar{x})(y_{t+k} - \bar{y})/N & k = 0, 1, \ldots, N-1 \\ \sum_{t=1-k}^{N} (x_t - \bar{x})(y_{t+k} - \bar{y})/N & k = -1, -2, \ldots, -(N-1) \end{cases} \tag{8.4}$$

and the **sample cross-correlation function** is

$$r_{XY}(k) = c_{XY}(k)/s_X s_Y, \tag{8.5}$$

where $s_X, s_Y$ are the sample standard deviations of observations on $x_t$ and $y_t$, respectively.

It can be shown that these estimators are asymptotically unbiased and consistent. However, it can also be shown that estimators at neighbouring lags are themselves autocorrelated. Furthermore, it can be shown that the variances of sample cross-correlations depend on the autocorrelation functions of the two components. In general, the variances will be inflated. Thus, even for moderately large values of $N$ up to about 200, or even higher, it is possible for two series, which are actually unrelated, to give rise to apparently 'large' cross-correlation coefficients, which are spurious, in that they arise solely from autocorrelations within the two series. Thus, if a test is required for non-zero correlation between two time series, then (at least) one of the series should first be filtered to convert it to (approximate) white noise. The same filter should then be applied to the second series before computing the cross-correlation function – see also Section 9.4.2.

For example, suppose that one series appears to be a first-order autoregressive process with sample mean $\bar{x}$ and estimated parameter $\widehat{\alpha}$. Then the filtered series is given by

$$x'_t = (x_t - \bar{x}) - \widehat{\alpha}(x_{t-1} - \bar{x}).$$

In other words, the new filtered series consists of the residuals from the fitted model. Applying the same filter to the second series, we get $y'_t = (y_t - \bar{y}) - \widehat{\alpha}(y_{t-1} - \bar{y})$.

If the two series are actually uncorrelated, the above procedure will not affect the expected value of the sample cross-correlation coefficients, but should reduce their variance. This makes it much easier to interpret the cross-correlations, or indeed to use them to fit a linear model as in Section 9.4.2.

Note that some authors (e.g. Brockwell and Davis, 1991, Chapter 11) recommend prewhitening *both* series, sometimes called double prewhitening, before calculating cross-correlations.

For two uncorrelated series, of which one is white noise, it can be shown that

$$E[r_{XY}(k)] \simeq 0, \qquad \text{Var}[r_{XY}(k)] \simeq 1/N.$$

Thus values outside the interval $\pm 2/\sqrt{N}$ are significantly different from zero, and this result can be used to test for zero cross-correlation.

### *8.1.3 Interpretation*

Sometimes, it is rather difficult to interpret a sample cross-correlation function. If the series are properly prewhitened, we have seen that it is easy to test whether any of the cross-correlation coefficients are significantly different from zero. Example 8.2 suggests that a significant peak in the estimated cross-correlation function at lag $d$ may indicate that one series is related to the other when delayed by time $d$. However, you are more likely to find a series of significant coefficients at neighbouring lags, and they are more difficult to interpret.

The interpretation is even more difficult, and indeed fraught with danger, if the prefiltering procedure, described in Section 8.1.2, is not used. For example, Coen et al. (1969) calculated cross-correlation functions between variables such as the (detrended) *Financial Times (FT)* share index and (detrended) U.K. car production, and this resulted in a fairly smooth, roughly sinusoidal function with 'large' coefficients at lags 5 and 6 months. Coen et al. used this information to set up a regression model to 'explain' the variation in the *FT* share index in terms of car production 6 months earlier. However, Box and Newbold (1971) have shown that the 'large' cross-correlation coefficients are spurious as the two series had not been properly filtered. Rather than having models with *independent* error terms, the appropriate models for the given series were close to being random walks. This meant there were very high autocorrelations within each series and this inflated the variances of the cross-correlations. When the series were properly modelled, it was found that cross-correlations were negligible.

Box and Newbold (1971) also presented some interesting simulations. They constructed two *independent* random walks and then computed the cross-correlations. The latter should have an expectation close to zero, but the high autocorrelations inflated the variance so that several spuriously high coefficients were observed. The general sinusoidal shape was similar to that found for the real data. The best general advice is 'Beware'!

## 8.2 The Cross-Spectrum

The cross-correlation function is the natural tool for examining the relationship between two time series in the time domain. This section introduces a complementary function, called the cross-spectral density function or cross-spectrum, which is the natural tool in the frequency domain.

By analogy with Equation (6.11), we will define the **cross-spectrum** of a discrete-time bivariate stationary process, measured at unit intervals of time, as the Fourier transform of the cross-covariance function, namely

$$f_{XY}(\omega) = \frac{1}{\pi}\left[\sum_{k=-\infty}^{\infty} \gamma_{XY}(k)e^{-i\omega k}\right] \tag{8.6}$$

over the range $0 < \omega < \pi$. The physical interpretation of the cross-spectrum is more difficult than for the autospectrum (see Priestley, 1981, p. 657). Indeed a physical understanding of cross-spectra will probably not become clear until we have studied linear systems.

Note that $f_{XY}(\omega)$ is a *complex* function, unlike the autospectrum, which is real. This is because $\gamma_{XY}(k)$ is not an even function.

The reader should note that many authors define the cross-spectrum in the range $(-\pi, \pi)$ by analogy with Equation (6.13) as

$$f_{XY}(\omega) = \frac{1}{2\pi} \sum_{k=-\infty}^{\infty} \gamma_{XY}(k) e^{-i\omega k}. \tag{8.7}$$

This definition has certain mathematical advantages, notably that it can handle complex-valued processes and that it has a simple inverse relationship of the form

$$\gamma_{XY}(k) = \int_{-\pi}^{\pi} e^{i\omega k} f_{XY}(\omega) d\omega \tag{8.8}$$

whereas Equation (8.6) does not have a simple inverse relationship. However, Equation (8.7) introduces negative frequencies, and for ease of understanding we prefer (8.6). In any case, (8.7) means that $f_{XY}(-\omega)$ is the complex conjugate[1] of $f_{XY}(\omega)$ and so provides no extra information. Authors who use Equation (8.7) only examine the cross-spectrum at positive frequencies.

We now describe several functions derived from the cross-spectrum, which are helpful in interpreting the cross-spectrum. From Equation (8.6), the real part of the cross-spectrum, called the **co-spectrum**, is given by

$$c(\omega) = \frac{1}{\pi} \sum_{k=-\infty}^{\infty} \gamma_{XY}(k) \cos \omega k = \frac{1}{\pi} \left\{ \gamma_{XY}(0) + \sum_{k=1}^{\infty} \left[ \gamma_{XY}(k) + \gamma_{YX}(k) \right] \cos \omega k \right\}. \tag{8.9}$$

The complex part of the cross-spectrum, with a minus sign attached, is called the **quadrature** spectrum and is given by

$$q(\omega) = \frac{1}{\pi} \sum_{k=-\infty}^{\infty} \gamma_{XY}(k) \sin \omega k = \frac{1}{\pi} \sum_{k=1}^{\infty} \left[ \gamma_{XY}(k) - \gamma_{YX}(k) \right] \sin \omega k \tag{8.10}$$

so that

$$f_{XY}(\omega) = c(\omega) - iq(\omega). \tag{8.11}$$

An alternative way of expressing the cross-spectrum is in the form

$$f_{XY}(\omega) = \alpha_{XY}(\omega) e^{i\phi_{XY}(\omega)}, \tag{8.12}$$

where

$$\alpha_{XY}(\omega) = \sqrt{c^2(\omega) + q^2(\omega)} \tag{8.13}$$

[1]If $z = a + ib$, then its complex conjugate is given by $\bar{z} = a - ib$.

is the **cross-amplitude** spectrum, and

$$\phi_{XY}(\omega) = \tan^{-1}[-q(\omega)/c(\omega)] \tag{8.14}$$

is the **phase** spectrum. From Equation (8.14), it appears that $\phi_{XY}(\omega)$ is undetermined by a multiple of $\pi$. However, if the cross-amplitude spectrum is required to be positive so that we take the positive square root in Equation (8.13), then the phase is actually undetermined by a multiple of $2\pi$ using the equality of Equations (8.11) and (8.12). This apparent non-uniqueness makes it difficult to evaluate the phase. However, when we consider linear systems in Chapter 9, we will see that there are physical reasons why the phase *is* often uniquely determined and does not need to be confined to the range $\pm\pi$. The phase is usually zero at $\omega = 0$ and it makes sense to treat it as a continuous function of $\omega$, as $\omega$ goes from 0 to $\pi$. Thus, if the phase grows to $+\pi$ say, then it can be allowed to continue to grow rather than revert to $-\pi$.

Another useful function derived from the cross-spectrum is the (squared) **coherency**, which is given by

$$C(\omega) = [c^2(\omega) + q^2(\omega)]/[f_X(\omega)f_Y(\omega)] = \alpha_{XY}^2(\omega)/f_X(\omega)f_Y(\omega) \tag{8.15}$$

where $f_X(\omega), f_Y(\omega)$ are the power spectra of the individual processes, $\{X_t\}$ and $\{Y_t\}$. It can be shown that

$$0 \leq C(\omega) \leq 1.$$

This quantity measures the square of the linear correlation between the two components of the bivariate process at frequency $\omega$ and is analogous to the square of the usual correlation coefficient. The closer $C(\omega)$ is to one, the more closely related are the two processes at frequency $\omega$.

Finally, we will define a function called the **gain** spectrum, which is given by

$$G_{XY}(\omega) = \sqrt{f_Y(\omega)C(\omega)/f_X(\omega)} = \alpha_{XY}(\omega)/f_X(\omega) \tag{8.16}$$

which is essentially the regression coefficient of the process $Y_t$ on the process $X_t$ at frequency $\omega$. A second gain function can also be defined by $G_{YX}(\omega) = \alpha_{XY}(\omega)/f_Y(\omega)$, where we divide by $f_Y$ rather than $f_X$. In the terminology of linear systems – see Chapter 9 – this definition corresponds to regarding $Y_t$ as the input and $X_t$ as the output.

By this point, the reader will probably be rather confused by all the different functions that have been introduced in relation to the cross-spectrum. Whereas the cross-correlation function is a relatively straightforward development from the autocorrelation function (in theory at least, if not always in practice), statisticians often find the cross-spectrum much harder to understand than the autospectrum. It is no longer possible to interpret the results as some sort of contribution to power (variance) at different frequencies. Usually *three* functions have to be plotted against frequency to describe the relationship between two series in the frequency domain. Sometimes the co-, quadrature and coherency spectra are most suitable. Sometimes the coherency, phase and cross-amplitude are more appropriate, while another possible trio is coherency, phase and gain. Each trio can be determined from any other trio. The physical interpretation of these functions will probably not become clear until we have studied linear systems in Chapter 9.

### *8.2.1 Examples*

This subsection derives the cross-spectrum and related functions for the two examples discussed in Section 8.1.1.

*Example 8.3* For Example 8.1, we may use Equation (8.6) to derive the cross-spectrum from the cross-covariances by

$$f_{XY}(\omega) = (0.5\sigma_Z^2 e^{-i\omega} + 0.5\sigma_Z^2 e^{-2i\omega})/\pi$$

for $0 < \omega < \pi$. Using (8.9), the co-spectrum is given by

$$c(\omega) = 0.5\sigma_Z^2(\cos\omega + \cos 2\omega)/\pi.$$

Using (8.10), the quadrature spectrum is given by

$$q(\omega) = 0.5\sigma_Z^2(\sin\omega + \sin 2\omega)/\pi.$$

Using (8.13), the cross-amplitude spectrum is given by

$$\alpha_{XY}(\omega) = \frac{0.5\sigma_Z^2}{\pi}\sqrt{(\cos\omega + \cos 2\omega)^2 + (\sin\omega + \sin 2\omega)^2}$$

which, after some algebra, gives

$$\alpha_{XY}(\omega) = \sigma_Z^2 \cos(\omega/2)/\pi.$$

Using (8.14), the phase spectrum is given by

$$\tan\phi_{XY}(\omega) = -(\sin\omega + \sin 2\omega)/(\cos\omega + \cos 2\omega).$$

Figure 8.1 shows the co-spectrum, the quadrature, the cross-amplitude, and the phase spectra. In order to evaluate the coherency, we need to find the power spectra of the two individual processes. Since $X_t = Z_t$ is white noise, it has constant spectrum $f_X(\omega) = \sigma_Z^2/\pi$. The second process, $Y_t$, is a form of moving average process and it can readily be shown that its spectrum is given by

$$f_Y(\omega) = 0.5\sigma_Z^2(1 + \cos\omega)/\pi = \sigma_Z^2 \cos^2(\omega/2)/\pi.$$

Thus, using Equation (8.15), the coherency spectrum is given by

$$C(\omega) = 1 \quad \text{for all } \omega \text{ in } (0, \pi).$$

This latter result may appear surprising at first sight. However, both $X_t$ and $Y_t$ are generated from the *same* noise process and this explains why there is perfect correlation between the components of the two processes at any given frequency.

Finally, using Equation (8.16), the gain spectrum is given by

$$G_{XY}(\omega) = \sqrt{f_Y(\omega)C(\omega)/f_X(\omega)} = \cos(\omega/2)$$

since the coherency is unity.

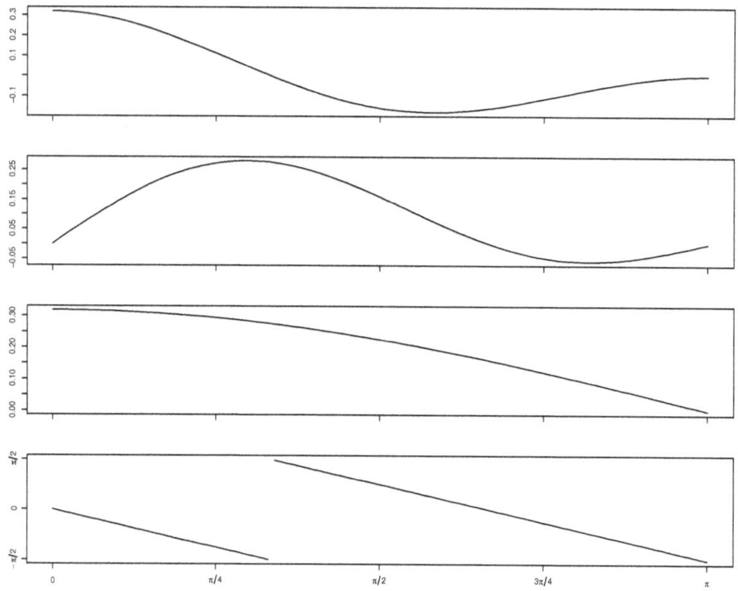

Figure 8.1 *The spectra $c(\omega)$, $q(\omega)$, $\alpha_{XY}(\omega)$ and $\phi_{XY}(\omega)$ for Example 8.3.*

*Example 8.4* For the bivariate process of Example 8.2, Equation (8.6) gives the cross-spectrum as $f_{XY}(\omega) = \sigma_Z^2 e^{-i\omega d}/\pi$, and this leads in turn to the following functions:

$$\begin{aligned} c(\omega) &= \sigma_Z^2 \cos \omega d/\pi \\ q(\omega) &= \sigma_Z^2 \sin \omega d/\pi \\ \alpha_{XY}(\omega) &= \sigma_Z^2/\pi \\ \tan \phi_{XY}(\omega) &= -\tan \omega d. \end{aligned} \tag{8.17}$$

Then, as the two autospectra are given by

$$\begin{aligned} f_X(\omega) &= \sigma_Z^2/\pi \\ f_Y(\omega) &= 2\sigma_Z^2/\pi \end{aligned}$$

we find

$$C(\omega) = 1/2.$$

Note that all the above functions are defined on $(0, \pi)$. The function of particular interest in this example is the phase, which, from Equation (8.17), is a straight line with slope $-d$ when $\phi_{XY}(\omega)$ is unconstrained and is plotted against $\omega$ as a continuous function starting with zero phase at zero frequency (see the left panel of Figure 8.2).

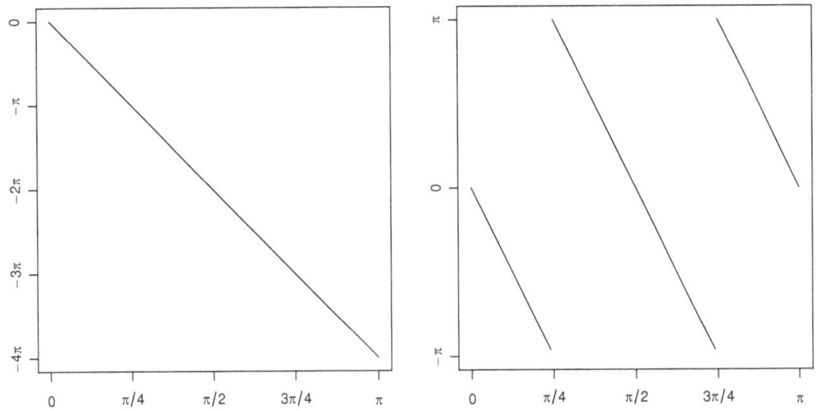


Figure 8.2 *The phase spectrum, $\phi_{XY}(\omega)$, for Example 8.4 with $d = 4$, with phase unconstrained (left) and phase constrained (right).*

If, however, the phase is constrained to lie within the interval $(-\pi, \pi)$ then a graph like the right panel of Figure 8.2 will result, where the slope of each line is $-d$.

This result is often used to help in the identification of relationships between time series. If the estimated phase approximates a straight line through the origin, then this indicates a delay between the two series equal to the slope of the line. More generally, the time delay between two recording sites will change with frequency, due, for example, to varying speeds of propagation. This is called the **dispersive** case, and can be recognized by changes in the slope of the phase function.

### 8.2.2 *Estimation*

As in univariate spectral analysis, there are two basic approaches to estimating a cross-spectrum. First, we can take a Fourier transform of the truncated sample cross-covariance function (or of the cross-correlation function to get a normalized cross-spectrum). This is analogous to Equation (7.21) for the univariate case. The estimated co-spectrum is given by

$$\widehat{c}(\omega) = \frac{1}{\pi}\left[\sum_{k=-M}^{M} \lambda_k c_{XY}(k)\cos\omega k\right] \tag{8.18}$$

where $M$ is the truncation point, and $\{\lambda_k\}$ is the lag window. The estimated quadrature spectrum is given by

$$\widehat{q}(\omega) = \frac{1}{\pi}\left[\sum_{k=-M}^{M} \lambda_k c_{XY}(k)\sin\omega k\right]. \tag{8.19}$$

Equations (8.18) and (8.19) are often used in the equivalent forms

$$\begin{aligned}
\widehat{c}(\omega) &= \frac{1}{\pi}\left\{\lambda_0 c_{XY}(0) + \sum_{k=1}^{M} \lambda_k [c_{XY}(k) + c_{XY}(-k)] \cos \omega k\right\} \\
\widehat{q}(\omega) &= \frac{1}{\pi}\left\{\sum_{k=1}^{M} \lambda_k [c_{XY}(k) - c_{XY}(-k)] \sin \omega k\right\}.
\end{aligned}$$

The truncation point $M$ and the lag window $\{\lambda_k\}$ are chosen in a similar way to that used in spectral analysis for a single series, with the Tukey and Parzen windows being most popular.

Having estimated the co- and quadrature spectra, estimates of the cross-amplitude spectrum, phase and coherency follow, in an obvious way, from Equations (8.13), (8.14) and (8.15). Estimates of the power spectra of the two individual series, namely, $\widehat{f}_X$ and $\widehat{f}_Y$, are needed in the latter case. We find

$$\begin{aligned}
\widehat{\alpha}_{XY}(\omega) &= \sqrt{\widehat{c}^2(\omega) + \widehat{q}^2(\omega)}, \\
\tan \widehat{\phi}_{XY}(\omega) &= -\widehat{q}(\omega)/\widehat{c}(\omega), \\
\widehat{C}(\omega) &= \widehat{\alpha}^2_{XY}(\omega)/\widehat{f}_X(\omega)\widehat{f}_Y(\omega).
\end{aligned}$$

When plotting the estimated phase spectrum, similar remarks apply as to the (theoretical) phase. Phase estimates are undetermined by a multiple of $2\pi$, but can usually be plotted as a continuous function, which is zero at zero frequency.

Before estimating the coherency, it may be advisable to **align** the two series. If this is not done, Jenkins and Watts (1968) have demonstrated that estimates of coherency will be biased if the phase changes rapidly. If the sample cross-correlation function has its largest value at lag $s$ say, then the two series are aligned by translating one series a distance $s$ so that the peak in the cross-correlation function of the aligned series is at zero lag.

The second approach to cross-spectral analysis is to smooth a function called the **cross-periodogram**. The univariate periodogram of a series $\{x_t\}$ can be written in the form

$$\begin{aligned}
I(\omega_p) &= \left(\sum x_t e^{i\omega_p t}\right)\left(\sum x_t e^{-i\omega_p t}\right)/N\pi \\
&= N(a_p^2 + b_p^2)/4\pi \qquad (8.20)
\end{aligned}$$

using Equations (7.17) and (7.10), where $\{a_p\}$, $\{b_p\}$ are the coefficients in the Fourier series representation of $\{x_t\}$. By analogy with Equation (8.20), we may define the cross-periodogram of two series $\{x_t\}$ and $\{y_t\}$ as

$$I_{XY}(\omega_p) = \left(\sum x_t e^{i\omega_p t}\right)\left(\sum y_t e^{-i\omega_p t}\right)/N\pi. \qquad (8.21)$$

After some algebra, it can be shown that the real and imaginary parts of $I_{XY}(\omega_p)$ are given by

$$N(a_{px}a_{py} + b_{px}b_{py})/4\pi \quad \text{and} \quad N(a_{px}b_{py} - a_{py}b_{px})/4\pi,$$

where $(a_{px}, b_{px})$, $(a_{py}, b_{py})$ are the Fourier coefficients of $\{x_t\}, \{y_t\}$, respectively at $\omega_p$. These real and imaginary parts may then be smoothed to get consistent estimates of the co- and quadrature spectral density functions by

$$\widehat{c}(\omega_p) = N \sum_{q=p-m^*}^{p+m^*} (a_{qx}a_{qy} + b_{qx}b_{qy})/4\pi m$$

$$\widehat{q}(\omega_p) = N \sum_{q=p-m^*}^{p+m^*} (a_{qx}b_{qy} - a_{qy}b_{qx})/4\pi m,$$

where $m = 2m^* + 1$ is chosen, as in the univariate case, so as to balance variance and resolution. These two equations are analogous to Equation (7.24) in the univariate case. The above estimates may then be used to estimate the cross-amplitude spectrum, phase and coherency as before.

The computational advantages of the second type of approach are clear. Once a periodogram analysis has been made of the two individual processes, nearly all the work has been done as the estimates of $c(\omega)$ and $q(\omega)$ only involve the Fourier coefficients of the two series. The disadvantage of the approach is that alignment is only possible if the cross-correlation function is calculated separately. This can be done directly or by the use of two (fast) Fourier transforms by an analogous procedure to that described in Section 7.4.5.

The properties of cross-spectral estimators are discussed, for example, by Jenkins and Watts (1968), Priestley (1981) and Bloomfield (2000). The following points are worth noting. Estimates of phase and cross-amplitude are imprecise when the coherency is relatively small. Estimates of coherency are constrained to lie between 0 and 1, and there may be a bias towards 1/2, which may be serious with short series. Finally, we note that rapid changes in phase may bias coherency estimates, which is another reason why alignment is generally a good idea.

### 8.2.3 *Interpretation*

Cross-spectral analysis is a technique for examining the relationship between two series in the frequency domain. The technique may be used for two time series that 'arise on a similar footing' and then the coherency spectrum is perhaps the most useful function. It measures the linear correlation between two series at each frequency and is analogous to the square of the ordinary product-moment correlation coefficient.

The other functions introduced in this chapter, such as the phase spectrum, are most readily understood in the context of linear systems, which will be discussed later in Chapter 9. We will therefore defer further discussion of how to interpret cross-spectral estimates until Section 9.3.

*Example 8.5*

In this example, we consider the total population (in millions) and birth rate (per 1,000) series for the United States from 1965 to 2015 (Figure 1.5, and explore the relationship between changes in these two series. Figure 8.3 shows the original and differenced values of the two series.

```
> pop<-read.csv("US_pop_birthrate.csv", header=T)
> ts.pop<-ts(pop[,2], start=1960)/10^6
> ts.birth<-ts(pop[,3], start=1960)
> ts.dpop<-diff(ts.pop)
> ts.dbirth<-diff(ts.birth)

> par(mfrow=c(2,2), mar=c(4,4,4,4))
> plot(ts.pop, xlab="Year", ylab="Population")
> plot(ts.birth, xlab="Year", ylab="Birth rates")
> plot(ts.dpop, xlab="Year", ylab="Population changes")
> plot(ts.dbirth, xlab="Year", ylab="Birth rate changes")
```

We first show periodograms of changes of population and changes of birth rates in Figure 8.4. The spectra of both series are large at low frequencies and small at middle and high frequencies. Besides, the shape of estimated spectra varies significantly when different smoothing parameters are used.

```
> # FIgure 8.4
> par(mfrow=c(2,1), mar=c(4,4,4,4))
> sp.dpop <- spec.pgram(ts.dpop,plot=F,sp=5)
> plot(sp.dpop$freq*2*pi,sp.dpop$spec/(2*pi),type="l",
    lty=1, xlab="Frequency", ylab="Spectrum")
> sp.dpop<-spec.pgram(ts.dpop,pl=F,sp=10)
> lines(sp.dpop$freq*2*pi, sp.dpop$spec/(2*pi),
    lty=2,lwd=2)
> sp.dpop <- spec.pgram(ts.dpop,pl=F,sp=25)
> lines(sp.dpop$freq*2*pi, sp.dpop$spec/(2*pi),
    lty=3,lwd=2)
> legend('topright',c("sp=5", "sp=10", "sp=25"),
    lty=c(1,2,3))

> sp.dbirth <- spec.pgram(ts.dbirth,plot=F)
> plot(sp.dbirth$freq*2*pi,sp.dbirth$spec/(2*pi),type="l",
    lty=1, xlab="Frequency", ylab="Spectrum")
> sp.dbirth<-spec.pgram(ts.dbirth, pl=F, sp=10)
> lines(sp.dbirth$freq*2*pi, sp.dbirth$spec/(2*pi),
    lty=2,lwd=2)
> sp.dbirth <- spec.pgram(ts.dbirth,pl=F,sp=25)
> lines(sp.dbirth$freq*2*pi, sp.dbirth$spec/(2*pi),
    lty=3,lwd=2)
> legend('topright',c("sp=5", "sp=10", "sp=25"),
    lty=c(1,2,3))
```

We then consider the cross-spectrum of the two series of changes. Figure 8.5

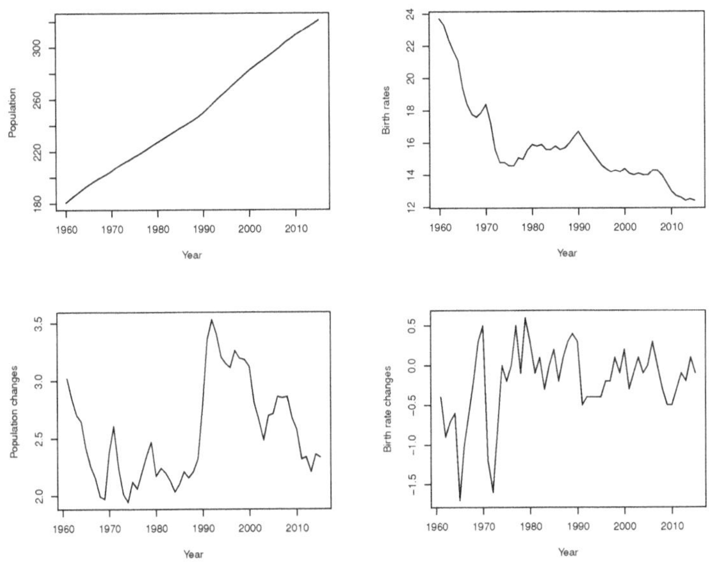


Figure 8.3 *The total population (Top left), the birth rates (Top right), the changes of population (Bottom left), and the changes of the birth rates (Bottom right) for the United States from 1965 to 2015.*

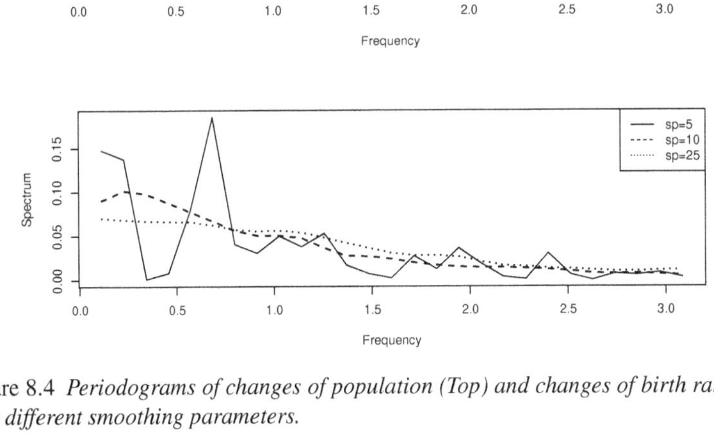


Figure 8.4 *Periodograms of changes of population (Top) and changes of birth rates (Bottom) with different smoothing parameters.*

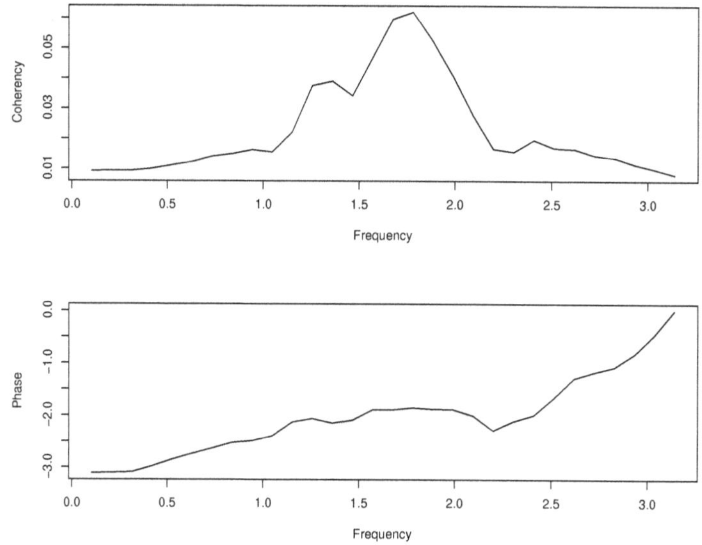


Figure 8.5 *The squared coherency (Top) and phase spectrum (Bottom) of the cross-spectrum.*

shows the squared coherency and the phase spectrum of the cross-spectrum for changes of populations and changes of birth rates. We find that the squared coherency is large at middle frequencies, suggesting that there is a significant linear relationship between two series at middle frequencies. We also note that the phase seems linear at both low and high frequencies, but with different slopes; on the other hand, the phase seems linear in the middle frequencies.

```
> # Figure 8.5
> par(mfrow=c(2,1), mar=c(4,4,4,4))
> p<-spec.pgram(xy,t=0,plot=F,sp=25)
> f<-p$freq*2*pi;
> plot(f,p$coh/(2*pi),type="l",xlab="Frequency",
   ylab="Coherency")
> plot(f,p$phase,type="l",xlab="Frequency", ylab="Phase")
```

## Exercises

**8.1** Show that the cross-covariance function of the stationary bivariate process $\{X_t, Y_t\}$ where

$$
\begin{aligned}
X_t &= Z_{1,t} + \beta_{11} Z_{1,t-1} + \beta_{12} Z_{2,t-1} \\
Y_t &= Z_{2,t} + \beta_{21} Z_{1,t-1} + \beta_{22} Z_{2,t-1}
\end{aligned}
$$

and $\{Z_{1,t}\}$, $\{Z_{2,t}\}$ are independent purely random processes with zero mean and variance $\sigma_Z^2$, is given by

$$\gamma_{XY}(k) = \begin{cases} (\beta_{11}\beta_{21} + \beta_{12}\beta_{22})\sigma_Z^2 & k = 0 \\ \beta_{21}\,\sigma_Z^2 & k = 1 \\ \beta_{12}\,\sigma_Z^2 & k = -1 \\ 0 & \text{otherwise.} \end{cases}$$

Hence evaluate the cross-spectrum.

**8.2** Define the cross-correlation function $\rho_{XY}(\tau)$ of a bivariate stationary process and show that $|\rho_{XY}(\tau)| \leq 1$ for all $\tau$.

Two first-order moving average processes

$$\begin{aligned} X_t &= Z_t + 0.4 Z_{t-1} \\ Y_t &= Z_t - 0.4 Z_{t-1} \end{aligned}$$

are formed from a purely random process $\{Z_t\}$, which has mean zero and variance $\sigma_Z^2$. Find the cross-covariance and cross-correlation functions of the bivariate process $\{X_t, Y_t\}$ and hence show that the cross-spectrum is given by

$$f_{XY}(\omega) = \sigma_Z^2\,(0.84 + 0.8\mathrm{i}\sin\omega)\,/\,\pi \qquad \text{for } 0 < \omega < \pi.$$

Evaluate the co-, quadrature, cross-amplitude, phase and coherency spectra.

Chapter 9

# Linear Systems

## 9.1 Introduction

An important problem in engineering and the physical sciences is that of identifying a model for a physical system (or process) given observations on the input and output to the system. For example, the yield from a chemical reactor (the output) depends in part on the temperature at which the reactor is kept (the input). Much of the literature assumes that the system can be adequately approximated over the range of interest by a linear model whose parameters do not change with time, although recently there has been increased interest in time-varying and non-linear systems. It turns out that the study of linear systems is useful, not only for examining the relationship between different time series, but also for examining the properties of linear filtering procedures such as many of the formulae for removing trend and seasonality.

This chapter confines attention to time-invariant linear systems. After defining such a system, Sections 9.2 and 9.3 look at their properties and show how to describe a linear system in the time and frequency domains, respectively. Then Section 9.4 discusses how to identify the structure of a linear system from observed data.

Much of the literature on linear systems (e.g. Bendat and Piersol, 2000) is written from an engineering viewpoint, and looks especially at such topics as control theory and digital communications (Glover and Grant, 1998, e.g.), as well as the identification of input/output systems. We naturally concentrate on more statistical issues and acknowledge, in particular, the major contribution of Box et al. (1994).

We generally denote the input and output series by $\{x_t\}$, $\{y_t\}$, respectively in discrete time, and by $x(t)$, $y(t)$, respectively in continuous time, though we sometimes use the latter notation for either type of series, as in the following definition.

*Definition of a Linear System.* Suppose $y_1(t), y_2(t)$ are the outputs corresponding to inputs $x_1(t), x_2(t)$, respectively. Then the system is said to be **linear** if, and only if, any linear combination of the inputs, say $\lambda_1 x_1(t) + \lambda_2 x_2(t)$, produces the same linear combination of the outputs, namely, $\lambda_1 y_1(t) + \lambda_2 y_2(t)$, where $\lambda_1, \lambda_2$ are any constants[1].

[1] An alternative, neater way of writing this, is to denote the transformation effected by the system by $\mathscr{L}$ say. Mathematicians would call this an *operator*. Then if $\mathscr{L}x_1(t) = y_1(t)$ and $\mathscr{L}x_2(t) = y_2(t)$, the operator is linear if $\mathscr{L}[\lambda_1 x_1(t) + \lambda_2 x_2(t)] = \lambda_1 \mathscr{L}x_1(t) + \lambda_2 \mathscr{L}x_2(t) = \lambda_1 y_1(t) + \lambda_2 y_2(t)$.


DOI: 10.1201/9781003646112-9

As two special cases of the above definition, we note that a linear system preserves addition and scalar multiplication, and some writers define linearity by specifying *two* conditions that correspond to these two special cases. For addition, we set $\lambda_1 = \lambda_2 = 1$, and get a condition sometimes called the **Principle of Superposition**, which says that the sum of the inputs gives rise to the sum of the outputs. For scalar multiplication, we set $\lambda_2 = 0$, and get a condition sometimes called the *homogeneity*, or *proportionality* or *scale-invariant* condition. Note that the latter condition means that if, for example, you double the input, then the output will also be doubled. We emphasize again that both these conditions are special cases of the more general definition given above.

We further confine attention to linear systems that are time-invariant. This term is defined as follows. If input $x(t)$ produces output $y(t)$, then the system is said to be **time-invariant** if a delay of time $\tau$ in the input produces the same delay in the output. In other words, $x(t-\tau)$ produces output $y(t-\tau)$, so that the input–output relation does not change with time. Generally speaking, any equation with constant coefficients defines a time-invariant system (though it need not be linear).

We only consider systems having one input and one output. The extension to several inputs and outputs is straightforward in principle, though more difficult in practice.

The reader may readily verify that the following equations both define a time-invariant linear system:

$$(1) \qquad y_t = 0.5x_t + 0.25x_{t-1},$$
$$(2) \qquad y_t = 0.5y_{t-1} + 0.25x_t,$$

where we note that (2) involves lagged values of the output. Differencing (or differentiation in continuous time) also gives a linear system. However, the reader should check that the equation

$$(3) y_t = 0.5x_t + 2,$$

which looks like a linear equation, does not in fact define a linear system[2], although it is time-invariant. Of course, an equation involving a non-linear function, such as $y_t = 0.5x_t^2$, is not linear either. Further note that an equation with time-varying coefficients, such as $y_t = 0.5tx_t$, will not be time-invariant, even if it is linear.

## 9.2 Linear Systems in the Time Domain

A time-invariant linear system may generally be written in the form

$$y(t) = \int_{-\infty}^{\infty} h(u)x(t-u)\, du \tag{9.1}$$

[2]Hint: You can apply the definition of linearity to two carefully chosen known input series such as $x_{1,t} = K$ for all $t$, where $K$ is a constant, and see what you get. Alternatively, write the equation in operator form as $\mathscr{L}x_t = [0.5x_t + 2]$, and then $\mathscr{L}[x_{1,t} + x_{2,t}] = [0.5(x_{1,t} + x_{2,t}) + 2] = \mathscr{L}x_{1,t} + \mathscr{L}x_{2,t} - 2$ and so does not satisfy the Principle of Superposition. Mathematicians would describe a non-linear transformation of this type as an *affine* transformation, which can readily be made linear by a suitable linear transformation, namely, $z_t = y_t - 2$.

in continuous time, or

$$y_t = \sum_{k=-\infty}^{\infty} h_k x_{t-k} \tag{9.2}$$

in discrete time. The weight function, $h(u)$ in continuous time or $\{h_k\}$ in discrete time, provides a description of the system in the time domain. This function is called the **impulse response function** of the system, for reasons that will become clear in Section 9.2.2.

It is obvious that Equations (9.1) and (9.2) define a linear system. Moreover, the fact that the impulse response functions do not depend on $t$, ensures that the systems are time invariant. A linear system is said to be **physically realizable** or **causal** if

$$h(u) = 0 \qquad u < 0$$

in continuous time, or

$$h_k = 0 \qquad k < 0$$

in discrete time.

We further restrict attention to **stable** systems for which any bounded input produces a bounded output, although control engineers are often concerned with controlling unstable systems. In discrete time, a sufficient condition for stability is that the impulse response function should satisfy

$$\sum_k |h_k| < C,$$

where $C$ is a finite constant. In continuous time, the above sum is replaced by an appropriate integral.

Engineers have been mainly concerned with continuous-time systems but are increasingly studying sampled-data control problems. Statisticians generally work with discrete data and so the subsequent discussion concentrates on the discrete-time case.

### *9.2.1 Some types of linear systems*

The linear filters introduced in Section 2.5.2 are examples of linear systems. For example, the simple moving average given by

$$y_t = (x_{t-1} + x_t + x_{t+1})/3$$

has impulse response function

$$h_k = \begin{cases} 1/3 & k = -1, 0, +1 \\ 0 & \text{otherwise.} \end{cases}$$

Note that this filter is not 'physically realizable', as defined above, although it can of course be used in practice as a mathematical smoothing device. The filter can be made physically realizable by defining a new output variable, $z_t$ say, where $z_t = y_{t-1}$.

Another general class of linear systems is those expressed as *linear differential equations* with constant coefficients in continuous time. For example, the equation

$$K\frac{\mathrm{d}y(t)}{\mathrm{d}t} + y(t) = x(t)$$

is a description of a linear system, where $K > 0$ for stability. In discrete time, the analogues of differential equations are *difference* equations given by

$$y_t + \alpha_1 \nabla y_t + \alpha_2 \nabla^2 y_t + \cdots = \beta_0 x_t + \beta_1 \nabla x_t + \beta_2 \nabla^2 x_t + \cdots \tag{9.3}$$

where $\nabla y_t = y_t - y_{t-1}$, and $\{\alpha_i\}, \{\beta_j\}$ are suitably chosen constants so as to make the system stable. Equation (9.3) can be rewritten as

$$y_t = a_1 y_{t-1} + a_2 y_{t-2} + \cdots + b_0 x_t + b_1 x_{t-1} + \cdots. \tag{9.4}$$

It is clear that Equation (9.4) can be rewritten in the form (9.2) by successive substitution, or, more elegantly, by using polynomial equations in the backward shift operator $B$. For example, if

$$y_t = \frac{1}{2} y_{t-1} + x_t$$

then we can write the equation as $(1 - \frac{1}{2}B)y_t = x_t$, or as $y_t = (1 - \frac{1}{2}B)^{-1} x_t$ so that we find

$$y_t = x_t + \frac{1}{2} x_{t-1} + \frac{1}{4} x_{t-2} + \cdots.$$

Thus the impulse response function is given by

$$h_k = \begin{cases} (\frac{1}{2})^k & k = 0, 1, \ldots, \\ 0 & k < 0. \end{cases}$$

The reader will notice that $h_k$ is defined for all positive $k$ giving an impulse response function of infinite order. Engineers call such a system an **infinite impulse response** (IIR) system. More generally, whenever the equation for the system includes lagged values of the output, as in Equation (9.4), the system will generally be IIR. However, if there are no lagged values of the output and a finite sum of input values at different lags, then engineers call the system a **finite impulse response** (FIR) system.

Two very simple FIR linear systems are given by

$$y_t = x_{t-d} \tag{9.5}$$

called **simple delay**, where the integer $d$ denotes the delay time, and

$$y_t = g x_t \tag{9.6}$$

called **simple gain**, where $g$ is a constant called the gain. The impulse response functions of Equations (9.5) and (9.6) are

$$h_k = \begin{cases} 1 & k = d \\ 0 & \text{otherwise} \end{cases}$$

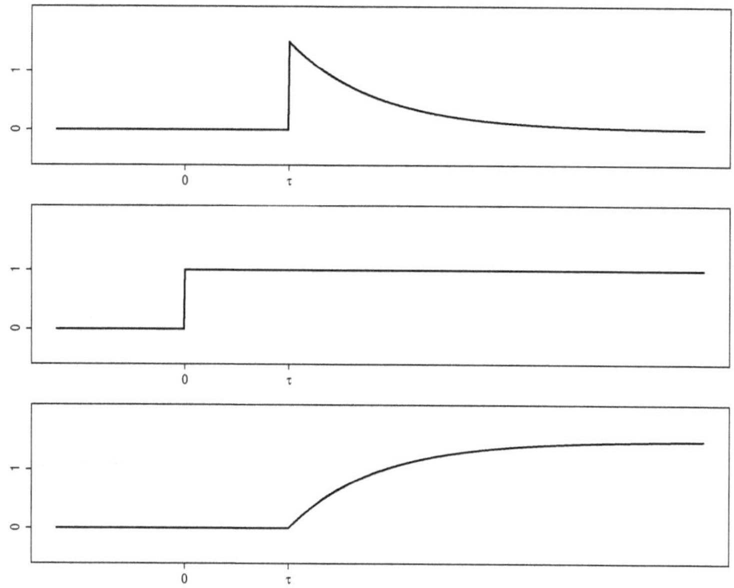


Figure 9.1 *A delayed exponential response system showing graphs of the impulse response function $h(u)$ (top), an input $x_t$ with a unit step change at time zero (middle), and the corresponding output $S_t$ (bottom).*

and

$$h_k = \begin{cases} g & k=0 \\ 0 & \text{otherwise} \end{cases}$$

respectively.

In continuous time, the impulse response functions of the corresponding equations for simple delay and simple gain, namely, $y(t) = x(t-\tau)$ and $y(t) = gx(t)$ can only be represented in terms of a function called the Dirac delta function, denoted by $\delta(u)$, which is mathematically tricky to handle – see Appendix B. The resulting impulse response functions are $\delta(u-\tau)$ and $g\,\delta(u)$, respectively.

An important class of impulse response functions, which often provides a reasonable approximation to physically realizable systems, is given by

$$h(u) = \begin{cases} [g\,\mathrm{e}^{-(u-\tau)/T}]/T & u > \tau \\ 0 & u < \tau. \end{cases}$$

A function of this type is called a **delayed exponential**, and depends on three constants, denoted by $g$, $T$ and $\tau$. The constant $\tau$ is called the **delay**. When $\tau = 0$, we have simple exponential response. The constant $g$ is called the **gain**, and represents the eventual change in output when a step change of unit size is made to the input. The constant $T$ governs the rate at which the output changes. The top panel of Figure 9.1 presents a graph of the impulse response function for a delayed exponential system together with an input showing a step change of unity at time zero and the corresponding output — see Section 9.2.3.

### 9.2.2 *The impulse response function: An explanation*

The impulse response function of a linear system describes how the output is related to the input as defined in Equations (9.1) or (9.2). The name 'impulse response' arises from the fact that the function describes the response of the system to an impulse input of unit size. For example, in discrete time, suppose that the input $x_t$ is zero for all $t$ except at time zero when it takes the value unity. Thus

$$x_t = \begin{cases} 0 & t \neq 0 \\ 1 & t = 0. \end{cases}$$

Then the output at time $t$ is given by

$$\begin{aligned} y_t &= \sum h_k x_{t-k} \\ &= h_t. \end{aligned}$$

Thus the output resulting from the unit impulse input is the same as the impulse response function, and this explains why engineers often prefer the description 'unit impulse response function'.

### 9.2.3 *The step response function*

An alternative, equivalent, way of describing a linear system in the time domain is by means of a function called the step response function, which is defined by

$$S(t) = \int_{-\infty}^{t} h(u)\, du$$

in continuous time, and

$$S_t = \sum_{k \leq t} h_k$$

in discrete time.

The name 'step response' arises from the fact that the function describes the response of the system to a unit step change in the input at time zero. For example, in discrete time, consider the input which is zero before time zero but unity thereafter. This can be represented mathematically by

$$x_t = \begin{cases} 0 & t < 0 \\ 1 & t \geq 0. \end{cases}$$

The middle panel of Figure 9.1 shows such an input $x_t$. Then the corresponding output is given by

$$y_t = \sum_k h_k x_{t-k} = \sum_{k \leq t} h_k = S_t,$$

so that the output is equal to the step response function.

Engineers sometimes use this relationship to measure the properties of a physically realizable system. The input is held steady for some time and then a unit step change is made to the input. The output is then observed and this provides an estimate of the step response function, and hence of its derivative, the impulse response function. A step change in the input may be easier to arrange in practice than an impulse.

The step response function for a delayed exponential system is given by

$$S(t) = g\,[1 - \mathrm{e}^{-(t-\tau)/T}] \qquad t > \tau$$

and the graph of $y(t)$ in the bottom panel of Figure 9.1 is also a graph of $S(t)$.

## 9.3 Linear Systems in the Frequency Domain

### *9.3.1 The frequency response function*

An alternative way of describing a time-invariant linear system is by means of a function, called the **frequency response function**, which is the Fourier transform of the impulse response function. It is defined by

$$H(\omega) = \int_{-\infty}^{\infty} h(u)\mathrm{e}^{-\mathrm{i}\omega u}\,\mathrm{d}u \qquad 0 \le \omega < \infty \tag{9.7}$$

in continuous time, and

$$H(\omega) = \sum_k h_k \mathrm{e}^{-\mathrm{i}\omega k} \qquad 0 \le \omega \le \pi \tag{9.8}$$

in discrete time. The frequency response function is sometimes given the alternative description of **transfer function**, but the former term is arguably more descriptive while the latter term is sometimes used in a different way – see below.

The frequency response and impulse response functions are equivalent ways of describing a linear system, in a somewhat similar way that the autocovariance and power spectral density functions are equivalent ways of describing a stationary stochastic process, one function being the Fourier transform of the other. We shall see that, for some purposes, $H(\omega)$ is much more useful than $h(u)$ or $h_k$. First, we prove the following theorem.

*Theorem 9.1* A sinusoidal input to a linear system gives rise, in the steady state, to a sinusoidal output at the *same* frequency. The amplitude of the sinusoid may change and there may also be a phase shift.

*Proof* The proof is given for continuous time, the extension to discrete time being straightforward. Suppose that the input to a linear system, with impulse response function $h(u)$, is given by

$$x(t) = \cos \omega t \qquad \text{for all } t.$$

Then the output is given by

$$y(t) = \int_{-\infty}^{\infty} h(u) \cos \omega(t-u) \, du. \tag{9.9}$$

Now $\cos(A-B) = \cos A \cos B + \sin A \sin B$, so we may rewrite Equation (9.9) as

$$y(t) = \cos \omega t \int_{-\infty}^{\infty} h(u) \cos \omega u \, du + \sin \omega t \int_{-\infty}^{\infty} h(u) \sin \omega u \, du. \tag{9.10}$$

As the two integrals do not depend on $t$, it is now obvious that $y(t)$ is a mixture of sine and cosine terms at the same frequency $\omega$. Thus the output is a sinusoidal perturbation at the same frequency $\omega$ as the input.

If we write

$$\begin{aligned} A(\omega) &= \int_{-\infty}^{\infty} h(u) \cos \omega u \, du, & (9.11) \\ B(\omega) &= \int_{-\infty}^{\infty} h(u) \sin \omega u \, du, & (9.12) \\ G(\omega) &= \sqrt{A^2(\omega) + B^2(\omega)}, & (9.13) \\ \tan \phi(\omega) &= -B(\omega)/A(\omega), & (9.14) \end{aligned}$$

then Equation (9.10) may be rewritten as

$$\begin{aligned} y(t) &= A(\omega) \cos \omega t + B(\omega) \sin \omega t, \\ \text{or as} \quad y(t) &= G(\omega) \cos[\omega t + \phi(\omega)]. & (9.15) \end{aligned}$$

Equation (9.15) shows that a cosine wave is amplified by a factor $G(\omega)$, which is called the **gain** of the system. The equation also shows that the cosine wave is shifted by an angle $\phi(\omega)$, which is called the **phase shift**. The above results have been derived for a particular frequency, $\omega$, but hold true for any fixed $\omega$ such that $\omega > 0$. However, note that both the gain and phase shift may vary with frequency.

Looking at Equation (9.14), the reader may note that the phase shift is not uniquely determined, because $\tan x = \tan(x + 2\pi)$. If we take the positive square root in Equation (9.13), so that the gain is required to be positive, then the phase shift is undetermined by a multiple of $2\pi$ (see also Sections 8.2 and 9.3.2). This is not usually a problem in practice as the analyst may choose to constrain the phase to lie within a range, such as $(-\pi, \pi)$, or allow it to be a continuous function.

We have so far considered an input cosine wave. By a similar argument it can be shown that an input sine wave, $x(t) = \sin \omega t$, gives an output $y(t) = G(\omega) \sin[\omega t + \phi(\omega)]$, so that there is the same gain and phase shift. More generally if we consider an input given by

$$x(t) = e^{i\omega t} = \cos \omega t + i \sin \omega t$$

then the output is given by

$$\begin{aligned} y(t) &= G(\omega)\{\cos[\omega t + \phi(\omega)] + i \sin[\omega t + \phi(\omega)]\} \\ &= G(\omega) e^{i[\omega t + \phi(\omega)]} \\ &= G(\omega) e^{i\phi(\omega)} x(t). & (9.16) \end{aligned}$$

Thus we still have gain $G(\omega)$ and phase shift $\phi(\omega)$.

We can now link these results back to the frequency response function, $H(\omega)$. Equations (9.7), (9.11) and (9.12) give us

$$\begin{aligned} H(\omega) &= \int_{-\infty}^{\infty} h(u)\mathrm{e}^{-\mathrm{i}\omega u}\,\mathrm{d}u \\ &= \int_{-\infty}^{\infty} h(u)(\cos\omega u - \mathrm{i}\sin\omega u)\,\mathrm{d}u \\ &= A(\omega) - \mathrm{i}B(\omega). \end{aligned}$$

This may be further rewritten, using Equations (9.13) and (9.14), in the form

$$H(\omega) = G(\omega)\,\mathrm{e}^{\mathrm{i}\phi(\omega)}. \tag{9.17}$$

Thus, when the input in Equation (9.16) is of the form $e^{\mathrm{i}\omega t}$, the output is obtained simply by multiplying by the frequency response function, and we have (in the steady-state situation) that

$$y(t) = H(\omega)x(t) = G(\omega)\,\mathrm{e}^{\mathrm{i}\phi(\omega)}x(t). \tag{9.18}$$

This completes the proof of Theorem 9.1. □

*Transients* The reader should note that Theorem 9.1 only applies in the **steady state** where it is assumed that the input sinusoid was applied at $t = -\infty$. If, in fact, the sinusoid is applied starting at say $t = 0$, then the output will take some time to settle to the steady-state form given by the theorem. The difference between the observed output and the steady-state output is called the **transient** component. If the system is stable, then the transient component tends to zero as $t \to \infty$. If the relationship between input and output is expressed as a differential equation (or a difference equation in discrete time), then the reader who knows how to solve such equations will recognize that the steady-state behaviour of the system corresponds to the particular integral of the equation, while the transient component corresponds to the complementary function.

It is easier to describe the transient behaviour of a linear system by using the **Laplace** transform[3] of the impulse response function. The Laplace transform also has the advantage of being defined for some unstable systems and so is sometimes preferred by engineers. However, statisticians customarily deal with steady-state behaviour for stable systems and so typically use Fourier transforms. We will continue this custom.

*Discussion of Theorem 9.1* Theorem 9.1 helps to explain the importance of the frequency response function. For inputs consisting of an impulse or step change it is easy to calculate the output using the (time-domain) impulse response function. However, for a sinusoidal input, it is much easier to calculate the output using the frequency response function. (Compare the convolution in the time domain as in

[3]The Laplace transform is defined in Appendix A.

Equation (9.9) with the simple multiplication in the frequency domain using Equation (9.18).) More generally for an input containing several sinusoidal perturbations, namely

$$x(t) = \sum_j A_j(\omega_j)e^{i\omega_j t}$$

it is easy to calculate the output using the frequency response function as

$$y(t) = \sum_j A_j(\omega_j)H(\omega_j)e^{i\omega_j t}$$

and this is one type of situation where linear systems are easier to study in the frequency domain.

*Further comments on the frequency response function* Returning to the definition of the frequency response function as given by Equations (9.7) and (9.8), note that some authors define $H(\omega)$ for negative as well as positive frequencies. However, for real-valued processes we need only consider $H(\omega)$ for $\omega > 0$.

Note that, in discrete time, $H(\omega)$ is only defined for frequencies up to the Nyquist frequency $\pi$ (or $\pi/\Delta t$ if there is an interval $\Delta t$ between successive observations). We have already introduced the Nyquist frequency in Section 7.2.1 and can apply similar ideas to a linear system measured at unit intervals of time. Given any sinusoidal input, with a frequency higher than $\pi$, we can find a corresponding sinusoid at a lower frequency in the range $(0, \pi)$, which gives identical readings at unit intervals of time. This alternative sinusoid will therefore give rise to an identical output when measured only at unit intervals of time.

We have already noted that $H(\omega)$ is sometimes called the frequency response function and sometimes the transfer function. We prefer the former term, as it is more descriptive, indicating that the function shows how a linear system responds to sinusoids at different frequencies. In any case the term 'transfer function' is used by some authors in a different way. Engineers sometimes use the term to denote the Laplace transform of the impulse response function – see, for example, Bendat and Piersol (2000, p. 30). For a physically realizable stable system, the Fourier transform of the impulse response function may be regarded as a special case of the Laplace transform. A necessary and sufficient condition for a linear system to be stable is that the Laplace transform of the impulse response function should have no poles in the right half-plane or on the imaginary axis. For an unstable system, the Fourier transform does not exist, but the Laplace transform does. However, we only consider stable systems, in which case the Fourier transform is adequate.

Further confusion can arise because Jenkins and Watts (1968) use the term 'transfer function' to denote the z-transform[4] of the impulse response function in the discrete-time case, while Box et al. (1994) use the term for a similar expression in connection with what they call transfer-function models – see Section 9.4.2. The z-transform is also used more generally by engineers for analysing discrete-time systems (e.g. Hayes, 1996, Chapter 2) but the Fourier transform is adequate for our purposes.

[4]The z-transform is defined in Appendix A.

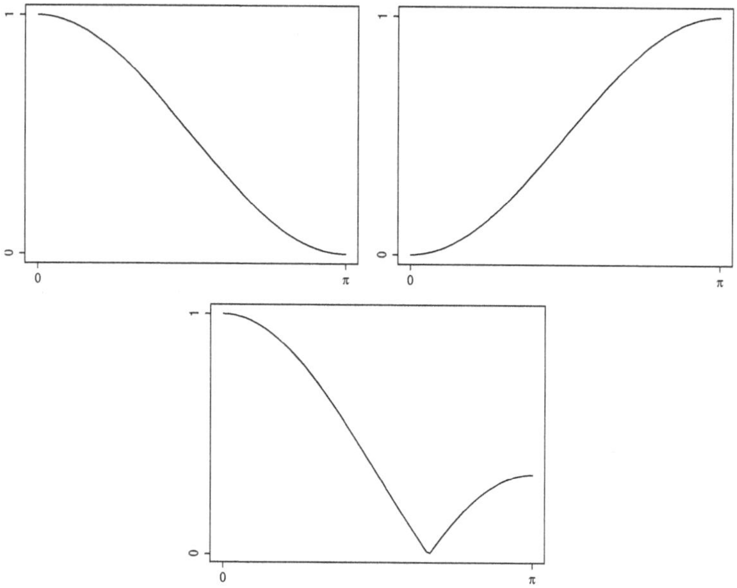


Figure 9.2 *Gain diagrams for, (a) a low-pass filter; (b) a high-pass filter; (c) a simple moving average of three successive observations.*

### *9.3.2 Gain and phase diagrams*

The frequency response function $H(\omega)$ of a linear system is a complex function that may be written in the form

$$H(\omega) = G(\omega)\mathrm{e}^{\mathrm{i}\phi(\omega)}$$

where $G(\omega)$, $\phi(\omega)$ are the gain and phase, respectively – see Equation (9.17). In order to understand the properties of a linear system, it is helpful to plot $G(\omega)$ and $\phi(\omega)$ against $\omega$ to obtain what are called the **gain diagram** and the **phase diagram**.

The gain may be calculated via Equation (9.13) or equivalently as $|H(\omega)|$, which may be obtained via $G^2(\omega) = |H(\omega)|^2 = H(\omega)\overline{H(\omega)}$ where $\overline{H(\omega)}$ denotes the complex conjugate[5] of $H(\omega)$. If $G(\omega)$ is 'large' for low values of $\omega$, but 'small' for high values of $\omega$, as in Figure 9.2(a), then we have what is called a **low-pass** filter. This description is self-explanatory in that, if the input is a mixture of variation at several different frequencies, only those components with a low frequency will 'get through' the filter. Conversely, if $G(\omega)$ is 'small' for low values of $\omega$, but 'large' for high values of $\omega$, then we have a **high-pass** filter as in Figure 9.2(b).

The phase shift $\phi(\omega)$ may be calculated from the real and imaginary parts of $H(\omega)$ using Equation (9.14). Effectively this means plotting the value of $H(\omega)$ in the complex plane and examining the result to see which quadrant it is in. The phase can

[5]If $z = a + ib$, then its complex conjugate is $\overline{z} = a - ib$.

be written directly as $\tan^{-1}[-B(\omega)/A(\omega)]$ or equivalently as $\arctan[-B(\omega)/A(\omega)]$. However, as noted earlier, plotting the phase diagram is complicated by the fact that the phase is not uniquely determined. If the gain is always taken to be positive, then the phase is undetermined by a multiple of $2\pi$ and is often constrained to the range $(-\pi, \pi)$. Unfortunately this may result in spurious discontinuities when the phase reaches the upper or lower boundary and engineers often prefer to plot the phase as a continuous unconstrained function, using the fact that $\phi(0) = 0$ provided $G(0)$ is finite. Even then, the phase may have a discontinuity when the gain becomes zero as in Example 9.1, and this can only be avoided by allowing the gain to go negative.

### *9.3.3 Some examples*

*Example 9.1* Consider the simple moving average

$$y_t = (x_{t-1} + x_t + x_{t+1})/3$$

which is a linear system with impulse response function

$$h_k = \begin{cases} 1/3 & k = -1, 0, +1 \\ 0 & \text{otherwise.} \end{cases}$$

The frequency response function of this filter is (using Equation (9.8))

$$H(\omega) = \frac{1}{3}e^{-i\omega} + \frac{1}{3} + \frac{1}{3}e^{i\omega} = \frac{1}{3} + \frac{2}{3}\cos\omega \qquad 0 < \omega < \pi.$$

This function happens to be real, not complex, and so the phase appears to be given by

$$\phi(\omega) = 0 \qquad 0 < \omega < \pi.$$

However, $H(\omega)$ is negative for $\omega > 2\pi/3$, and so if we adopt the convention that the gain should be positive, then we have

$$\begin{aligned} G(\omega) &= \left|\frac{1}{3} + \frac{2}{3}\cos\omega\right| \\ &= \begin{cases} \frac{1}{3} + \frac{2}{3}\cos\omega & 0 < \omega \le 2\pi/3 \\ -\frac{1}{3} - \frac{2}{3}\cos\omega & 2\pi/3 < \omega < \pi \end{cases} \end{aligned}$$

and

$$\phi(\omega) = \begin{cases} 0 & 0 < \omega \le 2\pi/3 \\ \pi & 2\pi/3 < \omega < \pi. \end{cases}$$

The gain is plotted in Figure 9.2(c) and is of low-pass type. This is to be expected as a moving average smooths out local fluctuations (high-frequency variation) and measures the trend (low-frequency variation). In fact it is arguably more sensible to allow the gain to go negative in $(2\pi/3, \pi)$ so that the phase is zero for all $\omega$ in $(0, \pi)$.

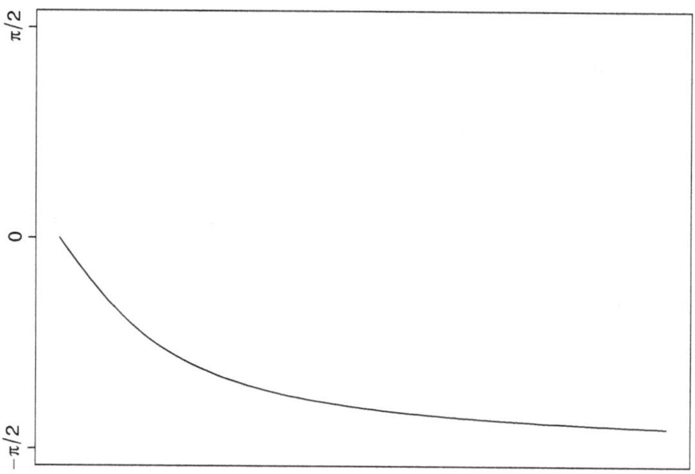


Figure 9.3 *Phase diagram for a simple exponential response system.*

*Example 9.2* A linear system showing simple exponential response has impulse response function

$$h(u) = [g\,\mathrm{e}^{-u/T}]/T \qquad u > 0.$$

Using Equation (9.7), the frequency response function is

$$H(\omega) = g(1 - \mathrm{i}\omega T)/(1 + \omega^2 T^2) \qquad \omega > 0.$$

Hence

$$\begin{aligned} G(\omega) &= g/\sqrt{(1+\omega^2 T^2)} \\ \tan\phi(\omega) &= -T\omega. \end{aligned}$$

As the frequency increases, $G(\omega)$ decreases so that the system is of low-pass type. As regards the phase, if we take $\phi(\omega)$ to be zero at zero frequency, then the phase becomes increasingly negative as $\omega$ increases until the output is completely out of phase with the input (see Figure 9.3).

*Example 9.3* Consider the linear system consisting of pure delay, so that

$$y(t) = x(t - \tau)$$

where $\tau$ is a constant. The impulse response function is given by

$$h(u) = \delta(u - \tau)$$

where $\delta$ denotes the Dirac delta function — see Appendix B. Then the frequency response function is given by

$$H(\omega) = \int_{-\infty}^{\infty} \delta(u-\tau)\,\mathrm{e}^{-\mathrm{i}\omega u}\,\mathrm{d}u = \mathrm{e}^{-\mathrm{i}\omega\tau}.$$

In fact $H(\omega)$ can be derived without using the rather difficult delta function by using Theorem 9.1. Suppose that input $x(t) = \mathrm{e}^{\mathrm{i}\omega t}$ is applied to the system. Then the output is $y(t) = \mathrm{e}^{\mathrm{i}\omega(t-\tau)} = \mathrm{e}^{-\mathrm{i}\omega\tau} \times$ input. Thus, by analogy with Equation (9.18), we have $H(\omega) = \mathrm{e}^{-\mathrm{i}\omega\tau}$. Then, using Equation (9.17), it is easy to see that this linear system has a constant gain equal to unity, namely, $G(\omega) = 1$, while the phase is given by $\phi(\omega) = -\omega\tau$.

### 9.3.4 General relation between input and output

So far, we have only considered sinusoidal inputs in the frequency domain. This section considers any type of input and shows that it is generally easier to work with linear systems in the frequency domain than in the time domain.

The general relation between input and output in continuous time is given by Equation (9.1), namely

$$y(t) = \int_{-\infty}^{\infty} h(u)x(t-u)\,\mathrm{d}u. \tag{9.19}$$

When $x(t)$ is not of a simple form, this integral may be hard to evaluate. Now consider the Fourier transform of the output, given by

$$\begin{aligned} Y(\omega) &= \int_{-\infty}^{\infty} y(t)\mathrm{e}^{-\mathrm{i}\omega t}\,\mathrm{d}u = \int_{-\infty}^{\infty}\int_{-\infty}^{\infty} h(u)x(t-u)\mathrm{e}^{-\mathrm{i}\omega t}\,\mathrm{d}u\,\mathrm{d}t \\ &= \int_{-\infty}^{\infty}\int_{-\infty}^{\infty} h(u)\mathrm{e}^{-\mathrm{i}\omega u}x(t-u)\,\mathrm{e}^{-\mathrm{i}\omega(t-u)}\,\mathrm{d}u\,\mathrm{d}t. \end{aligned}$$

However,

$$\int_{-\infty}^{\infty} x(t-u)\mathrm{e}^{-\mathrm{i}\omega(t-u)}\,\mathrm{d}t = \int_{-\infty}^{\infty} x(t)\mathrm{e}^{-\mathrm{i}\omega t}\,\mathrm{d}t$$

for all values of $u$ and is therefore the Fourier transform of $x(t)$, which we will denote by $X(\omega)$. Furthermore

$$\int_{-\infty}^{\infty} h(u)\,\mathrm{e}^{-\mathrm{i}\omega u}\,\mathrm{d}u = H(\omega)$$

so that

$$Y(\omega) = H(\omega)X(\omega). \tag{9.20}$$

Thus the integral in Equation (9.19) corresponds to a multiplication in the frequency domain provided that the Fourier transforms exist. A similar result holds in discrete time.

A more useful general relation between input and output, akin to Equation (9.20), can be obtained when the input $x(t)$ is a stationary process with a continuous power spectrum. This result will be given as Theorem 9.2.

*Theorem 9.2* Consider a stable time-invariant linear system with gain function $G(\omega)$. Suppose that the input $X(t)$ is a stationary process with continuous power spectrum $f_X(\omega)$. Then the output $Y(t)$ is also a stationary process, whose power spectrum $f_Y(\omega)$ is given by

$$f_Y(\omega) = G^2(\omega) f_X(\omega). \tag{9.21}$$

*Proof* The proof will be given for continuous time, but a similar proof yields the same result in discrete time. It is easy to show that a stationary input to a stable linear system gives rise to a stationary output, and this will not be shown here.

We denote the impulse response and frequency response functions of the system by $h(u), H(\omega)$, respectively. Thus $G(\omega) = |H(\omega)|$. By definition, the output is related to the input by

$$Y(t) = \int_{-\infty}^{\infty} h(u) X(t-u)\, \mathrm{d}u.$$

For mathematical convenience, we assume that the input has mean zero, in which case the output also has mean zero. It is straightforward to extend the proof to an input with non-zero mean.

Denote the autocovariance functions (acv.f.s) of $X(t), Y(t)$ by $\gamma_X(\tau), \gamma_Y(\tau)$, respectively. Then

$$\begin{aligned}
\gamma_Y(\tau) &= E[Y(t)Y(t+\tau)] \qquad \text{since } E[Y(t)] = 0 \\
&= E\left[\int_{-\infty}^{\infty} h(u)X(t-u)\,\mathrm{d}u \int_{-\infty}^{\infty} h(u')X(t+\tau-u')\,\mathrm{d}u'\right] \\
&= \int_{-\infty}^{\infty}\int_{-\infty}^{\infty} h(u)h(u')E[X(t-u)X(t+\tau-u')]\,\mathrm{d}u\,\mathrm{d}u'.
\end{aligned}$$

But

$$E[X(t-u)X(t+\tau-u')] = \gamma_X(\tau - u' + u).$$

Thus

$$\gamma_Y(\tau) = \int_{-\infty}^{\infty}\int_{-\infty}^{\infty} h(u)h(u')\gamma_X(\tau-u'+u)\,\mathrm{d}u\,\mathrm{d}u'. \tag{9.22}$$

The relationship between the acv.f.s of the input and the output in Equation (9.22) is not of a simple form. However, if we take Fourier transforms of both sides of Equation (9.22) by multiplying by $\mathrm{e}^{-\mathrm{i}\omega t}/\pi$ and integrating with respect to $\tau$ from $-\infty$ to $+\infty$, we find, using Equation (6.17), that the left-hand side is the spectrum of the output, namely

$$\frac{1}{\pi}\int_{-\infty}^{\infty} \gamma_Y(\tau)\mathrm{e}^{-\mathrm{i}\omega t}\,\mathrm{d}\tau = f_Y(\omega).$$

The right-hand side of Equation (9.22) requires some work to simplify. We find

$$\int_{-\infty}^{\infty}\int_{-\infty}^{\infty} h(u)h(u')\left[\frac{1}{\pi}\int_{-\infty}^{\infty}\gamma_X(\tau-u'+u)\mathrm{e}^{-\mathrm{i}\omega\tau}\,\mathrm{d}\tau\right]\mathrm{d}u\,\mathrm{d}u'$$

$$= \int_{-\infty}^{\infty}\int_{-\infty}^{\infty} h(u)\mathrm{e}^{\mathrm{i}\omega u}h(u')\mathrm{e}^{-\mathrm{i}\omega u'}\left[\frac{1}{\pi}\int_{-\infty}^{\infty}\gamma_X(\tau-u'+u)\mathrm{e}^{-\mathrm{i}\omega(\tau-u'+u)}\,\mathrm{d}\tau\right]\mathrm{d}u\,\mathrm{d}u'$$

However,

$$\frac{1}{\pi}\int_{-\infty}^{\infty}\gamma_X(\tau-u'+u)\mathrm{e}^{-\mathrm{i}\omega(t-u'+u)}\,\mathrm{d}\tau=\frac{1}{\pi}\int_{-\infty}^{\infty}\gamma_X(\tau)\mathrm{e}^{-\mathrm{i}\omega\tau}\,\mathrm{d}\tau=f_X(\omega)$$

for all $u, u'$, and (denoting the complex conjugate of $H(\omega)$ by $\overline{H(\omega)}$)

$$\int_{-\infty}^{\infty}h(u)\mathrm{e}^{\mathrm{i}\omega u}=\overline{H(\omega)}=G(\omega)\mathrm{e}^{-\mathrm{i}\phi(\omega)}.$$

Thus

$$f_Y(\omega)=\overline{H(\omega)}H(\omega)f_X(\omega)=G^2(\omega)f_X(\omega).$$

This completes the proof of Theorem 9.2. □

The relationship between the spectra of the input and the output of a linear system is very important and yet of a surprisingly simple form. Once again a result in the frequency domain – Equation (9.21) – is much simpler than the corresponding result in the time domain – Equation (9.22).

Theorem 9.2 can be used in various ways and, in particular, can be used to evaluate the spectra of various classes of stationary processes in a simpler manner to that used in Chapter 6. There the procedure was to evaluate the acv.f. of the process and then find its Fourier transform. This can be algebraically tedious, and so we give three examples showing how to use Equation (9.21) instead.

(1) *Moving average (MA) processes*

An MA process of order $q$ is given by

$$X_t=\beta_0Z_t+\beta_1Z_{t-1}+\cdots+\beta_qZ_{t-q},$$

where $Z_t$ denotes a purely random process with variance $\sigma_Z^2$. Usually $\beta_0$ is one, but it simplifies the algebra to include this extra coefficient. Comparing with Equation (9.2), we see that the MA equation may be regarded as specifying a linear system with $\{Z_t\}$ as input and $\{X_t\}$ as output. This system is stable and time-invariant and so, using Equation (9.8), has frequency response function

$$H(\omega)=\sum_{j=0}^{q}\beta_j\mathrm{e}^{-\mathrm{i}\omega j}.$$

As $\{Z_t\}$ is a purely random process, its spectrum is constant, namely,

$$f_Z(\omega)=\sigma_Z^2/\pi.$$

Thus, using Equation (9.21), the spectrum of $\{X_t\}$ is given by

$$f_X(\omega)=\left|\sum_{j=0}^{q}\beta_j\mathrm{e}^{-\mathrm{i}\omega j}\right|^2\sigma_Z^2/\pi.$$

For example, for the first-order MA process

$$X_t = Z_t + \beta Z_{t-1}, \tag{9.23}$$

we have

$$H(\omega) = 1 + \beta\, \mathrm{e}^{-i\omega}$$

and

$$\begin{aligned} G^2(\omega) &= |H(\omega)|^2 = (1+\beta \cos\, \omega)^2 + \beta^2 \sin^2\, \omega \\ &= 1 + 2\beta \cos\, \omega + \beta^2, \end{aligned}$$

so that $f_X(\omega) = (1 + 2\beta \cos\, \omega + \beta^2)\sigma_Z^2/\pi$ as already derived in Section 6.5.

This type of approach can also be used when $\{Z_t\}$ is not a purely random process. For example, suppose that the $\{Z_t\}$ process in Equation (9.23) is stationary with spectrum $f_Z(\omega)$. Then the spectrum of $\{X_t\}$ is given by

$$f_X(\omega) = (1 + 2\beta \cos\, \omega + \beta^2) f_Z(\omega).$$

(2) *Autoregressive (AR) processes*

The stationary first-order AR process

$$X_t = \alpha X_{t-1} + Z_t$$

with $|\alpha| < 1$, may be regarded as a linear system producing output $X_t$ from input $Z_t$. It may also be regarded as a linear system 'the other way round', producing output $Z_t$ from input $X_t$ by

$$Z_t = X_t - \alpha X_{t-1}.$$

In the latter form, the frequency response function is

$$H(\omega) = 1 - \alpha\, \mathrm{e}^{-i\omega}$$

and hence has (squared) gain given by

$$G^2(\omega) = 1 - 2\alpha \cos\, \omega + \alpha^2.$$

Thus, using Equation (9.21) with $Z_t$ as output, we have

$$f_Z(\omega) = (1 - 2\alpha \cos\, \omega + \alpha^2) f_X(\omega). \tag{9.24}$$

However, if $\{Z_t\}$ denotes a purely random process with spectrum $f_Z(\omega) = \sigma_Z^2/\pi$, then we know the spectrum of the 'output' and may rewrite Equation (9.24) to get the spectrum of the 'input'. We find

$$f_X(\omega) = \sigma_Z^2/\pi(1 - 2\alpha \cos\, \omega + \alpha^2)$$

which has already been obtained as Equation (6.23) by our earlier method. I hope the reader agrees that this method is easier, and it can readily be extended to handle higher-order AR processes.

(3) *Differentiation*

Consider the process that converts a continuous input $X(t)$ into a continuous output $Y(t)$ by differentiation, namely

$$Y(t) = \frac{\mathrm{d}X(t)}{\mathrm{d}t}. \tag{9.25}$$

It can readily be shown that this is a time-invariant linear system. A differentiator is of considerable mathematical interest, although in practice only approximations to it are physically realizable.

If the input is sinusoidal, say $X(t) = \mathrm{e}^{\mathrm{i}\omega t}$, then, by differentiating, we find the output is given by

$$Y(t) = \mathrm{i}\omega\, \mathrm{e}^{\mathrm{i}\omega t},$$

so that, using Equation (9.18), the frequency response function is given by

$$H(\omega) = \mathrm{i}\omega.$$

If the input is a stationary process, with spectrum $f_X(\omega)$, then it appears that the output has spectrum

$$f_Y(\omega) = |\mathrm{i}\omega|^2 f_X(\omega) = \omega^2 f_X(\omega) \tag{9.26}$$

However, this result assumes that the linear system in Equation (9.25) is stable, when in fact it is only stable for certain types of input processes. For example, it can be shown that the response to a unit step change is an unbounded impulse, which means the system is not stable. In order for the system to be stable, the variance of the output must be finite. Now

$$\mathrm{Var}[Y(t)] = \int_0^\infty f_Y(\omega)\mathrm{d}\omega = \int_0^\infty \omega^2 f_X(\omega)\mathrm{d}\omega.$$

However, using equation (6.18), we have

$$\gamma_X(k) = \int_0^\infty f_X(\omega) \cos \omega k \; \mathrm{d}\omega$$

and

$$\frac{\mathrm{d}^2\gamma_X(k)}{\mathrm{d}k^2} = -\int_0^\infty \omega^2 f_X(\omega) \cos \omega k \; \mathrm{d}\omega$$

so that

$$\mathrm{Var}[Y(t)] = -\left[\frac{\mathrm{d}^2\gamma_X(k)}{\mathrm{d}k^2}\right]_{k=0}.$$

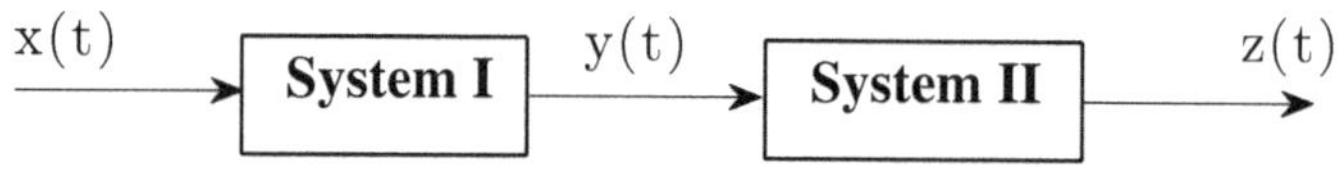


Figure 9.4 *Two linear systems in series.*

Thus $Y(t)$ has finite variance provided that $\gamma_X(k)$ can be differentiated twice at $k = 0$, and only then does Equation (9.26) hold.

### 9.3.5 *Linear systems in series*

The advantages of working in the frequency domain are also evident when we consider two or more *linear systems in series* (sometimes said to be *in cascade*). For example, Figure 9.4 shows two linear systems in series, where the input $x(t)$ to system I produces output $y(t)$, which in turn is the input to system II producing output $z(t)$. It is often of interest to evaluate the properties of the overall combined system, having $x(t)$ as input and $z(t)$ as output. It can readily be shown that the combined system is also linear, and we now find its properties. Denote the impulse response and frequency response functions of systems I and II by $h_1(\tau), h_2(\tau), H_1(\omega)$ and $H_2(\omega)$, respectively.

In the time domain, the relationship between $x(t)$ and $z(t)$ would be in the form of a double integral involving $h_1(u)$ and $h_2(u)$, which is rather complicated. However, in the frequency domain we can denote the Fourier transforms of $x(t), y(t), z(t)$ by $X(\omega), Y(\omega), Z(\omega)$, respectively, and use Equation (9.20). Then

$$Y(\omega) = H_1(\omega)X(\omega)$$

and

$$Z(\omega) = H_2(\omega)Y(\omega) = H_2(\omega)H_1(\omega)X(\omega).$$

Thus it is easy to see that the overall frequency response function of the combined system is

$$H(\omega) = H_1(\omega)H_2(\omega). \tag{9.27}$$

If

$$H_1(\omega) = G_1(\omega)\,e^{i\phi_1(\omega)}, \qquad H_2(\omega) = G_2(\omega)\,e^{i\phi_2(\omega)},$$

then

$$H(\omega) = G_1(\omega)G_2(\omega)\,e^{i[\phi_1(\omega)+\phi_2(\omega)]}. \tag{9.28}$$

Thus the overall gain is the *product* of the component gains, while the overall phase is the *sum* of the component phases.

This result may be immediately applied to the case where the input $x(t)$ is a stationary process with power spectrum $f_X(\omega)$. Generalizing Equation (9.21), we find

$$f_Z(\omega) = G_1^2(\omega)G_2^2(\omega)f_X(\omega).$$

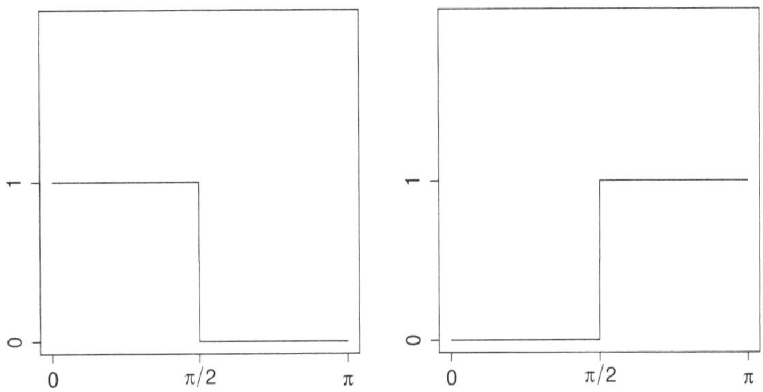


Figure 9.5 *Two types of ideal filters; (a) a low-pass filter or trend estimator; (b) a high-pass filter or trend eliminator.*

The above results are easily extended to the situation where there are $m$ linear systems in series with respective frequency response functions $H_1(\omega), \dots, H_m(\omega)$. The overall frequency response function is the product of the individual functions, namely

$$H(\omega) = H_1(\omega)H_2(\omega)\dots H_m(\omega).$$

### 9.3.6 *Design of filters*

We are now in a position to reconsider in more depth the properties of the filters introduced in Sections 2.5.2 and 2.5.3. Given a time series $\{x_t\}$, the filters for estimating or removing trend are of the general form

$$y_t = \sum_k h_k x_{t-k}.$$

This equation clearly defines a time-invariant linear system and its frequency response function is given by

$$H(\omega) = \sum_k h_k e^{-i\omega k}$$

with gain function $G(\omega) = |H(\omega)|$.

How do we set about choosing an appropriate filter for a time series? The design of a filter involves a choice of $\{h_k\}$ and hence of $H(\omega)$ and $G(\omega)$. Two types of 'ideal' filters are shown in Figure 9.5. Both have sharp cut-offs, the low-pass filter completely eliminating high-frequency variation and the high-pass filter completely eliminating low-frequency variation.

However, ideal filters of this type are impossible to achieve with a finite set of weights. Instead the smaller the number of weights used, the less sharp will generally be the cut-off property of the filter. For example, the gain diagram of a simple moving average of three successive observations is of low-pass type but has a much less

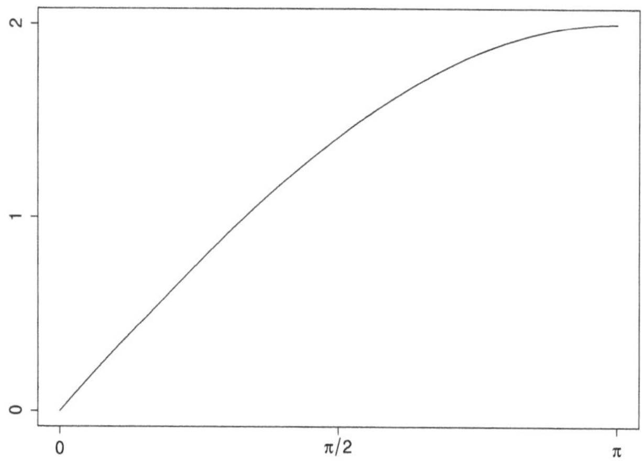


Figure 9.6 *The gain diagram $G(\omega)$ for the difference operator.*

sharp cut-off than the ideal low-pass filter (compare Figure 9.2(c) with Figure 9.5(a)). More sophisticated trend estimators, such as Spencer's 15-point moving average, have better cut-off properties.

As an example of a trend eliminator, consider first differencing, namely

$$y_t = x_t - x_{t-1}.$$

This has frequency response function

$$H(\omega) = 1 - e^{-i\omega}$$

and gain function

$$G(\omega) = \sqrt{2(1 - \cos\omega)}$$

which is plotted in Figure 9.6. This is indeed of high-pass type, as required, but the shape is a long way from the ideal filter in Figure 9.5(b). This should be borne in mind when working with first differences.

## 9.4 Identification of Linear Systems

We have so far assumed that the structure of the linear system under consideration is known. Given the impulse response function of a system, or equivalently the frequency response function, we can find the output corresponding to a given input. In particular, when considering the properties of filters for estimating or removing trend and seasonality, a formula for the 'system' is given when the filter is specified.

However, many problems concerning linear systems are of a completely different type. The structure of the system is *not* known *a priori* and the problem is to examine the relationship between input and output so as to infer the properties of the system. This procedure is called the **system identification**. For example, suppose we are interested in the effect of temperature on the yield from a chemical process. Here we have a physical system, which we assume, initially at least, is approximately linear over the range of interest. By examining the relationship between observations on temperature (the input) and yield (the output) we can infer the properties of the chemical process.

The identification process is straightforward if the input to the system can be controlled and if the system is 'not contaminated by noise'. In this case, we can simply apply an impulse or step change input, observe the output, and hence calculate the impulse response or step response function. Alternatively, we can apply sinusoidal inputs at a range of different frequencies and observe the corresponding amplitude and phase shift of the output (which should be sinusoidal at the same frequency as the input if the system really is linear). This enables us to estimate the gain and phase diagrams.

However, many systems are contaminated by noise as illustrated in Figure 9.7, where $N(t)$ denotes a noise process. This noise process need not be white noise but it is usually assumed that it is uncorrelated with the input process $X(t)$.

A further difficulty arises when the input can be observed but cannot be controlled. In other words one cannot make changes, such as a step change, to the input in order to see what happens to the output. For example, attempts have been made to treat the economy of a country as a linear system and to examine the relationship between observed variables like the retail price index (regarded as the input) and average wages (regarded as the output). However, price increases can only be controlled to a certain extent by governmental decisions, and this makes system identification difficult. Moreover, we will see later in Section 9.4.3 that there is an additional problem with this sort of data in that the output (e.g. wage increases) may in turn affect the input (e.g. price increases) and this is called a *feedback problem*.

When the system is affected by noise or the input is not controllable, more refined techniques are required to identify the system. We will describe two alternative approaches, one in the frequency domain and one in the time domain. Section 9.4.1 shows how cross-spectral analysis of input and output may be used to estimate the frequency response function of a linear system. Section 9.4.2 describes the Box–Jenkins approach to estimating the impulse response function of a linear system.

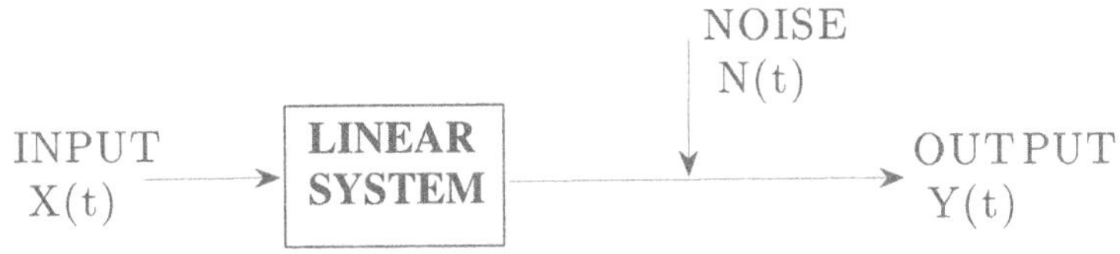


Figure 9.7 *A linear system with added noise.*

### 9.4.1 Estimating the frequency response function

Suppose that we have a system with added noise, as depicted in Figure 9.7. Although the structure of the system is unknown, it is often reasonable to assume that it is linear, time-invariant and stable, and that the noise is a stationary process that is uncorrelated with the input and has mean zero. Suppose that we have observations on the input and output over some time period. The input should be observations on a stationary process, in which case the output will also be stationary. Given these observations, we wish to estimate the frequency response function of the system. We will denote the (unknown) impulse response and frequency response functions of the system by $h(u)$, $H(\omega)$, respectively.

The reader may think that Equation (9.21), namely

$$f_Y(\omega) = G^2(\omega) f_X(\omega)$$

can be used to estimate the gain of the system, by estimating the spectra of input and output. However, this equation does not hold in the presence of noise and does not in any case give information about the phase of the system. Instead we derive a relationship involving the cross-spectrum of input and output.

In continuous time, the output $Y(t)$ is given by

$$Y(t) = \int_0^\infty h(u) X(t-u) \mathrm{d}u + N(t). \tag{9.29}$$

Note that we are only considering physically realizable systems, so that $h(u)$ is zero for $u < 0$. For mathematical convenience, we assume $E[X(t)] = 0$ so that $E[Y(t)] = 0$, but the following results also hold if $E[X(t)] \neq 0$. Multiplying through Equation (9.29) by $X(t-\tau)$ and taking expectations, we have

$$E[Y(t)X(t-\tau)] = \int_0^\infty h(u) E[X(t-u)X(t-\tau)] \mathrm{d}u + E[N(t)X(t-\tau)].$$

The last term on the right-hand side is zero as $N(t)$ is assumed to be uncorrelated with the input $X(t)$. Remembering that the processes have mean zero, the equation may be rewritten as

$$\gamma_{XY}(\tau) = \int_0^\infty h(u) \gamma_X(\tau - u) \mathrm{d}u, \tag{9.30}$$

where $\gamma_{XY}$ is the cross-covariance function of $X(t)$ and $Y(t)$, and $\gamma_X$ is the autocovariance function of $X(t)$. Equation (9.30) is called the Wiener–Hopf integral equation. Given $\gamma_{XY}$ and $\gamma_X$, this equation can, in principle, be solved to give the impulse response function $h(u)$. However, it is often easier to work with the corresponding relationship in the frequency domain.

First, we revert to discrete time and note that the discrete-time analogue of Equation (9.30) is

$$\gamma_{XY}(\tau) = \sum_{k=0}^{\infty} h_k \gamma_X(\tau - k). \tag{9.31}$$

Take Fourier transforms of both sides of this equation by multiplying by $\mathrm{e}^{-\mathrm{i}\omega\tau}/\pi$ and summing from $\tau = -\infty$ to $+\infty$. Then we find

$$\begin{aligned} f_{XY}(\omega) &= \sum_{\tau=-\infty}^{\infty}\sum_{k=0}^{\infty} h_k \mathrm{e}^{-\mathrm{i}\omega k}\gamma_X(\tau-k)\,\mathrm{e}^{-\mathrm{i}\omega(\tau-k)}/\pi \\ &= \sum_{k=0}^{\infty} h_k \mathrm{e}^{-\mathrm{i}\omega k} f_X(\omega) \\ &= H(\omega) f_X(\omega), \end{aligned} \tag{9.32}$$

where $f_{XY}$ is the cross-spectrum of input and output and $f_X$ is the (auto)spectrum of the input. Thus, once again, a convolution in the time domain corresponds to a multiplication in the frequency domain, and Equation (9.32) is much simpler than (9.30).

Estimates of $f_{XY}(\omega)$ and $f_X(\omega)$ can now be used to estimate $H(\omega)$ using Equation (9.32). Denote the estimated spectrum of the input by $\hat{f}_X(\omega)$, and the estimate cross-spectrum by $\hat{f}_{XY}(\omega)$. Then

$$\hat{H}(\omega) = \hat{f}_{XY}(\omega)/\hat{f}_X(\omega).$$

In practice, we normally use the equation

$$H(\omega) = G(\omega)\mathrm{e}^{\mathrm{i}\phi(\omega)}$$

and estimate the gain and phase separately. We have

$$\begin{aligned} \hat{G}(\omega) &= |\hat{H}(\omega)| = |\hat{f}_{XY}(\omega)/\hat{f}_X(\omega)| \\ &= |\hat{f}_{XY}(\omega)|/\hat{f}_X(\omega) \quad \text{since } \hat{f}_X(\omega) \text{ is real} \\ &= \hat{\alpha}_{XY}(\omega)/\hat{f}_X(\omega) \end{aligned} \tag{9.33}$$

where $\alpha_{XY}(\omega)$ is the cross-amplitude spectrum (see Equation (8.13)).

We also find

$$\tan\hat{\phi}(\omega) = -\hat{q}(\omega)/\hat{c}(\omega) \tag{9.34}$$

where $q(\omega)$, $c(\omega)$ are the quadrature and co-spectra, respectively (see Equation (8.14)). Thus, having estimated the cross-spectrum, Equations (9.33) and (9.34) enable us to estimate the gain and phase of the linear system, whether or not there is added noise.

We can also use cross-spectral analysis to estimate the properties of the noise process. The discrete-time version of Equation (9.29) is

$$Y_t = \sum_{k=0}^{\infty} h_k X_{t-k} + N_t. \tag{9.35}$$

For mathematical convenience, we again assume that $E(N_t) = E(X_t) = 0$ so that $E(Y_t) = 0$. If we multiply both sides of (9.35) by $Y_{t-m}$, we find

$$Y_t Y_{t-m} = \left(\sum h_k X_{t-k} + N_t\right)\left(\sum h_k X_{t-m-k} + N_{t-m}\right).$$

Taking expectations we find

$$\gamma_Y(m) = \sum_k \sum_j h_k h_j \gamma_X(m-k+j) + \gamma_N(m)$$

since $\{X_t\}$ and $\{N_t\}$ are assumed to be uncorrelated. Taking Fourier transforms of both sides of this equation, we find

$$f_Y(\omega) = H(\omega)\overline{H(\omega)} f_X(\omega) + f_N(\omega).$$

However,

$$H(\omega)\overline{H(\omega)} = G^2(\omega) = C(\omega) f_Y(\omega)/f_X(\omega)$$

where $C(\omega)$ denotes the *coherency* — see Equations (8.15) and (8.16). Thus

$$f_N(\omega) = f_Y(\omega)[1 - C(\omega)]. \tag{9.36}$$

Thus an estimate of $f_N(\omega)$ is given by

$$\hat{f}_N(\omega) = \hat{f}_Y(\omega)[1 - \hat{C}(\omega)]. \tag{9.37}$$

Equation (9.36) also enables us to see that if there is no noise, so that there is a pure linear relation between $X_t$ and $Y_t$, then $f_N(\omega) = 0$ and $C(\omega) = 1$ for all $\omega$. On the other hand if $C(\omega) = 0$ for all $\omega$, then $f_Y(\omega) = f_N(\omega)$ and the output is not linearly related to the input. This confirms the point mentioned in Chapter 8 that the coherency $C(\omega)$ measures the linear correlation between input and output at frequency $\omega$.

The results of this section not only show us how to identify a linear system by cross-spectral analysis but also give further guidance on the interpretation of functions derived from the cross-spectrum, particularly the gain, phase and coherency. Examples are given, for example, by Bloomfield (2000, Chapter 10) and Jenkins and Watts (1968).

In principle, estimates of the frequency response function of a linear system may be transformed to give estimates of the impulse response function, e.g. Jenkins and Watts (1968, p. 444) and Box et al. (1994, Appendix A11.1), but we do not recommend this. For instance, Example 8.4 appears to indicate that the sign of the phase may be used to indicate which series is 'leading' the other. However, for more complicated lagged models of the form given by Equation (9.35), it becomes increasingly difficult to make inferences from phase estimates. For such *time-domain* models, it is usually better to try to estimate the relationships directly (perhaps after pre-whitening the series), rather than via spectral estimates. One approach is the so-called Box–Jenkins approach, described in the next subsection.

### 9.4.2 *The Box–Jenkins approach*

This section gives a brief introduction to the method proposed by Box and Jenkins (1970, Chapters 10 and 11)[6] for identifying a physically realizable linear system, in the time domain, in the presence of added noise.

[6]There is virtually no change in Chapters 10 and 11 in Box et al. (1994).

The input and output series are both differenced $d$ times until both are stationary and are also mean-corrected. The modified series will be denoted by $\{X_t\}$, $\{Y_t\}$, respectively. We want to find the impulse response function $\{h_k\}$ of the system, where

$$Y_t = \sum_{k=0}^{\infty} h_k X_{t-k} + N_t. \tag{9.38}$$

The 'obvious' way to estimate $\{h_k\}$ is to multiply through Equation (9.38) by $X_{t-m}$ and take expectations to give

$$\gamma_{XY}(m) = h_0\gamma_X(m) + h_1\gamma_X(m-1) + \cdots \tag{9.39}$$

assuming that $N_t$ is uncorrelated with the input. If we assume that the weights $\{h_k\}$ are effectively zero beyond say $k = K$, then the first $K+1$ equations of type (9.39) for $m = 0, 1, \ldots, K$, can be solved for the $K+1$ unknowns $h_0, h_1, \ldots, h_K$, on substituting estimates of $\gamma_{XY}$ and $\gamma_X$. Unfortunately these equations do not, in general, provide good estimators for the $\{h_k\}$, and, in any case, assume knowledge of the truncation point $K$. The basic trouble, as already noted in Section 8.1.2, is that autocorrelation within the input and output series will increase the variance of cross-correlation estimates.

Box and Jenkins (1970) therefore propose two modifications to the above procedure. First, they suggest 'prewhitening' the input before calculating the sample cross-covariance function. Second, they propose an alternative form of Equation (9.35), which will in general require fewer parameters by allowing the inclusion of lagged values of the output[7]. They represent the linear system by the equation

$$\begin{aligned} Y_t - \delta_1 Y_{t-1} - \cdots - \delta_r Y_{t-r} \\ = \omega_0 X_{t-b} - \omega_1 X_{t-b-1} - \cdots - \omega_s X_{t-b-s}. \end{aligned} \tag{9.40}$$

This is rather like Equation (9.4), but is given in the notation used by Box and Jenkins (1970, Chapter10) and involves an extra parameter $b$, which is called the **delay** of the system. The delay can be any non-negative integer. Using the backward shift operator $B$, Equation (9.40) may be written as

$$\delta(B)Y_t = \omega(B)X_{t-b} \tag{9.41}$$

where

$$\delta(B) = 1 - \delta_1 B - \cdots - \delta_r B^r$$

and

$$\omega(B) = \omega_0 - \omega_1 B - \cdots - \omega_s B^s.$$

Box and Jenkins (1970) describe Equation (9.41) as a **transfer function** model, which is a potentially misleading description in that the term 'transfer function' is

[7] This is similar to the idea that the general linear process, represented by an MA model of possibly infinite order, can often be parsimoniously approximated by a mixed ARMA model of low order – see Section 3.4.5.

sometimes used to describe some sort of transform of the impulse response function. Indeed Box and Jenkins (1970, Chapter 10) describe the generating function of the impulse response function, namely

$$h(B) = h_0 + h_1 B + h_2 B^2 + \cdots$$

as the **transfer function**[8].

The Box–Jenkins procedure begins by fitting an ARMA model to the (differenced) input. Suppose this model is of the form (see Section 3.4.5)

$$\phi(B)X_t = \theta(B)\alpha_t$$

where $\{\alpha_t\}$ denotes a purely random process, in the notation of (Box and Jenkins, 1970, Chapter 11). Thus we can transform the input to a white noise process by

$$\theta^{-1}(B)\phi(B)X_t = \alpha_t.$$

Suppose we apply the same transformation to the output, to give

$$\theta^{-1}(B)\phi(B)Y_t = \beta_t$$

and then calculate the cross-covariance function of the filtered input and output, namely, $\{\alpha_t\}$ and $\{\beta_t\}$. It turns out that this function gives a better estimate of the impulse response function, since if we write

$$h(B) = h_0 + h_1 B + h_2 B^2 + \cdots$$

so that

$$Y_t = h(B)X_t + N_t$$

then we find

$$\begin{aligned} \beta_t &= \theta^{-1}(B)\phi(B)Y_t = \theta^{-1}(B)\phi(B)[h(B)X_t + N_t] \\ &= h(B)\alpha_t + \theta^{-1}(B)\phi(B)N_t. \end{aligned}$$

If we now evaluate the cross-covariance function of the two derived series, namely, $\{\alpha_t\}$ and $\{\beta_t\}$, then we find

$$\gamma_{\alpha\beta}(m) = h_m \operatorname{Var}(\alpha_t) \tag{9.42}$$

since $\{\alpha_t\}$ is a purely random process, and $N_t$ is uncorrelated with $\{\alpha_t\}$. Equation (9.42) is of a much simpler form than Equation (9.39) and will give more reliable estimates than those obtained by solving equations of type (9.39). This is partly because the sample cross-covariances of the derived series will have lower variances than those for the original series because there is less autocorrelation in the two series.

[8]Note that Box and Jenkins (1970) use the notation $\{v_k\}$ for the impulse response function. Box and Jenkins also use $x_t$ and $y_t$ for the differenced input and output, respectively, but we retain the convention of using capital letters to denote random variables and lower case letters for observed values.

Observed values of the derived prewhitened series can be found by calculating $\hat{\alpha}_t = \hat{\phi}(B)\hat{\theta}^{-1}(B)x_t$ where $\hat{\phi}$, $\hat{\theta}$ denote estimates of $\phi$ and $\theta$ when fitting an ARMA model to the input, and $x_t$ denotes the observed value of the input at time $t$. The same estimated transform is applied to the observed output values, and the sample cross-covariance function of $\hat{\alpha}_t$ and $\hat{\beta}_t$, namely, $c_{\alpha\beta}(k)$, is then computed. The observed variance of $\hat{\alpha}_t$, namely, $s_{\alpha}^2$, is also computed. Then an estimate of $h_k$ is given by

$$\hat{h}_k = c_{\alpha\beta}(k)/s_{\alpha}^2. \tag{9.43}$$

Box and Jenkins (1970) give the theoretical impulse response functions for various models given by Equation (9.40) and go on to show how the shape of the estimated impulse response function given by Equation (9.43) can be used to suggest appropriate values for the integers $r, b$ and $s$ in Equation (9.40). They then show how to obtain least squares estimates of $\delta_1, \delta_2, \ldots, \omega_0, \omega_1, \ldots$, given values of $r, b$ and $s$. These estimates can in turn be used to obtain refined estimates of $\{h_k\}$ if desired.

Box and Jenkins go on to show how a transfer function model, with added noise, can be used for forecasting and control. Several successful case studies have been published (e.g. Jenkins, 1979; Jenkins and McLeod, 1982) and the method looks potentially useful. However, it should be noted that the main example discussed by Box and Jenkins (1970, Chapter 11), using some gas furnace data, has been criticized by Young (1984) and Chatfield (1977, p. 504) on a number of grounds, including the exceptionally high correlation between input and output, which means that virtually any identification procedure will give good results.

Finally, it is worth noting that a somewhat similar method to the Box–Jenkins approach has been independently developed in the control engineering literature (Astrom and Bohlin, 1966; Astrom, 1970) and is called the **Astrom–Bohlin approach**. This method also involves prewhitening and a model similar to Equation (9.40), but the control engineering literature does not discuss identification and estimation procedures in the same depth as the statistical literature. One difference in the Astrom–Bohlin approach is that non-stationary series may be converted to stationarity by high-pass filtering methods other than differencing.

The reader should note that we have implicitly assumed throughout this subsection that the output does not affect the input — in other words there is no feedback, as discussed in Section 9.4.3. If there is a suspicion that feedback may be present, then it may be advisable to use alternative methods or try to fit the more general multivariate (vector) ARMA model as discussed later in Section 13.4. In fact we will see that a transfer function model can be regarded as a special case of the vector AR model.

At this point the reader may be wondering whether it is better to adopt a frequency-domain approach using cross-spectral analysis or to fit the sort of time-domain parametric model considered in this subsection. Our view is that it is unwise to attempt to make general pronouncements on the relative virtues of time-domain and frequency-domain methods. The two approaches are *complementary* rather than rivals, and it may be helpful to try both. It depends in part on the context, on the nature of the data and on what sort of model is desired.

### 9.4.3 *Systems involving feedback*

A system of the type illustrated in Figure 9.7 is called an **open-loop** system, and the procedures described in the previous two sections are appropriate for data collected under these conditions. However, data are often collected from systems where some form of **feedback control** is being applied, and then we have what is called a **closed-loop** system as illustrated in Figure 9.8. For example, when trying to identify a full-scale industrial process, it could be dangerous, or an unsatisfactory product could be produced, if some form of feedback control is not applied to keep the output somewhere near target. Similar problems arise in an economic context. For example, attempts to find a linear relationship showing the effect of price changes on wage changes are bedevilled by the fact that wage changes will in turn affect prices.

The problem of identifying systems in the presence of feedback control is discussed, for example, by Gustavsson et al. (1977) and Priestley (1983). The comments by Granger and Newbold (1986, Section 7.3) on 'Causality and Feedback' are also relevant. The key message is that open-loop procedures may not be applicable to data collected in a closed-loop situation. The problem can be explained more clearly in the frequency domain. Assuming that all processes are stationary, let $f_{XY}(\omega)$ denote the cross-spectrum of $X(t)$ and $Y(t)$ in Figure 9.8, and let $f_X(\omega), f_N(\omega), f_V(\omega)$ denote the spectra of $X(t)$, $N(t)$ and $V(t)$, respectively. Then if $H(\omega)$ and $J(\omega)$ denote the frequency response functions of the system and controller, respectively, it can be shown that

$$f_{XY}/f_X = (Hf_V + \bar{J}f_N)/(f_V + J\bar{J}f_N), \tag{9.44}$$

where all terms are functions of frequency, and $\bar{J}$ is the complex conjugate of $J$. Only if $f_N \equiv 0$ or $J \equiv 0$ is the ratio $f_{XY}/f_X$ equal to $H$ as is the case for an open-loop system (Equation (9.32)). Thus the estimate of $H$ provided by $\hat{f}_{XY}/\hat{f}_X$ will be poor unless $f_N/f_V$ is small. In particular, if $f_V \equiv 0, \hat{f}_{XY}/\hat{f}_X$ will provide an estimate of $J^{-1}$ and *not* of $H$.

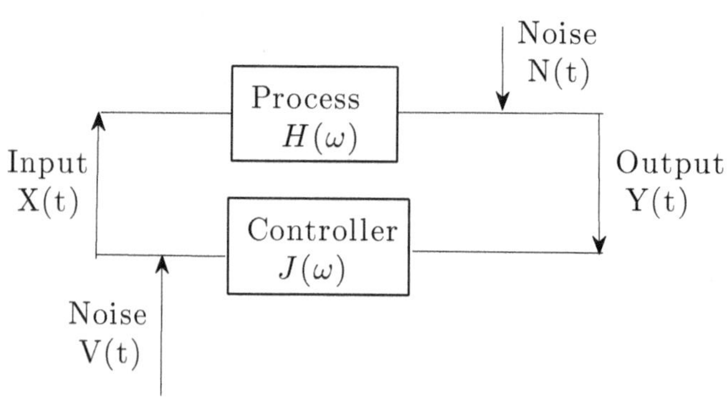


Figure 9.8 *A closed-loop system.*

Similar remarks apply to an analysis in the time domain. The time-domain equivalent of Equation (9.44) is given by Box and MacGregor (1974).

The above problem is not specifically discussed by Box et al. (1994), although it is quite clear from the remarks in their Section 11.6 that their methods are only intended for use in open-loop systems. However, some confusion could be created by the fact that Box et al. (1994, Section 13.2) do discuss ways of choosing optimal feedback control, which is quite a different problem. Having identified a system in open loop, they show how to choose feedback control action so as to satisfy some chosen criterion.

Unfortunately, open-loop identification procedures have sometimes been used for a closed-loop system where they are not appropriate. Tee and Wu (1972) studied a paper machine while it was already operating under manual control and proposed a control procedure that has been shown to be worse than the existing form of control (Box and MacGregor, 1974). In marketing, several studies have looked at the relationship between advertising expenditure and sales of products such as dishwashing liquid and coffee. However, expenditure on advertising is often chosen as a result of changes in sales levels, so that any conclusions obtained by an open-loop analysis are open to doubt.

What then can be done if feedback is present? Box and MacGregor (1974) suggest one possible approach in which the analyst deliberately adds an independent noise sequence on top of the noise $V(t)$. Alternatively, one may have some knowledge of the noise structure or of the controller frequency response function. Akaike (1968) claims that it is possible to identify a system provided only that instantaneous transmission of information does not occur in both system and controller, and an example of his, rather complicated, procedure is given by Otomo et al. (1972).

We have left to last the final, and perhaps most important concern, namely, whether feedback is present within a particular system. Sometimes it is clear from the context whether feedback is present. For example, if one wanted to study the relationship between average (ambient) temperature and sales of a seasonal product, like ice cream, then it is clear that sales cannot possibly affect temperature. Thus one has an open-loop system and open-loop procedures, such as the Box–Jenkins approach in Section 9.4.2, can be used. However, it is not always clear from the context alone as to whether feedback is present, particularly in economics and marketing. However, some contextual information may still be available. For example, economists know that prices typically affect wages, and wages in turn affect prices. If contextual information is not available, then the analyst may have to rely on data analysis to give some indication. For example, if significantly large cross-correlation coefficients between (prewhitened) input and output are observed at a zero or positive lag, then feedback may be present. However, this sort of inference is generally rather difficult to make in practice. The key point to remember is that, if feedback is present, then cross-correlation and cross-spectral analysis of the raw data may give misleading results if analyzed in an open-loop way. Thus it is best to be conservative and allow for the possible presence of feedback if unsure about the true position.

### Exercises

**9.1** Which of the following equations define a time-invariant linear system?

(a) $y_t = 2x_t$
(b) $y_t = 0.7x_t - 3$
(c) $y_t = 0.5x_t + 0.3x_{t-1}$
(d) $y_t = 0.5y_{t-1} + 0.3x_t$
(e) $y(t) = 1.5t\,x_t$
(f) $y(t) = \frac{\mathrm{d}}{\mathrm{d}t}x(t)$
(g) $y(t) = 1/x(t)$
(h) $y(t) = x(t-\tau)$ for $\tau > 0$

**9.2** Find the impulse response function, the step response function, the frequency response function, the gain and the phase shift for the following linear systems (or filters):

(a) $y_t = \frac{1}{2}x_{t-1} + x_t + \frac{1}{2}x_{t+1}$
(b) $y_t = \frac{1}{5}(x_{t-2} + x_{t-1} + x_t + x_{t+1} + x_{t+2})$
(c) $y_t = \nabla x_t$
(d) $y_t = \nabla^2 x_t$

where in each case $t$ is integer valued. Plot the gain and phase shift for filters (a) and (c). Which of the filters are low-pass and which high-pass?

If filters (a) and (b) are joined in series, find the frequency response function of the combined filter.

**9.3** Find the frequency response functions of the following linear systems in continuous time:

(a) $y(t) = gx(t-\tau)$
(b) $y(t) = \frac{g}{T}\int_0^\infty \mathrm{e}^{-u/T}x(t-u)\,\mathrm{d}u$

where $g$, $T$ and $\tau$ are positive constants.

**9.4** Consider the AR(1) process, given by $X_t = \alpha X_{t-1} + Z_t$, where $Z_t$ denotes a purely random process with zero mean and constant variance $\sigma_Z^2$ and $|\alpha| < 1$ for stationarity. If we regard the model equation as a linear system with $X_t$ denoting the input and $Z_t$ denoting the output, find the frequency response function of the system. Hence find the power spectrum of $X_t$, knowing the spectrum of the 'output' is constant – see Section 9.3.4 if you get stuck.
Now consider the one-parameter second-order AR process

$$X_t = \alpha X_{t-2} + Z_t$$

where the lagged value of $X$ is two steps before. Show that the process is second-order stationary if $|\alpha| < 1$ (as for the AR(1) process), and find the autocovariance and autocorrelation functions.

Show that the power spectral density function of the process is given by

$$f(\omega) = \sigma_Z^2/\pi(1 - 2\alpha\cos 2\omega + \alpha^2) \qquad 0 < \omega < \pi$$

using two different methods: (a) by transforming the autocovariance function; (b) by using the approach of Section 9.3.4.

Suppose now that $\{Z_t\}$ is any stationary process with power spectrum $f_Z(\omega)$. What then is the power spectrum of $\{X_t\}$ as defined by the above AR(1) model?

**9.5** If $\{X_t\}$ is a stationary time series in discrete time with power spectral density function $f(\omega)$, show that the smoothed time series

$$Y_t = \sum_{p=0}^{k} a_p X_{t-p}$$

where the $a_p$ are real constants, is a stationary process with power spectral density function

$$\left[\sum_{q=0}^{k}\sum_{p=0}^{k} a_p a_q \cos(p-q)\omega\right] f(\omega).$$

In particular, if $a_p = 1/3$ for $p = 0, 1, 2$, show that the power spectrum of $Y_t$ is

$$f(\omega)[1-\cos 3\omega]/9(1-\cos\omega).$$

(Hint: Use Equation (9.21) and the trigonometric relation $\cos A \cos B = \frac{1}{2}[\cos(A+B)+\cos(A-B)]$.)

**9.6** Show that the power spectral density function of the ARMA(1,1) process

$$X_t = \alpha X_{t-1} + Z_t + \beta Z_{t-1}$$

is given by $f_X(\omega) = \sigma_Z^2(1+2\beta\cos\omega+\beta^2)/\pi(1-2\alpha\cos\omega+\alpha^2)$ for $0<\omega<\pi$, using the approach of Section 9.3.4. It may help to let $Y_t = Z_t + \beta Z_{t-1}$. (This power spectrum may be shown to be equivalent to the normalized spectrum in Exercise 6.7 after some algebra.)

More generally, for the MA process $X_t = \theta(B)Z_t$, show that the frequency response function of the filter $Z_t \to X_t$ is $H(\omega) = \theta(\mathrm{e}^{-\mathrm{i}\omega})$, so that the spectrum of $X_t$ is given by $\theta(\mathrm{e}^{-\mathrm{i}\omega})\theta(\mathrm{e}^{\mathrm{i}\omega})\sigma_Z^2/\pi$ for $0<\omega<\pi$. Hence show that the spectrum of the general ARMA process $\phi(B)X_t = \theta(B)Z_t$ is given by $\theta(\mathrm{e}^{-\mathrm{i}\omega})\,\theta(\mathrm{e}^{\mathrm{i}\omega})\,\sigma_Z^2\,/\,\pi\,\phi(\mathrm{e}^{-\mathrm{i}\omega})\,\phi(\mathrm{e}^{\mathrm{i}\omega})$. Check this result on the above ARMA(1,1) process.

Chapter 10

# State-Space Models and the Kalman Filter

A general class of models, arousing much interest in many areas of application, is that of state-space models. They were originally developed by control engineers, particularly for applications requiring continuous updating of the current position. An example, from the field of navigation systems, is 'controlling the position of a space rocket'. However, state-space models have also found increasing use in many types of time-series problems, including parameter estimation, smoothing and prediction.

This chapter introduces state-space models for the time-series analyst, as well as describing the Kalman filter, which is an important general method of handling state-space models. Essentially, Kalman filtering is a method of signal processing, which provides optimal estimates of the current state of a dynamic system. It consists of a set of equations for recursively estimating the current state of a system and for finding variances of these estimates. More details may be found, for example, in Harvey (1989), Janacek and Swift (1993) and Durbin and Koopman (2001).

## 10.1 State-Space Models

When a scientist or engineer tries to measure any sort of signal, it will typically be contaminated by noise, so that the actual observation is given (in words) by

$$\text{OBSERVATION} \quad = \quad \text{SIGNAL} \quad + \quad \text{NOISE.} \tag{10.1}$$

The corresponding equation that is more familiar to statisticians is written as

$$\text{DATA} \quad = \quad \text{FIT} \quad + \quad \text{RESIDUAL.}$$

Both these equations can be intuitively helpful in understanding the key distinction between 'explained' and 'unexplained' variation. However, as is customary, we prefer to use Equation (10.1) here because it seems more natural for introducing state-space models.

The key step to defining the class of state-space models (in their basic linear form) is as follows. We assume that Equation (10.1) holds, but further assume that the signal is a linear combination of a set of variables, called **state variables**, which constitute what is called the **state vector** at time $t$. This vector describes the state of


DOI: 10.1201/9781003646112-10

the system at time $t$ and is sometimes called the 'state of nature'. The state variables can take many forms and the reader should wait to examine the various examples given below before expecting to fully understand this concept.

Although the above jargon derives from control engineering, it should be emphasized that the ideas are equally applicable in many other scientific areas. For example, in economics, the observation could be an economic variable, such as the unemployment rate, and the state variables could then include such (unobserved) quantities as the current true underlying level and the current seasonal factor (if any).

It is an unfortunate complication that there is no standard notation for state-space models and the Kalman filter. We confine attention to the case of a univariate observed time series and denote the observation at time $t$ by $X_t$. We denote the $(m \times 1)$ state vector at time $t$ by $\boldsymbol{\theta}_t$, and write Equation (10.1) as

$$X_t = \boldsymbol{h}_t^T \boldsymbol{\theta}_t + n_t, \tag{10.2}$$

where the $(m \times 1)$ column vector $\boldsymbol{h}_t$ is assumed to be a known vector[1] and $n_t$ denotes the observation error.

The state vector $\boldsymbol{\theta}_t$ which is of prime importance, cannot usually be observed directly (i.e. is unobservable). The state variables are typically model parameters of some sort, such as regression coefficients in a regression model (see Section 10.1.6) or parameters describing the state of a system in a rather different way (see Sections 10.1.1–10.1.3). Thus, the analyst will typically want to use the observations on $X_t$ to make inferences about $\boldsymbol{\theta}_t$. However, in some applications (see Section 10.1.4), the state vector will at least be partially known, but the state-space formulation and the Kalman filter may still be useful for making predictions and handling missing values, quite apart from estimating model parameters.

Although $\boldsymbol{\theta}_t$ may not be directly observable, it is often reasonable to assume that we know how it changes through time, and we denote the updating equation by

$$\boldsymbol{\theta}_t = G_t \boldsymbol{\theta}_{t-1} + \boldsymbol{w}_t, \tag{10.3}$$

where the $(m \times m)$ matrix $G_t$ is assumed known, and $\boldsymbol{w}_t$ denotes a $(m \times 1)$ vector of deviations such that $\boldsymbol{w}_t^T = (w_{1,t}, w_{2,t}, \ldots, w_{m,t})$.

The pair of equations in (10.2) and (10.3) constitute the general form of the (univariate) **state-space model**. Equation (10.2) is called the **observation** (or **measurement**) equation, while (10.3) is called the **transition** (or **state** or **system**) equation.

The 'errors' in the observation and transition equations are generally assumed to be serially uncorrelated and also to be uncorrelated with each other at all time periods. We may further assume that $n_t$ is $\mathrm{N}(0, \sigma_n^2)$ while $\boldsymbol{w}_t$ is multivariate normal with zero mean vector and a known variance–covariance matrix[2] denoted by $W_t$. Note

[1] We use the superscript $^T$ throughout to denote the transpose of a matrix or vector. Here $\boldsymbol{h}_t$ is a column vector, and so its transpose $\boldsymbol{h}_t^T$ is a row vector.

[2] This is a square symmetric matrix, of size $(m \times m)$, that specifies the variances of each element of $\boldsymbol{w}_t$ on its diagonal and the covariances between pairs of elements of $\boldsymbol{w}_t$ as the off diagonal terms. Thus the $(i, j)$th element of $W_t$ is given by $\mathrm{Cov}\{w_{i,t}, w_{j,t}\}$.

that if $w_t$ is independent of $\theta_t, \theta_{t-1}, \theta_{t-2}, \ldots$, then the sequence $\{\theta_t\}$ is said to have the Markov property in that $\theta_t$ depends on $\theta_{t-1}$ but not on earlier values.

The state-space model can readily be generalized to the case where $X_t$ is a vector by making $h_t$ a matrix of appropriate size and by making $n_t$ a vector of appropriate length. It is also possible to add terms involving known linear combinations of explanatory (or exogenous) variables to the right-hand side of Equation (10.2).

The application of state-space models to engineering problems, such as controlling a dynamic system, is fairly clear. There the equations of motion of a system are often assumed to be known *a priori*, as are the properties of the system disturbances and measurement errors, although some model parameters may have to be estimated from data. Neither the equations nor the 'error' statistics need be constant as long as they are known functions of time. However, at first sight, state-space models may appear to have little connection with earlier time-series models. Nevertheless it can be shown, for example, that it is possible to put many types of time-series models into a state-space form. They include regression and autoregressive moving average (ARMA) models as well as various trend-and-seasonal models for which exponential smoothing methods are thought to be appropriate.

In fact, the class of state-space models actually covers a very wide collection of models, often appearing under different names in different parts of the literature. For example, the so-called **unobserved components models**, used widely by econometricians, are of a state-space form. Bayesian forecasting (see Section 10.1.5) relies on a class of models, called **dynamic linear models**, which are essentially a state-space representation, while some models with time-varying coefficients can also be represented in this way. Moreover, Harvey (1989) has described a general class of trend-and-seasonal models, called **structural models**, which involve the classical decomposition of a time series into trend, seasonality and irregular variation, but which can also be represented as state-space models. We pay particular attention to these important models.

Note that the model decomposition must be additive in order to get a linear state-space model. If, for example, the seasonal effect is thought to be multiplicative, then logarithms must be taken in order to fit a structural model, although this implicitly assumes that the 'error' terms are also multiplicative. A key feature of structural models (and more generally of linear state-space models) is that the observation equation involves a *linear* function of the state variables and yet does not restrict the model to be constant through time. Rather it allows local features, such as trend and seasonality, to be updated through time using the transition equation.

Several examples of state-space models are presented in the following subsections, starting with the random walk plus noise model that involves just one state variable.

### *10.1.1 The random walk plus noise model*

Suppose that the observation equation is known to be given by

$$X_t = \mu_t + n_t \tag{10.4}$$

where the unobservable local level $\mu_t$ is assumed to follow a random walk given by

$$\mu_t = \mu_{t-1} + w_t. \tag{10.5}$$

Here, Equation (10.5) is the transition equation, and the state vector $\theta_t$ consists of a single state variable, namely, $\mu_t$. Thus $\theta_t$ is a scalar, rather than a vector, while $h_t$ and $G_t$ are also constant scalars, namely, unity. The model involves two error terms, namely, $n_t$ and $w_t$, which are usually assumed to be independent and normally distributed with zero means and respective variances $\sigma_n^2$ and $\sigma_w^2$. The ratio of these two variances, namely, $\sigma_w^2/\sigma_n^2$, is called the **signal-to-noise ratio** and is an important quantity in determining the features of the model. In particular, if $\sigma_w^2 = 0$, then $\mu_t$ is a constant and the model reduces to a trivial, constant-mean model.

The state-space model defined by Equations (10.4) and (10.5) is usually called the **random walk plus noise model**, but has also been called the **local level** model and the **steady** model. Note that no trend term is included. While relatively simple, the model is very important since it can be shown that simple exponential smoothing produces optimal forecasts, not only for an ARIMA$(0,1,1)$ model (see Chapter 5) but also for the above model (see Section 10.2 and Exercise 10.1). The reader can readily explore the relation with the ARIMA$(0,1,1)$ model by taking first differences of $X_t$ in Equation (10.4) and using Equation (10.5) to show that the first differences are stationary and have the same autocorrelation function as an MA$(1)$ model. It can also be shown that the random walk plus noise model and the ARIMA$(0,1,1)$ model give rise to the same forecast function. Thus the random walk plus noise model is an alternative to the ARIMA$(0,1,1)$ model for describing data showing no long-term trend or seasonality but some short-term correlation.

### *10.1.2 The linear growth model*

The linear growth model is specified by these three equations

$$\begin{aligned} X_t &= \mu_t + n_t, \\ \mu_t &= \mu_{t-1} + \beta_{t-1} + w_{1,t}, \\ \beta_t &= \beta_{t-1} + w_{2,t}. \end{aligned} \tag{10.6}$$

The first equation is the observation equation, while the next two are transition equations. The state vector $\theta_t^T = (\mu_t, \beta_t)$ has two components, which can naturally be interpreted as the local level $\mu_t$, and the local trend (or growth rate) $\beta_t$. Note that the latter state variable does not actually appear in the observation equation. Comparing with the general state-space form in Equations (10.2) and (10.3), the reader may readily verify that $h_t^T = (1,0)$ and $G_t = \begin{bmatrix} 1 & 1 \\ 0 & 1 \end{bmatrix}$ are both constant through time.

The title 'linear growth model' is reasonably self-explanatory, although it might be clearer to add the adjective 'local'. If the trend term $\beta_t$ is constant, then the current level $\mu_t$ changes linearly through time. However, the trend (or growth rate) may also evolve through time. Of course if $w_{1,t}$ and $w_{2,t}$ have zero variance, then the trend is constant (or deterministic) and we have what is called a **global** linear trend model.

However, this situation is arguably unlikely to occur in practice and the modern preference is to use the above **local** linear trend model where the trend *is* allowed to change. In any case the global model is a special case of Equation (10.6) and so it seems more sensible to fit the latter, more general model.

The reader may easily verify that the second differences of $X_t$ in Equation (10.6) are stationary and have the same autocorrelation function as an MA(2) model. In fact it can be shown that two-parameter exponential smoothing (where level and trend are updated) is optimal for an ARIMA$(0,2,2)$ model and also for the above linear growth model (e.g. Abraham and Ledolter, 1986). It is arguably easier to get a variety of trend models from special cases of a general structural state-space model than from the Box–Jenkins ARIMA class of models.

### *10.1.3 The basic structural model*

There are various ways of incorporating seasonality into a state-space model. An important example is the following model specified by four equations:

$$\begin{aligned} X_t &= \mu_t + i_t + n_t, \qquad (10.7)\\ \text{where} \quad \mu_t &= \mu_{t-1} + \beta_{t-1} + w_{1,t} \\ \beta_t &= \beta_{t-1} + w_{2,t} \\ i_t &= -\sum_{j=1}^{s-1} i_{t-j} + w_{3,t} \end{aligned}$$

Here, $\mu_t$ denotes the local level, $\beta_t$ denotes the local trend, $i_t$ denotes the local seasonal index and $s$ denotes the number of periods in 1 year (or season). The model also incorporates four separate 'error' terms that are all assumed to be additive and to have mean zero. Note that the fourth equation in (10.7) assumes that the expectation of the sum of the seasonal effects over 1 year is zero. In this model, the state vector has $s+2$ components, namely, $\mu_t, \beta_t, i_t, i_{t-1}, \ldots, i_{t-s+1}$. The model is similar in spirit to that implied by the additive Holt–Winters method (see Section 5.2.3). The latter depends on three smoothing parameters, which correspond in some sense to the three error variance ratios, namely, $\sigma_1^2/\sigma_n^2$, $\sigma_2^2/\sigma_n^2$ and $\sigma_3^2/\sigma_n^2$, where $\sigma_n^2 = \text{Var}(n_t)$, and $\sigma_i^2 = \text{Var}(w_{i,t})$ for $i = 1,2,3$.

The above model is called the **basic structural model** by Harvey (1989), who discusses the properties of the model in some detail. Harvey (1989) also discusses various extensions of the model, such as the incorporation of explanatory variables. Andrews (1994) gives some encouraging empirical results as regards the forecasting ability of structural models.

An alternative class of seasonal state-space models are described by Ord et al. (1997) and have the feature that they incorporate a single source of error. Some special cases of this class of models are found to be models for which exponential smoothing is optimal (Chatfield et al., 2001).

*10.1.4 State-space representation of an AR(2) process*

We illustrate the connection between state-space models and ARIMA models by considering the AR(2) model. The latter can be written as

$$X_t = \phi_1 X_{t-1} + \phi_2 X_{t-2} + Z_t. \tag{10.8}$$

Consider the (rather artificial) state vector at time $t$ defined by $\theta_t^T = (X_t, \phi_2 X_{t-1})$. Then the observation equation may be written (trivially) as

$$X_t = (1,0)\,\theta_t$$

with $h_t^T = (1,0)$ and $\sigma_n^2 = 0$, while Equation (10.8) may be written as the first line of the transition equation

$$\theta_t = \begin{bmatrix} \phi_1 & 1 \\ \phi_2 & 0 \end{bmatrix} \theta_{t-1} + \begin{bmatrix} 1 \\ 0 \end{bmatrix} Z_t \tag{10.9}$$

since $\theta_{t-1}^T = (X_{t-1}, \phi_2 X_{t-2})$.

This looks (and is!) a rather contrived piece of mathematical trickery, and we normally prefer to use Equation (10.8), which appears more natural than Equation (10.9). (In contrast the state-space linear growth model in Equation (10.6) may well appear more natural than an ARIMA$(0,2,2)$ model.) However, Equation (10.9) does replace two-stage dependence with two equations involving one-stage dependence, and also allows us to use the general results relating to state-space models, such as the recursive estimation of parameters, should we need to do so. For example, the Kalman filter provides a general method of estimation for ARIMA models (e.g. Kohn and Ansley, 1986). However, it should also be said that the approach to identifying state-space models is generally quite different from that for ARIMA models in that more knowledge about model structure is typically assumed *a priori* (see Section 10.1.7).

Note that the state-space representation of an ARMA model is not unique and it may be possible to find many equivalent representations. For example, we could find alternative state-space representations of Equation (10.8) using the state vector $\theta_t^T = (X_t, X_{t-1})$. In such case, the observation equation is still $X_t = (1,0)\theta_t$ with $\mathbf{h}_t = (1,0)$ and $\sigma_n^2 = 0$, but the transition equation becomes

$$\theta_t = \begin{bmatrix} \phi_1 & \phi_2 \\ 1 & 0 \end{bmatrix} \theta_{t-1} + \begin{bmatrix} 1 \\ 0 \end{bmatrix} Z_t. \tag{10.10}$$

We may also consider using the (more useful?) state vector $\theta_t^T = [X_t, \widehat{X}_t(1)]$, where $\widehat{X}_t(1)$ is the optimal one-step-ahead forecast at time $t$ — see Exercise 10.4. Of course, in this case, the state vector *can* be observed directly, and the problem is no longer one of estimating $\theta_t$. However, the state-space formulation may still be useful for other purposes, such as making predictions. For the other two forms of the state vector, the first component can be observed directly, but the second component contains unobserved parameters and so does need to be estimated.

### *10.1.5 Bayesian forecasting*

Bayesian forecasting (West and Harrison, 1997) is a general approach to forecasting that includes a variety of methods, such as regression and exponential smoothing, as special cases. It relies on a model, called the **dynamic linear model**, which is closely related to the general class of state-space models. The Bayesian formulation means that the Kalman filter is regarded as a way of updating the (prior) probability distribution of $\theta_t$ when a new observation becomes available to give a revised (posterior) distribution. The Bayesian approach also enables the analyst to consider the case where several different models are entertained and it is required to choose a single model to represent the process. Alternatively, when there are several plausible candidate models, the approach allows the analyst to compute some sort of combined forecast. For example, when the latest observation appears to be an outlier, one could entertain the possibility that this represents a step change in the process, or that it arises because of a single intervention, or that it is a 'simple' outlier with no change in the underlying model. The respective probabilities of each model being 'true' are updated after each new observation.

An expository introduction to the Bayesian approach, together with case studies and computer software, is given by Pole et al. (1994). Further developments are described by West and Harrison (1997). The approach has some staunch adherents, while others find the avowedly Bayesian approach rather intimidating. This author has no practical experience with the approach. Fildes (1983) suggested that the method is generally not worth the extra complexity compared with alternative, simpler methods. However, some of the examples in Pole et al. (1994) and in West and Harrison (1997) are persuasive, especially for short series with prior information.

### *10.1.6 A regression model with time-varying coefficients*

This example demonstrates how state-space models can be used to generalize familiar constant-parameter models to the situation where the model parameters are allowed to change through time. Suppose that the observed variable $X_t$ is known to be linearly related to a known explanatory variable $u_t$ by

$$X_t = a_t + b_t u_t + n_t.$$

If the parameters $a_t$ and $b_t$ are constant, then we have the familiar linear regression model, but we suppose instead that the regression coefficients $a_t$ and $b_t$ are allowed to evolve through time according to a random walk. The two regression coefficients may be taken as the components of the (unobservable) state vector $\theta_t$, while the values of the explanatory variable $u_t$ are put into the known vector $h_t$. Thus writing $\theta_t^T = [a_t, b_t]$ and $h_t^T = [1, u_t]$, the model may be written in in state-space form as

$$\begin{aligned} X_t &= h_t^T \theta_t + n_t, \\ \theta_t &= \theta_{t-1} + w_t, \end{aligned}$$

where the transition matrix $G_t$ is constant, namely, the constant $(2 \times 2)$ identity matrix given by $G_t = \begin{bmatrix} 1 & 0 \\ 0 & 1 \end{bmatrix}$.

Of course if the elements of $w_t$ have zero variance, then $\theta_t$ is constant, say $\theta_t^T = (a,b)$, and we are back to the usual linear regression model with constant coefficients. In this case, the transition equation is a trivial identity and there is little point in using the state-space format. The advantage of Equation (10.11) is that it covers a much more general class of models, including ordinary linear regression as a special case, but also allowing the model parameters to change through time. As well as covering a wider class of possibilities, the state-space formulation means that we can apply the general theory relating to state-space models.

### *10.1.7 Model building*

An important difference between state-space modelling in time-series applications and in some engineering problems is that the structure and properties of a time series will usually not be assumed known *a priori*. In order to apply state-space theory, we need to know $h_t$ and $G_t$ in the model equations and also to know the variances and covariances of the disturbance terms, namely, $\sigma_n^2$ and $W_t$. The choice of a suitable state-space model (i.e. the choice of suitable values for $h_t$ and $G_t$) may be accomplished using a variety of aids including external knowledge and a preliminary examination of the data. For example, Harvey (1989) claims that the basic structural model (see Section 10.1.3) can describe many time series showing trend and seasonality, but, to use the standard basic structural model, the analyst must, for example, check that the seasonal variation really is additive. If it is not additive, the analyst should consider transforming the data or trying an alternative model. In other words, the use of a state-space model does not take away the difficult problem of finding a suitable model for a given set of data. As usual, model fitting is easy, but model building can be hard.

Another problem in time-series applications is that the error variances are generally not known *a priori*. This can be dealt with by guesstimating them, and then updating them in an appropriate way, or, alternatively, by estimating them from a set of data over a suitable fit period.

## 10.2 The Kalman Filter

In state-space modelling, the prime objective is usually to estimate the signal in the presence of noise. In other words we want to estimate the $(m \times 1)$ state vector $\theta_t$, which cannot usually be observed directly. The Kalman filter provides a general method for doing this. It consists of a set of equations that allow us to update the estimate of $\theta_t$ when a new observation becomes available. We will see that this updating procedure has two stages, called the prediction stage and the updating stage.

Suppose we have observed a univariate time series up to time $(t-1)$, and that $\widehat{\theta}_{t-1}$ is the 'best' estimator for $\theta_{t-1}$ based on information up to this time. Here 'best' is defined as the minimum mean square error estimator. Further, suppose that we have evaluated the $(m \times m)$ variance–covariance matrix[3] of $\widehat{\theta}_{t-1}$, which we denote

[3] See footnote on p. 204.

by $P_{t-1}$. The first stage, called the **prediction stage**, is concerned with forecasting $\theta_t$ from data up to time $(t-1)$, and we denote the resulting estimator in an obvious notation by $\widehat{\theta}_{t|t-1}$. Considering Equation (10.3), where $w_t$ is still unknown at time $t-1$, the obvious estimator for $\theta_t$ is given by

$$\widehat{\theta}_{t|t-1} = G_t \widehat{\theta}_{t-1} \tag{10.11}$$

with variance–covariance matrix

$$P_{t|t-1} = G_t P_{t-1} G_t^T + W_t. \tag{10.12}$$

Equations (10.11) and (10.12) are called the **prediction equations**. Equation (10.12) follows from standard results on variance–covariance matrices for vector random variables (e.g. Chatfield and Collins, 1980, Equation (2.9)).

When the new observation at time $t$, namely, $X_t$, has been observed, the estimator for $\theta_t$ can be modified to take account of this extra information. At time $(t-1)$, the best forecast of $X_t$ is given by $h_t^T \widehat{\theta}_{t|t-1}$ so that the prediction error is given by

$$e_t = X_t - h_t^T \widehat{\theta}_{t|t-1}.$$

This quantity can be used to update the estimate of $\theta_t$ and of its variance–covariance matrix. It can be shown that the best way to do this is by means of the following equations:

$$\widehat{\theta}_t = \widehat{\theta}_{t|t-1} + K_t e_t \tag{10.13}$$

and

$$P_t = P_{t|t-1} - K_t h_t^T P_{t|t-1} \tag{10.14}$$

where

$$K_t = P_{t|t-1} h_t / [h_t^T P_{t|t-1} h_t + \sigma_n^2] \tag{10.15}$$

is called the Kalman gain matrix. In the univariate[4] case, $K_t$ is just a vector of size $(m \times 1)$. Equations (10.13) and (10.14) constitute the second **updating stage** of the Kalman filter and are called the **updating equations**.

We will not attempt to derive the updating equations or to demonstrate the optimality of the Kalman filter. However, we note that the results may be found via least squares theory or using a Bayesian approach. A clear introduction to the Kalman filter is given by Meinhold and Singpurwalla (1983), while more detailed accounts are given by Harvey (1989, 1993, Chapter 4), Aoki (1990) and Durbin and Koopman (2001).

A major practical advantage of the Kalman filter is that the calculations are recursive, so that, although the current estimates are based on the whole past history of measurements, there is no need for an ever-expanding memory. Rather the new estimate of the signal is based solely on the previous estimate and the latest observation. A second advantage of the Kalman filter is that it converges fairly quickly when

[4]The observation $X_t$ is univariate, but remember that the state vector $\theta_t$ is $(m \times 1)$.

there is a constant underlying model, but can also follow the movement of a system where the underlying model is evolving through time.

The Kalman filter equations look rather complicated at first sight, but they may readily be programmed in their general form and reduce to much simpler equations in certain special cases. For example, consider the random walk plus noise model of Section 10.1.1 where the state vector $\theta_t$ consists of just one state variable, the current level $\mu_t$. After some algebra (e.g. Abraham and Ledolter, 1986), it can be shown that the Kalman filter for this model in the steady-state case (as $t \to \infty$) reduces to the simple recurrence relation

$$\widehat{\mu}_t = \widehat{\mu}_{t-1} + \alpha e_t, \tag{10.16}$$

where the smoothing constant $\alpha$ is a (complicated) function of the signal-to-noise ratio $\sigma_w^2/\sigma_n^2$ (see Exercise 10.1). Equation (10.16) is, of course, *simple exponential smoothing*. When $\sigma_w^2$ tends to zero, so that $\mu_t$ is a constant, we find that $\alpha$ tends to zero as would intuitively be expected, while as $\sigma_w^2/\sigma_n^2$ becomes large, then $\alpha$ approaches unity.

As a second example, consider the linear regression model with time-varying coefficients in Section 10.1.6. Abraham and Ledolter (1983, Section 8.3.3) show how to find the Kalman filter for this model. In particular, it is easy to demonstrate that, when $W_t$ is the zero matrix, so that the regression coefficients are constant, then $G_t$ is the identity matrix, while $P_{t|t-1} = P_{t-1}$. Then the Kalman filter reduces to the equations

$$\begin{aligned} \widehat{\theta}_t &= \widehat{\theta}_{t-1} + K_t e_t, \\ P_t &= P_{t-1} - K_t h_t^T P_{t-1}, \end{aligned}$$

where

$$\begin{aligned} e_t &= X_t - h_t^T \widehat{\theta}_{t-1}, \\ K_t &= P_{t-1} h_t [h_t^T P_{t-1} h_t + \sigma_n^2]^{-1}. \end{aligned}$$

Abraham and Ledolter (1983, Section 8.3.3) demonstrate that these equations are the same as the 'well-known' updating equations for recursive least squares provided that starting values are chosen in an appropriate way.

In order to initialize the Kalman filter, we need estimates of $\theta_t$ and $P_t$ at the start of the series. This can be done by *a priori* guesswork, relying on the fact that the Kalman filter will rapidly update these quantities so that the initial choices become dominated by the data. Alternatively, one may be able to estimate the $(m \times 1)$ vector $\theta_t$ at time $t = m$ by least squares from the first $m$ observations, since if we can write

$$X_0 = M\theta_m + e,$$

where $X_0^T = (X_m, X_{m-1}, \ldots, X_1)$, $M$ is a known non-singular $(m \times m)$ matrix and $e$ is an $m$-vector of independent 'error' terms, then

$$\widehat{\theta}_m = M^{-1} X_0 \tag{10.17}$$

is the least squares estimate of $\theta_m$ (since $M$ is a square matrix). An example is given in Exercise 10.2.

Once a model has been put into state-space form, the Kalman filter can be used to provide recursive estimates of the signal, and they in turn lead to algorithms for various other calculations, such as making predictions and handling missing values. For example, *forecasts* may readily be obtained from the state-space model using the latest estimate of the state vector. Given data to time $N$, the best estimate of the state vector is written as $\widehat{\theta}_N$ and the $h$-step-ahead forecast is given by

$$\widehat{X}_N(h) = h_{N+h}^T \widehat{\theta}_{N+h} = h_{N+h}^T G_{N+h} G_{N+h-1} \ldots G_{N+1} \widehat{\theta}_N,$$

where we assume $h_{N+h}$ and future values of $G_t$ are known. Of course if $G_t$ is a constant, say $G$, then

$$\widehat{X}_N(h) = h_{N+h}^T G^h \widehat{\theta}_N. \tag{10.18}$$

If future values of $h_t$ or $G_t$ are *not* known, then they must themselves be forecasted or otherwise guesstimated.

The Kalman filter is applied to state-space models that are linear in the parameters. In practice many time-series models, such as multiplicative seasonal models, are non-linear. Then it may be possible to apply a filter, called the **extended Kalman filter**, by making a locally linear approximation to the model. Applications to data where the noise is not necessarily normally distributed are also possible, but we will not pursue these more advanced topics here (see, e.g. Durbin and Koopman, 2001).

As an example, we will evaluate the Kalman filter for the linear growth model of Section 10.1.2. Suppose that, from data up to time $(t-1)$, we have estimates $\widehat{\mu}_{t-1}$ and $\widehat{\beta}_{t-1}$ of the level and trend. At time $(t-1)$ the best forecasts of $w_{1,t}$ and $w_{2,t}$ are both zero so that the best forecasts of $\mu_t$ and $\beta_t$ in Equation (10.6) are clearly given by

$$\widehat{\mu}_{t|t-1} = \widehat{\mu}_{t-1} + \widehat{\beta}_{t-1}$$

and

$$\widehat{\beta}_{t|t-1} = \widehat{\beta}_{t-1}.$$

These agree with Equation (10.11). When $X_t$ becomes available, we can find the prediction error, namely, $e_t = X_t - \widehat{\mu}_{t|t-1}$, and this can then be inserted into Equation (10.13) to give the following pair of scalar equations, namely

$$\widehat{\mu}_t = \widehat{\mu}_{t|t-1} + k_{1,t}\, e_t = \widehat{\mu}_{t-1} + \widehat{\beta}_{t-1} + k_{1,t}\, e_t,$$

and

$$\widehat{\beta}_t = \widehat{\beta}_{t|t-1} + k_{2,t}\, e_t = \widehat{\beta}_{t-1} + k_{2,t}\, e_t,$$

where $k_{1,t}$, $k_{2,t}$ are the elements of the Kalman gain 'matrix' (here a $2 \times 1$ vector) $K_t$, which can be evaluated after some algebra. It is interesting to note that these two equations are of similar form to those in Holt's two-parameter (non-seasonal) version of exponential smoothing (see Section 5.2.3). There the level and trend are denoted by $L_t, T_t$, respectively, and we have, for example, that

$$L_t = \alpha X_t + (1-\alpha)(L_{t-1} + T_{t-1}) = L_{t-1} + T_{t-1} + \alpha e_t,$$

where $e_t = X_t - [L_{t-1} + T_{t-1}]$. In the steady state as $t \to \infty$, it can be shown that $k_{1,t}$ tends to a constant, which corresponds to the smoothing parameter $\alpha$. This demonstrates that the forecasting method called Holt's (two-parameter) exponential smoothing is optimal for the linear growth model.

An intuitively reasonable way to initialize the two state variables from the first two observations is to take $\widehat{\mu}_2 = X_2$ and $\widehat{\beta}_2 = X_2 - X_1$ (see Exercise 10.2).

## Exercises

**10.1** Consider the random walk plus noise model in Section 10.1.1, and denote the signal-to-noise ratio $\sigma_w^2/\sigma_n^2$ by $c$. Show that the first-order autocorrelation coefficient of $(1-B)X_t$ is $-1/(2+c)$ and that higher-order autocorrelations are all zero.

For the ARIMA$(0,1,1)$ model

$$(1-B)X_t = Z_t + \theta Z_{t-1}$$

show that the first-order autocorrelation coefficient of $(1-B)X_t$ is $\theta/(1+\theta^2)$ and that higher-order autocorrelations are all zero. Thus the two models have equivalent autocorrelation properties when $\theta/(1+\theta^2) = -1/(2+c)$. Hence show that the invertible solution, with $|\theta| < 1$, is $\theta = \frac{1}{2}[(c^2+4c)^{1/2} - c] - 1$.

Applying the Kalman filter to the random walk plus noise model, we find (after some algebra) that, in the steady state (as $t \to \infty$ and $P_t \to$constant), we have

$$\widehat{\mu}_t = \widehat{\mu}_{t-1} + \alpha e_t$$

and

$$\alpha = 1 + \theta = \frac{1}{2}[(c^2+4c)^{1/2} - c]$$

and this is simple exponential smoothing. Now the ARIMA model is invertible provided that $-1 < \theta < 1$, suggesting that $0 < \alpha < 2$. However, the random walk plus noise model restricts $\alpha$ to the range $0 < \alpha < 1$ (and hence $-1 < \theta < 0$) and physical considerations suggest that this is generally a more sensible model. Do you agree?

**10.2** Consider the following special case of the linear growth model:

$$\begin{aligned} X_t &= \mu_t + n_t \\ \mu_t &= \mu_{t-1} + \beta_{t-1} \\ \beta_t &= \beta_{t-1} + w_t \end{aligned}$$

where $n_t, w_t$ are independent normal with zero means and respective variances $\sigma_n^2, \sigma_w^2$. Show that the initial least squares estimator of the state vector at time $t = 2$, in terms of the observations $X_1$ and $X_2$, is $[\widehat{\mu}_2, \widehat{\beta}_2] = [X_2,\ X_2 - X_1]$ with variance–covariance matrix

$$P_2 = \begin{bmatrix} \sigma_n^2 & \sigma_n^2 \\ \sigma_n^2 & 2\sigma_n^2 + \sigma_w^2 \end{bmatrix}.$$

If $\sigma_w^2 = 0$, so that we have ordinary linear regression with constant coefficients, and a third observation $X_3$ becomes available, apply the Kalman filter to show that the estimator of the state vector at time $t = 3$ is given by

$$[\widehat{\mu}_3, \widehat{\beta}_3] = \left[\frac{5}{6}X_3 + \frac{1}{3}X_2 - \frac{1}{6}X_1,\ (X_3 - X_1)/2\right].$$

Verify that these are the same results that would be obtained by ordinary least squares regression.

**10.3** Find a state-space representation of (a) the MA(1) process $X_t = Z_t + \beta Z_{t-1}$, (b) the MA(2) process.
(Hint for (a): Try $\theta_t^T = [X_t, \widehat{X}_t(1)] = [X_t, \beta Z_t]$.)

**10.4** Find a state-space representation of the AR(2) process in Equation (10.8) based on the state vector $\theta_t^T = [X_t, \widehat{X}_t(1)]$, where the optimal one-step-ahead predictor at time $t$ is given by $\widehat{X}_t(1) = \phi_1 X_t + \phi_2 X_{t-1}$, and show that $G = \begin{bmatrix} 0 & 1 \\ \phi_2 & \phi_1 \end{bmatrix}$ with $w_t^T = (1, \phi_1)Z_t$.

**10.5** Find a state-space representation of the AR($p$) process

$$X_t = \phi_1 X_{t-1} + \cdots + \phi_p X_{t-p} + Z_t$$

based on the state vector $\theta_t^T = (X_t, \ldots, X_{t-p})$ Note that the state vector is observable in this case – a rather unusual case. Show that

$$G = \begin{bmatrix} \phi_1 & \phi_2 & \cdots & \phi_{p-1} & \phi_p \\ 1 & 0 & \cdots & 0 & 0 \\ \cdots & \cdots & \cdots & \cdots & \cdots \\ 0 & 0 & \cdots & 1 & 0 \end{bmatrix}.$$

Chapter 11

# Non-Linear Models

Most of this book (like most of the time-series statistical literature) is concerned with *linear* methods and models. However, there is no reason why real-life generating processes should all be linear, and the assumption of linearity is often made more for mathematical and computational convenience rather than because it is really believed to be true. Nowadays, there is growing interest in non-linear models combined with a greater computational facility for fitting them, and so this chapter provides an introduction to various types of univariate non-linear models. A brief bibliography is given at the end of the chapter.

## 11.1 Introduction

This section motivates the need for non-linear models and attempts the difficult task of distinguishing between linear and non-linear models.

### *11.1.1 Why non-linearity?*

Figure 11.1 displays a famous time series giving the average number of sunspots recorded in successive months. The data are listed, for example, by Andrews and Herzberg (1985), and the (updated) series may be obtained in R. Two representations of the data from 1749 to 1983 are given, with different vertical axes. Close inspection of either graph reveals that there is regular cyclic behaviour with a period of approximately 11 years. At first sight, the upper graph seems a more natural way to display the data, but the graph goes up and down so rapidly that it is really only possible to see where the maxima and minima occur. The lower graph uses a smaller height, and it[1] enables us to see that the series generally increases at a faster rate than it decreases. Behaviour like this cannot be explained by a linear model and is quite different from the asymmetry arising from a linear model with an asymmetric error distribution. Figure 11.1 can be reproduced by the following R script.

[1]This time plot provides an excellent example of the care that is needed in drawing graphs, particularly as regards the choice of scales (see Sections 2.3 and 14.4). One representation may be 'good' for one purpose (e.g. displaying the maxima) while a second representation may be 'good' for a different purpose (e.g. looking at the slope of increases and decreases). The plotting of the yearly version of the sunspot data is discussed by Cleveland (1994) together with additional relevant examples and advice.

DOI: 10.1201/9781003646112-11

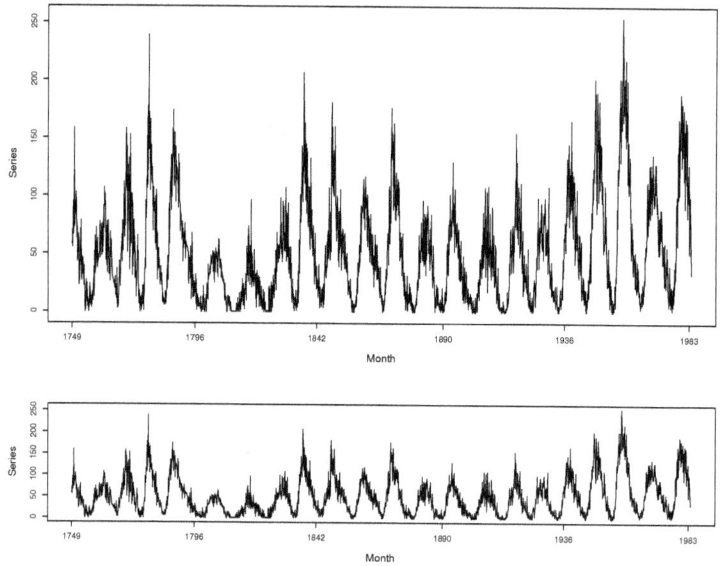


Figure 11.1 *Average monthly sunspot numbers from 1860 to 1983. The lower graph has a shortened vertical axis, which enables the rise and fall of the graph to be more easily assessed.*

```
> library(tseries)
> m<-rbind(1,1,2)
> layout(m)
> par(mar=c(4,4,4,4), cex.lab=1.2)
> plot(sunspots, xlab="Year", ylab="Series", xaxt="n")
> x.pos<-c(1749, 1796, 1842, 1890, 1936, 1983)
> axis(1, x.pos, x.pos)
> plot(sunspots, xlab="Year", ylab="Series", xaxt="n")
> x.pos<-c(1749, 1796, 1842, 1890, 1936, 1983)
> axis(1, x.pos, x.pos)
```

Another famous time series records the annual numbers of lynx trapped in the Mackenzie River district of Canada between 1822 and 1934 (e.g. Hand et al., 1994, Data Set 109). These data are shown in Figure 11.2 and also show asymmetric cyclic behaviour but with the series *falling* faster than it rises. Asymmetric behaviour can also arise in studying the economy since the relationships between economic variables tend to be different when the economy is moving into a recession rather than when coming out of a recession – downturns are often steeper and more short-lived than upturns. Thus 'it seems to be generally accepted that the economy is nonlinear' Granger and Teräsvirta (1993, p. 1). As yet another example, the amount of water flowing down a river tends to increase sharply after a heavy storm and then to tail off gradually.

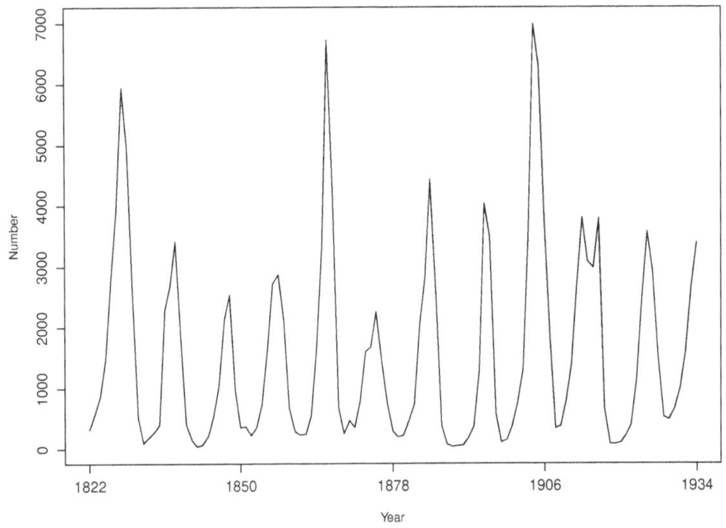


Figure 11.2 *Annual numbers of lynx trapped in the Mackenzie River district of Canada between 1821 and 1934.*

For seasonal series with a fixed cycle length, it may be possible to model asymmetric behaviour with a non-sinusoidal seasonal component, but when the cycle length is not fixed as in the above examples, a non-linear model is much more compelling for describing series with properties such as 'going up faster than coming down'. Non-linear models are also needed to describe data where the variance changes through time — see Section 11.3 and Figure 11.3.

Non-linear models can also be used to explain, and give forecasts for, data exhibiting regular cyclic behaviour. As such they provide an interesting alternative to the use of harmonic components, especially if the behaviour is asymmetric. For some non-linear models, if the noise process is 'switched off', then the process will converge asymptotically to a strictly periodic form called a **limit cycle** (Priestley, 1988; Tong, 1990).

Questions about non-linearity also arise when we consider transforming a variable using a non-linear transformation such as the Box–Cox transformation (see Section 2.4). Data may be transformed for a variety of reasons such as to make the data more normally distributed or to achieve constant variance. However, if we are able to fit a linear model to the transformed data, this will imply that a non-linear model is appropriate for the original data. In particular, a series that shows multiplicative seasonality can be transformed to additive seasonality by taking logs and can then be handled using linear methods. However, the multiplicative model for the original data will be non-linear.

Before embarking on a non-linear analysis, it is sensible to check that the data really are non-linear (Darbellay and Slama, 2000). There are various technical procedures described in the literature for assessing and testing different aspects of non-linearity. A recent survey, with further references, is given by Tsay (2001), and the so-called BDS test, in particular, is briefly discussed later in this chapter. However, tests for non-linearity often have poor power, and the simplest, and arguably the most important, tool (as in the rest of time-series analysis) is a careful inspection of the time plot. Behaviour such as that seen in Figure 11.1 can be self-evident *provided the scales are chosen carefully*. It should also be noted that tests for non-linearity can have difficulty in distinguishing between data from a non-linear model and data from a linear model to which outliers have been added. While some non-linear models can give rise to occasional sharp spikes, the same is true of a linear model with occasional outliers. Here again, a careful inspection of the time plot can be both crucial and fruitful, especially when allied to expert contextual knowledge as to when, where and why unusual observations might occur. This highlights the close connection between non-linearity and non-normality. If, for example, a time series exhibits more 'spikes' up than down, then it is often not clear if this is due to non-linearity, non-normality, or both.

### *11.1.2 What is a linear model?*

The first point to make is that there is no clear consensus as to exactly what is meant by a linear stochastic time-series model and hence no consensus as to what is meant by a *non-linear* model. In much of statistical methodology, the term **general linear model** is used to describe a model that is linear in the parameters but that could well involve non-linear functions of the explanatory variables in the so-called design matrix. In contrast a linear model (or linear system) in time series would certainly exclude non-linear functions of lagged or explanatory variables. To complicate matters further, an MA process, which is certainly a linear process (see Chapter 3) and which, at first sight, appears to be linear in the parameters, is actually regarded as non-linear in the parameters in that the one-step-ahead errors (on which least-squares estimation is based) are non-linear functions of the parameters. This means that explicit analytic formulae for estimators may not be available (see Section 4.3.1) and Box and Jenkins (1970) use the term **non-linear estimation** to describe procedures for minimizing a sum-of-squares function when numerical methods (such as hill-climbing) have to be used. This chapter only discusses the use of the term 'non-linear' as applied to models and forecasting methods.

The most obvious example of a linear model is the **general linear process** (see Section 3.11), which arises when the value of a time series, say $X_t$, can be expressed as a linear function of the present and past values of a purely random process, say $Z_t$. This class of models includes stationary autoregressive (AR), moving average (MA), and ARMA models. The linearity of the process is clear when the model is viewed as a **linear system** (see Chapter 9) for converting the sequence of $Z_t$s into a sequence of $X_t$s. In addition the state-space model defined by Equations (10.2) and (10.3) is generally regarded as linear provided the disturbances are normally distributed, $h_t$

is a constant known vector and $G_t, W_t$ are constant known matrices (or at least are non-stochastic, so that if they change through time, they do so in a predetermined way).

A linear forecasting **method** is one where the $h$-steps-ahead forecast at time $N$ can be expressed as a linear function of the observed values up to, and including, time $N$. This applies to exponential smoothing, and the additive (though not the multiplicative) version of Holt–Winters. It also applies to minimum mean square error (MMSE) forecasts derived from a stationary ARMA model with known parameters, as would be expected for a general linear process. However, note that when the model parameters have to be estimated from the data (as is normally the case), the MMSE forecasts will *not* be linear functions of past data.

The status of (non-stationary) ARIMA models is not so obvious. Apart from the non-stationarity (which means they can't be expressed as a general linear process), they look linear in other respects. An ARI model, for example, can be regarded as a linear system by treating $\{X_t\}$ and $\{Z_t\}$ as if they were the input and output, respectively, although it is really the other way round. Moreover MMSE forecasts from ARIMA models (assuming known model parameters) will be linear functions of past data. This suggests that it might be possible to define a linear model as any model for which MMSE forecasts are linear functions of observed data. However, while this is a necessary condition, it is not sufficient, because some models give linear prediction rules while exhibiting clear non-linear properties in other respects. A further complication is that it is possible to have models that are **locally linear**, but **globally non-linear** (see Sections 2.5 and 10.1.2). Thus it appears that it may not be feasible to define linearity precisely, but rather that it is possible to move gradually away from linearity towards non-linearity.

### *11.1.3 What is a non-linear model?*

A non-linear model could be defined by exclusion as any model that is not linear. However, this is not helpful since (1) linear models have not been exactly defined, and (2) there is a blurred borderline between linearity and non-linearity. For example, some long-memory models (see Section 13.3) are more linear than non-linear, while for non-stationary models, the non-stationarity property is often more important than whether the model is linear. In any case, as noted earlier, a variable, which can be described by a linear model, becomes non-linear after applying a non-linear transformation.

This chapter therefore restricts attention to certain classes of models that are conventionally regarded as non-linear models, even though some of them have some linear characteristics while some excluded models have non-linear characteristics.

### *11.1.4 What is white noise?*

When examining the properties of non-linear models, it can be very important to distinguish between independent and uncorrelated random variables. In Section 3.4, **white noise** (or a **purely random process**) was defined to be a sequence of independent and identically distributed (i.i.d.) random variables. This is sometimes called **strict white noise** (SWN), and the phrase **uncorrelated white noise** (UWN) is used when successive values are merely uncorrelated, rather than independent. Of course if successive values follow a normal (Gaussian) distribution, then zero correlation implies independence so that Gaussian UWN is SWN. However, with non-linear models, distributions are generally non-normal and zero correlation need not imply independence.

In reading the literature, it is also helpful to understand the idea of a martingale difference. A series of random variables $\{X_t\}$ is called a **martingale** if $\mathrm{E}[X_{t+1} | \text{data to time} t]$ is equal to the observed value of $X_t$, say $x_t$. Then a series $\{Y_t\}$ is called a **martingale difference** (MD) if $\mathrm{E}[Y_{t+1} | \text{ data to time } t] = 0$. This last result follows by letting $\{Y_t\}$ denote the first differences of a martingale, namely, $Y_t = X_t - X_{t-1}$. An MD is like UWN except that it does not need to have constant variance. Of course a Gaussian MD with constant variance is SWN.

UWN and MDs have known linear, second-order properties. For example, they have constant mean and zero autocorrelations. However, the definitions say nothing about the non-linear properties of such series. In particular, although $\{X_t\}$ may be UWN or an MD, the series of squared observations $\{X_t^2\}$ need not be. Only if $\{X_t\}$ is SWN, will $\{X_t^2\}$ be UWN. There are many tests for linearity whose power depends on the particular type of non-linearity envisaged for the alternative hypothesis (e.g. see Brock and Potter, 1993; Tsay, 2010, Section 4.2). These tests generally involve looking at the properties of moments of $\{X_t\}$, which are higher than second order, particularly at the autocorrelation function of $\{X_t^2\}$, which involves fourth-order moments.

Some analysts also like to look at **polyspectra**, which are the frequency domain equivalent of this and involve taking the Fourier transform of higher order moments of the process. The simplest example is the **bispectrum**, which is the transform of third-order terms of the general form $X_t X_{t-j} X_{t-k}$. For linear Gaussian processes, all polyspectra (including the bispectrum) are identically zero for order three or more and so bispectra have been used as part of test procedures for normality and for linearity. The sunspots data and the lynx data, for example, are both found to be non-linear and non-normal using tests based either on the bispectrum or on time-domain statistics. This author has little experience with polyspectra and will not discuss them here. It appears that they can be useful as a descriptive tool for the analyst interested in frequency domain characteristics of data, but they can be difficult to interpret and may not be best suited for use in tests of linearity.

The rest of this chapter introduces some stochastic time-series models with a non-linear structure. Attention is restricted to some important classes of models that are of particular theoretical and practical importance.

## 11.2 Non-Linear Autoregressive Processes

An obvious way to generalize the (linear) AR model of order $p$ is to assume that

$$X_t = f(X_{t-1}, X_{t-2}, \ldots, X_{t-p}) + Z_t, \tag{11.1}$$

where $f$ is some non-linear function and $Z_t$ denotes a purely random process. This is called a **non-linear autoregressive** (NLAR) model of order $p$. Note that the 'error' term is assumed to be additive. A more general error structure is possible, perhaps with $Z_t$ incorporated inside the $f$ function, but will not be considered here.

For simplicity consider the case $p = 1$. Then we can rewrite Equation (11.1) as

$$X_t = \phi(X_{t-1})X_{t-1} + Z_t \tag{11.2}$$

where $\phi$ is some non-constant function. It can be shown that a sufficient condition for Equation (11.2) to describe a stable model is that $\phi$ must satisfy the constraint that $|\phi(x)| < 1$, at least when $|x|$ is large. A model such as

$$X_t = \alpha X_{t-1}^2 + Z_t,$$

which does not satisfy this condition, will generally be explosive (unless $Z_t$ is constrained to an appropriate finite interval), and thus will not be much use for describing real data, even though a comparable quadratic model, with $X_{t-1}^2$ replaced by the square of some explanatory variable, might seem plausible in a regression, rather than autoregression, context. Clearly non-linearity has to be introduced in a more subtle way in a time-series context.

Several variations of an AR model have been proposed, which allow the parameter(s) of the AR model to change through time, either deterministically (unusual) or stochastically or determined in some way by past data. Considering the stochastic option first, and taking the first-order case as an example, we could let

$$X_t = \alpha_t X_{t-1} + Z_t$$

where

$$\alpha_t = \gamma + \beta\alpha_{t-1} + \varepsilon_t$$

and $\gamma$, $\beta$ are constants with $\{\varepsilon_t\}$ being an i.i.d. sequence independent of the $\{Z_t\}$ sequence. Thus the parameter of the AR(1) process for $X_t$ itself follows an AR(1) process. Such models are called **time-varying parameter models** (e.g. see Nicholls and Pagan, 1985). In the case when $\beta = 0$ in Equation (11.2), the model reduces to what is sometimes called a **random coefficient model**. While such models appear intuitively plausible at first sight – we all know the world is changing – they can be tricky to handle and it is hard to distinguish between the constant and time-varying parameter cases. It is also not immediately obvious if the model is linear or non-linear. The model may appear to be linear in Equation (11.2), but appears non-linear if Equations (11.2) and (11.2) are combined, with $\beta = 0$ for simplicity, as

$$X_t = \gamma X_{t-1} + Z_t + \varepsilon_t X_{t-1}.$$

The last term is non-linear. The 'best' point forecast at time $t$ is $\hat{\gamma} x_t$. This looks linear at first sight, but of course $\hat{\gamma}$ will involve past data leading to a non-linear function (though a similar situation arises for the constant-parameter AR model when the parameter(s) have to be estimated from the data). A more convincing argument for non-linearity is that, when Equation (11.2) is expressed as a state-space model (as in Equation 10.2), it is not a linear model since $h_t = \alpha_t$ changes stochastically through time. Moreover it can be shown that the width of prediction intervals depends in a non-linear way on the latest value of the series.

Another general possibility arises if we assume that the function $f$ in Equation (11.1) is piecewise linear. This means that it consists of two or more linear functions defined over different regions of the lagged values of $X_t$. Consider, for example, the NLAR model of order 1, namely

$$X_t = f(X_{t-1}) + Z_t. \tag{11.3}$$

If we assume that $f$ is piecewise linear, then the parameters of $f$ could, for example, depend on whether the value of $X_{t-1}$ is larger or smaller than a critical value, customarily called a threshold. Thus we effectively allow the model parameters to be determined by past data. This leads to the idea of a threshold AR model, which will be considered in the next subsection.

## 11.3 Threshold Autoregressive Models

Following from Equation (11.3), a threshold AR model is a piecewise linear model where the parameters of an AR model are determined by the values taken by one or more of the lagged values of the time series. Consider, for example, the model

$$X_t = \begin{cases} \alpha^{(1)} X_{t-1} + Z_t & \text{if } X_{t-1} < \gamma, \\ \alpha^{(2)} X_{t-1} + Z_t & \text{if } X_{t-1} \geq \gamma, \end{cases} \tag{11.4}$$

where $\alpha_{(1)}$, $\alpha_{(2)}$, $\gamma$ are constants and $\{Z_t\}$ denotes SWN. In one sense, this is an AR(1) model, but the AR parameter depends on whether $X_{t-1}$ exceeds the value $r$ called the **threshold**. Below $r$ the AR parameter is $\alpha_{(1)}$, but above $r$ it is $\alpha_{(2)}$. This feature makes the model non-linear and it is an example of a large class of models called **threshold autoregressive** (TAR) models.

Tong (1990) calls a TAR model **self-exciting** when the choice from the various sets of possible parameter values is determined by just one of the past values, say $X_{t-d}$ where $d$ is the *delay*. In Equation (11.4), the choice is determined by the value of $X_{t-1}$ and so the model is indeed self-exciting. Model (11.4) can be readily be extended to higher order autoregressions and to more than two thresholds depending on the values of one or more past data values. In particular, a $k$-regime self-exciting TAR (SETAR) model with threshold variable $X_{t-d}$ if it satisfies

$$\begin{aligned} X_t &= \alpha_0^{(j)} + \alpha_1^{(j)} X_{t-1} + \cdots + \alpha_p^{(j)} X_{t-p} + Z_t^{(j)}, \\ &\qquad \text{if } \gamma_{j-1} \leq X_{t-d} < \gamma_j, \end{aligned} \tag{11.5}$$

in the $j$th regime, where $k$ and $d$ are positive integers, $j = 1, \ldots, k$, $\gamma_j$ are real numbers such that $-\infty = \gamma_0 < \gamma_1 < \cdots < \gamma_k = \infty$, and $\{Z_t^{(j)}\}$ are i.i.d. sequences with mean 0 and variance $\sigma_j^2$ and are mutually independent for different $j$.

Some theory for threshold models is given by Tong (1990). Threshold models are piecewise linear in that they are linear in a particular subset of the sample space, and can be thought of as providing a piecewise linear approximation to some general non-linear function as in Equation (11.1). Threshold models sometimes give rise to periodic behaviour with a limit cycle and it can be fruitful to demonstrate such a property by plotting $X_t$ against $X_{t-1}$, or more generally by plotting $X_t$ against $X_{t-k}$. Such a graph is sometimes called a **phase diagram** (actually a discrete-time version of such a diagram). Such graphs are useful, not just for identifying TAR models, but more generally in assessing the general form of a lagged relationship, particularly whether it is linear or non-linear.

Estimation and forecasting for TAR models are, perhaps inevitably, rather more difficult than for linear AR models and will not be covered here. Estimation usually involves some sort of iterative procedure, while the difficulties involved in forecasting can be illustrated for the first-order model in Equation (11.4). Suppose we have data up to time $N$ and that $x_N$ happens to be larger than the threshold $r$. Then it is obvious from the model that $\hat{x}_N(1) = E[X_{N+1}|\text{data to time } N] = \alpha_{(2)}x_N$, and so the one-step-ahead forecast is easy to find. However, finding the two-steps-ahead forecast for the above model is much more complicated, as it will depend on whether the next observation happens to exceed the threshold. The expectation that arises is algebraically intractable and some sort of integration or approximation will need to be made to evaluate expectations over future error terms and the corresponding thresholds.

An interesting application of threshold models was made by Chappell et al. (1996) in regard to exchange rates within the European Union. These rates are supposed to stay within prescribed bounds. A threshold model led to improved forecasts as compared with a random walk model. A second example, using economic data, is given by Tiao and Tsay (1994). Although the latter authors found little improvement in forecasts using a threshold model, the modelling process required to fit a non-linear model led to greater insight into economic relationships, particularly that the economy behaves in a different way when it is going into, or coming out of, recession. Tsay (1998) has extended threshold models to the multivariate case, and gives details on testing, estimation and modelling together with some examples.

*Example 11.1*

We consider an example of analysing the average monthly sunspot numbers from 1860 to 1983 (see Figure 11.1) with SETAR models. Denote $Y_t$ the sunspot number at year $t$. We first transform the data by taking logarithm,

$$\widetilde{Y}_t = \log(Y_t + 1).$$

The reason for adding 1 to $Y_t$ is that the sunspot numbers at certain years are zero. We note that the solar cycle or solar magnetic activity cycle is nearly 11-years, so we consider the sunspot numbers are approximately periodic with 11-year cycles. Accordingly, we use the `R` function `decompose` to single out the seasonal effect in $\widetilde{Y}_t$. Figure 11.3 shows the estimated trend, seasonal effect $S_t$, and the remainder series. Figure 11.4 shows the deseasoned series $X_t := \widetilde{Y}_t - S_t$, and its sample ac.f.'s and partial sample ac.f.'s. Figures 11.3 and 11.4 can be reproduced by the following `R` script.

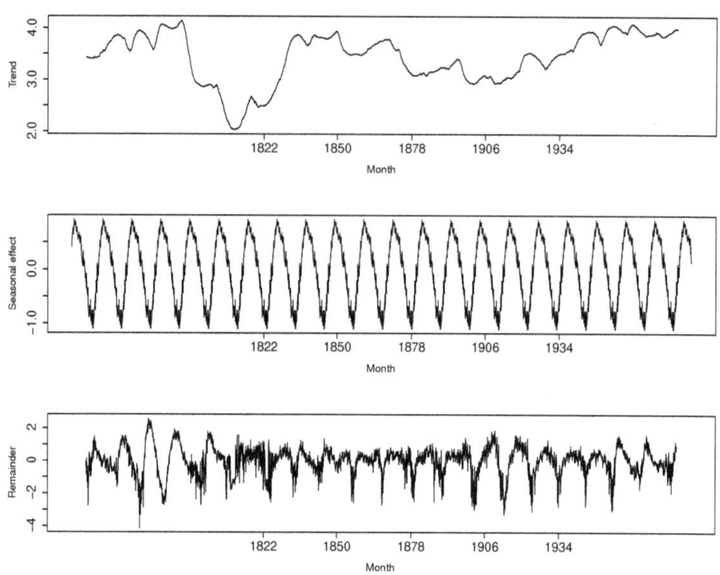


Figure 11.3 *The decomposition of the transformed sunspots series $\widetilde{Y}_t$ (Top: Estimated trend $m_t$; Middle: Estimated seasonal effect $S_t$; Bottom: The remainder series).*

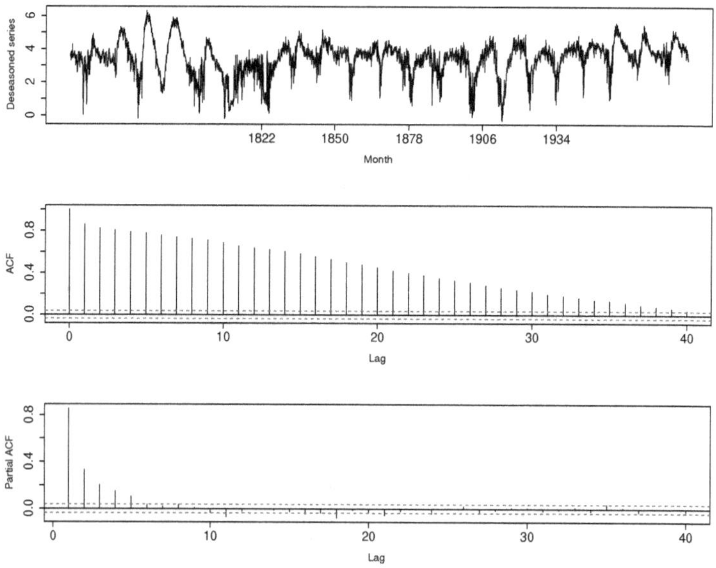


Figure 11.4 *The deseasoned series $X_t = \widetilde{Y}_t - S_t$ (top), and its sample ac.f. (middle) and sample partial ac.f. (bottom).*

```
> # Decompose the transformed series with 11-year cycles
> ss<-ts(log(sunspots+1), frequency=12*11, start=c(1749,1))
> ss.de<-decompose(ss)
> ss.trend<-ts(ss.de$trend, frequency=12, start=c(1749,1))
> ss.random<-ts(ss.de$random, frequency=12, start=c(1749,1))
> ss.se<-ts(ss.de$seasonal, frequency=12, start=c(1749,1))
> x.pos<-c(1749, 1796, 1842, 1890, 1936, 1983)

> # Figure 11.3
> par(mfrow=c(3,1), mar=c(4,4,4,4), cex.axis=1.5, cex.lab=1.2)
> plot(ss.trend, xlab="Year", ylab="Trend", xaxt="n")
> axis(1, x.pos, x.pos)
> plot(ss.se, xlab="Year", ylab="Seasonal effect", xaxt="n")
> axis(1, x.pos, x.pos)
> plot(ss.random, xlab="Year", ylab="Remainder", xaxt="n")
> axis(1, x.pos, x.pos)

> # Figure 11.4
> ss.nose<-ts(ss-ss.de$seasonal, frequency=12, start=c(1749,1))
> par(mfrow=c(3,1), mar=c(4,4,4,4), cex.axis=1.5, cex.lab=1.2)
> plot(ss.nose, xlab="Year", ylab="Deseasoned series", xaxt="n")
> axis(1, x.pos, x.pos)
> acf(as.numeric(ss.nose), ylab="ACF", main="", lag=40)
> pacf(as.numeric(ss.nose), ylab="Partial ACF", main="", lag=40)
```

The sample ac.f.'s and partial ac.f.'s in Figure 11.4 indicates that the series $X_t$ can be approximately considered as an AR process. To fit nonlinear AR models to a time series, we may use the package `tsDyn` in R. For comparison purposes, we first use the model selection procedure in Section 4.8 to find a good ARMA model for $X_t$.

```
> ssnose.arma.aic<-matrix(0,10,10)
> for (i in 0:9) for (j in 0:9) {
  tmp.fit<-arima(ss.nose, order=c(i,0,j))
  ssnose.arma.aic[i+1,j+1]<-tmp.fit$aic
 }
> which(ssnose.arma.aic==min(ssnose.arma.aic), TRUE)
     row col
[1,]  10   7
```

The above code shows that an ARMA(9, 6) model gives the minimum AIC among ARMA$(p,q)$ with $0 \leq p,q \leq 10$. Obviously, ARMA(9, 6) is not a good choice in practice as the model contains too many parameters. In such a case, it is worthwhile to try a nonlinear time series model, for example, a TAR model. The following R script shows how to fit a SETAR(2) model to $X_t$ using function `setar` in `tsDyn`.

```
> library(tsDyn)
> fit.setar<-setar(ss.nose, m=2, mL=2, mH=2, thDelay=1)
> summary(fit.setar)

Non linear autoregressive model

SETAR model ( 2 regimes)
Coefficients:
Low regime:
  const.L    phiL.1    phiL.2
0.5325051 0.4889504 0.3329734

High regime:
  const.H    phiH.1    phiH.2
0.1210850 0.7154626 0.2441025

Threshold:
-Variable: Z(t) = + (0) X(t)+ (1)X(t-1)
-Value: 3.277
Proportion of points in low regime: 34.6%  High regime: 65.4%

Residuals:
      Min        1Q    Median        3Q       Max
-2.933373 -0.215514  0.024918  0.251950  2.421657

Fit:
residuals variance = 0.2403,  AIC = -4007, MAPE = 23.28%

Coefficient(s):

         Estimate  Std. Error  t value  Pr(>|t|)
const.L  0.532505    0.051617  10.3166 < 2.2e-16 ***
phiL.1   0.488950    0.021724  22.5078 < 2.2e-16 ***
phiL.2   0.332973    0.024634  13.5168 < 2.2e-16 ***
const.H  0.121085    0.085173   1.4216    0.1552
phiH.1   0.715463    0.031474  22.7317 < 2.2e-16 ***
phiH.2   0.244103    0.036852   6.6238 4.179e-11 ***
---
Signif. codes:  0 '***' 0.001 '**' 0.01 '*' 0.05 '.' 0.1 ' ' 1

Threshold
Variable: Z(t) = + (0) X(t) + (1) X(t-1)

Value: 3.277
```

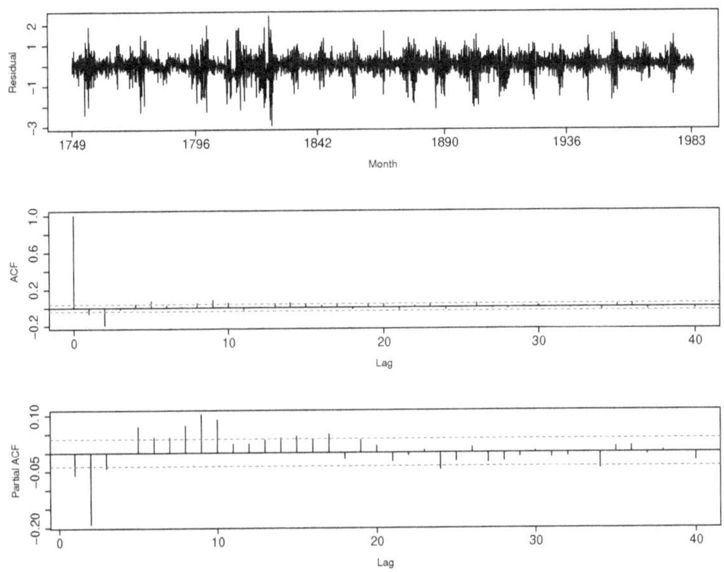


Figure 11.5 *The residual series in the fitted SETAR(2) model (top), and its sample ac.f. (middle) and sample partial ac.f. (bottom).*

```
> # Figure 11.5
> fit.setar.resid<-ts(fit.setar$resid, frequency=12,
    start=c(1749,3))
> par(mfrow=c(3,1), mar=c(4,4,4,4), cex.axis=1.5, cex.lab=1.2)
> plot(fit.setar.resid, xlab="Year", ylab="Residual",
    type="l", xaxt="n")
> axis(1, x.pos, x.pos)
> acf(fit.setar$resid, xlab="Lag",ylab="ACF", main="", lag=40)
> pacf(fit.setar$resid, xlab="Lag",ylab="Partial ACF",
    main="", lag=40)
```

The fitted SETAR(2) model is

$$X_t = \begin{cases} 0.5325_{(.0516)} + 0.4890_{(.0217)} X_{t-1} + 0.3330_{(.0246)} X_{t-2} + Z_t, \\ \qquad\qquad\qquad\qquad \text{if } X_{t-1} \le 3.277, \\ 0.1211_{(.0852)} + 0.7155_{(.0315)} X_{t-1} + 0.2441_{(.0369)} X_{t-2} + Z_t, \\ \qquad\qquad\qquad\qquad \text{if } X_{t-1} > 3.277, \end{cases}$$

where $Z_t \sim N(0, 0.2403)$. Figure 11.5 shows the residual series, and its sample ac.f.'s and partial ac.f.'s. Comparing the sample ac.f.'s and partial ac.f.'s of $X_t$ and $Z_t$, we note that serial correlations of $Z_t$ are significantly smaller than those of $X_t$.

## 11.4 Smooth Transition Autoregressive Models

A key feature of TAR models is the discontinuous nature of the AR relationship as the threshold is passed. Those who believe that nature is generally continuous may prefer an alternative model such as the **smooth transition autoregressive** (STAR) model where there is a smooth continuous transition from one linear AR model to another, rather than a sudden jump. For example, consider the model

$$X_t = a_0 + \sum_{j=1}^{p} a_j X_{t-j} + (b_0 + \sum_{j=1}^{p} b_j X_{t-j}) I(X_{t-d}) + Z_t, \tag{11.6}$$

where $d$ is the delay parameter, and $I$ is a smooth function with sigmoid characteristics. One example is the logistic function

$$I(x) = 1/[1+\exp[\gamma(r-x)]]$$

which depends on two parameters, namely, $r$, which is comparable to the threshold, and $1/\gamma$, which controls the speed of the switch from one model to the other. Of course, a STAR model reduces to a simple threshold model by choosing $I$ to be an indicator function taking the value zero below a threshold and one above it. Note that the conditional mean of a STAR model is a weighted sum of two conditional means,

$$E(X_t|X_{t-1}, X_{t-2}, \ldots) = \mu_{t1} + I(X_{t-d})\mu_{t2},$$

where

$$\mu_{t1} = a_0 + \sum_{j=1}^{p} a_j X_{t-j}, \qquad \mu_{t2} = b_0 + \sum_{j=1}^{p} b_j X_{t-j}.$$

A favourable property of the STAR model versus the SETAR model is that the conditional mean function is differentiable.

A STAR model can be estimated using the `R` function `lstar` in package `tsDyn`; see details in the following example.

*Example 11.2*

Consider the average monthly sunspot numbers $Y_t$ from 1860 to 1983 (see Figure 11.1). In Example 11.1 of Section 11.3, we first transformed the series, then removed the seasonal effect from the transformed series and obtained the series $X_t$ (see the top panel of Figure 11.4). Instead of fitting a TAR model to the deseasoned series $X_t$, we can also try to fit a STAR model to $X_t$. In particular, the following `R` script fits a STAR(2) model to the deseasoned series $X_t$ in Section 11.3. We shall note that the model representation in `lstar` is

$$X_t = \left(\alpha_0 + \sum_{j=1}^{p} \alpha_j X_{t-j}\right) I(X_{t-d}) + \left(\beta_0 + \sum_{j=1}^{p} \beta_j X_{t-j}\right) [1 - I(X_{t-d})] + Z_t,$$

which is different from the representation (11.6).

```
> library(tsDyn)
> fit.star<-lstar(ss.nose, m=2, d=1)
> summary(fit.star)

Non linear autoregressive model

LSTAR model
Coefficients:
Low regime:
  const.L    phiL.1    phiL.2
0.7016242 0.3207036 0.3535763

High regime:
    const.H      phiH.1      phiH.2
-0.58144876  0.35282417 -0.06858923

Smoothing parameter: gamma = 100

Threshold
Variable: Z(t) = + (1) X(t) + (0) X(t-1)

Value: 2.423

Residuals:
      Min         1Q    Median        3Q       Max
-2.960203 -0.215721  0.029487  0.254014  2.357019

Fit:
residuals variance = 0.2394,  AIC = -4015, MAPE = 23.27%

Coefficient(s):
          Estimate  Std. Error  t value  Pr(>|z|)
const.L   0.701624    0.071471   9.8168 < 2.2e-16 ***
phiL.1    0.320704    0.043619   7.3523 1.947e-13 ***
phiL.2    0.353576    0.033351  10.6018 < 2.2e-16 ***
const.H  -0.581449    0.088855  -6.5438 5.998e-11 ***
phiH.1    0.352824    0.051362   6.8693 6.451e-12 ***
phiH.2   -0.068589    0.041905  -1.6368    0.1017
gamma   100.000008  404.860071   0.2470    0.8049
th        2.422947    0.044916  53.9440 < 2.2e-16 ***
---
Signif. codes:  0 '***' 0.001 '**' 0.01 '*' 0.05 '.' 0.1 ' ' 1

Non-linearity test of full-order LSTAR model against
full-order AR model
```

```
 F = 25.696 ; p-value = 8.7349e-12

Threshold
Variable: Z(t) = + (1) X(t) + (0) X(t-1)

> # Figure 11.6
> fit.star.resid<-ts(fit.star$resid, frequency=12,
    start=c(1749,3))
> par(mfrow=c(3,1), mar=c(4,4,4,4), cex.axis=1.5, cex.lab=1.2)
> plot(fit.star.resid, xlab="Year", ylab="Residual",
    type="l", xaxt="n")
> axis(1, x.pos, x.pos)
> acf(fit.star$resid, xlab="Lag",ylab="ACF", main="", lag=40)
> pacf(fit.star$resid, xlab="Lag",ylab="Partial ACF",
    main="", lag=40)
```

The estimated model is given by

$$\begin{aligned} X_t = &\Big(0.7016_{(.0715)} + 0.3207_{(.0436)} X_{t-1} + 0.3536_{(.0333)} X_{t-2}\Big) I(X_{t-1}) \\ &+\Big(-0.5814_{(.0889)} + 0.3528_{(.0513)} X_{t-1} - 0.0686_{(.0429)} X_{t-2}\Big)\big[1 \\ &\qquad -I(X_{t-1})\big] + Z_t, \end{aligned}$$

where

$$I(x) = \Big[1 + \exp\Big(100.0_{(404.9)} \cdot (2.4230_{(.0449)} - x)\Big)\Big]^{-1}.$$

Note that the large standard error of $\gamma$ is very large. Actually, other studies also show that the standard errors of $\gamma$ and $s$ are often very large; see Teräsvirta (1994). We also refer to it for further discussion. Figure 11.6 shows the residual series, and its sample ac.f.'s and partial ac.f.'s. Note that those plots are quite similar to those in Figures 11.5, as the difference between residual series in the fitted SETAR(2) and STAR(2) models is quite small. For example, the following R script shows that the mean and standard deviations of the difference between two residual series are quite small.

```
> resid.diff<-fit.star.resid-fit.setar.resid
> c(mean(resid.diff), sd(resid.diff))
[1] 2.106738e-14 5.829785e-02
```

## 11.5 Bilinear Models

A class of non-linear models, called the **bilinear** class, may be regarded as a plausible non-linear extension of the ARMA model, rather than of the AR model. Bilinear models incorporate cross-product terms involving lagged values of the time series and of the innovation process. The model may also incorporate ordinary AR and MA

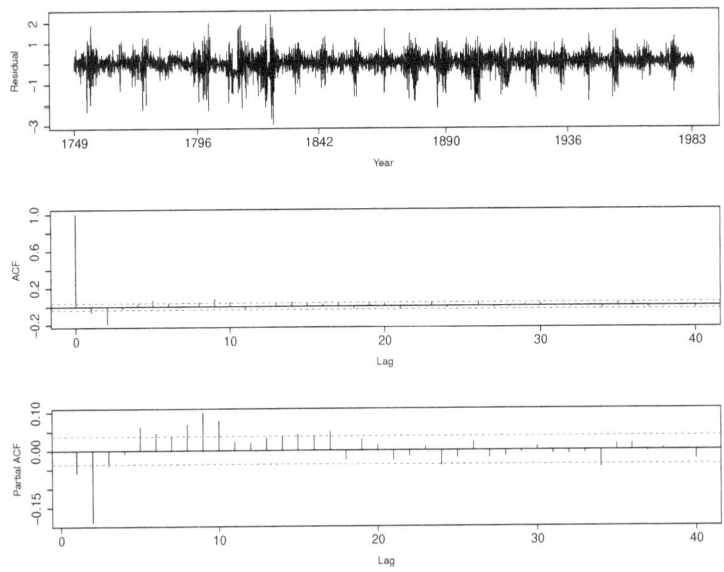


Figure 11.6 *The residual series in the fitted STAR(2) model (top), and its sample ac.f. (middle) and sample partial ac.f. (bottom).*

terms. Denoting the time series by $\{X_t\}$ and the innovation process by $\{Z_t\}$, a simple example could be

$$X_t = \alpha X_{t-1} + \beta Z_{t-1} X_{t-1} + Z_t, \tag{11.7}$$

where $\alpha$ and $\beta$ are constants. As well as the innovation term $Z_t$, this model includes one AR term plus one cross-product term involving $Z_{t-1}$ and $X_{t-1}$. It is this cross-product term that is the non-linear term and makes this a bilinear model.

One natural way for such a model to arise is to consider an AR(1) model where the AR parameter is not constant but is itself subject to innovations. If the perturbations in the AR parameter are such that the model for $X_t$ is given by

$$X_t = (\alpha + \beta Z_{t-1}) X_{t-1} + Z_t \tag{11.8}$$

then this leads to Equation (11.7).

A bilinear model could appear to be uncorrelated white noise (UWN) when examined in the usual way by inspecting, and perhaps testing, the sample ac.f. For example, consider the model

$$X_t = \beta Z_{t-1} X_{t-2} + Z_t. \tag{11.9}$$

It can be shown (after some horrid algebra) that $\rho(k) = 0$ for all $k \neq 0$. Thus, the series has the second-order properties of UWN, even though it has a clear structure. If, instead, we examine the series $\{X_t^2\}$, then its ac.f. turns out to be of similar form to that of an ARMA(2,1) model. Thus Equation (11.9) is certainly not SWN.

This example re-emphasizes the earlier remarks that there is no point in looking at second-order properties of $\{X_t\}$ and hoping they will indicate any non-linearity, because they won't! Rather, the search for non-linearity must rely on specially tailored procedures. As noted earlier, one general approach is to look at the properties of the $\{X_t^2\}$ series. If both $\{X_t\}$ and $\{X_t^2\}$ appear to be UWN, then $\{X_t\}$ can reasonably be treated as SWN.

Bilinear models are interesting theoretically but are perhaps not particularly helpful in providing insight into the underlying generating mechanism; hence, they have been little used in practice. They have, for example, been found to give a good fit to the famous sunspots data (see Figure 11.1), but subsequent analysis revealed that they gave poor long-term forecasts. The forecasts diverged and were not able to predict the periodic behaviour observed, because bilinear models are not designed to reflect such behaviour.

## 11.6 Regime-Switching Models

Several other classes of non-linear models have been introduced in the literature and we refer briefly to two of them. **State-dependent models** are described by Priestley (1988). They include threshold and bilinear models as special cases and can be thought of as a locally linear ARMA model. We say no more about them here.

A second class of non-linear models, called **regime-switching models**, has been widely applied in econometrics. The key feature is that the generating mechanism is different at different points in time and may be non-linear. When the model changes, it is said to *switch between regimes*. The time points at which the regime changes may be known in advance, or the regimes may change according to a Markov process. In particular, let $s_t$ be a random variable that takes integer values $\{1,2,\dots,K\}$. Assume that the prior probability of $s_0$ taking value $k$ is $\pi_k$, i.e.

$$P(s_0 = k) = \pi_k, \qquad k \in \{1,\dots,K\}.$$

Then obviously, $\pi_1 + \cdots + \pi_K = 1$. Suppose that

$$P(s_t = j \mid s_{t-1} = i, s_{t-2} = l, \dots) = P(s_t = j \mid s_{t-1} = i) = p_{ij}; \qquad (11.10)$$

that is, the probability that $s_t$ equals some value $j$ depends on the past only through the most recent value $s_{t-1}$. Such a process is called an $K$-state **Markov chain** with transition probabilities $\{p_{ij}\}_{i,j=1,\dots,K}$. Note that $p_{ij}$ is the probability that regime $i$ will be followed by regime $j$, and hence

$$p_{i1} + \cdots + p_{iK} = 1.$$

The transition probabilities $\{p_{ij}\}_{i,j=1,\dots,K}$ are often summarized in a $K \times K$ matrix, called the **transition matrix**:

$$P = \begin{pmatrix} p_{11} & p_{12} & \cdots & p_{1K} \\ p_{21} & p_{22} & \cdots & p_{2K} \\ \vdots & \vdots & \ddots & \vdots \\ p_{K1} & p_{K2} & \cdots & p_{KK} \end{pmatrix}.$$

Then a regime switching autoregressive model with order $p$ can be written in the following form

$$X_t = \alpha_{s_t,1}X_{t-1} + \cdots + \alpha_{s_t,p}X_{t-p} + Z_t, \tag{11.11}$$

where the regime $s_t$ is the outcome of an unobserved $K$-state Markov chain with $s_t$ independent of $Z_t$ for all $t$. A simple illustration is provided by the following model for sales at time $t$ in terms of demand and supply, namely

$$\text{sales}_t = \text{minimum } [\ \text{demand}_t, \text{supply}_t\ ].$$

Here the two regimes are (1) demand exceeds supply; and (2) supply exceeds demand. While the above equation is deterministic, as well as non-linear, separate equations are needed to model demand and supply and these are likely to be stochastic. The reader is referred to Harvey (1993, Section 8.6) for an introduction to these models.

The estimation of regime-switching models is usually challenging. For the regime-switching AR model (11.10), one can still find an explicit expression for its likelihood which can be maximized numerically. Comparing to the estimation of model parameters, the more interesting quantities in regime-switching models is the conditional probability that the regime at $t$ is $i$ given the observed data, i.e.

$$P(s_t = i \mid \{X_t\}).$$

The above probabilities, called *smoothed probabilities*, can be calculated based on conditional probabilities $P(s_t = i \mid s_{t-1} = j, \{X_t\})$; see details in Hamilton (1994, Section 22.4). The idea of regime-switching can also be extended to other time series models. For example, one may consider regime-switching ARCH or GARCH model; see ARCH and GARCH models in Sections 12.3 and 12.4. For those regime-switching models, the estimation becomes very difficult as the computation of the likelihood may involve path integrals. For illustration purpose, we provide an example of regime-switching AR models below.

*Example 11.3*

Consider again the average monthly sunspot numbers $Y_t$ from 1860 to 1983 (see Figure 11.1). In Section 11.3, we first transformed the series, then removed the seasonal effect from the transformed series and obtained the series $X_t$ (see the top panel of Figure 11.4). Note that the series seems to fluctuate between two regimes; this indicates that a two-state regime-switching model may be appropriate for interpreting the data. Specifically, we consider an AR(2) model with $K = 2$ regimes.

$$X_t = \begin{cases} \alpha_{1,0} + \alpha_{1,1}X_{t-1} + \alpha_{1,2}X_{t-2} + Z_{1,t}, & \text{if } s_t = 1, \\ \alpha_{2,0} + \alpha_{2,1}X_{t-1} + \alpha_{2,2}X_{t-2} + Z_{2,t}, & \text{if } s_t = 2, \end{cases}$$

where the noises $Z_{1,t} \sim N(0, \sigma_1^2)$ and $Z_{2,t} \sim N(0, \sigma_2^2)$. The two-state prior and transition probability matrix can be expressed as

$$P(s_0 = 1) = \pi \in [0, 1], \qquad P(s_0 = 2) = 1 - \pi.$$

$$P = \begin{pmatrix} \phi & 1-\phi \\ 1-\psi & \psi \end{pmatrix}.$$

This model can be estimated by using function `fit.MSAR` in the `R` package `NHMSAR`; see below.

```
> library("NHMSAR")
> data<-array(as.numeric(ss.nose),c(length(ss.nose),1,1))
> theta.init<-init.theta.MSAR(data,M=2,order=2,label="HH")
> fit.rs<-fit.MSAR(data, theta.init)
> fit.rs$theta   # estimated model parameters
$A
                A1        A2
Regime1 0.6406985 0.3069456
Regime2 0.4652951 0.2517079

$A0
              A01
Regime1 0.2022228
Regime2 0.7489775

$sigma

Regime1 0.07480023
Regime2 0.59640904

$prior

Regime1 1.000000e+00
Regime2 1.438698e-21

$transmat
           Regime1    Regime2
Regime1 0.98122641 0.01877359
Regime2 0.04352881 0.95647119

attr(,"NbComp")
[1] 1
attr(,"NbRegimes")
[1] 2
attr(,"order")
[1] 2
attr(,"label")
[1] "HH"
attr(,"n_par")
[1] 8
```

```
attr(,"emis.linear")
[1] FALSE
attr(,"class")
[1] "MSAR"
```

Therefore, the estimated model becomes

$$X_t = \begin{cases} 0.2022 + 0.6407X_{t-1} + 0.3069X_{t-2} + Z_{1,t}, & \text{if } s_t = 1, \\ 0.7490 + 0.4653X_{t-1} + 0.2517X_{t-2} + Z_{2,t}, & \text{if } s_t = 2, \end{cases}$$

$$Z_{1,t} \sim N(0, 0.0748^2), \qquad Z_{2,t} \sim N(0, 0.5964^2).$$

with the prior and transition probabilities

$$\pi_1 = 1, \qquad \pi_2 = 1 - \pi = 0,$$

$$P = \begin{pmatrix} 0.9812 & 0.0188 \\ 0.0435 & 0.9565 \end{pmatrix}.$$

The estimated conditional probabilities can be extracted from the fitted model. Figure 11.7 shows the estimated conditional probabilities for each regime at each time point, which can be reproduced using the following R script.

```
> fit.rs.smprob<-ts(fit.rs$smoothedprob, frequency=12,
    start=c(1749,2))
> x.pos<-c(1749, 1796, 1842, 1890, 1936, 1983)
> par(mfrow=c(2,1), mar=c(2,4,2,2), cex.axis=1.5,
    cex.lab=1.5)
> plot(fit.rs.smprob[,1], xlab="", ylab="", main="",
    type="l", xaxt="n")
> axis(1, x.pos, x.pos)
> plot(fit.rs.smprob[,2], xlab="", ylab="", main="",
    type="l", xaxt="n")
> axis(1, x.pos, x.pos)
```

After the model is estimated, we can compute the fitted values and residual as follows. Figure 11.8 shows the residual series, and its sample ac.f. and sample partial ac.f.

```
> # Compute the fitted values and residuals
> Xt<-as.numeric(ss.nose)
> n<-length(Xt)
> theta.regime1<-c(fit.rs$theta[2]$A0[1],fit.rs$theta[1]$A[1,])
> theta.regime2<-c(fit.rs$theta[2]$A0[2],fit.rs$theta[1]$A[2,])
> values.in.regime<-cbind(1, Xt[2:(n-1)], Xt[1:(n-2)]
      )%*%cbind(theta.regime1, theta.regime2)
> fit.rs.fitted<-rowSums(values.in.regime*fit.rs$smoothedprob)
> fit.rs.resid<- Xt[-c(1,2)]-fit.rs.fitted
```

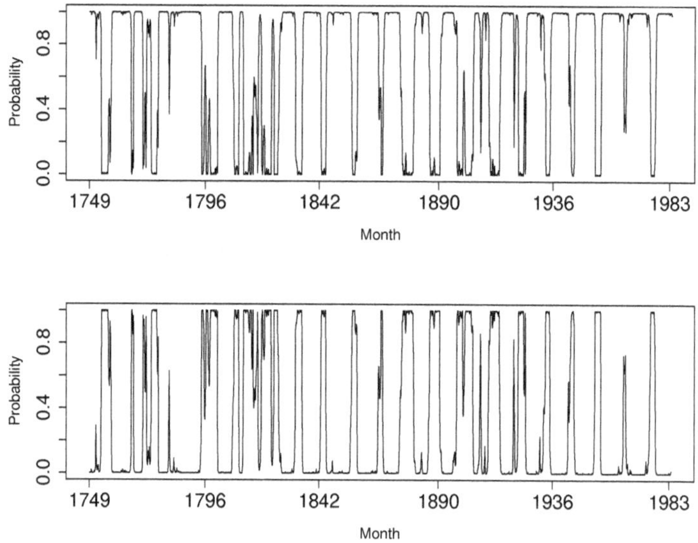


Figure 11.7 *Estimated conditional probabilities* $\widehat{P}(s_t = 1|\{X_t\})$ *(top) and* $\widehat{P}(s_t = 2|\{X_t\})$ *(bottom).*

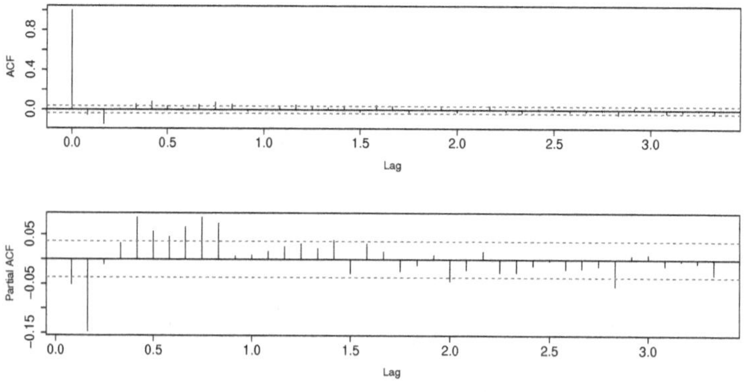


Figure 11.8 *The residual series in the fitted two-state regime switching model (top), and its sample ac.f. (middle) and sample partial ac.f. (bottom).*

```
> # Figure 11.8
> fit.rs.resid<-ts(fit.rs.resid, frequency=12, start=c(1749,3))
> par(mfrow=c(3,1), mar=c(4,4,4,4), cex.axis=1.5, cex.lab=1.2)
> plot(fit.rs.resid, xlab="Year", ylab="Residual",
    type="l", xaxt="n")
> axis(1, x.pos, x.pos)
> acf(fit.rs.resid, xlab="Lag", ylab="ACF", main="", lag=40)
> pacf(fit.rs.resid, xlab="Lag", ylab="Partial ACF",
    main="", lag=40)
```

## 11.7 Chaos

The topic of **chaos** has attracted much attention in recent years, especially from applied mathematicians. Chaotic behaviour arises from certain types of non-linear models, and a loose definition is 'apparently random behaviour that is generated by a purely deterministic, non-linear system'. A non-technical overview is given by Gleick (1987) while Kantz and Schreiber (1997) provide a readable introduction from the perspective of mathematical physics. Chan and Tong (2001) and Isham (1993) provide a statistical perspective, while further helpful material is given by Tong (1990, Chapter 2).

If a chaotic deterministic system can appear to behave as if it were 'random', the statistician then has the task of deciding what is meant by random and further has to try to decide whether an apparently random time series has been generated by a stochastic model, by a chaotic (non-linear) deterministic model or by some combination of the two. It would be of great interest to scientists if fluctuations, previously thought to be random, turned out to have a deterministic explanation. Unfortunately, distinguishing between the different possibilities can be difficult in practice

The main idea is well illustrated by the famous example of chaos called the **logistic map**, sometimes alternatively called the **quadratic** map. This is an example of what mathematicians would call a *difference equation* or *map*, but which statisticians would probably regard as a deterministic time series.

Suppose a time series is generated by the (deterministic) equation

$$x_t = kx_{t-1}(1 - x_{t-1})$$

for $t = 1, 2, 3, \ldots$ with $x_0 \in (0, 1)$. Then, provided $0 < k \leq 4$, the series will stay within the range $(0, 1)$. For low values of $k$, the deterministic nature of the series will generally be self-evident. For $0 < k < 1$, it can easily be seen that the series will always decline to zero, whatever starting value is given. For $1 \leq k \leq 3$, it can be shown that the series will always converge to the value $x_t = (1 - 1/k)$ for large $t$, and this value is called a stable (or fixed) point, or, in the jargon of chaos theory, an **attractor** (although such a series is not in fact chaotic). Note that the stable point $(1 - 1/k)$ is the point of intersection of the 45° line and the quadratic curve, $x_t =$

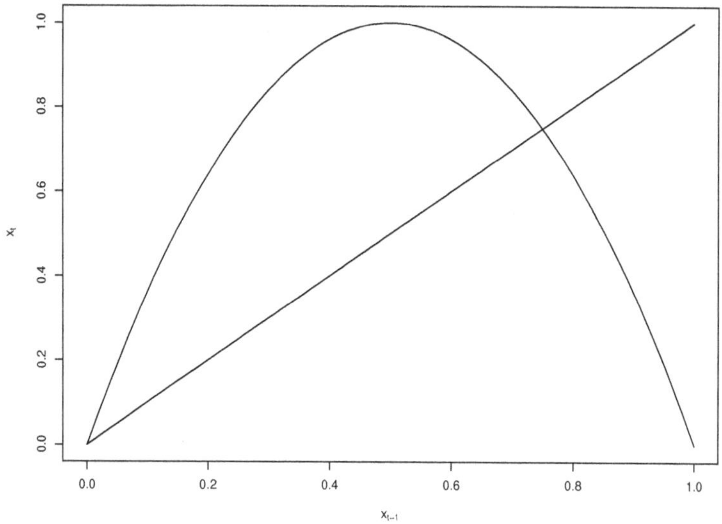


Figure 11.9 *A plot of $x_t$ (y-axis) versus $x_{t-1}$ (x-axis) for the logistic map with $k = 4$. The 45° line (where $x_t = x_{t-1}$) is also shown.*

$kx_{t-1}(1 - x_{t-1})$ for $1 \leq k \leq 3$. For $3 < k < 3.57$, the series exhibits cyclic behaviour whose period depends on $k$ but as $k$ approaches 4, the series looks more and more chaotic. Indeed when $k = 4$, it can be shown that the series has a flat spectrum and has the second-order properties of UWN. The series 'jumps around' all over the interval $(0, 1)$ and the deterministic nature of the series may not be readily apparent in the time plot. This is an example of chaotic behaviour. Of course the deterministic nature of the series can readily be demonstrated in this case by plotting $x_t$ versus $x_{t-1}$ as the points will lie on a quadratic curve, as illustrated in Figure 11.9.

A chaotic system has the property that a small change in initial conditions will generally magnify through time rather than die out. This is exemplified in the so-called **butterfly effect**, whereby a butterfly flapping its wings could produce an effect that is eventually transformed into a tropical storm! The sensitivity to initial conditions (the rate at which a small perturbation is magnified) is measured by a quantity called the **Lyapunov exponent**. This will not be defined here, but values greater than zero indicate divergence. Moreover some starting values lead to greater instability than others (e.g. Yao and Tong, 1994) so that the width of 'error bounds' on predictions will depend on the latest value from which forecasts are to be made.

A chaotic series also has the property that a sequence of $m$ values will generally lie in a restricted area of $m$-space. For example, for the logistic map with $k = 4$, the values of paired observations of the form $(x_t, x_{t-1})$ lie on the quadratic curve illustrated in Figure 11.4 rather than anywhere in the unit square. This leads to the concept of the **dimension** of a chaotic system. Whereas random processes generally have infinite dimension, chaotic deterministic behaviour can take place in a much lower dimension. The dimension of a time series has been defined in several different ways, but none of the definitions is very accessible and so none will be given here. Note that non-integer values of dimension are possible as in the so-called **fractal** behaviour, which will also not be discussed here, except to note that fractals are **self-similar** in that they look to have the same properties regardless of the scale at which they are studied.

We have seen that a time series generated by the logistic map, with $1 \leq k \leq 3$, will always converge to the value $x_t = (1 - 1/k)$ for large $t$, and that this value can be called an attractor. However, for chaotic behaviour when $k = 4$, there is no convergence and the so-called domain of attraction is just the whole unit interval, meaning that values may be found anywhere in the range $(0, 1)$ for large $t$. However, as chaotic behaviour leads to successive values lying in a restricted subspace, it is sometimes found that the domain of attraction has a strange geometric form (e.g. fractals), and this is called a **strange attractor**.

As another example of the sort of strange behaviour that can result, Bartlett (1990) reproduces an example of a process that is random when considered as going forward through time but that becomes a non-linear chaotic deterministic series *when time is reversed*. The fascinating topic of time reversibility is reviewed by Lawrance (1991). It can be shown that a stationary (Gaussian) ARMA model *is* reversible in that the properties of data generated by such a model look the same in either direction. In contrast a non-linear mechanism typically gives data, which are not time reversible. For example, the sunspots data in Figure 11.1 would look different if plotted backwards in time as the series would then decrease faster than it rises.

The study of chaos leads to much fascinating mathematics both with difference equations in discrete time (as considered here) or with non-linear differential equations in continuous time. However, are these results of any use to statisticians? One obvious question is whether it is possible to forecast a chaotic series. Because of sensitivity to initial conditions, long-term forecasting is not possible. However, for short-term forecasting of low-dimensional chaotic series, some progress is possible if we know the model (Berliner, 1991). Unfortunately, the analyst does *not* generally know the model *a priori* and so the next important question is whether it is possible to identify a chaotic model and, related to this, whether it is possible to distinguish between chaotic and stochastically random series. In economics, we would like to go further and disentangle the systematic component of a chaotic model from the 'noise' that will inevitably affect the system, either in an additive way, so that $x_t = f(x_{t-1}) +$ noise, or as measurement error, so that we actually observe $y_t = x_t +$ noise even though the system equation for the $\{x_t\}$ series remains deterministic. Unfortunately it seems to be difficult to tell whether a series is (deterministically) chaotic or stochastic or some combination of the two, though some progress is being made. For long financial series, a test called the BDS test (Brock and Potter, 1993; Granger and Teräsvirta, 1993, Chapter 6) is often used to test the null hypothesis that a series is (linear) i.i.d. against a (non-linear) chaotic alternative, but it is not clear when this test is superior to other tests for non-linearity. More generally the extent to which a non-linear deterministic process retains its properties when corrupted by noise is also unclear. It is, however, worth noting that classical (linear) methods for smoothing series, which are designed to separate the signal from the noise, can actually make things worse for chaotic series (Kantz and Schreiber, 1997, Example 1.1). Work continues on these difficult questions. Sadly, it does appear to be the case that very large samples are needed to identify attractors in high-dimensional chaos.

A few years ago, there were high hopes that the use of chaotic models might lead to improved economic forecasts. Sadly, this has not yet occurred. For example, Granger (1992) says that it seems unlikely that the stock market could obey a simple deterministic model, while Granger and Teräsvirta (1993, p. 36) and Brock and Potter (1993) both say that there is strong evidence for nonlinearity in economic data but weak evidence that they are also chaotic. On the other hand, May (1987) has argued that chaos is likely to be pervasive in biology and genetics. In general, the analyst will find it difficult to apply models with chaotic properties to real time-series data, but research is continuing (e.g. Tong, 1995) and the position may well change in the future. In any case, the study of chaotic models is fascinating and may contribute to our understanding of random behaviour in time-series modelling. It may even cause us to re-examine the meaning of the word 'random'. Even if a system under study is regarded in principle as deterministic, the presence of chaotic effects with unknown initial conditions (as will usually be the case in practice) means that prediction becomes difficult or impossible. Moreover a system may appear deterministic at the microscopic level, but appear stochastic at the macroscopic level. Putting this in a different way, it can be argued that whether a system is 'random' depends not on its intrinsic properties, but on how it appears to an observer in the given context with available knowledge. These are deep waters!!

## 11.8 Concluding Remarks

The real world generally changes through time and often behaves in a non-linear way. Thus alternatives to linear models with constant parameters should often be considered. There are several ways that the need for a non-linear model may be indicated, namely

- Looking at the time plot and noting asymmetry, changing variance, etc.
- Plotting $x_t$ against $x_{t-1}$, or more generally against $x_{t-k}$ for $k = 1, 2, \ldots$, and looking for strange attractors, limit cycles, etc.
- Looking at the properties of $\{x_t^2\}$ as well as at those of $\{x_t\}$
- Taking account of context, background knowledge, known theory, etc.

Non-linear models often behave quite differently from linear models. In general, they are harder to handle and require considerable technical and numerical skill. In particular, it is generally much more difficult to construct forecasts from a non-linear, rather than linear, model, for more than one step ahead. In the latter case, analytic formulae are generally available, but non-linear forecasting becomes increasingly difficult for longer lead times where some sort of numerical approach will generally be needed to calculate conditional expectations. Fortunately, specialist computer packages, such as STAR (see Tong, 1990, p. xv), are becoming available to do this.

Two general references on non-linear forecasting are Lin and Granger (1994) and Chatfield (2001, Section 4.2.4).

An additional feature of non-linear models, which may be unexpected at first, is that the width of prediction intervals need not increase with the lead time. This may happen, for example, for data exhibiting multiplicative seasonality, where prediction intervals tend to be narrower near a seasonal trough rather than near a seasonal peak. Another peculiarity of non-linear models, even with normal errors, is that the distribution of the forecast error is *not* in general normal, and may even be bimodal or have some other unexpected shape. In the bimodal case, a sensible prediction interval may comprise, not a single interval, but two disjoint intervals. This seems most peculiar at first sight, but second thoughts remind us of situations where we might expect a high or low outcome but not an intermediate result. Sales of a new fashion commodity could be like this. In such circumstances it could be particularly misleading to give a single point forecast by calculating the conditional expectation.

It is difficult to give advice on how to choose an appropriate non-linear model. As always contextual information and background knowledge are vital, but the only statistical advice I can give is that periodic behaviour suggests trying a threshold model, while, with a very long series, some analysts believe that it may be worth trying a neural net. It is also worth remembering that alternatives to linear models include, not only non-linear models but also models where the parameters change through time in a pre-determined way and models that allow a sudden change in structure. While the latter may be regarded as non-linear, it may be more helpful to think of them as non-stationary (see Section 13.2).

Non-linear models are mathematically interesting and sometimes work well in practice. The fitting procedure is more complicated than for linear models, but may lead to greater insight, and it should be remembered that the prime motivation for modelling is often to improve understanding of the underlying mechanism. This being so, it would be a bonus if non-linear models gave better forecasts as well. Sadly, it appears from the literature that gains in forecasting accuracy are often (usually?) modest if they exist at all, and may not by themselves compensate for the additional effort required to compute them. Thus even when the data appear to exhibit clear non-linear properties, it may still be safer to use a linear model for forecasting purposes.

## 11.9 Bibliography

The literature on non-linear models is growing rapidly and many references have been given throughout this chapter. Alternative introductions are given by Franses (1998, Chapter 8), Granger and Newbold (1986, Chapter 10) and Harvey (1993, Chapter 8). More detailed accounts are given by Priestley (1981, Chapter 11) and by Tong (1990). The advanced text by Granger and Teräsvirta (1993) is concerned with economic relationships and extends discussion to multivariate non-linear models. Fan and Yao (2003) introduced various parametric and nonparametric techniques for nonlinear time series analysis and prediction.

## Exercise

It is difficult to set exercises on non-linear models that are mathematically and practically tractable. The reader may like to try the following exercise on the logistic map.

**11.1** Consider the logistic map with $k = 4$, namely

$$x_t = 4x_{t-1}(1 - x_{t-1})$$

This series is non-linear, deterministic and chaotic. Given the starting value $x_0 = 0.1$, evaluate the first three terms of the series. Change $x_0$ to $0.11$ and repeat the calculations. Show that $x_3$ changes from 0.289 to 0.179 (to 3 decimal places). Thus a change of 0.01 in the starting value leads to a change of 0.110 in the value of $x_3$ and this provides a simple demonstration that a chaotic series is very sensitive to initial conditions.

**11.2** Use the following models to analyse the daily returns of the adjusted closing prices of the S&P500 index series, shown in Figure 1.2.

(a) Non-linear autoregressive models;

(b) Threshold autoregressive models;

(c) Smooth transition autoregressive models;

(d) Regime-switching models with two or three regimes;

(e) Neural network autoregressive models.

Chapter 12

# Volatility Models

The non-linear models introduced in the previous chapter could be described as having structural non-linearity. They allow improved point forecasts of the observed variable to be made when the true model is known. This chapter considers various classes of non-linear models of a completely different type, which are primarily concerned with modelling *changes in variance* or *volatility*. They do not generally lead to better point forecasts of the measured variable, but may lead to better estimates of the (local) variance. This, in turn, allows more reliable prediction intervals to be computed and hence a better assessment of risk.

Volatility models have many applications in economics and finance. In options trading, volatility shows fluctuation levels of the market price of the underlying asset and is one of key factors that determine option prices. In risk management, volatility models provide a simple approach to calculating the value at risk of a financial position. Volatility also plays an important role in asset allocation and portfolio optimization. Given the increasing interest in this topic, this chapter introduces various types of univariate volatility models. A brief bibliography is given at the end of the chapter.

## 12.1 Structure of a Model for Asset Returns

Suppose we have a time series from which any trend and seasonal effects have been removed and from which linear (short-term correlation) effects may also have been removed. We denote this derived series by $\{Y_t\}$, to distinguish it from the original observed series, $\{P_t\}$. $Y_t$ might be the first differences (or the percentage changes) of a financial time series such as the natural log of a share price $P_t$, i.e.

$$Y_t = \log P_t - \log P_{t-1}, \quad \text{or} \quad Y_t = (\frac{P_t}{P_{t-1}} - 1) \times 100\%$$

and in this context is often called the **return** or the **growth rate** of a series. For illustration purposes, let $P_t$ be the adjusted closing prices of the Standard & Poor's 500 (S&P500) at the $t$th trading day, one can calculate its daily returns $Y_t$ at each day; Figure 1.2 shows the daily return series from January 4, 1995, to December 30, 2016.

The basic idea in volatility modelling is that the return series $\{Y_t\}$ has very few serial correlations, but it is a dependent series. To see this, let $Y_t$ denote the daily

DOI: 10.1201/9781003646112-12

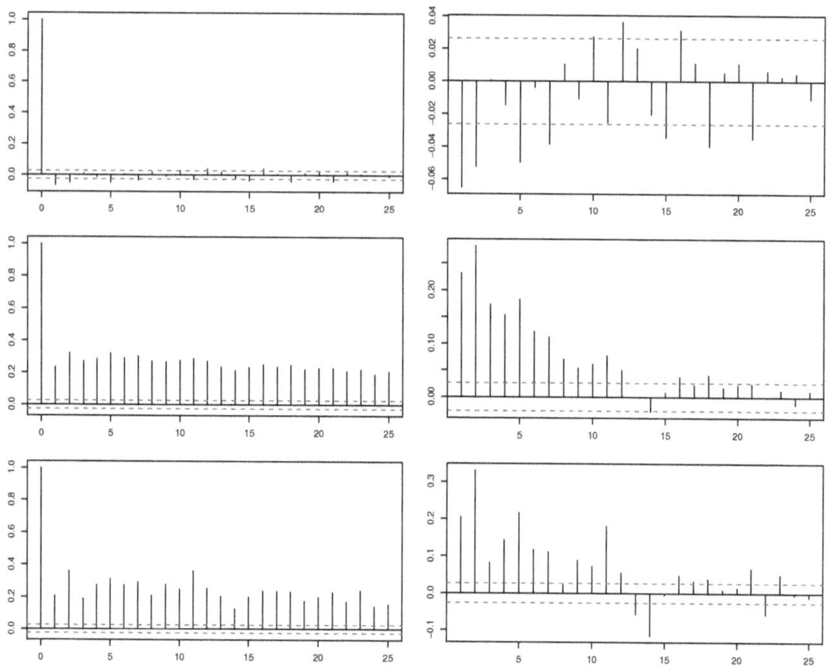

Figure 12.1 *Sample ac.f. (left) and sample partial ac.f. (right) of various functions of the daily returns, $Y_t$, of adjusted closing prices of S&P500 index from January 4, 1995, to December 30, 2016. Top: Original series $Y_t$; Middle: Absolute values of $Y_t$; Bottom: Squared values of $Y_t$.*

returns $Y_t$ of the S&P500 index shown in Figure 1.2, we calculate sample ACFs and sample PACF of $Y_t$, $|Y_t|$, and $Y_t^2$, respectively, and show them in Figure 12.1. Note that the sample ACFs of the returns $Y_t$ suggest no significant serial correlations except for small ones at lags 1, 2, and 5. However, the sample ACFs of the absolute and squared returns, $|Y_t|$ and $Y_t^2$, show strong dependence over all lags. From this example, we find an important feature of the return series; that is, the returns may seem serially uncorrelated, but it is dependent.

To focus our discussion on volatility of a return series, we may further assume that $Y_t$ is the series of residuals from a regression or innovations in a linear time series model. In particular, suppose that $X_t$ follow an ARMA$(p,q)$ model described in Section 3.8,

$$\phi(B)X_t = \theta(B)Y_t,$$

where $\phi(B)$ and $\theta(B)$ are polynomials of $B$ with orders $p$ and $q$, respectively. Denote $\mathscr{F}_t$ the set of observed data up to time $t$, i.e. $\{X_1,\ldots,X_t\}$; the observation $X_t$ can be expressed as

$$X_t = \mu_t + Y_t, \tag{12.1}$$

where $\mu_t$ is the mean of $X_t$ conditional on observed data $\mathscr{F}_{t-1}$ and given by

$$\mu_t = E(X_t|\mathscr{F}_{t-1}) = \phi(B)X_t - (\theta(B)-1)Y_t, \tag{12.2}$$

and the innovation series $Y_t$ has mean 0 and conditional variance

$$\sigma_t^2 = \mathrm{Var}(X_t|\mathscr{F}_{t-1}) = \mathrm{Var}(Y_t|\mathscr{F}_{t-1}). \tag{12.3}$$

Volatility models discussed in the rest of this chapter deal with the evolution of $\sigma_t^2$ over time.

## 12.2 Historic Volatility

For illustration purpose, we may assume the conditional mean of the process $X_t$ is zero, i.e. $X_t = Y_t$. Let $\sigma_t$ be the volatility of $Y_t$ at time $t$; then motivated by Equation (12.3), $\sigma_t^2$ at time $t-1$ can be simply estimated by the sample variance based on the most recent $k$ observations:

$$\widehat{\sigma}_t^2 = \frac{1}{k-1}\sum_{i=1}^{k}(Y_{t-i}-\bar{Y})^2, \tag{12.4}$$

where $\bar{Y} = \sum_{i=1}^{k} Y_{t-i}/k$. In finance, if $\sigma_t$ in Equation (12.4) is in the daily basis, then it can be converted to the annual volatility by $\sqrt{A}\sigma$, where the annualizing factor $A$ is the number of trading days, usually taken as around 252. The volatility estimate $\widehat{\sigma}_t$ given by (12.4) is called the *k-day historic volatility*. Figure 12.2 displays historic volatilities with two different window sizes $k$. Both estimates of $\sigma_t$ capture important changes in the volatility process. In particular, they all show big volatilities during the 2008–2009 financial crisis. Note that for larger window size $k$, the time course of historic volatility is smoother.

Historic volatility uses equal weights for the observations in the moving window of returns. Since the most recent observations may have more contributions to levels of volatility, the conditional variance at time $t$ may be approximated by

$$\widehat{\sigma}_t^2 = \sum_{i=1}^{k}\alpha_i(Y_{t-i}-\overline{Y})^2, \tag{12.5}$$

where $\sum_{i=1}^{k}\alpha_i = 1$. One variant of Equation (12.5) is to include a long-run variance rate $V$ in the weighted sum, leading to

$$\widehat{\sigma}_t^2 = \gamma V + \sum_{i=1}^{k}\alpha_i(Y_{t-i}-\overline{Y})^2, \tag{12.6}$$

where $\gamma+\sum_{i=1}^{k}\alpha_i = 1$. Another variant of (12.5) is the *exponentially weighted moving average* (EWMA), which is a special case of Equation (12.5) and assumes that the weights $\alpha_i$ decrease exponentially fast. Specifically, let $k=\infty$, $u_j = 0 = \widehat{\sigma}_j$ for $j \le 0$, and $\alpha_i = (1-\lambda)\lambda^{i-1}$ for some $0<\lambda<1$; Equation (12.5) becomes

$$\widehat{\sigma}_t^2 = \gamma V + (1-\lambda)\sum_{i=1}^{\infty}\lambda^{i-1}(Y_{t-i}-\overline{Y})^2, \tag{12.7}$$

which can be recursively represented as

$$\widehat{\sigma}_t^2 = (1-\lambda)\gamma V + \lambda\widehat{\sigma}_{t-1}^2 + (1-\lambda)(Y_{t-1}-\overline{Y})^2. \tag{12.8}$$

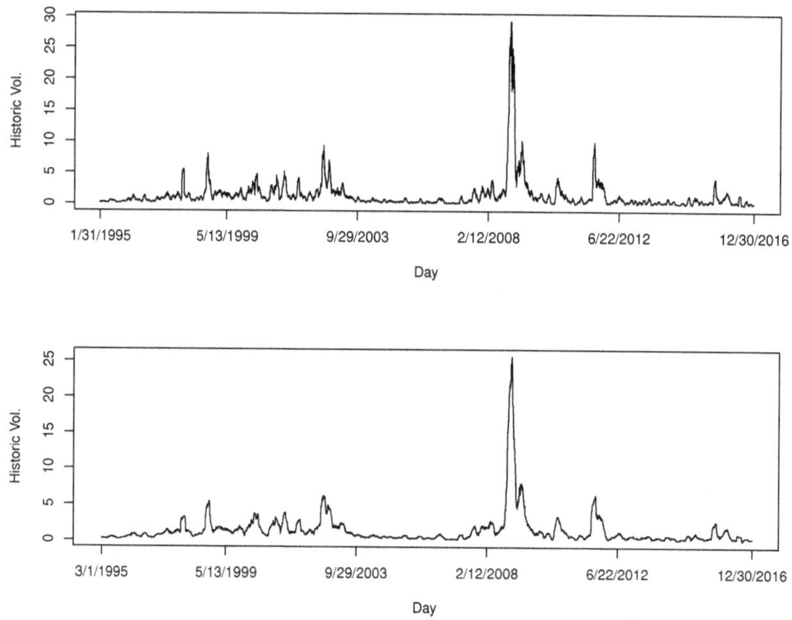


Figure 12.2 *Historic volatility estimated with different window sizes. Top:* $k = 20$*; bottom:* $k = 40$.

## 12.3 Autoregressive Conditional Heteroskedastic (ARCH) Models

To better describe the idea, we may represent $Y_t$ having mean zero in the form

$$Y_t = \sigma_t \varepsilon_t, \tag{12.9}$$

where $\{\varepsilon_t\}$ denotes a sequence of i.i.d. random variables with zero mean and unit variance, and $\sigma_t$ may be thought of as the local conditional standard deviation of the process. The $\varepsilon_t$ may have a normal distribution but this assumption is not necessary for much of what follows. For example, one may assume that $\varepsilon_t$ follow a standardized Student $t$-distribution. Let $x_\nu$ be a Student $t$-distribution with $\nu > 2$ degrees of freedom. Then $\varepsilon_t = x_\nu / \sqrt{\nu/(\nu-2)}$ has a *standardized Student t-distribution* with variance 1 and probability density function

$$p(\varepsilon) = \frac{\Gamma((\nu+1)/2)}{\Gamma(\nu/2)\sqrt{(\nu-2)\pi}} \left(1 + \frac{\varepsilon^2}{\nu-2}\right)^{-(\nu+1)/2}.$$

In any case the unconditional distribution of $Y_t$ generated by a non-linear model will not generally be normal but rather fat-tailed (or leptokurtic). Suppose we additionally assume that the square of $\sigma_t$ depends on the most recent value of the derived series by

$$\sigma_t^2 = \omega + \sum_{j=1}^{k} \alpha_j Y_{t-j}^2, \tag{12.10}$$

where the parameters $\omega$ and $\alpha_1, \ldots, \alpha_k$ are nonnegative and satisfy

$$\alpha_1 + \cdots + \alpha_k < 1$$

to ensure that $\sigma_t^2$ is non-negative. A model for $Y_t$ satisfying Equations (12.9) and (12.10) is called an **autoregressive conditionally heteroscedastic model** of order $k$ (ARCH($k$)). The adjective 'autoregressive' arises because the value of $\sigma_t^2$ depends on past values of the derived series, albeit in squared form. Note that Equation (12.10) does not include an 'error' term and so does not define a stochastic process.

Since (12.10) is similar to the autoregressive model in Chapter 3, for $Y_t^2$ to be covariance stationary, the roots of the equation $1 - \alpha_1 z - \cdots - \alpha_k z^k = 0$ are required to lie outside the unit circle. The unconditional variance $V$ of $Y_t$ is given by

$$V = \frac{\omega}{1 - \alpha_1 - \cdots - \alpha_k}. \tag{12.11}$$

*ARCH(1) model*

To understand the ARCH models, we examine the first-order case, i.e. $k = 1$. Then

$$Y_t = \sigma_t \varepsilon_t, \qquad \sigma_t^2 = \omega + \alpha_1 Y_{t-1}^2,$$

where $\omega > 0$ and $\alpha_1 > 0$. It is clear that the unconditional mean of $Y_t$ is zero, i.e.

$$E(Y_t) = E(E(Y_t|\mathscr{F}_{t-1})) = E(\sigma_t E(\varepsilon_t)) = 0.$$

The unconditional variance of $Y_t$ can be obtained as

$$\text{Var}(Y_t) = E(Y_t^2) = E(\omega + \alpha_1 Y_{t-1}^2) = \omega + \alpha_1 E(Y_{t-1}^2).$$

Given that $Y_t$ is stationary, we have $\text{Var}(Y_t) = \text{Var}(Y_{t-1}) = E(Y_{t-1}^2)$; hence

$$\text{Var}(Y_t) = E(Y_t^2) = \omega/(1 - \alpha_1).$$

One nice property of ARCH models is that the tail distribution of $Y_t$ is heavier than that of a normal distribution. To see that, we need to first compute the unconditional kurtosis of $Y_t$. Assume that $Z_t$ are i.i.d. standard normal random variables; we note that

$$\begin{aligned} E(Y_t^4) &= E[E(\sigma_t^4 \varepsilon_t^4|\mathscr{F}_{t-1})] = 3E[E(Y_t^2|\mathscr{F}_{t-1})]^2 \\ &= 3E[\omega + \alpha_1 Y_{t-1}^2]^2 = 3E[\omega^2 + 2\omega\alpha_1 E(Y_{t-1}^2) + \alpha_1^2 E(Y_{t-1}^4)]. \end{aligned}$$

If $Y_t$ is fourth-order stationary with $E(Y_t^4)$ being a constant, we can solve the above equation for $E(Y_t^4)$ and obtain that

$$E(Y_t^4) = \frac{3\omega(1+\alpha_1)}{(1-\alpha_1)(1-3\alpha_1^2)}.$$

Therefore, the unconditional kurtosis of $Y_t$ is given by

$$\kappa := \frac{E(Y_t^4)}{[\mathrm{Var}(Y_t)]^2} = 3\frac{1-\alpha_1^2}{1-3\alpha_1^2}.$$

When the fourth moment of $Y_t$ is positive, we see that $1 - 3\alpha_1^2 > 0$ and then the unconditional kurtosis of $Y_t$ is larger than 3, which is the unconditional kurtosis of a normal distribution. This shows that the tail distribution of $Y_t$ is heavier than that of a normal distribution; hence, the innovation process $Y_t$ in a Gaussian ARCH(1) model tends to generate more 'outliers' than a Gaussian white noise process.

These properties also hold for general ARCH models with higher orders, but the argument become more complicated.

*Estimation and prediction*

Parameters in ARCH models can be estimated by maximizing their likelihood, which usually needs to be done numerically.

It can be shown that ARCH models are martingale differences (MDs), so that knowledge of the value of $\sigma_t$ does not lead to improved point forecasts of $Y_{t+h}$. The value of modelling $\{\sigma_t\}$ lies in getting more reliable bounds for prediction intervals for $Y_{t+h}$ and in assessing risk more generally. When the derived series $\{Y_t\}$ has mean zero, the point forecast of $Y_{t+1}$ is zero and prediction intervals are typically calculated using the appropriate percentage point of the standard normal distribution, even when there are doubts about the normality assumption (though alternative heavy-tailed distributions could be used).

Consider an ARCH($k$) model. At the forecast origin $N$, the 1-step-ahead forecast for $\sigma_{N+1}^2$ is

$$\widehat{\sigma}_N^2(1) = E(\sigma_{N+1}^2|\mathscr{F}_N) = \omega + \alpha_1 Y_N^2 + \cdots + \alpha_k Y_{N+1-k}^2,$$

and the $l$-step-ahead forecast for $\sigma_{N+l}^2$ is

$$\widehat{\sigma}_N^2(l) = E(\sigma_{N+l}^2|\mathscr{F}_N) = \omega + \sum_{i=1}^{k} \alpha_i \widehat{\sigma}_N^2(l-k),$$

where $\widehat{\sigma}_N^2(l-k) = E(Y_{N+l-k}^2|\mathscr{F}_N)$ if $l > k$, and $\widehat{\sigma}_N^2(l-k) = Y_{N+l-k}^2$ if $l \le k$. Thus, the one-step-ahead $100(1-\alpha')\%$ prediction interval for $Y_{t+1}$ is typically taken to be of the form $0 \pm z_{\alpha'/2}\hat{\sigma}_t$, where $z_{\alpha'/2}$ denotes the value of the standard normal distribution for which the probability of being exceeded is $\alpha'/2$.

*Example 12.1*

We now apply the modelling procedure to build a simple ARCH model for the daily returns, $Y_t$, of adjusted closing prices of the S&P500 index. Figure 1.2 shows the daily return series from January 4, 1995, to December 30, 2016. The sample ACF and PACF of the squared returns in the middle and bottom panels of Figure 12.1 show the existence of conditional heteroscedasticity. We consider an ARCH(2) model with the following specification for the daily return series

$$X_t = \mu + Y_t, \qquad Y_t = \sigma_t \varepsilon_t, \qquad \sigma_t^2 = \omega + \alpha_1 Y_{t-1}^2 + \alpha_2 Y_{t-2}^2.$$

We use the R function garchFit in the R package fGarch to estimate the model. Assuming that $\varepsilon_t$ are i.i.d. standard normal, we obtain the fitted model

$$\begin{aligned} X_t &= 0.0807_{(.0125)} + Y_t, \qquad Y_t = \sigma_t \varepsilon_t, \\ \sigma_t^2 &= 0.6580_{(.0216)} + 0.1965_{(.0202)} Y_{t-1}^2 + 0.3523_{(.0246)} Y_{t-2}^2, \end{aligned}$$

where the standard errors of the parameters are given in the parentheses; see some of the R output below. Figure 12.3 shows the estimated volatility $\widehat{\sigma}_t$, the standardized residual $\widehat{\varepsilon}_t$, and the same ac.f.'s and sample partial ac.f.'s of $\widehat{\varepsilon}_t$ and $\widehat{\varepsilon}_t^2$, respectively. Note that both the output and the plot show that the estimated residuals still have conditional heteroscedasticities, and hence indicates that the ARCH(2) model is not adequate.

```
> library("fGarch")
> fit1<-garchFit(~garch(2,0), data=sp500[,2])
> summary(fit1)

Title: GARCH Modelling

Call: garchFit(formula = ~garch(2, 0), data = sp500[, 2])

Conditional Distribution: norm

Coefficient(s):
     mu     omega    alpha1    alpha2
0.08072  0.65795  0.19649  0.35226

Std. Errors: based on Hessian

Error Analysis:
        Estimate  Std. Error  t value Pr(>|t|)
mu       0.08072     0.01252    6.447 1.14e-10 ***
omega    0.65795     0.02160   30.464  < 2e-16 ***
alpha1   0.19649     0.02021    9.720  < 2e-16 ***
```

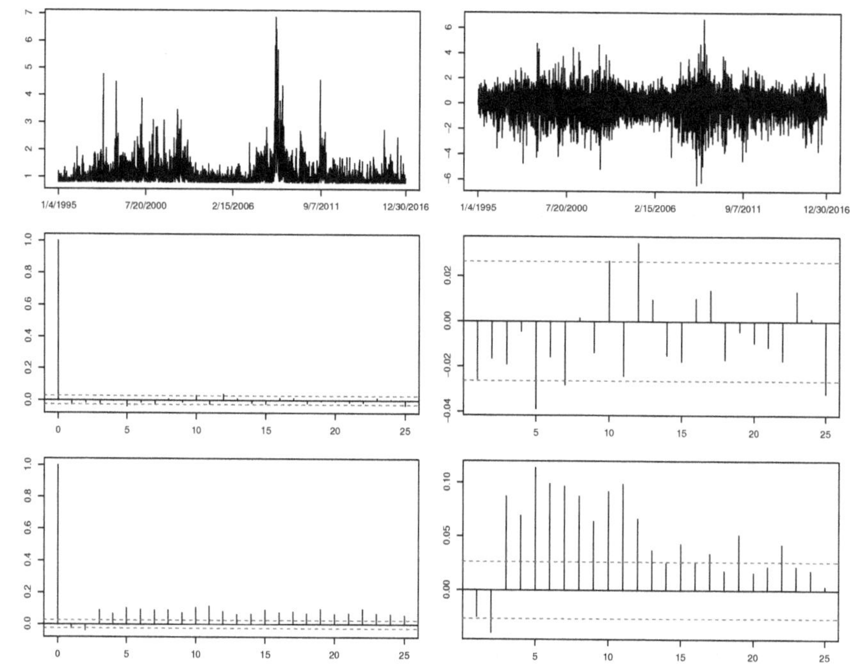

Figure 12.3 *Estimated volatility $\widehat{\sigma}_t$ (top left), the standardized residual $\widehat{\varepsilon}_t$ (top right), and the same ac.f.'s and sample partial ac.f.'s of $\widehat{\varepsilon}_t$ (middle) and $\widehat{\varepsilon}_t^2$ (bottom), respectively, in the ARCH(2) model for daily returns of adjusted closing prices of the S&P500 index from January 4, 1995, to December 30, 2016.*

```
alpha2    0.35226     0.02456    14.343   < 2e-16 ***
---
Signif. codes:  0 '***' 0.001 '**' 0.01 '*' 0.05 '.' 0.1 ' ' 1

Log Likelihood: -8184.593    normalized:  -1.47763

Standardised Residuals Tests:
                                 Statistic p-Value
 Jarque-Bera Test   R    Chi^2  1909.616  0
 Shapiro-Wilk Test  R    W      NA        NA
 Ljung-Box Test     R    Q(10)  25.34687  0.004725394
 Ljung-Box Test     R    Q(15)  40.03293  0.0004483369
 Ljung-Box Test     R    Q(20)  43.53819  0.001734486
 Ljung-Box Test     R^2  Q(10)  363.8295  0
 Ljung-Box Test     R^2  Q(15)  563.6726  0
 Ljung-Box Test     R^2  Q(20)  730.354   0
 LM Arch Test       R    TR^2   430.027   0
```

```
Information Criterion Statistics:
     AIC      BIC      SIC     HQIC
2.956704 2.961485 2.956703 2.958371

> # Plot the result
> fit1.vol<-volatility(fit1)
> fit1.resid.st<-residuals(fit1)/volatility(fit1)
> x.pos<-c(seq(1,n,1400),n)
> par(mfrow=c(3,2), mar=c(2,2,2,2))
> plot(fit1.vol, type="l", xlab="", ylab="", xaxt="n")
> axis(1, x.pos, sp500$Date[x.pos])
> plot(fit1.resid.st, type="l",xlab="", ylab="", xaxt="n")
> axis(1, x.pos, sp500$Date[x.pos])
> acf(fit1.resid.st, 25, ylab="", main="")
> acf(fit1.resid.st, 25, type="partial", ylab="", main="")
> acf(fit1.resid.st^2, 25, ylab="", main="")
> acf(fit1.resid.st^2, 25, type="partial", ylab="", main="")
```

## 12.4 Generalized ARCH Models

The ARCH model has been generalized to allow the variance to depend on past values of $\sigma_t^2$ as well as on past values of $Y_t^2$. A derived variable satisfying Equation (12.9) is said to follow a **generalized ARCH** (or GARCH) model of order $(h,k)$ when the local conditional variance is given by

$$\sigma_t^2 = \omega + \sum_{j=1}^{h} \beta_j \sigma_{t-j}^2 + \sum_{i=1}^{k} \alpha_i Y_{t-i}^2, \tag{12.12}$$

where $\omega \geq 0$ and $\alpha_i, \beta_j \geq 0$ for all $i, j$.

The GARCH$(h,k)$ model (12.12) can be considered as an ARMA model of volatility with martingale difference innovations: Let $\beta_i = 0$ if $i > h$, $\alpha_j = 0$ if $j > k$, and $\eta_t = Y_t^2 - \sigma_t^2$. Note that

$$E(\eta_t) = E(\sigma_t^2 \varepsilon_t^2) - E(\sigma_t^2) = E(\varepsilon_t^2 E(\sigma_t^2 | \mathscr{F}_{t-1})) - E(\sigma_t^2) = 0.$$

Then $\eta_t$ is a martingale difference. Moreover,

$$Y_t^2 = \omega + \sum_{j=1}^{\max(h,k)} (\alpha_j + \beta_j) Y_{t-j}^2 + \eta_t - \sum_{i=1}^{h} \beta_i \eta_{t-i}. \tag{12.13}$$

Hence the same invertibility and stationarity assumptions of ARMA models apply to GARCH models. For example, to ensure that $Y_t$ is covariance stationary, it is required that all roots of $1 - \sum_{j=1}^{\max(h,k)} (\alpha_j + \beta_j) z^j = 0$ lie outside the unit circle. Since the $\alpha$'s and $\beta$'s are usually assumed to be nonnegative, then we have

$$\sum_{j=1}^{k} \alpha_j + \sum_{i=1}^{h} \beta_i < 1. \tag{12.14}$$

Under (12.14), the unconditional variance of $u_t$ is given by

$$E(u_t^2) = \frac{\omega}{1 - \sum_{i=1}^{h} \beta_i - \sum_{j=1}^{k} \alpha_j}. \tag{12.15}$$

*GARCH(1,1) model*

To understand properties of GARCH models, we examine the GARCH(1, 1) case,

$$Y_t = \sigma_t \varepsilon_t, \qquad \sigma_t^2 = \omega + \alpha Y_{t-1}^2 + \beta \sigma_{t-1}^2, \tag{12.16}$$

where $\omega > 0$ and $\alpha, \beta > 0$. Note that The GARCH(1, 1) model is closely related to the EWMA estimate (12.8).

The model implies that a large $Y_{t-1}^2$ or $\sigma_{t-1}^2$ will lead to a large $\sigma_t^2$, which in turn will give rise to a large $Y_t^2 = \sigma_t^2 \varepsilon_t^2$. This is consistent with the volatility clustering observed in financial time series; hence, the GARCH(1, 1) model is often used to fit financial time series. Moreover, even when $\varepsilon_t$ is standard normal, GARCH(1, 1) can still be highly leptokurtic since

$$\frac{E(Y_t^4)}{[\text{Var}(Y_t)]^2} = \frac{3[1-(\alpha+\beta)^2]}{1-(\alpha+\beta)^2 - 2\alpha^2} > 3 \tag{12.17}$$

when $(\alpha+\beta)^2 + 2\alpha^2 < 1$.

Recall that $\eta_t = Y_t^2 - \sigma^2$ and let $\lambda = \alpha + \beta$; the GARCH(1, 1) model (12.16) can be expressed as

$$(1 - \lambda B)\sigma_t^2 = \omega + \alpha \eta_{t-1},$$

where $B$ is the backward shift operator. Therefore, when $Y_t$ is covariance stationary, we have

$$\begin{aligned} \sigma_t^2 &= \frac{1}{1-\lambda B}(\omega + \alpha \eta_{t-1}) \\ &= \frac{\omega}{1-\lambda} + \alpha(1 + \lambda B + \cdots + \lambda^i B^i + \ldots)\eta_{t-1} \\ &= \frac{\omega}{1-\lambda} + \alpha(\eta_{t-1} + \lambda \eta_{t-2} + \cdots + \lambda^i \eta_{t-1-i}). \end{aligned}$$

Note that as $\lambda$ approaches 1, the weight of $\eta_{t-i}$ in the above equation also approaches 1, indicating the persistent effect of events that occurred a long time ago on current volatility.

*Prediction of volatilities*

Similar to ARCH models, GARCH models do not affect point forecasts of the original observed variable, and it is therefore rather difficult to make a fair comparison of the forecasting abilities of different models for changing variance. Thus the modelling aspect (understanding the changing structure of a series), and the assessment of risk, are both more important than their ability to make point forecasts.

Replacing time index $t$ by $t+1$ in (12.16) and taking expectations on both sides yields the one-step-ahead forecast of $\sigma^2_{N+1}$ at the forecast origin $N$,

$$\widehat{\sigma}^2_N(1) = \omega + \alpha Y_N^2 + \beta \sigma_N^2,$$

where $\sigma_N$ can be estimated by fitting the model (12.16) to observations $Y_1, \ldots, Y_N$. To obtain a $k$-step-ahead forecast of $\sigma^2_{N+k}$, let $\sigma^2 = \omega/(1-\alpha-\beta)$. The GARCH(1, 1) model (12.16) can be written in the form

$$\sigma_t^2 - \sigma^2 = \alpha(Y_{t-1}^2 - \sigma^2) + \beta(\sigma_{t-1}^2 - \sigma^2). \tag{12.18}$$

Replacing time index $t$ by future time $N+k$ in (12.18) yields

$$\sigma^2_{N+k} - \sigma^2 = \alpha(Y^2_{N+k-1} - \sigma^2) + \beta(\sigma^2_{N+k-1} - \sigma^2).$$

This suggests the $k$-step-ahead forecast of $\sigma^2_{N+k}$ is

$$\widehat{\sigma}^2_N(k) := E(\sigma^2_{N+k}|\mathscr{F}_N) = \sigma^2 + \lambda^{k-1}\{\alpha(Y_N^2 - \sigma^2) + \beta(\sigma_N^2 - \sigma^2)\}. \tag{12.19}$$

Since $\sigma^2 = \omega/(1-\lambda)$, then

$$\widehat{\sigma}^2_N(k) = \omega(1-\lambda^k)/(1-\lambda) + (\alpha Y_N^2 + \beta\sigma_N^2)\lambda^{k-1}. \tag{12.20}$$

*Example 12.2*

We now consider a GARCH(1, 1) model with the following specification for the daily return series of the S&P500 index

$$X_t = \mu + Y_t, \qquad Y_t = \sigma_t \varepsilon_t, \qquad \sigma_t^2 = \omega + \alpha Y_{t-1}^2 + \beta \sigma_{t-1}^2.$$

We still use the R function `garchFit` to estimate the model. Assuming that $\varepsilon_t$ are i.i.d. standard normal, we obtain the fitted model

$$\begin{aligned} X_t &= 0.0740_{(.0111)} + Y_t, \qquad Y_t = \sigma_t\varepsilon_t, \\ \sigma_t^2 &= 0.0189_{(.0029)} + 0.1005_{(.0085)} Y_{t-1}^2 + 0.8857_{(.0092)}\sigma_{t-1}^2, \end{aligned}$$

where the standard errors of the parameters are given in the parentheses; see the R output below. Figure 12.4 shows the estimated volatility $\widehat{\sigma}_t$, the standardized residual $\widehat{\varepsilon}_t$, and the same ac.f.'s and sample partial ac.f.'s of $\widehat{\varepsilon}_t$ and $\widehat{\varepsilon}_t^2$, respectively. Comparing it to Figure 12.3, the standardized residuals in the fitted GARCH(1,1) model show little conditional heteroscedasticity, indicating the model is adequate.

```
> fit2<-garchFit(~garch(1,1), data=sp500[,2])
> summary(fit2)

Title: GARCH Modelling
```

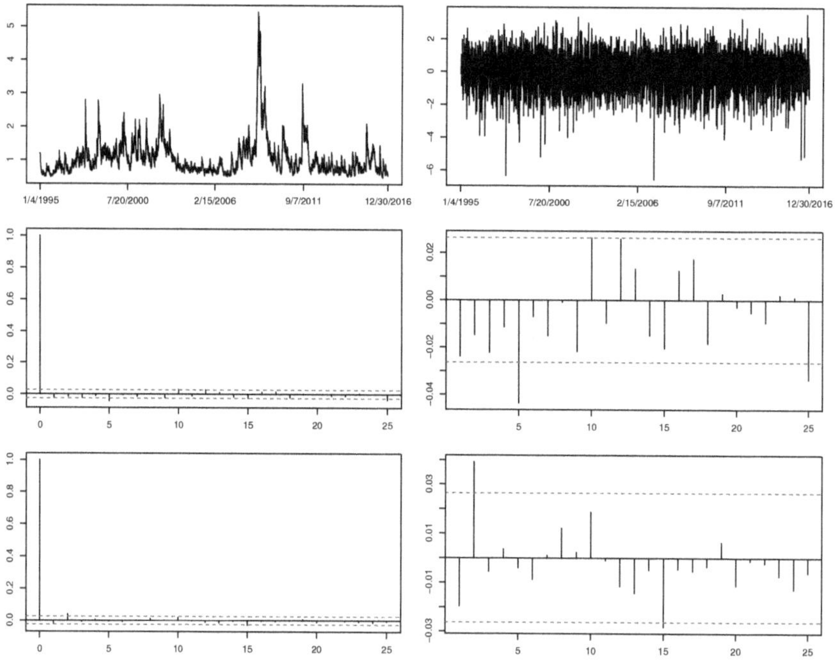

Figure 12.4 *Estimated volatility* $\widehat{\sigma}_t$ *(top left), the standardized residual* $\widehat{\varepsilon}_t$ *(top right), and the same ac.f.'s and sample partial ac.f.'s of* $\widehat{\varepsilon}_t$ *(middle) and* $\widehat{\varepsilon}_t^2$ *(bottom), respectively, in the GARCH(1, 1) model for daily returns of adjusted closing prices of the S&P500 index from January 4, 1995, to December 30, 2016.*

```
Call: garchFit(formula = ~garch(1, 1), data = sp500[, 2])

Conditional Distribution: norm

Coefficient(s):
      mu      omega     alpha1      beta1
0.073951  0.018870  0.100541  0.885716

Std. Errors: based on Hessian

Error Analysis:
        Estimate  Std. Error  t value Pr(>|t|)
mu      0.073951    0.011127    6.646 3.01e-11 ***
omega   0.018870    0.002920    6.464 1.02e-10 ***
alpha1  0.100541    0.008528   11.790  < 2e-16 ***
beta1   0.885716    0.009208   96.186  < 2e-16 ***
---
```

```
Signif. codes:  0 '***' 0.001 '**' 0.01 '*' 0.05 '.' 0.1 ' ' 1

Log Likelihood:
 -7703.91    normalized:  -1.390849

Description:
 Sun Nov 12 11:59:15 2017 by user:

Standardised Residuals Tests:
                                Statistic p-Value
 Jarque-Bera Test   R    Chi^2  692.0824  0
 Shapiro-Wilk Test  R    W      NA        NA
 Ljung-Box Test     R    Q(10)  25.4232   0.004598514
 Ljung-Box Test     R    Q(15)  34.91386  0.002529427
 Ljung-Box Test     R    Q(20)  39.16812  0.006353632
 Ljung-Box Test     R^2  Q(10)  14.55688  0.1490715
 Ljung-Box Test     R^2  Q(15)  21.04238  0.1354807
 Ljung-Box Test     R^2  Q(20)  22.33248  0.3227451
 LM Arch Test       R    TR^2   14.79585  0.2527914

Information Criterion Statistics:
     AIC      BIC      SIC     HQIC
2.783141 2.787922 2.783140 2.784808
```

## 12.5 The ARMA-GARCH Models

The linear time series models in Chapter 3 can be combined with GARCH to model the dynamics of asset returns and their volatilities. Assuming that $X_t$ follows an ARMA model with GARCH innovations yields the following ARMA$(p,q)$-GARCH$(h,k)$ model for $(X_t, \sigma_t)$:

$$X_t = \mu + \sum_{i=1}^{p} \phi_i X_{t-i} + Y_t + \sum_{j=1}^{q} \psi_j Y_{t-j}, \qquad Y_t = \sigma_t \varepsilon_t,$$
$$\sigma_t^2 = \omega + \sum_{j=1}^{k} \alpha_j Y_{t-j}^2 + \sum_{i=1}^{h} \beta_i \sigma_{t-i}^2. \tag{12.21}$$

The $\varepsilon_t$ in (12.21) are i.i.d. standard normal or standardized Student-$t$ random variables. The stationarity conditions for the ARMA part and the GARCH part are the same as those in Sections 3.8 and 12.4. The second equation in (12.21) can be replaced by other volatility models so that different aspects of volatilities can be characterized.

Given the model parameters, the two equations in (12.21) can be used to obtain one-step-ahead forecasts of the conditional mean and conditional variance of $r_t$.

Specifically, from Sections 5.3 and 12.4, we have the one-step-ahead forecasts at the forecast origin $N$

$$\widehat{X}_N(1) = \mu + \sum_{i=1}^{p} \phi_i X_{N+1-i} + \sum_{j=1}^{q} \psi_j Y_{N+1-j}, \tag{12.22}$$

$$\widehat{\sigma}_N^2(1) = \omega + \sum_{j=1}^{k} \alpha_j Y_{N+1-j}^2 + \sum_{j=1}^{h} \beta_i \sigma_{N+1-i}^2. \tag{12.23}$$

The conditional distribution of $X_{N+1}$ given the current and past observations up to time $N$ is $N(\widehat{X}_N(1), \widehat{\sigma}_N^2(1))$.

The $k$-step-ahead forecast $\widehat{X}_N(k)$ can be evaluated by using the method for the ARMA model $r_t$ described in Section 5.3.1. To evaluate the $k$-step-ahead forecast $\widehat{\sigma}_N^2(k)$, we first use the MA($\infty$) representation

$$X_t = \mu + \sum_{i=0}^{\infty} \psi_i Y_{t-i}$$

with $\psi_0 = 1$ (see Section 5.3.1). Therefore the $k$-step-ahead forecast of $X_{N+k}$ can be expressed as

$$\widehat{X}_N(k) = \mu + \sum_{i=k}^{\infty} \psi_i Y_{t+k-i},$$

and the corresponding forecast error is

$$e_N(k) = X_{N+k} - \widehat{X}_N(k) = \sum_{i=1}^{k} \psi_{k-i} Y_{t+i}.$$

From this and the property $E(Y_{N+i}^2 | \mathscr{F}_{N+i-1}) = \sigma_{N+i}^2$, which has been used to derive (12.19), it follows that the variance of the forecast error is

$$\text{Var}(e_N(k)) = \sum_{i=1}^{k} \psi_{k-i}^2 E(\sigma_{N+i}^2 | \mathscr{F}_N) = \sum_{i=1}^{k} \psi_{k-i}^2 \sigma_N^2(i), \tag{12.24}$$

in which $\sigma_N^2(i)$ are given by (12.19).

*Example 12.3*

We now consider an ARMA(1, 1)-GARCH(1, 1) model with the following specification for the daily return series of the S&P500 index

$$\begin{aligned} X_t &= \mu + \phi X_{t-1} + Y_t + \psi Y_{t-1}, \\ Y_t &= \sigma_t \varepsilon_t, \\ \sigma_t^2 &= \omega + \alpha Y_{t-1}^2 + \beta \sigma_{t-1}^2. \end{aligned}$$

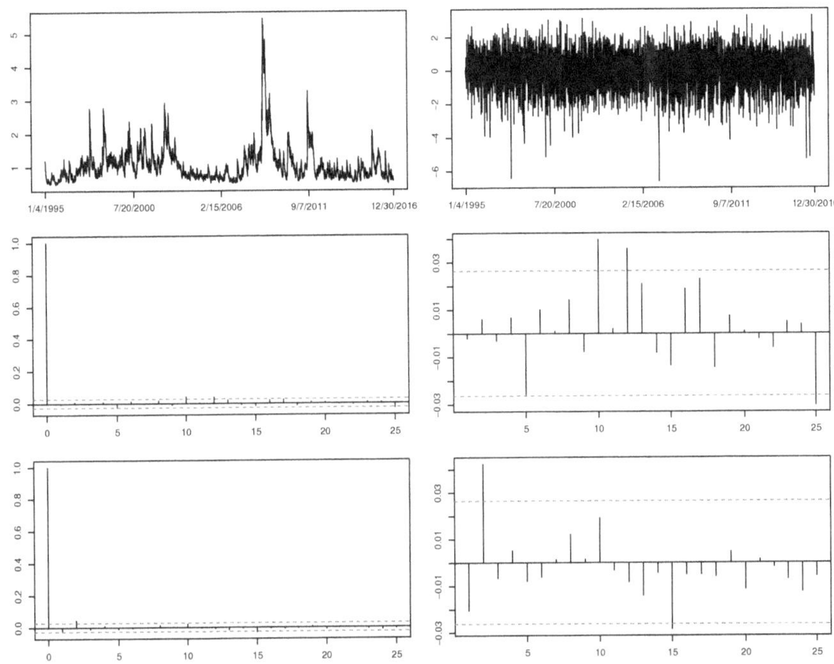

Figure 12.5 *Estimated volatility $\widehat{\sigma}_t$ (top left), the standardized residual $\widehat{\varepsilon}_t$ (top right), and the same ac.f.'s and sample partial ac.f.'s of $\widehat{\varepsilon}_t$ (middle) and $\widehat{\varepsilon}_t^2$ (bottom), respectively, in the ARMA(1,1)-GARCH(1, 1) model for daily returns of adjusted closing prices of the S&P500 index from January 4, 1995, to December 30, 2016.*

We still use the R function garchFit to estimate the model. Assuming that $\varepsilon_t$ are i.i.d. standard normal, we obtain the fitted model

$$\begin{aligned} r_t &= 0.0066_{(.0042)} + 0.9116_{(.0540)} X_{t-1} + Y_t - 0.9373_{(.0456)}, \\ Y_t &= \sigma_t \varepsilon_t, \\ \sigma_t^2 &= 0.0185_{(.0029)} + 0.0994_{(.0084)} Y_{t-1}^2 + 0.8871_{(.0091)} \sigma_{t-1}^2, \end{aligned}$$

where the standard errors of the parameters are given in the parentheses; see the R output below. Figure 12.5 shows the estimated volatility $\widehat{\sigma}_t$, the standardized residual $\widehat{\varepsilon}_t$, and the same ac.f.'s and sample partial ac.f.'s of $\widehat{\varepsilon}_t$ and $\widehat{\varepsilon}_t^2$, respectively.

```
> fit3<-garchFit(~arma(1,1)+garch(1,1), data=sp500[,2])
> fit3.vol<-volatility(fit3)
> fit3.resid.st<-residuals(fit3)/volatility(fit3)

> summary(fit3)
```

```
Title: GARCH Modelling

Call: garchFit(formula = ~arma(1, 1) + garch(1, 1),
          data = sp500[, 2])

Mean and Variance Equation: data ~ arma(1, 1) + garch(1, 1)

Conditional Distribution: norm

Coefficient(s):
       mu        ar1        ma1      omega     alpha1      beta1
0.0066356 0.9115801 -0.9372960 0.0184682 0.0993799 0.8871178

Std. Errors: based on Hessian

Error Analysis:
        Estimate  Std. Error  t value Pr(>|t|)
mu      0.006636    0.004237    1.566    0.117
ar1     0.911580    0.054019   16.875  < 2e-16 ***
ma1    -0.937296    0.045582  -20.563  < 2e-16 ***
omega   0.018468    0.002859    6.461 1.04e-10 ***
alpha1  0.099380    0.008421   11.801  < 2e-16 ***
beta1   0.887118    0.009078   97.719  < 2e-16 ***
---
Signif. codes:  0 '***' 0.001 '**' 0.01 '*' 0.05 '.' 0.1 ' ' 1

Log Likelihood: -7693.559    normalized:  -1.38898

Standardised Residuals Tests:
                                Statistic p-Value
 Jarque-Bera Test   R    Chi^2  750.2105  0
 Shapiro-Wilk Test  R    W      NA        NA
 Ljung-Box Test     R    Q(10)  15.42737  0.1172401
 Ljung-Box Test     R    Q(15)  26.54412  0.0326778
 Ljung-Box Test     R    Q(20)  32.5605   0.03767997
 Ljung-Box Test     R^2  Q(10)  16.5621   0.08463268
 Ljung-Box Test     R^2  Q(15)  22.95862  0.0850223
 Ljung-Box Test     R^2  Q(20)  24.22309  0.232783
 LM Arch Test       R    TR^2   16.26175  0.1795395

Information Criterion Statistics:
     AIC      BIC      SIC     HQIC
2.780126 2.787296 2.780124 2.782626
```

## 12.6 Other ARCH Type Models

Many other types of ARCH models have been proposed. We introduce two of them in this section. We shall note that, it is difficult to choose between different models for changing variance using the data alone and so it is advisable, as usual, to use the context and any background theory, to supplement the results of the exploratory analysis of the data.

### 12.6.1 *The integrated GARCH model*

In many empirical analyses of financial time series, it is found that the volatilities of asset returns are highly persistent, since the $\alpha+\beta$ values in the fitted GARCH(1, 1) models are very close to 1. If $\alpha+\beta=1$, the GARCH(1, 1) model becomes an **integrated GARCH** model

$$\sigma_t^2 = \omega + \beta\sigma_{t-1}^2 + (1-\beta)u_{t-1}^2,$$

More generally, the integrated GARCH model, IGARCH($h$, $k$), is of the form

$$\sigma_t^2 = \omega + \sum_{j=1}^{h} \beta_j \sigma_{t-j}^2 + \sum_{i=1}^{k} \alpha_i Y_{t-i}^2, \tag{12.25}$$

with $\sum_{j=1}^{k}\alpha_j + \sum_{i=1}^{h}\beta_i = 1$.

Note that the above condition implies that IGARCH($h$, $k$) models have infinite unconditional variance, which can be seen by letting $\sum_{i=1}^{h}\beta_i + \sum_{j=1}^{k}\alpha_j$ approach 1 in (12.15), hence IGARCH models cannot be covariance stationary. However, IGARCH models are in fact strictly stationary. Another interesting property of IGARCH models is that, the $k$-step-ahead forecast of $\sigma_{N+k}^2$ at forecast origin $N$ can be very simple. In particular, letting $\lambda \to 1$ in (12.20) yields the following forecast of $\sigma_{N+k}^2$

$$\sigma_N^2(k) = E(\sigma_{t+k}^2|\mathscr{F}_t) = k\omega + \beta\sigma_t^2 + (1-\beta)u_t^2. \tag{12.26}$$

### 12.6.2 *The exponential GARCH model*

As indicated in many empirical studies, a stylized fact of the volatility of asset returns is that the volatility response to a large positive return is considerably smaller than that of a negative return of the same magnitude. The GARCH model, which is defined by $\sigma_t^2$ and $Y_{t-j}^2$, cannot incorporate this leverage effect. To accommodate the asymmetry, Nelson (1991) proposed the **exponential GARCH** model, EGARCH($h,k$), that has the form

$$Y_t = \sigma_t\varepsilon_t, \quad \log(\sigma_t^2) = \omega + \sum_{i=1}^{h} \beta_i \log(\sigma_{t-i}^2) + \sum_{j=1}^{k} f_j(\varepsilon_{t-j}), \tag{12.27}$$

where the $\varepsilon_t$ are i.i.d. with mean 0 and $f_j(\varepsilon) = \alpha_j \varepsilon + \gamma_j(|\varepsilon| - E|\varepsilon|)$. Note that the random variable $f_j(\varepsilon_t)$ is the sum of two zero-mean random variables $\alpha_j \varepsilon_t$ and $\gamma_j(|\varepsilon_t| - E|\varepsilon_t|)$. We can rewrite $f_j(\varepsilon_t)$ as

$$f_j(\varepsilon_t) = \begin{cases} (\alpha_j + \gamma_j)\varepsilon_t - \gamma_j E|\varepsilon_t|, & \text{if } \varepsilon_t \geq 0, \\ (\alpha_j - \gamma_j)\varepsilon_t - \gamma_j E|\varepsilon_t|, & \text{if } \varepsilon_t < 0, \end{cases}$$

which shows the asymmetry of the volatility response to positive and negative returns. Since (12.27) represents $\log(\sigma_t^2)$ in ARMA form with innovations $f_j(\varepsilon_{t-j})$, $\sigma_t^2$ and therefore $Y_t$ also are stationary if the roots of $1 - \beta_1 z - \cdots - \beta_h z^h = 0$ lie outside the unit circle; see Section 3.7.2.

## 12.7 Stochastic Volatility Models

The reader will notice that the formulae for $\sigma_t^2$ in all GARCH type models introduced in this chapter are essentially deterministic in that there is no 'error' term in either equation. An alternative to ARCH or GARCH models is to assume that $\sigma_t$ follows a stochastic process. This is usually done by modelling the logarithm of $\sigma_t^2$ or of $\sigma_t$ to ensure that $\sigma_t^2$ remains positive. A simple example is to assume that $\log(\sigma_t^2) = h_t$, say, follows an AR process with an 'error' component that is independent of the $\{\varepsilon_t\}$ in the innovation series $\{Y_t\}$, that is,

$$\begin{aligned} Y_t &= \sigma_t \varepsilon_t, \qquad \sigma_t^2 = e^{h_t}, \\ h_t &= \phi_0 + \phi_1 h_{t-1} + \cdots + \phi_p h_{t-p} + \eta_t, \end{aligned} \tag{12.28}$$

which has AR($p$) dynamics for $\log \sigma_t^2$. Models of this type are called **stochastic volatility** or **stochastic variance models**. The $\varepsilon_t$ and $\eta_t$ in (12.28) are assumed to be independent normal random variables with $\varepsilon_t \sim N(0,1)$ and $\eta_t \sim N(0, \sigma^2)$. A complication of the SV model is that unlike in usual AR($p$) models, the $h_t$ in (12.28) is an unobserved state undergoing AR($p$) dynamics, while the observations are $u_t$ such that $u_t | h_t \sim N(0, e^{h_t})$. The likelihood function of $\theta = (\sigma, \phi_0, \ldots, \phi_p)^T$, based on a sample of $n$ observations $u_1, \ldots, u_n$, involves $n$-fold integrals, making it prohibitively difficult to compute the MLE by numerical integration for usual sample sizes.

Although the likelihood function is more difficult to handle, the model ties in more naturally with other finance models and is easier to generalize to the multivariate case. Moreover it seems intuitively more reasonable to assume that $\sigma_t$ changes stochastically through time rather than deterministically, especially when one sees the sudden changes in volatility that can occur in the financial market as a result of a special event like a war involving oil-producing countries. More details may be found in Harvey (1993, Section 8.4). Taylor (1994) suggests that a judicious combination of both ARCH and stochastic volatility models may provide more satisfactory results than a single model.

## 12.8 Bibliography

Further information about ARCH and GARCH models is given by Bollerslev et al. (1992, 1994), (Enders, 1995, Chapter 3), (Franses, 1998, Chapter 7), Gourieroux (1997), Shephard (1996), and Lai and Xing (2008, Chapters 6 and 9).

### Exercises

**12.1** Consider an ARCH(1,1) model

$$Y_t = \sigma_t \varepsilon_t, \qquad \sigma_t^2 = 0.1 + 0.5\sigma_{t-1}^2,$$

where $\varepsilon_t$ are i.i.d. standard Gaussian random variables. What are the unconditional variance and unconditional kurtosis of $Y_t$?

**12.2** Use the following R script to simulate an ARCH(1,1) process given in Exercise 12.1.

```
> set.seed(10)
> n<-500
> eps<-rnorm(n)
> sigma2<-rep(n)
> sigma2[1]<-0.1+0.5*1
> for (i in 2:n)  sigma2[i]<-0.1+0.5*sigma2[i-1]
> vol<-sqrt(sigma2)
> y<-vol*eps
```

(a) Plot the series $Y_t$ and $\sigma_t$.

(b) Plot the sample ac.f. and sample ac.f. of $Y_t$, $|Y_t|$, $Y_t^2$, and $\sigma_t^2$.

**12.3** Show that the kurtosis of the series generated by the GARCH(1, 1) model (12.16) is given by (12.17).

**12.4** Consider an AR(1)-GARCH(1,1) model

$$X_t = \theta X_{t-1} + Y_t, \quad Y_t = \sigma_t \varepsilon_t, \quad \sigma_t^2 = \omega + \alpha Y_{t-1}^2 + \beta \sigma_{t-1}^2.$$

(a) What are the 1-step and 2-step ahead forecasts of $X_{n+1}$ at the forecast origin $X_n$?

(a) What are the 1-step and 2-step ahead forecasts of $\sigma_{n+1}^2$ at the forecast origin $X_n$?

**12.5** Consider the daily returns, $Y_t$, of adjusted closing prices of the S&P500 index, which has been analysed in Examples 12.1–12.3. Fit a GARCH(2, 1) model with the following specification for $Y_t$

$$X_t = \mu + Y_t, \quad Y_t = \sigma_t \varepsilon_t, \quad \sigma_t^2 = \omega + \alpha_1 Y_{t-1}^2 + \beta_1 \sigma_{t-1}^2 + \beta_2 \sigma_{t-2}^2.$$

(a) Compare the maximized log-likelihood and AIC of a GARCH(2,1) model with that of an ARCH(2) model (Example 12.1), a GARCH(1,1) model (Example 12.2), and an ARMA(1,1)-GARCH(1,1) model.

(b) What are the 1-step and 2-step ahead forecasts of $\sigma_{n+1}^2$ at the forecast origin $X_n$?

Chapter 13

# Multivariate Time Series Modelling

## 13.1 Introduction

Observations are often taken simultaneously on two or more time series. For example, in meteorology we might observe temperature, air pressure and rainfall at the same site for the same sequence of time points. In economics, many different measures of economic activity are typically recorded at regular intervals. Examples include the retail price index, the gross domestic product and the level of unemployment. Given multivariate data like these, it may be helpful to develop a multivariate model to describe the interrelationships among the series.

Section 5.3.1 briefly discussed some multivariate forecasting procedures, focusing particularly on the use of multiple regression models. This chapter takes a more detailed look at some multivariate time-series models, giving particular attention to vector autoregressive (VAR) models.

The enormous improvement in computing capability over recent years has made it much easier to fit a given multivariate model from a computational point of view. However, the overall model-building process is still much more difficult for multivariate than univariate models. In particular, there are typically (far) more parameters to estimate than in the univariate case. Moreover, the pool of candidate models is much wider, and this can be seen as a strength – more possibilities – or as a weakness – it is harder to find the 'right' model. As regards the latter point, it appears that multivariate models are more vulnerable to misspecification than simpler univariate models, and this emphasizes the importance of getting sufficient background information so as to understand the context and identify all relevant explanatory variables before starting the modelling process. As always, it is vital to ask appropriate questions and formulate the problem carefully. An iterative approach to model building (see Section 4.8) is generally required, and the use to which the model will be put should be considered as well as the goodness of fit. There is always tension between seeking a parsimonious model (so that fewer parameters need to be estimated) while ensuring that important effects are not mistakenly omitted from the model.

With multivariate time-series data, the modelling process is complicated by the need to model the serial dependence *within* each series, as well as the interdependence *between* series. Information about the latter is provided by the cross-correlation function, which was introduced in Section 8.1. However, as noted in Section 8.1.3, the interpretation of cross-correlations is difficult and choosing


DOI: 10.1201/9781003646112-13

appropriate candidate models is not easy. There also tend to be more problems with the data in the multivariate case. Having more variables to measure means that there is more chance of mistakes in the data. Moreover, multivariate data are sometimes unsuitable for fitting multivariate models, perhaps because some explanatory variables have been held more or less constant in the past.

While univariate models can be very useful for describing short-term correlation effects, for forecasting large numbers of series, and as a benchmark in comparative forecasting studies, it is clear that multivariate models should also have much to offer in gaining a better understanding of the underlying structure of a given system and (hopefully) in getting better forecasts. Sadly, as noted in Section 5.4.2, the latter does not always happen. While multivariate models can usually be found that give a better *fit* than univariate models, there are a number of reasons why better forecasts need not necessarily result (though of course they sometimes do). We have seen that multivariate models, being more complicated, are generally more difficult to fit than univariate ones, while multivariate data may leave much to be desired. Furthermore, multivariate forecasts may require values of explanatory variables that are not yet available and so must themselves be forecast. If this cannot be done very accurately, then poor forecasts of the response variable may also result. Overall, the analyst should be prepared for the possibility that multivariate forecasts are not always as good as might be expected.

### *13.1.1 One equation or many?*

One basic question is whether the model should involve a single equation or several equations. In multiple regression, for example, the model explains the variation in a single **response** variable, say $y$, in terms of the variation in one or more **predictor**, or **explanatory**, variables, say $x_1, x_2, \ldots$. This is done with the aid of a single equation. For a single-equation model to be appropriate, there must be only one response variable of interest (e.g. forecasts are only required for this one variable), and there should be no suggestion that the value of the response variable could itself affect the predictor variables. In other words, the regression equation assumes there is an **open-loop** system (see Figure 9.7). If the relationship between a single predictor variable and a single response variable can be modelled by a regression equation, some people would say that there is a **causal relationship** between the variables, though in practice it may be difficult to decide whether there is a direct link or if the link comes via relationships with a third, possibly unobserved, variable.

A completely different situation arises when the 'outputs' affect the 'inputs' so that there is **feedback** in a **closed-loop** system (see Figure 9.8). For example, in economics, we know that a rise in prices will generally lead to a rise in wages, which will in turn lead to a further rise in prices. Then a regression model is not appropriate. Instead, a model with more than one equation will be needed to satisfactorily model the system.

One type of model with more than one equation is the econometric simultaneous equation model (see Section 5.3.2). This model comprises a number of equations, which need not be linear, and which are generally constructed using economic theory. By including relevant policy variables, they may be used to evaluate alternative economic strategies, as well as improving understanding of the system (the economy) and producing forecasts (although the latter may not be the prime objective). Models of this type are typically constructed by econometricians and so we do not attempt to describe the modelling process in this time-series text, as the problems are usually more of an economic nature than statistical. We will, however, make the following general remarks.

We have already contrasted the differing viewpoints of econometricians and statisticians, while emphasizing the complementary nature of their skills. It is certainly true that economic models based on theory need to be validated with real data, but a statistical data-based approach to modelling the economy will not get very far by itself without some economic guidelines. This is because an unrestricted analysis will find it difficult to select a sensible model from a virtually infinite number of choices. However, we note the following three general points in regard to building models for economic data:

1. Economic data is naturally affected by feedback, and this always makes model building difficult. For a physical system, such as a chemical reactor, where the feedback is well-controlled, there may not be enough information in the available data to identify the structure of the system. However, this is not a major problem in that known perturbations can be superimposed on the system in order to see what effect they have. However, it is generally less easy to control an economy than something like a chemical reactor, and efforts to control the economy are often made in a fairly subjective way. As a result, the amount of information in economic data may be less than one would like. Furthermore, it is difficult to carry out experiments on the economy in the same sort of way that perturbations can be added to a physical system.
2. The economy has a complex, non-linear structure, which may well be changing through time, and yet data sets are often quite small.
3. Statistical inference is usually carried out conditionally on an assumed model and focuses on uncertainty due to sampling variation and having to estimate model parameters. However, specification errors, arising from choosing the wrong model, are often more serious, particularly in economics.

The main time-series alternative to econometric simultaneous equation models is VAR models and they will be covered in Section 13.3.

### *13.1.2 The cross-correlation function*

A key tool in modelling multivariate time-series data is the cross-correlation function, which was defined in Section 8.1 in the context of a (bivariate) linear system. Before continuing, it may be helpful to redefine this function for an $m$-variate multivariate process, say $\{X_t\}$, where $X_t^T = (X_{1t}, X_{2t}, \ldots, X_{mt})$. Analogous to the univariate case, we begin by defining cross-covariances.

Let $\mu_t$ denote the vector of **mean** values of $X_t$ at time $t$, so that its $i$th component is $\mu_{it} = E(X_{it})$. Let $\Gamma(t, t+k)$ denote the **cross-covariance matrix** of $X_t$ and

$X_{t+k}$, so that its $(i,j)$th element is the cross-covariance coefficient of $X_{it}$ and $X_{j,t+k}$. A multivariate process is said to be **second-order stationary** if the mean and the cross-covariance matrices at different lags do not depend on time. Then $\mu_t$ will be a constant, say $\mu$, while $\Gamma(t,t+k)$ will be a function of the lag $k$ only, say $\Gamma(k)$. Then the $(i,j)$th element of $\Gamma(k)$, say $\gamma_{ij}(k)$, is given by

$$\gamma_{ij}(k) = \text{Cov}(X_{it}, X_{j,t+k}) = E[(X_{it} - \mu_i)(X_{j,t+k} - \mu_j)], \tag{13.1}$$

see Equation (8.1). In the stationary case, the set of cross-covariance matrices, $\Gamma(k)$ for $k = 0, \pm 1, \pm 2, \ldots$, is called the **covariance matrix function**.

It has rather different properties to the (auto)covariance function in univariate time series in that it is not an even function of lag. Actually, since

$$\gamma_{ij}(k) = \text{Cov}(X_{it}, X_{j,t+k}) = \text{Cov}(X_{j,t+k}, X_{it}) = \gamma_{ji}(-k),$$

we have

$$\Gamma(k) = \Gamma^T(-k), \qquad k = 0, \pm 1, \pm 2, \ldots. \tag{13.2}$$

Note that the diagonal terms, $\gamma_{ii}(k)$, which are auto- rather than cross-covariances, still have the property of being an even function of lag.

Given the covariance matrix function, it is easy to standardize any particular element of any matrix (by dividing by the product of the standard deviations of the two relevant series) to find the corresponding cross-correlation and hence construct the set of $(m \times m)$ cross-correlation matrices, $R(k)$ for $k = 0, \pm 1, \pm 2, \ldots$, called the **correlation matrix function** of the process. Thus the $(i,j)$th element of $R(k)$ is given by

$$\rho_{ij}(k) = \text{Corr}(X_{it}, X_{j,t+k}) = \gamma_{ij}(k)/\sigma_i\sigma_j \tag{13.3}$$

where $\sigma_i$, the standard deviation of $X_{it}$, can also be expressed as $\sqrt{\gamma_{ii}(0)}$. Note that $\rho_{ij}(k)$ is the correlation coefficient between $X_{it}$ and $X_{j,t+k}$. When $k > 0$, this correlation coefficient measures the linear dependence of $X_{j,t+k}$ on $X_{it}$, which occurs after time $t$. Consequently, if $\rho_{ij}(k) \neq 0$ and $k > 0$, the series $X_{it}$ leads the series $X_{jt}$ at lag $k$. Furthermore, using (13.2), we obtain that

$$R(k) = R^T(-k), \qquad k = 0, \pm 1, \pm 2, \ldots. \tag{13.4}$$

Due to the above property, it suffices in practice to consider the cross-correlation matrices $R(k)$ for $k \geq 0$.

Assuming that the same number of observations, say $N$, have been collected on the $m$ variables over the same time period, the sample cross-covariances and cross-correlations may be calculated by a natural extension of the formulae given in Section 8.1.2. For example, the sample cross-covariance coefficient of $X_i$ and $X_j$ at lag $k$ is given by

$$c_{ij}(k) = \begin{cases} \sum_{t=1}^{N-k} (x_{it} - \bar{x}_i)(x_{j,t+k} - \bar{x}_j)/N & k = 0, 1, \ldots, N-1 \\ \sum_{t=1-k}^{N} (x_{it} - \bar{x}_i)(x_{j,t+k} - \bar{x}_j)/N & k = -1, \ldots, -(N-1) \end{cases} \tag{13.5}$$

and the **sample cross-correlation coefficient** of $X_i$ and $X_j$ at lag $k$ is given by

$$r_{ij}(k) = c_{ij}(k)/s_i s_j \tag{13.6}$$

where $s_i = \sqrt{c_{ii}(0)}$ denotes the sample standard deviation of observations on the $i$th variable.

### *13.1.3 Initial data analysis*

In time-series analysis, the first step should normally be to plot the data. With multivariate data, this step is no less important. A time plot for each variable will indicate the presence of trend, seasonality, outliers and discontinuities. For stationary series, it will also be helpful to calculate the autocorrelation function (ac.f.) for each series in order to suggest an appropriate univariate model for each series. It may also be helpful to calculate the cross-correlation function for all meaningful pairs of variables, but the reader should refer back to Section 8.1.3, where the difficulties in interpreting cross-correlations are explained. In brief, cross-correlation variances are inflated by autocorrelation within the series and so some prefiltering is usually desirable to avoid spuriously large values.

With a good interactive graphics package, it can also be fruitful to scale all the series to have zero mean and unit variance, and then to plot pairs of variables on the same graph. One series can be moved backwards or forwards in time so as to make the visible characteristics of the series agree as closely as possible. If the cross-correlations between two series are generally negative around zero lag, then it may be necessary to turn one series 'upside-down' in order to get a good match. This type of approach may be helpful in alerting the analyst to the possible presence of linear relationships, perhaps with an in-built delay mechanism. However, the approach suffers from similar dangers as those involved in interpreting cross-correlations.

With a large number of variables, it can be very difficult to build a good multivariate time-series model, especially when many of the cross-correlations between predictor variables are 'large'. Then it may be fruitful to make some sort of multivariate transformation of the data (e.g. by using principal component analysis) so as to reduce the effective dimensionality of the data. This type of approach will not be considered here (see, e.g. Peña and Box, 1987).

*Example 13.1. Analysis of daily returns*

We show in Figure 13.1 daily percentage log returns of the Dow Jones Industrial Average, the S&P500, and the Nasdaq Composite indices from January 2, 2015, to November 30, 2017. These three market indices, labelled as $X_1, X_2$, and $X_3$, respectively, characterize the performance of the U.S. stock market from different perspectives; hence, they should be highly correlated. To see their inter-dependence, we show in Figure 13.2 their sample cross-correlations at lag $k = 0, 1, \ldots, 25$. We find that the concurrent inter-dependence of $X_1, X_2$ and $X_3$ are very strong, while the lead-lag effect among these series is relatively weak. Figures 13.1 and 13.2 can be reproduced using the following `R` script.

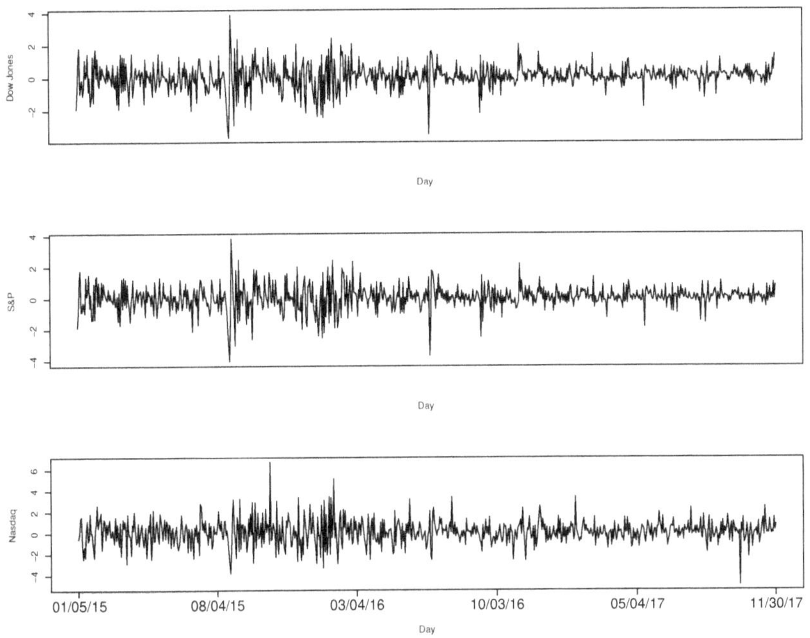


Figure 13.1 *Daily percentage log returns of adjusted closing prices of the Dow Jones Industrial Average (top), the S&P500 (middle), and the Nasdaq Composite indices from January 2, 2015, to November 30, 2017.*

```
> ret<-read.table("../data/indexret.txt", header=T)

> # Figure 13.1
> par(mfrow=c(3,1), mar=c(4,4,4,4))
> plot(ret$dji,type="l",xlab="Day",ylab="Dow Jones",xaxt="n")
> plot(ret$sp,type="l",xlab="Day",ylab="S&P",xaxt="n")
> plot(ret$nasdaq,type="l",xlab="Day",ylab="Nasdaq",xaxt="n")
> axis(1, x.pos, x.label, cex.axis=1.5)

> # a function of computing sample cross-correlation
> cross.corr<-function(dat1, dat2, k){
  dat.n<-length(dat1)
  cor(dat1[1:(dat.n-k)], dat2[(1+k):dat.n])
  }
> ret.ccor<-array(0, c(3,3,26))
> for (i in 1:3) for (j in 1:3) for (k in 1:26) {
     ret.ccor[i,j,k]<-cross.corr(ret[,i+1],ret[,j+1],k-1)
  }
```

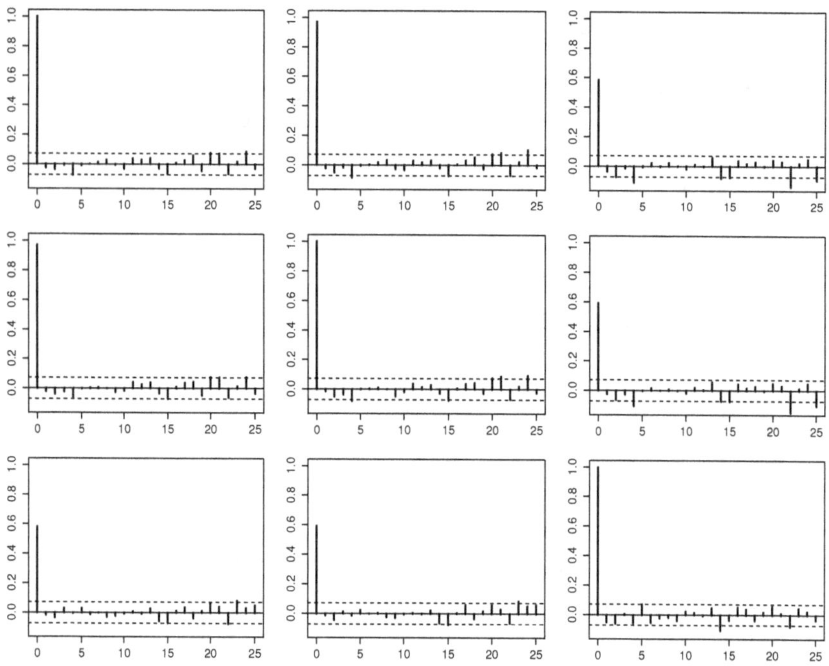

Figure 13.2 *Sample cross-correlations* $\widehat{\rho}_{ij}(k)$ $(i, j = 1,2,3)$, *of* $X_i$ *and* $X_j$ *at lag* $k = 0, 1, \ldots, 25$. $X_i$, $i = 1,2,3$, *are the daily percentage log returns of Adjusted closing prices of the Dow Jones Industrial Average, the S&P500 , and the Nasdaq Composite indices, respectively. The panel* $(i, j)$ *shows the sample cross-correlation coefficients of* $X_i$ *and* $X_j$.

```
> # Figure 13.2
> par(mfrow=c(3,3), mar=c(2,2,2,2))
> for (i in 1:3) for (j in 1:3) {
   plot(seq(0,25), ret.ccor[i,j,], type="h", xlab="",
     ylab="", ylim=c(-0.12, 1), lwd=2, cex.axis=1.3)
   abline(0,0)
   abline(1.96/sqrt(nrow(ret)), 0, lty=2)
   abline(-1.96/sqrt(nrow(ret)), 0, lty=2)
}
```

## 13.2 Single Equation Models

Section 13.1 said that it may be appropriate to model multivariate time-series data with a single equation when (1) there is a single response variable and several explanatory variables, and (2) there is no feedback from the response variable to the explanatory variables. The most obvious type of model to use is a multiple regression model, but the difficulties in fitting such models to time-series data have already been discussed in Section 5.3.1. We extend that discussion by considering the following simple model, namely

$$Y_t = a + bX_{t-d} + e_t \tag{13.7}$$

where $a, b, d$ are constants and $e_t$ denotes an error term (which may be autocorrelated).

If $d$ is an integer greater than zero in Equation (12.5), then $X_t$ is said to be a **leading indicator** for $Y_t$. Given data on $X$ and $Y$ until time $t$, this model enables forecasts of $Y_t$ to be made directly for up to $d$ steps ahead. However, to forecast more than $d$ steps ahead, the required value of $X_t$ will probably not be available and must itself be forecasted. In this sort of situation, a multivariate model is only able to give 'good' forecasts when forecasts of explanatory variables can be made (much) more accurately than those of the response variable.

Although Equation (13.7) appears to be a simple linear regression model at first sight, we said that the error terms may be autocorrelated and this makes the model non-standard. Furthermore, depending on the physical context, it may not be possible to control the values of $X_t$ or there may even be feedback from $Y_t$ to $X_t$, although this is not always immediately apparent. For all these reasons, it is often safer to fit an alternative class of models, which may include Equation (13.7) as a special case.

For an open-loop causal relationship between a single explanatory variable and a response variable, it is often worth trying to fit a member of the class of **transfer function models**. The latter were introduced in Section 9.4.2 and have the general form

$$Y_t = h(B)X_{t-d} + N_t, \tag{13.8}$$

where $h(B) = h_0 + h_1B + h_2B^2 + \cdots$ is a polynomial in the backward shift operator, $B$, $d$ denotes a non-negative integer and $N_t$ denotes noise (which may be autocorrelated). If $d > 0$, then $X_t$ is said to be a **leading indicator** for $Y_t$.

As noted in Section 9.4.2, Equation (13.8) can sometimes be parsimoniously rewritten in the form

$$\delta(B)Y_t = \omega(B)X_t + e_t, \tag{13.9}$$

where $\delta(B)$, $\omega(B)$ are low-order polynomials in $B$ such that $\omega(B) = \delta(B)h(B)$, while $e_t = \delta(B)N_t$ is usually assumed to follow some sort of autoregressive moving average (ARMA) process. Note that Equation (13.9) may include lagged values of $Y_t$ as well as of $X_t$. Models of this type can be fitted using the tools summarized in Section 9.4.2 and are fully described by Box et al. (1994, Chapter 11).

The **dynamic regression** models of Pankratz (1991) are of a somewhat similar type to those given above. Note that econometricians may refer to a model of the type described by Equation (13.8) as a **distributed lag model**, and when a polynomial lag function such as $h(B)$ is written as a *ratio* of polynomials as in the transfer function model of Equation (13.9), then econometricians may use the term **rational distributed lag**.

## 13.3 Vector Autoregressive Models

There are many situations where a single-equation model is inappropriate for multivariate time-series data. For example, there may be more than one response variable of interest, or the data may have been generated by a closed-loop system. In the latter case, it no longer makes sense to talk about an 'input' (an explanatory variable) and an 'output' (a response variable). More generally, there are many situations where there are two (or more) variables that, to a greater or lesser extent, 'arise on an equal footing', and which are all interrelated. Modelling such variables is often called **multiple time-series modelling**, and this section introduces arguably the most important class of models for this purpose.

### *13.3.1 VAR(1) models*

With $m$ variables, a natural way to represent them is by means of a $(m \times 1)$ vector $X_t$ where $X_t^T = (X_{1t}, \ldots, X_{mt})$. For simplicity, we initially restrict attention to the case $m = 2$. For stationary series, we may, without loss of generality, assume the variables have been scaled to have zero mean. In the latter case, a simple model would allow the values of $X_{1t}$ and $X_{2t}$ to depend linearly on the values of both series at time $(t-1)$. The resulting model for the two series would then consist of two equations, namely

$$\left.\begin{array}{lcl} X_{1t} & = & \phi_{11}X_{1,t-1} + \phi_{12}X_{2,t-1} + \varepsilon_{1t} \\ X_{2t} & = & \phi_{21}X_{1,t-1} + \phi_{22}X_{2,t-1} + \varepsilon_{2t}, \end{array}\right\} \tag{13.10}$$

where $\{\phi_{ij}\}$ are constants. The two 'error' terms $\varepsilon_{1t}$ and $\varepsilon_{2t}$ are usually both assumed to be white noise but are often allowed to be correlated contemporaneously. In other words, $\varepsilon_{1t}$ could be correlated with $\varepsilon_{2t}$ but not with past values of either $\varepsilon_{1t}$ or $\varepsilon_{2t}$.

Note that if coefficients $\phi_{12} = \phi_{21} = 0$, then $X_{1t}$ and $X_{2t}$ are not dynamically correlated. In such case, both $X_{1t}$ and $X_{2t}$ follow a univariate AR(1) process, respectively, and hence can be analysed separately. If one of $\phi_{12}$ and $\phi_{21}$ is not zero, say $\phi_{12} = 0$, but $\phi_{21} \neq 0$, then Equation (13.10) reduces to

$$\left.\begin{array}{lcl} X_{1t} & = & \phi_{11}X_{1,t-1} + \varepsilon_{1t} \\ X_{2t} & = & \phi_{21}X_{1,t-1} + \phi_{22}X_{2,t-1} + \varepsilon_{2t}. \end{array}\right\} \tag{13.11}$$

We notice that if $\phi_{12}$ is zero, then $X_{1t}$ does not depend on the lagged value of $X_{2t}$. This means that, while $X_{2t}$ depends on the lagged value of $X_{1t}$, there is no feedback from $X_{2t}$ to $X_{1t}$. Put another way, this means any causality goes in one direction only and we can think of $X_{1t}$ and $X_{2t}$ as the input and output, respectively. The first equation in (13.10) is univariate while the model for $X_{2t}$ could, in fact, be rewritten in the form of a transfer function model.

Equation (13.10) can be rewritten in vector form as

$$X_t = \Phi X_{t-1} + \varepsilon_t, \tag{13.12}$$

where $\varepsilon_t^T = (\varepsilon_{1t}, \varepsilon_{2t})$ and

$$\Phi = \begin{pmatrix} \phi_{11} & \phi_{12} \\ \phi_{21} & \phi_{22} \end{pmatrix}.$$

Equation (13.12) looks like an AR(1) model except that $X_t$ (and $\varepsilon_t$) are now vectors instead of scalars. Since $X_t$ depends on $X_{t-1}$, it is natural to call this model a **vector autoregressive model** of order 1 (VAR(1)). Equation (13.12) can be further rewritten as

$$(I - \Phi B)X_t = \varepsilon_t \tag{13.13}$$

where $B$ denotes the backward shift operator, $I$ is the $(2 \times 2)$ identity matrix and $\Phi B$ represents the operator matrix

$$\begin{pmatrix} \phi_{11}B & \phi_{12}B \\ \phi_{21}B & \phi_{22}B \end{pmatrix}.$$

The stationarity of $\mathbf{X}_t$ can be extended from the argument for univariate $X_t$. In particular, the necessary and sufficient condition for the stationarity of (13.12) or (13.13) is that the roots of the determinant of $I - \Phi B$ lie outside the unit circle.

### *13.3.2 VAR(p) models*

We can readily generalize the above model from two to $m$ variables and from first-order autoregression to $p$th order. In general, a VAR model of order $p$ (VAR($p$)) can be written in the form

$$\Phi(B)X_t = \varepsilon_t, \tag{13.14}$$

where $X_t$ is a $(m \times 1)$ vector of observed variables, and $\Phi$ is a matrix polynomial of order $p$ in the backward shift operator $B$ such that

$$\Phi(B) = I - \Phi_1 B - \cdots - \Phi_p B^p,$$

where $I$ is the $(m \times m)$ identity matrix and $\Phi_1, \Phi_2, \ldots, \Phi_p$ are $(m \times m)$ matrices of parameters.

Note that, if the variables can be ordered in such a way that each $\Phi_i$ matrix is lower triangular (meaning that all coefficients above the diagonal are zero), then a transfer-function model may be regarded as a special case of the VAR model in Equation (13.14). The last component of $X_t$ is then the output (or response variable) in an open-loop system. In contrast, in a closed-loop system, the 'outputs' feedback to affect the 'inputs' and the general VAR model may then be appropriate to describe the behaviour of the mutually dependent variables.

We restrict attention to stationary processes, and hence, without loss of generality, we may assume the variables have been scaled to have zero mean. This explains why there is no constant on the right-hand side of Equation (13.14). The condition for stationarity is that the roots of the equation,

$$\text{determinant}\{\Phi(x)\} = |I - \Phi_1 x - \Phi_2 x^2 - \cdots - \Phi_p x^p| = 0,$$

should lie outside the unit circle. Note that this condition reduces to the familiar condition for stationarity in the univariate case when $m = 1$.

In Equation (13.14), we have used $\varepsilon_t$ to denote $m$-dimensional white noise. Although we introduced bivariate white noise in the VAR(1) example above, we need to define white noise more generally in $m$ dimensions. Let $\varepsilon_t^T = (\varepsilon_{1t}, \varepsilon_{2t}, \ldots, \varepsilon_{mt})$ denote an $(m \times 1)$ vector of random variables. This multivariate time series will be called **multivariate white noise** if it is stationary with zero mean vector 0, and if the values of $\varepsilon_t$ at different times are uncorrelated. Then the $(m \times m)$ matrix of the cross-covariances of the elements of $\varepsilon_t$ with the elements of $\varepsilon_{t+j}$ is given by

$$\mathrm{Cov}(\varepsilon_t, \varepsilon_{t+j}) = \begin{cases} \Gamma_0 & j = 0, \\ 0_m & j \neq 0, \end{cases}$$

where $\Gamma_0$ denotes a $(m \times m)$ symmetric positive-definite matrix and $0_m$ denotes an $(m \times m)$ matrix of zeroes. This means that each component of $\varepsilon_t$ behaves like univariate white noise. Notice that the covariance matrix at lag zero, namely, $\Gamma_0$, does not need to be diagonal, as an innovation at a particular time point could affect more than one measured variable at that time point. Thus we do allow the components of $\varepsilon_t$ to be contemporaneously correlated.

The mathematics for a VAR model involves matrix polynomials that may look rather strange at first. To get a better feel for them, the reader is advised to look again at the first-order polynomial example included in Equation (13.13) and write out the scalar equations it represents.

Given the matrix $\Gamma_0$, describing the contemporaneous covariances of the white noise, it is possible in principle to evaluate the covariance matrix function, and hence the correlation matrix function of a VAR process. In practice, the algebra is usually horrid and it is only possible to find simple analytic functions in some simple cases. One gets a generalized matrix form of the Yule–Walker equations, which can be difficult to solve. Consider, for simplicity, the VAR(1) model in Equation (13.12). Multiply through on the right-hand side by $X_{t-k}^T$ and take expectations. When $k > 0$, we get

$$\Gamma(k) = \Gamma(k-1)\Phi^T,$$

but when $k = 0$, we get

$$\Gamma(0) = \Gamma(-1)\Phi^T + \Sigma_0 = \Gamma(1)^T\Phi^T + \Sigma_0.$$

These equations are only easy to solve when $\Phi$ is diagonal, in which case one has $m$ independent univariate AR(1) processes – see Exercise 13.3.

The definition of a VAR model in Equation (13.14) does not attempt to describe features such as trend and seasonality. It is possible to add deterministic terms to the right-hand side of Equation (13.14) to account for a non-zero mean, for trend and for seasonality (e.g. by including seasonal dummy variables). However, for seasonal data, it is usually easier to deseasonalize the data before attempting to model them, especially if the aim is to produce seasonally adjusted figures and forecasts. One simple possibility is to use seasonal differencing. As regards trend, non-seasonal (first) differencing may be employed to remove trend. However, the use of differencing is also not without problems, particularly if **co-integration** is present (see Section 13.6). Thus, fitting VAR models to real data is not easy and will be discussed separately in Section 13.5.

## 13.4 Vector ARMA Models

As in the univariate case, the VAR model may be generalized to include moving average (MA) terms. Building on Equation (13.14), this is done in an 'obvious' way by writing

$$\Phi(B)X_t = \Theta(B)\varepsilon_t \tag{13.15}$$

where

$$\Theta(B) = I + \Theta_1 B + \cdots + \Theta_q B^q$$

is a matrix polynomial of order $q$ in the backward shift operator $B$ and $\Theta_1, \Theta_2, \ldots, \Theta_q$ are $(m \times m)$ matrices of parameters. Then $X_t$ is said to follow a **vector ARMA** (VARMA) model of order $(p, q)$. Equation (13.15) is a natural generalization of the univariate ARMA model, and reduces to the familiar univariate ARMA model when $m = 1$. For stationary models, satisfying the stationarity condition given in Section 13.3, we can impose a condition for invertibility analogous to that in the univariate case. This requires that the roots of the equation,

$$\text{determinant}\{\Theta(x)\} = |I + \Theta_1 x + \Theta_2 x^2 + \cdots + \Theta_q x^q| = 0,$$

should lie outside the unit circle. This condition reduces to the usual univariate invertibility condition when $m = 1$.

If $\Phi(B)$ includes a factor of the form $I(1 - B)$, then the model is not stationary but rather acts on the first differences of the components of $X_t$. By analogy with the univariate case, such a model is called a **vector ARIMA** (VARIMA) model. Note that it may not be optimal in practice to difference each component of $X_t$ in the same way. Moreover the possible presence of co-integration (see Section 13.6) also needs to be considered before differencing multivariate data.

Forecasts can readily be computed for VAR, VARMA and VARIMA models by a natural extension of methods employed for univariate ARIMA models. Generally speaking, minimum mean square error (MMSE) forecasts can be obtained by replacing future values of white noise with zeroes while future values of $X_t$ are replaced with MMSE forecasts. Present and past values of $X_t$ and of $\varepsilon_t$ are replaced by the observed values and the (one-step-ahead forecast) residuals, respectively. Details will not be given here, but see Exercise 13.5.

One problem with VARMA (or VARIMA) models is that there may be different, but equivalent (or exchangeable) ways of writing what is really the same model. There are various ways of imposing constraints on the parameters involved in Equation (13.15) to ensure that a model is **identifiable**, meaning that the model is unique, but the conditions are complicated and will not be given here. There are further problems involved in identifying and fitting a model with MA components – see Section 13.5 – and, analogous to the univariate case, VARMA models are generally (much) harder to handle than VAR models.

Finally, we note that VARMA models can be generalized by adding terms, involving additional exogenous variables, to the right-hand side of Equation (13.15) and such a model is sometimes abbreviated as a VARMAX model.

## 13.5 Fitting VAR and VARMA Models

There are various approaches to the identification of VARMA models. They involve assessing the orders $p$ and $q$ of the model, estimating the parameter matrices in Equation (13.15) and estimating the variance–covariance matrix of the 'noise' components. We do not give details here, but rather refer the reader, for example, to Priestley (1981), Lütkepohl (1993) and Reinsel (1997). A recent survey of VARMA models is given by Tiao (2001).

Identification of a VARMA model is inevitably a difficult and complicated process because of the large number of model parameters that may need to be estimated. The number of parameters increases quadratically with $m$ and can become uncomfortably large when the lag length is more than one or two. This suggests that some constraints need to be placed on the model. One possibility is to use external knowledge or a preliminary analysis of the data to identify coefficient matrices where most of the parameters can *a priori* be taken to be zero. Such matrices are called *sparse* matrices. However, even with some sparse matrices included in the model, VARMA models are still difficult to fit, and so many analysts restrict attention to VAR models, which they hope will give an adequate approximation to VARMA models. Even then, there is still a danger of overfitting, and fitted VAR models do not always provide an approximation to real-life multivariate data that is as parsimonious and useful as AR models are for univariate data.

Because of the dangers of overfitting, a technique called **Bayesian vector autoregression** (BVAR) may be used to fit VAR models, in preference to using ordinary least squares. This approach can be used whether or not the analyst has a Bayesian philosophy. The technique essentially aims to prevent overfitting by shrinking parameters higher than first-order towards zero. The usual prior that is used for the parameters, called the **Minnesota prior**, has mean values, which assume *a priori* that every series is expected to be a random walk. Other priors have also been tried (e.g. Kadiyala and Karlsson, 1993). A tutorial paper showing how to select an appropriate BVAR model is given by Spencer (1993).

One important tool in VARMA model identification is the matrix of cross-correlation coefficients. The case of two-time series has already been considered in Section 8.1, and some difficulties involved in interpreting a sample cross-correlation function were noted. The author admits that he has typically found it difficult to interpret cross-correlation (and cross-spectral) estimates, even when there are only two variables. The analysis of three or more series is in theory a natural extension, but in practice is much more difficult and should only be attempted by analysts with substantial experience in univariate ARIMA model building. As previously noted, the interpretation of cross-correlations is complicated by the possible presence of autocorrelation within the individual series and by the possible presence of feedback between the series, and it is now generally recognized that series should be filtered or prewhitened before looking at cross-correlations.

A number of studies have been published, which suggest that, when carefully applied, the use of VARMA, and more especially of VAR models, can lead to improved forecasts as compared with univariate and other multivariate models. For example, the results in Boero (1990) suggest that a BVAR model is better than a large-scale econometric model for short-term forecasting, but not for long-term forecasts where the econometric model can benefit from judgemental interventions by the model user and may be able to pick up non-linearities not captured by (linear) VAR models. As is usually the case, different approaches and models are complementary. As a second example, Bidarkota (1998) found that a bivariate ARMA model gave better out-of-sample forecasts of real U.S. interest rates than a univariate unobserved components model. There is evidence that unrestricted VAR models do not forecast as well as when using Bayesian vector autoregression (e.g. Kadiyala and Karlsson, 1993), presumably because unrestricted models may incorporate spuriously many parameters.

Of course, multivariate models are often constructed to help describe and understand the measured system, rather than (just) to produce forecasts. We provide below an example of analysing the relationship among three macroeconomic variables.

*Example 13.2. Analysis of macro-economic series*

Figure 13.3 shows the time series plots of the U.S. *gross domestic product* (GDP), the civilian unemployment rate, and *consumer price index* (CPI) for all urban consumers from the first quarter of 1948 to the third quarter of 2017. Denote these series as $W_{1t}, W_{2t}$, and $W_{3t}$, respectively, and $\mathbf{W}_t = (W_{1t}, W_{2t}, W_{3t})^T$. Note that all $W_i$'s displays some extent of nonstationarity; we transform the data by calculating the rate of changes or difference for each series. Specifically, we obtain the change rate of GDP, $X_{1t} = (W_{1t} - W_{1,t-1})/W_{1,t-1} \times 100\%$, the difference of the unemployment rate, i.e. $X_{2t} = W_{2t} - W_{2,t-1}$, and a measure of inflation, $X_{3t} = (W_{3t} - W_{3,t-1})/W_{3,t-1} \times 100\%$. The transformed series $\mathbf{X}_t = (X_{1t}, X_{2t}, X_{3t})^T$ are shown in Figure 13.4. Figures 13.3 and 13.4 can be reproduced by the following R command.

```
> data<-read.table("data/macrots.txt", header=T)
> x.pos<- c(1, 70, 140, 210, 279)
> x.label<-c("1948.Q1", "1965.Q2", "1982.Q4", "2000.Q2",
"2017.Q3")

> # Figure 13.3
> par(mfrow=c(3,1), mar=c(4,4,4,4), cex.axis=1.5, cex.lab=1.2)
> plot(data$gdp,type="l",xlab="Quarter", ylab="GDP", xaxt="n")
> axis(1, x.pos, x.label)
> plot(data$unrate,type="l",xlab="Quarter", ylab="Unem. rate",
   xaxt="n")
> axis(1, x.pos, x.label)
> plot(data$cpi,type="l",xlab="Quarter", ylab="CPI", xaxt="n")
> axis(1, x.pos, x.label)

> gdp.rate<-diff(data$gdp)/data$gdp[1:(nrow(data)-1)]*100
> unrate.diff<-diff(data$unrate)
> cpi.rate<-diff(data$cpi)/data$cpi[1:(nrow(data)-1)]*100
> data2<-data.frame(data[-1,1], gdp.rate, unrate.diff, cpi.rate)
```

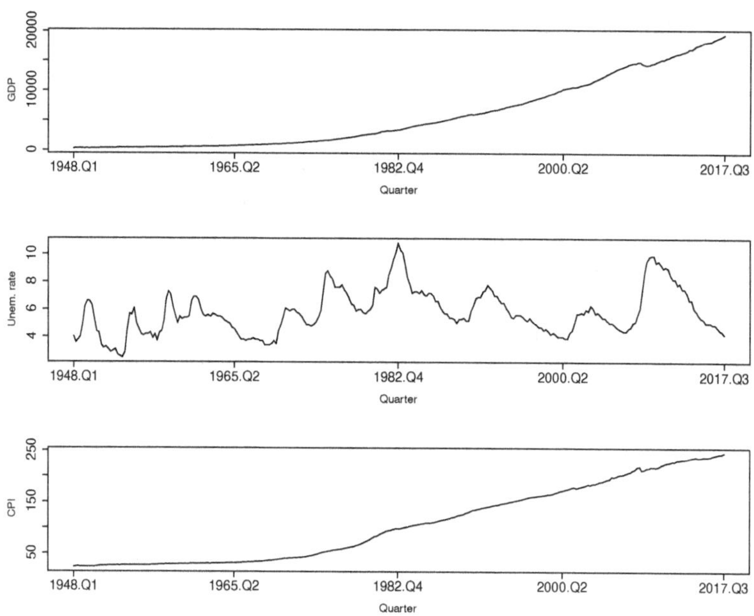


Figure 13.3 *Time series plot of GDP (top), unemployment rates (middle), and CPI (bottom).*

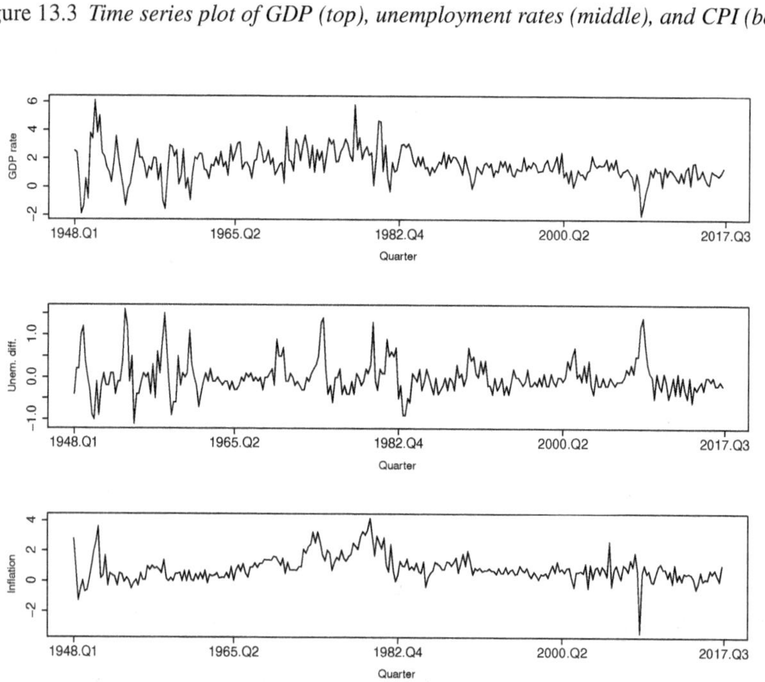


Figure 13.4 *Time series plot of the differenced series* $\mathbf{X}_t$.

```
> colnames(data2)<-c("quarter", "gdprate", "unemdiff",
"cpirate")

> # Figure 13.4
> par(mfrow=c(3,1), mar=c(4,4,4,4), cex.axis=1.5, cex.lab=1.2)
> plot(data2$gdprate,type="l",xlab="Quarter",
    ylab="GDP rate", xaxt="n")
> axis(1, x.pos, x.label)
> plot(data2$unemdiff,type="l",xlab="Quarter",
    ylab="Unem. diff.", xaxt="n")
> axis(1, x.pos, x.label)
> plot(data2$cpirate,type="l",xlab="Quarter",
    ylab="Inflation",xaxt="n")
> axis(1, x.pos, x.label)
```

To see the cross-sectional dependence of these three series, we show in Figure 13.5 the scatter plot of $X_i$ versus $X_j$, for $i, j = 1,2,3$. Note that the figure shows an obvious concurrent regression relationship between $X_{1t}$ and $X_{2t}$, $X_{1t}$ and $X_{3t}$, respectively. To see if there are any lead or lag effects among $X_{it}$'s, we show the sample cross-correlations in Figure 13.6. Figures 13.5 and 13.6 can be generated by the following R command.

```
> par(mar=c(2,2,2,2), cex.axis=1.5)
> plot(data2[,-1])

> par(mfrow=c(3,3), mar=c(2,2,2,2))
for (i in 1:3) for (j in 1:3) {
  plot(seq(0,25), data2.ccor[i,j,], type="h", xlab="", ylab="",
        ylim=c(-0.12, 1), lwd=2, cex.axis=1.3)
  abline(0,0)
  abline(1.96/sqrt(nrow(data2)), 0, lty=2)
  abline(-1.96/sqrt(nrow(data2)), 0, lty=2)
}
```

Given the concurrent and cross-sectional dependence of $\mathbf{X}_t$ in Figures 13.5 and 13.6, we may first try a single equation model for the series. For example, we consider the following equation

$$X_{1t} - \alpha X_{1,t-1} = \beta_0 + \beta_1 X_{2t} + \beta_2 X_{2,t-1} + Z_t,$$

where $Z_t \sim N(0, \sigma^2)$. This equation can be estimated as a linear regression via the following R script.

```
> n<-nrow(data2)
> attach(data2)
> fit1<-lm(gdprate[2:n]~gdprate[1:(n-1)]+unemdiff[2:n]
      +unemdiff[1:(n-1)])
```

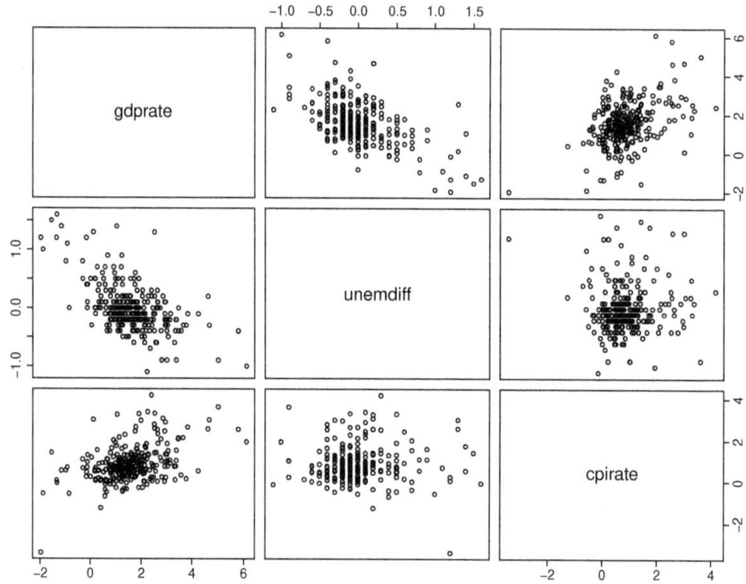


Figure 13.5 *Scatter plots of $X_i$ and $X_j$, $i, j = 1, 2, 3$.*

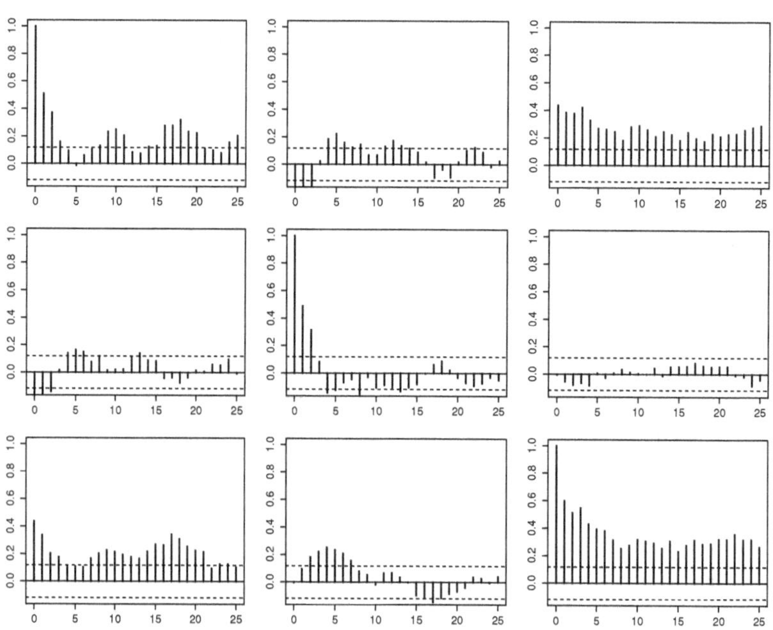

Figure 13.6 *Sample cross-correlations $\widehat{\rho}_{ij}(k)$ $(i, j = 1, 2, 3)$, of $\mathbf{X}_t$.*

```
> summary(fit1)

Call:
lm(formula = gdprate[2:n] ~ gdprate[1:(n - 1)] + unemdiff[2:n] +
    unemdiff[1:(n - 1)])

Residuals:
    Min      1Q  Median      3Q     Max
-2.1327 -0.5507 -0.1327  0.4315  3.6300

Coefficients:
                    Estimate Std. Error t value Pr(>|t|)
(Intercept)          0.95175    0.10117   9.408  < 2e-16 ***
gdprate[1:(n - 1)]   0.39013    0.05594   6.974 2.32e-11 ***
unemdiff[2:n]       -1.29118    0.14231  -9.073  < 2e-16 ***
unemdiff[1:(n - 1)]  0.27662    0.16110   1.717   0.0871 .
---
Signif. codes:  0 '***' 0.001 '**' 0.01 '*' 0.05 '.' 0.1 ' ' 1

Residual standard error: 0.8418 on 273 degrees of freedom
Multiple R-squared:  0.4357, Adjusted R-squared:  0.4295
F-statistic: 70.26 on 3 and 273 DF,  p-value: < 2.2e-16
```

Note that, at the 95% significance level, the estimated model is

$$X_{1t} = 0.9518_{(.1012)} + 0.3901_{(.0560)} X_{1,t-1} - 1.2912_{(.1423)} X_{2,t} + Z_t,$$

where $\widehat{\sigma}^2 = 0.8418^2$. This suggests that the changes in unemployment rates depend on the past values of the GDP growth rate. Similarly, we may fit a single equation model for $X_{1t}$ versus $X_{3t}$.

To get a full spectrum of relationships among the GDP growth rate, changes of unemployment rate, and the inflation rate, we consider a VAR model for $\mathbf{X}_t$. Specifically, we fit the following VAR(1) model to $\mathbf{X}_t$,

$$\mathbf{X}_t = \Phi \mathbf{X}_{t-1} + \varepsilon_t,$$

where $\text{Cov}(\varepsilon_t, \varepsilon_t) = \Gamma_0$, and $\text{Cov}(\varepsilon_t, \varepsilon_{t+j}) = 0$ for $j \neq 0$. This can be done by calling the function `VARMA` in the `R` package `MTS`. The following shows the `R` command and output of the estimation procedure.

```
> library("MTS")
> fit.var1<-VARMA(data2[,-1], p=1, q=0, include.mean=FALSE)

Number of parameters:  9
initial estimates:  0.7327 0.1979 0.251 -0.077 0.3654
```

```
      0.1182 0.2389 0.2681 0.5174
Par. lower-bounds:  0.6185 -0.1504 0.0784 -0.1156 0.2477
      0.0599 0.1651 0.0429 0.4058
Par. upper-bounds:  0.8469 0.5463 0.4237 -0.0384 0.483
      0.1765 0.3127 0.4933 0.629
Final Estimates:  0.711549 0.1507602 0.2785961 -0.06924344
      0.3811354 0.1073311 0.2308635 0.2543724 0.5149042

Coefficient(s):
           Estimate  Std. Error  t value Pr(>|t|)
gdprate     0.71155     0.05652   12.590  < 2e-16 ***
unemdiff    0.15076     0.17319    0.870 0.384027
cpirate     0.27860     0.08492    3.281 0.001035 **
gdprate    -0.06924     0.01926   -3.595 0.000324 ***
unemdiff    0.38114     0.05902    6.458 1.06e-10 ***
cpirate     0.10733     0.02894    3.709 0.000208 ***
gdprate     0.23086     0.03713    6.217 5.07e-10 ***
unemdiff    0.25437     0.11379    2.235 0.025393 *
cpirate     0.51490     0.05580    9.228  < 2e-16 ***
---
Signif. codes:  0 '***' 0.001 '**' 0.01 '*' 0.05 '.' 0.1 ' ' 1
---
Estimates in matrix form:
AR coefficient matrix
AR( 1 )-matrix
        [,1]  [,2]  [,3]
[1,]  0.7115 0.151 0.279
[2,] -0.0692 0.381 0.107
[3,]  0.2309 0.254 0.515

Residuals cov-matrix:
           [,1]         [,2]         [,3]
[1,]  1.0499988 -0.161464568 0.209923395
[2,] -0.1614646  0.121935341 0.002619803
[3,]  0.2099234  0.002619803 0.453308001
----
aic=  -3.137723
bic=  -3.020283
```

The estimated VAR(1) model is given by

$$\begin{pmatrix} X_{1t} \\ X_{2t} \\ X_{3t} \end{pmatrix} = \begin{pmatrix} 0.7115 & 0.1508 & 0.2786 \\ -0.0692 & 0.3811 & 0.1073 \\ 0.2309 & 0.2544 & 0.5149 \end{pmatrix} \begin{pmatrix} X_{1,t-1} \\ X_{2,t-1} \\ X_{3,t-1} \end{pmatrix} + \begin{pmatrix} \varepsilon_{1t} \\ \varepsilon_{2t} \\ \varepsilon_{3t} \end{pmatrix}$$

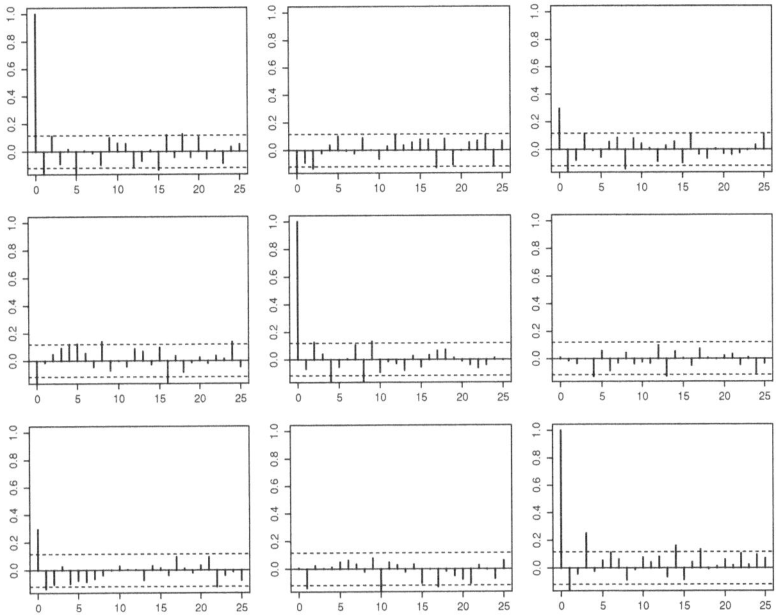

Figure 13.7 *Sample cross-correlations $\widehat{\rho}_{ij}(k)$ $(i, j = 1,2,3)$, of residual $\varepsilon_t$.*

with

$$\Gamma_0 = \begin{pmatrix} 1.0500 & -0.1615 & 0.2099 \\ -0.1615 & 0.1219 & 0.0026 \\ 0.2099 & 0.0026 & 0.4533 \end{pmatrix}.$$

To check if cross-covariance of residuals are zeros or not, we show in Figure 13.7 the sample cross-correlations of $\varepsilon_t$. Comparing the sample cross-correlations of $\mathbf{X}_t$ in Figure 13.6, the sample cross-correlations of $\varepsilon_t$ at lag $k \neq 0$ are significantly smaller. Actually, the majority of them are within the 95% confidence bands, which indicates that most of serial correlation among $\varepsilon_t$ is gone.

We may further consider fitting a VARMA model to the series $\mathbf{X}_t$, and it can also be done by calling the function `VARMA` in the `R` package `MTS`. Specifically, we can use the following command to fit a VARMA(1,1) model to $\mathbf{X}_t$.

```
fit.varma11<-VARMA(data2[,-1], p=1, q=1, include.mean=FALSE)
```

We omit the `R` output here and leave it to the interested readers for further exploration.

## 13.6 Co-Integration

Modelling multivariate time-series data is complicated by the presence of non-stationarity, particularly with economic data. One possible approach is to difference each series until it is stationary and then fit a VARMA model. However, this does not always lead to satisfactory results, particularly if different degrees of differencing are appropriate for different series or if the structure of the trend is of intrinsic interest in itself (and, in particular, assessing whether the trend is deterministic or stochastic). An alternative approach, much used in econometrics, is to look for what is called **co-integration**.

As a simple example, we might find that $X_{1t}$ and $X_{2t}$ are both non-stationary but that a particular linear combination of the two variables, say $(X_{1t} - kX_{2t})$ *is* stationary. Then the two variables are said to be co-integrated. If we now build a model for these two variables, there is no need to take first differences of both observed series, but rather the constraint implied by the stationary linear combination $(X_{1t} - kX_{2t})$ needs to be incorporated in the model.

A more general definition of co-integration is as follows. A series $\{X_t\}$ is said to be integrated of order $d$, written $I(d)$, if it needs to be differenced $d$ times to make it stationary. If two series $\{X_{1t}\}$ and $\{X_{2t}\}$ are both $I(d)$, then any linear combination of the two series will usually be $I(d)$ as well. However, if a linear combination exists for which the order of integration is less than $d$, say $(d-b)$, then the two series are said to be co-integrated of order $(d,b)$, written $CI(d,b)$. If this linear combination can be written in the form $\alpha^T X_t$, where $X_t^T = (X_{1t}, X_{2t})$, then the vector $\alpha$ is called a **co-integrating vector**.

Consider again the example given earlier in this section, where $X_{1t}$ and $X_{2t}$ are both non-stationary but $(X_{1t} - kX_{2t})$ is stationary. If $X_{1t}$ and $X_{2t}$ are both $I(1)$, then $d = b = 1$, $X_t$ is $CI(1,1)$ and a co-integrating vector is $\alpha^T = (1, -k)$.

In a non-stationary vector ARIMA model, there is nothing to constrain the individual series to make them 'move together' in some sense, yet the laws of economics suggest that there are bound to be long-run equilibrium forces that will prevent some economic series from drifting too far apart. This is where the notion of co-integration comes in. The constraint(s) implied by co-integration enable the analyst to fit a more realistic multivariate model.

In an introductory text, it is not appropriate to give further details here. Some useful references on co-integration include Banerjee et al. (1993), Dhrymes (1997), Engle and Granger (1991), and Johansen (2001), but there have been many other contributions to the subject dealing with topics such as tests for co-integration, error-correction models and ways of describing 'common trends'. An amusing non-technical introduction to the concept of co-integration is given by Murray (1994). We recommend that co-integration should always be considered when attempting to model multivariate economic data.

## 13.7 Multivariate Volatility Models

All sections in this chapter so far focus on the multivariate linear time series, which can be considered as an extension of univariate time series models in Chapter 3. We next consider an multivariate extension of the univariate volatility models in Chapter 12. Multivariate volatilities have many important applications in finance and other disciplines, and how to model and analyse multivariate volatilities has been an interesting topics during the last two decades.

Consider a multivariate time series $\mathbf{X}_t$. Similar to the univariate model (12.1), we write the series as

$$\mathbf{X}_t = \mu_t + \mathbf{Y}_t, \tag{13.16}$$

where $\mathbf{Y}_t = (Y_{1t}, \ldots, Y_{mt})^T$ is the innovation of the series at time $t$ with $E(\mathbf{Y}_t) = \mathbf{0}$, $\mu_t$ is the mean of $\mathbf{X}_t$ conditional on observed data up to time $t-1$, $\mathscr{F}_{t-1}$, and given by

$$\mu_t = E(\mathbf{X}_t | \mathscr{F}_{t-1}). \tag{13.17}$$

The process $\mu_t$ is assumed to follow the conditional expectation of a multivariate time series model in Sections 13.4 and 13.5. For simplicity, we may assume that $\mu_t$ follows

$$\mu_t = \sum_{i=1}^{p} \Phi_i \mathbf{X}_{t-i} + \sum_{j=1}^{q} \Theta_j \mathbf{Y}_{t-j}. \tag{13.18}$$

The conditional covariance matrix of $\mathbf{Y}_t$ given $\mathscr{F}_{t-1}$ is an $m \times m$ positive-definite matrix $\Sigma_t$ defined as $\Sigma_t = \mathrm{Cov}(\mathbf{Y}_t | \mathscr{F}_{t-1})$.

The major difficulty of extending univariate volatility models to their multivariate analogue is the curse of dimensionality, because there are $m(m+1)/2$ quantities in $\Sigma_t$ for an $m$-dimensional series. One way to overcome this difficulty is to consider some relatively simple multivariate volatility models. In this section, we introduce two kinds of volatility models. One is the exponentially weighted covariance matrix, and the other is called the BEKK model.

### *13.7.1 Exponentially weighted estimate*

Given the information set $\mathscr{F}_{t-1} = \{\mathbf{Y}_1, \ldots, \mathbf{Y}_{t-1}\}$, the covariance matrix of the innovation can be estimated by their sample covariance matrix, i.e.

$$\widehat{\Sigma}_t = \frac{1}{k-1} \sum_{i=1}^{k} (\mathbf{Y}_{t-i} - \bar{\mathbf{Y}})^{\otimes 2}, \tag{13.19}$$

where $\mathbf{a}^{\otimes 2} := \mathbf{a}\mathbf{a}^T$. This estimate put equal weights for all observations in the moving window of $\mathbf{Y}_t$'s. To emphasize the contribution of most recent innovations, the conditional covariance matrix at time $t$ can be approximated by

$$\widehat{\Sigma}_t = \sum_{i=1}^{k} \alpha_i (\mathbf{Y}_{t-i} - \bar{\mathbf{Y}})^{\otimes 2}, \tag{13.20}$$

where $0 < \alpha_1 < \alpha_2 < \cdots < \alpha_k < 1$ and $\sum_{i=1}^{k} \alpha_i = 1$. One specification of (13.20) is the *exponentially weighted moving average* (EWMA) estimate of the covariance matrix. Specifically, let $k = \infty$ and $\alpha_i = (1-\lambda)\lambda^{i-1}$ for some $0 < \lambda < 1$; Equation (13.20) becomes

$$\widehat{\Sigma}_t = (1-\lambda)\mathbf{Y}_{t-1}^{\otimes 2} + \lambda \widehat{\Sigma}_{t-1}. \tag{13.21}$$

When $\mathbf{Y}_1, \ldots, \mathbf{Y}_n$ are assumed to follow a multivariate normal distribution with mean $\mu$ and covariance matrix $\Sigma_t$, $\mu_t$ in (13.18) is a function of parameters $\Phi_i$'s and $\Theta_j$'s; hence, the log likelihood function of the data $\{\mathbf{X}_1, \ldots, \mathbf{X}_n\}$ can be derived, and maximized by numerical methods.

### 13.7.2 BEKK models

The BEKK model can be considered as a full extension of the univariate GARCH(1, 1) model. For an $m$-dimensional innovation process $\mathbf{Y}_t$, the BEKK$(h,k)$ volatility model takes the form

$$\Sigma_t = \mathbf{A}\mathbf{A}^T + \sum_{i=1}^{h} \mathbf{A}_i(\mathbf{Y}_{t-i}\mathbf{Y}_{t-i}^T)\mathbf{A}_i^T + \sum_{j=1}^{k} \mathbf{B}_j\Sigma_{t-j}\mathbf{B}_j^T, \tag{13.22}$$

where $\mathbf{A}$ is a lower triangular matrix such that $\mathbf{A}\mathbf{A}^T$ is positive-definite, $\mathbf{A}_i$ and $\mathbf{B}_j$ are $m \times m$ matrices. The model contains $m^2(h+k)+m(m+1)/2$ parameters, which increases rapidly with $h$ and $k$. It is difficult to study the property of a high-order BEKK model, so we focus only on an example for the BEKK(1,1) model.

*Example 13.3*

We consider the three macro-economic time series discussed in Section 13.5. In particular, we consider fitting a BEKK(1, 1) model to the transformed series $\mathbf{X}_t$; see Figure 13.4. For illustration purposes, we study the volatilities of GDP growth rates and inflation. Denote $\mathbf{Y}_t = (Y_{1t}, Y_{2t}) := (X_{1t}, X_{3t})$; we fit the following 2-dimensional BEKK(1,1) to $\mathbf{Y}_t$:

$$\Sigma_t = \mathbf{A}\mathbf{A}^T + \mathbf{A}_1(\mathbf{Y}_{t-1}\mathbf{Y}_{t-1}^T)\mathbf{A}_1^T + \mathbf{B}\Sigma_{t-1}\mathbf{B}^T. \tag{13.23}$$

The estimation can be carried out by calling the function `BEKK11` in the `R` package `MTS`. The following script and output in `R` show how to estimate the model parameters.

```
> library("MTS")
> fit<-BEKK11(data2[,-c(1,3)], include.mean=F)

Initial estimates:  1.113999 0.3721313 0.7685585 0.1 0.02
    0.02 0.1 0.8 0.1 0.1 0.8
Lower limits:  0.2227999 0.07442627 0.1537117 1e-06 -0.5
    -0.5 1e-06 1e-06 -0.5 -0.5 1e-06
Upper limits:  1.225399 0.4093445 0.8454144 0.999999 0.5
    0.5 0.999999 0.999999 0.5 0.5 0.999999

Coefficient(s):
        Estimate   Std. Error  t value   Pr(>|t|)
A011  0.28744139  0.13290378  2.16278   0.030558 *
A021  0.12166368  0.11316513  1.07510   0.282331
A022  0.15371171  0.06343682  2.42307   0.015390 *
A11   0.43739948  0.08634034  5.06599 4.0628e-07 ***
A21   0.10413559  0.03614344  2.88118   0.003962 **
A12   0.00772987  0.18349600  0.04213   0.966399
A22   0.45216477  0.07440824  6.07681 1.2260e-09 ***
```

```
B11   0.89702033  0.06957063 12.89366 < 2.22e-16 ***
B21  -0.01202102  0.02645793 -0.45434   0.649581
B12  -0.03443632  0.15553538 -0.22141   0.824777
B22   0.79452952  0.06505837 12.21256 < 2.22e-16 ***
---
Signif. codes:  0 '***' 0.001 '**' 0.01 '*' 0.05 '.' 0.1 ' ' 1
```

The volatility equation of the fitted BEKK(1, 1) model is

$$
\begin{aligned}
\begin{bmatrix} \sigma_{11,t} & \sigma_{12,t} \\ \sigma_{21,t} & \sigma_{22,t} \end{bmatrix} &= \begin{bmatrix} 0.287 & 0 \\ 0.122 & 0.154 \end{bmatrix} \begin{bmatrix} 0.287 & 0.122 \\ 0 & 0.154 \end{bmatrix} \\
&+ \begin{bmatrix} 0.437 & 0.008 \\ 0.104 & 0.452 \end{bmatrix} \begin{bmatrix} Y_{1,t-1}^2 & Y_{1,t-1}Y_{2,t-1} \\ Y_{1,t-1}Y_{2,t-1} & Y_{2,t-1}^2 \end{bmatrix} \begin{bmatrix} 0.437 & 0.104 \\ 0.008 & 0.452 \end{bmatrix} \\
&+ \begin{bmatrix} 0.897 & -0.034 \\ -0.012 & 0.795 \end{bmatrix} \begin{bmatrix} \sigma_{11,t-1} & \sigma_{12,t-1} \\ \sigma_{21,t-1} & \sigma_{22,t-1} \end{bmatrix} \begin{bmatrix} 0.897 & -0.012 \\ -0.034 & 0.795 \end{bmatrix},
\end{aligned}
$$

where seven elements are significant at 95% level. Figure 13.8 shows the estimated volatilities and the time-varying correlations of the BEKK(1,1) model. Note that the time-varying correlations of the BEKK(1, 1) model remain high during the most periods. Figure 13.8 can be reproduced by the following R command.

```
> x.pos<- c(1, 70, 140, 210, 279)
> x.label<-c("1948.Q1", "1965.Q2", "1982.Q4", "2000.Q2",
"2017.Q3")
> est.corr<-fit$Sig[,2]/sqrt(fit$Sig[,1]*fit$Sig[,4])

> par(mfrow=c(3,1), mar=c(4,4,4,4), cex.axis=1.5)
> plot(sqrt(fit$Sig[,1]), type="l", xlab="Quarter",
   ylab="Volatiltiy.GDP", xaxt="n")
> axis(1, x.pos, x.label)
> plot(sqrt(fit$Sig[,4]), type="l", xlab="Quarter",
   ylab="Volatility.inflation", xaxt="n")
> axis(1, x.pos, x.label)
> plot(est.corr, type="l", xlab="Quarter",
   ylab="Correlation", xaxt="n")
> axis(1, x.pos, x.label)
```

## 13.8 Bibliography

The beginner may find it helpful to read Chapter 5 of Chatfield (2001), Chapters 7 and 8 of Granger and Newbold (1986) or Chapters 13 and 14 of Wei (1990). A thorough treatment of VAR and VARMA models is provided by Lütkepohl (1993), while Reinsel (1997) covers similar material in a somewhat terser, more mathematical style. Multivariate versions of structural modelling, of Bayesian forecasting and

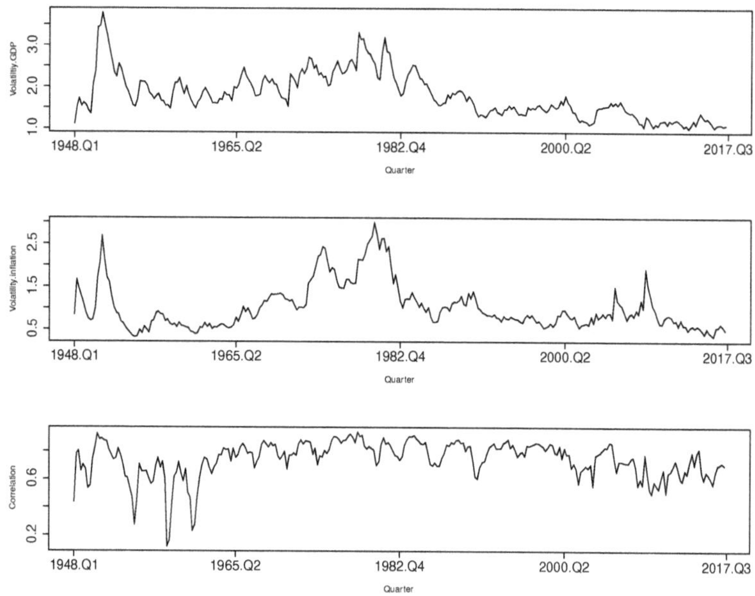


Figure 13.8 *Estimated volatilities (standard error) and time-varying correlations of a BEKK(1, 1) model for the quarterly GDP growth rate and the quarterly inflation rate from 1948 to 2017. Top: GDP growth rate volatility; Middle: Inflation rate volatility; Bottom: Time-varying correlations.*

of non-linear modelling have not been covered in this introductory chapter and the reader is referred to Harvey (1989), to West and Harrison (1997), to Granger and Teräsvirta (1993), and to Lai and Xing (2008, Chapter 9), respectively. Tsay (2015) gives a comprehensive summary of the basic concepts and ideas of analysing multivariate dependent data and provides the `R` package `MTS` for all the examples of multivariate time series analysis used in the book.

## Exercises

**13.1** Express the following bivariate model in vector form and say what sort of model it is.

$$\begin{aligned} X_{1t} &= X_{1,t-1} + 0.5X_{2,t-1} + \varepsilon_{1t} \\ X_{2t} &= 0.2X_{1,t-1} + 0.7X_{2,t-1} + \varepsilon_{2t} \end{aligned}$$

Is the model stationary and invertible?

**13.2** Consider the VAR(1) model for two variables as given by Equation (13.12). Determine whether the model is stationary for the following coefficient matrices at lag one;

(a) $\Phi = \begin{pmatrix} 0.3 & 0.3 \\ 0.3 & 0.3 \end{pmatrix}$; (b) $\Phi = \begin{pmatrix} 0.7 & 0.7 \\ 0.7 & 0.7 \end{pmatrix}$;

(c) $\Phi = \begin{pmatrix} 0.7 & 0 \\ 0 & 0.3 \end{pmatrix}$; (d) $\Phi = \begin{pmatrix} 0.9 & 0.5 \\ -0.1 & 0.3 \end{pmatrix}$.

**13.3** Find the covariance matrix function $\Gamma(k)$, and the correlation matrix function $P(k)$ for model (c) in Exercise 13.2, assuming that $\varepsilon_t$ denotes bivariate white noise with covariance matrix $\Gamma_0 = \begin{pmatrix} \sigma_1^2 & 0 \\ 0 & \sigma_2^2 \end{pmatrix}$ at lag zero.

**13.4** Consider the bivariate VARMA$(0,1)$ model from Equation (13.15) with $\Phi(B) = I$ and $\Theta(B) = I + \Theta_1 B$ where $\Theta_1 = \begin{pmatrix} 0.6 & 0.4 \\ 0.2 & 0.4 \end{pmatrix}$. Is the model stationary and invertible?

**13.5** Given data up to time $N$, find the one- and two-step-ahead forecasts for the bivariate VAR$(1)$ model in Exercise 13.2 (d).

**13.6** Consider the three dimensional series $\mathbf{X}_t = (X_{1t}, X_{2t}, X_{3t})$ in obtained in Example 13.2, where $X_{1t}$ is the growth rate of GDP, $X_{2t}$ is the change of the unemployment rate, and $X_{3t}$ is the inflation rate. Fit a VARMA(1,1) model to the series $\mathbf{X}_t$.

**13.7** Consider the three dimensional series $\mathbf{X}_t = (X_{1t}, X_{2t}, X_{3t})$ in obtained in Example 13.2, where $X_{1t}$ is the growth rate of GDP, $X_{2t}$ is the change of the unemployment rate, and $X_{3t}$ is the inflation rate.

(a) Fit a BEKK(1,1) model to the series $\mathbf{Y}_t := (X_{1t}, X_{2t})$.

(b) Fit a BEKK(1,1) model to the series $\mathbf{Y}_t := (X_{2t}, X_{3t})$.

Chapter 14

# Predictive AI for Time Series

Traditional time series models – such as ARIMA, exponential smoothing, and state-space models – rely on linear assumptions and fixed structures. While effective for simple seasonal or stationary patterns, they often fail to capture the nonlinear dynamics, complex interactions, and long-range dependencies found in real-world data. Abrupt regime changes, multiscale temporal behaviour, and interdependencies among multiple variables pose further challenges for classical approaches.

Deep learning offers a powerful alternative to overcome these limitations. Neural networks, often referred to as *artificial neural networks* (ANNs), are universal function approximators capable of modelling complex nonlinear relationships. They can also scale efficiently to large, high-dimensional datasets, making them well suited for modern applications in finance, healthcare, IoT, and climate science. Modern architectures provide remarkable flexibility: *Recurrent neural networks* (RNNs) and their variants – *Long short-term memory* (LSTM) and *gated recurrent units* (GRUs) – capture temporal dependencies; *convolutional neural networks* (CNNs) extract local temporal features; and *Transformers* use self-attention to model both short- and long-range dependencies in a parallel and scalable way.

This chapter introduces the foundations of neural networks and key architectures for predictive time series modelling. We discuss how these models can be designed, trained, and interpreted to improve forecasting accuracy, handle complex temporal dynamics, and uncover meaningful patterns in sequential data.

## 14.1 Artificial Neural Networks

*Artificial neural networks* (ANNs) provide the foundational framework for modern deep learning. They are computational models capable of capturing complex, nonlinear relationships between input variables and outputs. This section presents the essential building blocks of ANNs, including *layers*, *neurons*, *activation functions*, and the *training process*, establishing the foundation for understanding more advanced architectures such as *recurrent* and *convolutional* neural networks.

### *14.1.1 Feed-forward neural network*

An ANN is composed of a series of *layers*: an *input layer* that receives the explanatory variables, one or more *hidden layers* that process information through

 DOI: 10.1201/9781003646112-14

interconnected *neurons* or *nodes*, and an *output layer* that produces the final predictions. Each neuron applies a weighted transformation to its inputs and passes the result through a nonlinear activation function, enabling the network to represent highly flexible mappings. By iteratively adjusting these weights and biases during training, ANNs can approximate a wide range of functional relationships, enabling applications such as regression, classification, and time series forecasting.

A *feed-forward neural network* (FNN), also known as a *multilayer perceptron*, is a type of ANN in which information flows strictly from the input layer through the hidden layers to the output layer, without forming cycles. The term *feed-forward* highlights the absence of feedback connections, meaning that the network's outputs are never fed back as inputs. FNNs serve as the foundation of many deep learning architectures and are widely applied across tasks such as regression, classification, and sequential data modelling.

To understand how FNNs process information, it is helpful to examine the building block of the network – the *neuron*. A neuron is a function $g : \mathbb{R}^d \to \mathbb{R}$ that maps a vector of inputs $x \in \mathbb{R}^d$ to a scalar output:

$$g(\mathbf{x}) = \phi(\alpha'\mathbf{x} + \beta), \tag{14.1}$$

where $\alpha \in \mathbb{R}^d$ is a vector of *weights*, $\beta$ is a *bias*, and $\phi : \mathbb{R} \to \mathbb{R}$ is a continuous *activation function* that introduces nonlinearity into the network. Common active functions include the *rectified linear unit* (ReLU): $\text{ReLU}(x) = \max(0, x)$. the *sigmoid* function $\phi(x) = 1/(1+e^{-x})$, the *hyperbolic tangent* $\phi(x) = (e^x - e^{-x}/(e^x + e^{-x})$, and the *sign function* $\phi(x) = \text{sign}(x)$.

While a single neuron produces a scalar output, combining multiple neurons yields a *layer function* $h : \mathbb{R}^d \to \mathbb{R}^m$

$$h(\mathbf{x}) = (g_1(\mathbf{x}), \ldots, g_m(\mathbf{x}))', \tag{14.2}$$

where each $g_i$ is a neuron with its own weights $\alpha_i = (\alpha_{i1}, \ldots, \alpha_{id})'$ and bias $\beta_i$. Denoting the weights collectively as a $d \times m$ matrix of parameters $\mathbf{A} = \left[\alpha_1; \ldots; \alpha_m\right] = (\alpha_{ij})_{1 \le i \le d, 1 \le j \le m}$ and biases as $\beta = (\beta_1, \ldots, \beta_m)'$, this can be written compactly as

$$h(\mathbf{x}) = \bar{g}(\mathbf{A}'\mathbf{x} + \beta),$$

where $\bar{g} : \mathbb{R}^m \to \mathbb{R}^m$ is the vectorial activation function and applies the activation function element-wise.

A full feed-forward neural network is constructed by stacking $K$ layers:

$$h(\mathbf{x}) = h_K \circ h_{K-1} \circ \cdots \circ h_1(\mathbf{x}), \qquad K \ge 1, \tag{14.3}$$

where each layer $h_k : \mathbb{R}^{d_{k-1}} \to \mathbb{R}^{d_k}$ has its own weight matrix $\mathbf{A}_k$ and bias vector $\beta_k$. The input vector $\mathbf{x}$ forms the *input layer* and the final layer output is the *output layer*, and all intermediate layers are *hidden layers*. The number of neurons $d_k$ in each layer is called its *width*, while the total number of layers $k$ is the network *depth*. Together, $\{d_1, \ldots, d_K\}$ and $K$ define the *architecture* of the network. Networks with multiple hidden layers are commonly referred to as *deep neural networks* (DNNs).

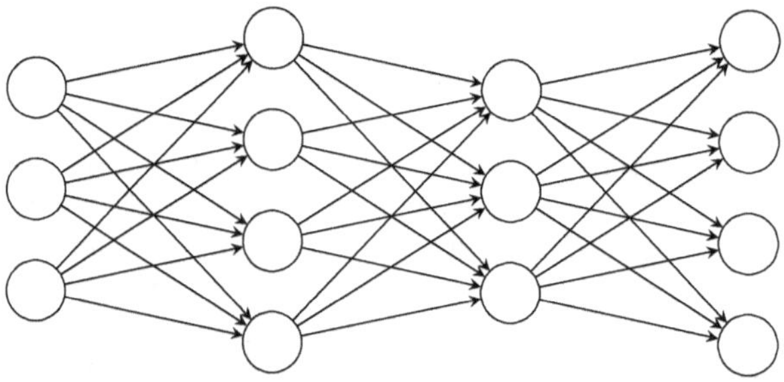

Figure 14.1 *A feed-forward neural network with two hidden layers. The network has 3 input neurons, two hidden layers with 4 and 3 neurons, and 4 output neurons.*

Figure 14.1 shows a FNN with $d = 3$ input neurons, two hidden layers with $d_1 = 4$ and $d_2 = 3$ neurons, and $d_3 = 4$ output neurons.

One of the most important theoretical results about ANNs is the *universal approximation property*, which states that a neural network with a single hidden layer, given enough neurons, can approximate any continuous function to arbitrary accuracy (Hornik et al., 1989). This highlights the remarkable flexibility of neural networks in capturing complex functional relationships. In practice, however, determining the optimal number of neurons in hidden layers is nontrivial, and practitioners often rely on empirical methods such as trial and error or cross-validation. Despite these challenges, ANNs remain a fundamental and versatile tool for tasks including regression, classification, pattern recognition, and time series prediction.

### *14.1.2 Training neural networks*

Consider a training set consisting of $n$ input-output pairs $(\mathbf{x}_i, \mathbf{y}_i)$, $i = 1, \ldots, n$. The FNN $h(\mathbf{x})$, defined in (14.3), is used to approximate the underlying relationship between the inputs $\mathbf{x}$ and the responses $\mathbf{y}$. To evaluate how well the network's predictions match the observed outputs, we define a *loss function* $\ell(h(\mathbf{x}), \mathbf{y})$ that measures the prediction error for each sample. The network is trained by finding the set of parameters $\theta$ that minimizes the *average loss* across all training examples:

$$L(\theta) = \frac{1}{n} \sum_{i=1}^{n} \ell(h(\mathbf{x}_i), \mathbf{y}_i). \tag{14.4}$$

This process enables the network to learn from the data and produce predictions that closely approximate the true underlying function. The optimization of $L(\theta)$ is typically carried out using the *back-propagation* algorithm, which efficiently computes the gradients of the loss with respect to all network parameters and allows iterative updates via gradient descent.

In particular, recall that the function of the $k$th layer is given by $h_k(\mathbf{x}) = \bar{g}(\mathbf{A}_k'\mathbf{x} + \boldsymbol{\beta}_k)$, where $\mathbf{A}_k = (\alpha_{k,ij})$ is a $d_{k-1} \times d_k$ weight matrix, $\boldsymbol{\beta}_k = (\beta_{k,1}, \ldots, \beta_{k,d_k})'$ is a $d_k$-

dimensional bias vector, and $\bar{g}$ is an activation function. Let $\mu_k = d_k(d_{k-1}+1)$ be the total number of parameters in the $k$th layer, and let $\theta_k$ be a vector containing all the parameters of $\mathbf{A}_k$ and $\beta_k$, i.e.

$$\theta_k = (\alpha_{k,11}, \alpha_{k,21}, \ldots, \alpha_{k,d_{k-1}d_k}, \beta_{k,1}, \ldots, \beta_{k,d_k})'.$$

The total number of parameters in the network is $\mu = \mu_1 + \cdots + \mu_k$, and the full parameter vector is $\theta = (\theta_1', \ldots, \theta_K')'$. Given the loss function $L(h(\mathbf{x}), \mathbf{y})$, the gradient with respect to all parameters can be expressed as

$$\nabla L(\theta) = \left( \left( \frac{\partial L}{\partial \theta_1} \right)', \ldots, \left( \frac{\partial L}{\partial \theta_K} \right)' \right)', \tag{14.5}$$

To calculate this gradient, consider each layer function $h_k$ for a single training sample $(\mathbf{x}_i, \mathbf{y}_i)$. Using the recursion $h_{k+1} = \bar{g}(\mathbf{A}_{k+1}' h_k + \beta_{k+1})$, the derivative of the loss with respect to the parameters of layer $k$ is

$$\begin{aligned} \frac{\partial \ell}{\partial \theta_k} &= \left( \frac{\partial h_k}{\partial \theta_k} \right) \left( \frac{\partial \ell}{\partial h_k} \right) = \left( \frac{\partial h_k}{\partial \theta_k} \right) \left( \frac{\partial h_{k+1}}{\partial h_k} \frac{\partial \ell}{\partial h_{k+1}} \right) \\ &= \frac{\partial (\mathbf{A}_k' h_{k-1} + \beta_k)}{\partial \theta_k} \left( \frac{d\bar{g}}{d\mathbf{x}} \Big|_{\mathbf{x} = \mathbf{A}_k h_{k-1} + \beta_k} \right) \left( \mathbf{A}_{k+1} \frac{d\bar{g}}{d\mathbf{x}} \Big|_{\mathbf{x} = \mathbf{A}_{k+1} h_k + \beta_{k+1}} \frac{\partial \ell}{\partial h_{k+1}} \right), \end{aligned} \tag{14.6}$$

where $\partial(\mathbf{A}_k h_{k-1} + \beta_k)/\partial\theta_k$ is a $d_k(d_{k-1}+1) \times d_k$ matrix. Summing over all training samples yields $\partial L/\partial \theta_k$, and concatenating these vectors across layers produces the full gradient $\nabla L(\theta)$ as in (14.5).

The central idea of back-propagation is the *recursive formula*

$$\frac{\partial \ell}{\partial h_k} = \frac{\partial h_{k+1}}{\partial h_k} \frac{\partial \ell}{\partial h_{k+1}}, \tag{14.7}$$

which propagates derivatives backward from the output layer ($k = K-1$) to the input layer ($k = 1$). This recursive computation allows the gradient to be obtained with *computational complexity* $O(\mu)$ per sample. By contrast, if one were to compute the derivative directly as

$$\frac{\partial \ell}{\partial h_k} = \frac{\partial \ell}{\partial h_K} \prod_{j=k}^{K-1} \frac{\partial h_{j+1}}{\partial h_j},$$

the total complexity would be $O(K\mu)$, which is substantially higher.

While back-propagation is computationally efficient, training deep networks presents several practical challenges:

- Vanishing and exploding gradients: In very deep networks, gradients can become extremely small or large, destabilizing learning. Solutions include careful weight initialization, normalization techniques such as batch normalization, and activation functions like ReLU.
- Network architecture selection: Depth, width, and connectivity patterns strongly affect performance, yet no universal rule exists; practitioners rely on heuristics, cross-validation, and experimentation.

- Computational cost: Training large networks requires substantial computational resources, often necessitating GPUs or TPUs for efficient parameter updates.

To mitigate overfitting, where the network learns noise rather than meaningful patterns, two common approaches are employed:

- Early stopping: Training is halted when performance on a validation set deteriorates, even if training loss continues to decrease.
- Regularization: The objective function is augmented with a penalty term that constrains model complexity, encouraging better generalization. Common methods include weight decay and dropout.

Beyond these classical approaches, modern neural networks incorporate additional improvements:

- Advanced activation functions: ReLU and its variants mitigate vanishing gradient issues.
- Normalization and regularization: Techniques like batch normalization and dropout stabilize training and reduce overfitting.
- Specialized architectures: CNNs exploit spatial structure, RNNs and LSTMs capture sequential dependencies, and transformers utilize attention mechanisms for long-range dependencies.
- Optimizers: Adaptive algorithms such as Adam and RMSProp accelerate convergence and improve robustness.

Collectively, these advancements have transformed neural networks into versatile models capable of handling complex tasks in vision, language, and decision-making domains, with ongoing research focusing on scalability, interpretability, and integration of domain knowledge.

### *14.1.3 Time series forecasting*

In time series analysis, neural networks (NNs) are typically trained by minimizing the sum of squared one-step-ahead forecast errors:

$$S = \sum_t \left(\hat{x}_{t-1}(1) - x_t\right)^2,$$

where $\hat{x}_{t-1}(1)$ denotes the forecast of $x_t$ from the previous time step. To avoid overfitting, the data is usually split into a *training set* for fitting the network and a *test set* for evaluating out-of-sample performance.

Training involves nonlinear optimization via back-propagation, which may require many iterations and can converge to local minima due to the large number of parameters. Regularization and model selection criteria, such as the Akaike Information Criterion (AIC), are commonly used to encourage simpler models and prevent overfitting.

The choice of network architecture, activation functions, and input variables affects forecasting performance. NNs can capture nonlinear patterns, seasonal effects, abrupt changes, and long-term dependencies, but they act as black-box models with limited interpretability. Combining domain knowledge with careful model design is therefore recommended.

Empirical evidence is mixed. For long series with clear nonlinear structure, NNs can outperform traditional methods, as in the Santa Fe competition. For short economic or financial series, simpler models like linear regression or random walk often yield more reliable forecasts. Overall, the success of NNs depends on data characteristics, preprocessing, and careful model selection.

*Example 14.1. FNN for time series prediction*

We demonstrate a simple FNN for predicting a univariate time series. The series is generated as

$$X_t = \sin(t) + Z_t, \qquad Z_t \text{ i.i.d. } N(0, \sigma_Z^2), \qquad \sigma_Z = 0.2.$$

Supervised learning data is created using a sliding window of 10 previous time steps to predict the next value. The dataset is split into 80% training and 20% testing.

The FNN has two hidden layers with 64 and 32 ReLU neurons and a single output neuron. It is trained with the Adam optimizer using mean squared error loss for 50 epochs, with a small validation set to monitor overfitting. Predictions on the test set are aligned with the original time steps.

The results are visualized in Figure 14.2. The top row shows the full time series, while the bottom row contains the training/validation loss and the predicted versus true values on the test set. The model successfully captures the underlying pattern, and the loss plot shows convergence without overfitting, illustrating that even a simple FNN can learn sequential dependencies in time series data.

## 14.2 Recurrent Neural Networks (RNNs)

While FNNs provide a flexible framework for modelling complex nonlinear relationships, they are inherently limited when dealing with sequential or time-dependent data. FNNs process each input independently and lack mechanisms to retain information from previous time steps, making them less suitable for tasks such as time series forecasting, speech recognition, or natural language processing. *Recurrent neural networks* (RNNs) extend the feed-forward architecture by introducing feedback connections that allow information to persist across time steps. This enables RNNs to model temporal dependencies and maintain a form of memory about past inputs.

However, basic RNNs can suffer from vanishing or exploding gradient problems, which limit their ability to capture long-term dependencies. To address these issues, more advanced recurrent units such as *Long Short-Term Memory* (LSTM) and *Gated Recurrent Unit* (GRU) networks have been developed. LSTMs and GRUs incorporate gating mechanisms to selectively retain or discard information, improving the learning of long-term patterns in sequences. In this section, we introduce the basic structure of RNNs, discuss their training dynamics, and present LSTM and GRU architectures, highlighting their advantages and practical applications in time series modelling.

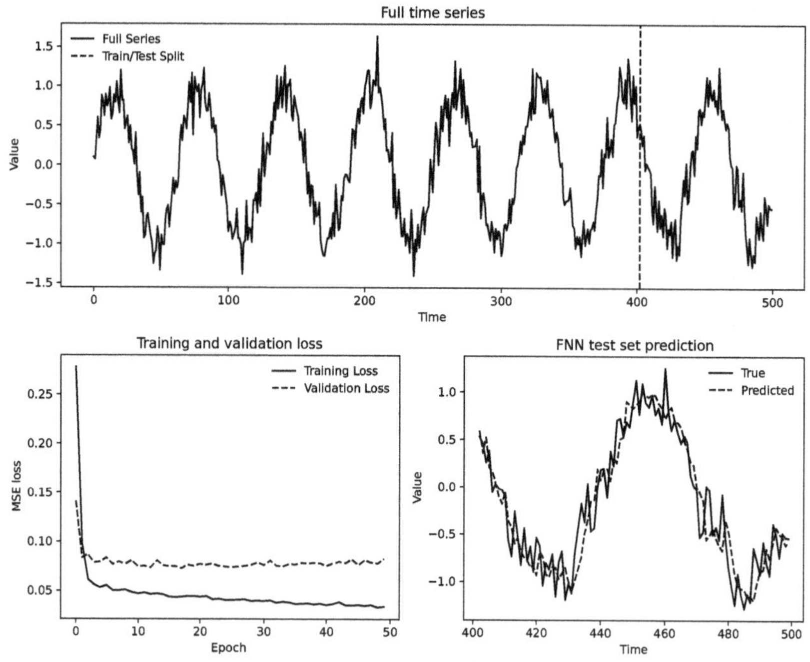


Figure 14.2 *Top panel: synthetic time series. Bottom left: training and validation loss. Bottom right: FNN time series predictions (Example 14.1).*

### *14.2.1 Basics of RNNs for sequential data*

A RNN is designed to model sequential data by processing inputs step by step while maintaining a hidden state that acts as a memory of past information. Unlike feed-forward networks, where inputs are treated independently, RNNs explicitly capture temporal dependencies by updating their hidden representation at each time step.

Formally, let $\{x_1, x_2, \ldots, x_T\}$ denote an input sequence of length $T$, where each $x_t \in \mathbb{R}^m$ represents the input vector at time $t$. The RNN maintains a hidden state $h_t \in \mathbb{R}^d$ that evolves according to the recursive relation

$$h_t = \sigma_h\big(W_h h_{t-1} + W_x x_t + b_h\big), \quad t = 1, \ldots, T. \tag{14.8}$$

Here, $h_0$ is typically initialized as the zero vector, although in some implementations it can be treated as a trainable parameter to improve flexibility. The matrix $W_h \in \mathbb{R}^{d\times d}$ governs the recurrent transformation applied to the previous hidden state $h_{t-1}$, while the matrix $W_x \in \mathbb{R}^{d\times m}$ maps the current input $x_t$ into the hidden state space. The term $b_h \in \mathbb{R}^d$ is a bias vector that ensures additional flexibility in the transformation. Finally, the nonlinear activation function $\sigma_h(\cdot)$, often chosen as tanh or the rectified linear unit (ReLU), introduces nonlinearity into the update rule, allowing the network to capture complex temporal patterns that cannot be represented by a purely linear recurrence. Figure 14.3 illustrates how an RNN processes a sequence, where each

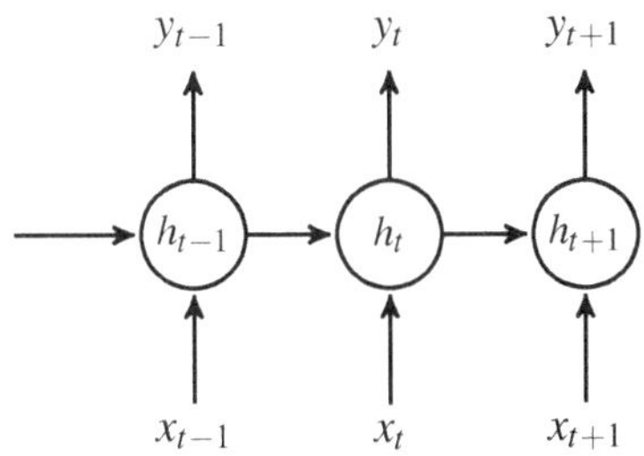


Figure 14.3 *Unrolled RNN across three time steps. Each hidden state $h_t$ depends on the previous hidden state $h_{t-1}$ and the current input $x_t$, while producing an output $y_t$.*

cell updates its hidden state from the previous state and current input to produce an output.

Through this recursive definition, the hidden state $h_t$ serves as a compact summary of all past inputs $\{x_1, \ldots, x_t\}$ up to time $t$. This mechanism enables the RNN to model dependencies across time, making it a natural tool for problems such as speech recognition, time series forecasting, and natural language processing, where current observations are strongly influenced by historical context.

The output at each time step is obtained by mapping the hidden state to an output space:

$$y_t = \sigma_y\big(W_y h_t + b_y\big), \tag{14.9}$$

where $W_y \in \mathbb{R}^{p \times d}$ and $b_y \in \mathbb{R}^p$ are trainable parameters, and $\sigma_y$ is an activation function appropriate for the task (e.g. softmax for classification, identity for regression).

The recurrent relation of an RNN can be expressed more explicitly by 'unrolling' the network across time. This unrolled view illustrates how the hidden state at time $t$ depends recursively on all previous hidden states and inputs:

$$h_t = \sigma_h\Big(W_h\,\sigma_h\big(W_h \cdots \sigma_h(W_h h_0 + W_x x_1 + b_h) + W_x x_2 + b_h \cdots\big) + W_x x_t + b_h\Big), \tag{14.10}$$

where $h_0$ is the initial hidden state, typically set to zero or treated as a trainable parameter. In this form, $h_t$ is a highly nonlinear function of the entire input sequence $\{x_1, \ldots, x_t\}$, allowing the network to capture complex temporal dependencies.

At each time step, the hidden state $h_t$ integrates information from the previous state $h_{t-1}$ and the current input $x_t$ through the nonlinear activation $\sigma_h$. The hidden state thus acts as a memory mechanism, accumulating and transforming information from all past inputs. The depth of this computation grows linearly with time, as each step introduces another nonlinear transformation.

This unrolled representation is particularly useful for understanding training via *backpropagation through time* (BPTT). By unfolding the RNN across all time steps, we can compute gradients of a loss function with respect to each weight and bias, considering contributions from all time steps. Despite this, very long sequences may lead to *vanishing* or *exploding gradients*, which are addressed by techniques such as gradient clipping or gated architectures like LSTM and GRU.

In supervised learning for sequential data, the objective is to adjust the network parameters $\theta = \{W_x, W_h, W_y, b_h, b_y\}$ so that the predicted outputs $\{y_1, \ldots, y_T\}$ match the target outputs $\{z_1, \ldots, z_T\}$. This is achieved using a loss function over all time steps:

$$\mathscr{L}(\theta) = \frac{1}{T} \sum_{t=1}^{T} \ell(y_t, z_t), \tag{14.11}$$

where $\ell(\cdot,\cdot)$ can be *mean squared error* (MSE) for regression,

$$\ell(y_t, z_t) = \|y_t - z_t\|^2,$$

or *cross-entropy* for classification,

$$\ell(y_t, z_t) = -\sum_k z_{t,k} \log y_{t,k}.$$

Training minimizes $\mathscr{L}(\theta)$ with respect to $\theta$, typically using *stochastic gradient descent* (SGD) or adaptive variants like Adam or RMSProp. Because hidden states depend recursively on previous inputs, gradients must account for contributions from all preceding steps. For the recurrent weight $W_h$, the gradient is

$$\frac{\partial \mathscr{L}}{\partial W_h} = \sum_{t=1}^{T} \frac{\partial \ell(y_t, z_t)}{\partial h_t} \frac{\partial h_t}{\partial W_h}, \tag{14.12}$$

where $\partial h_t / \partial W_h$ incorporates contributions from all prior hidden states.

RNNs can suffer from *vanishing* or *exploding gradients* when backpropagating through many time steps. The sensitivity of $h_t$ to a past state $h_{t-k}$ is

$$\frac{\partial h_t}{\partial h_{t-k}} = \prod_{i=1}^{k} \sigma_h'(a_{t-i}) W_h, \tag{14.13}$$

where $a_t = W_h h_{t-1} + W_x x_t + b_h$ and $\sigma_h'$ is the derivative of the activation. If the spectral radius of $W_h$ is less than 1, gradients vanish exponentially; if greater than 1, gradients explode, destabilizing training.

To address these issues, advanced RNN architectures such as LSTM and GRU introduce gating mechanisms that regulate the flow of information and gradients. Additional practical techniques include gradient clipping, careful weight initialization, and orthogonal or unitary weight matrices. These innovations enable the network to capture long-term dependencies effectively, making LSTM and GRU standard choices for applications such as language modelling, speech recognition, and time series forecasting.

### *14.2.2 LSTM for capturing long-term dependencies*

As discussed above, standard RNNs maintain hidden states to capture past inputs but often fail to learn long-term dependencies due to *vanishing* and *exploding gradients* during backpropagation. While techniques like gradient clipping or careful initialization can help, they do not fully resolve these issues for long sequences.

The LSTM network addresses these limitations with a memory cell and gating mechanisms that regulate information flow. The *input gate* controls new information added to the cell, the *forget gate* decides which past information to retain, and the *output gate* determines what to pass to the hidden state. By managing information explicitly, LSTMs enable reliable gradient propagation over long sequences, making them highly effective for tasks such as time series forecasting, speech recognition, and natural language processing.

Each LSTM cell comprises a memory cell $c_t$ and the three gates $i_t$, $f_t$, and $o_t$. Given the input $x_t$ and the previous hidden and cell states $(h_{t-1}, c_{t-1})$, the cell updates are defined as

$$f_t = \sigma\big(W_f x_t + U_f h_{t-1} + b_f\big), \tag{14.14a}$$
$$i_t = \sigma\big(W_i x_t + U_i h_{t-1} + b_i\big), \tag{14.14b}$$
$$\tilde{c}_t = \tanh\big(W_c x_t + U_c h_{t-1} + b_c\big), \tag{14.14c}$$
$$c_t = f_t \odot c_{t-1} + i_t \odot \tilde{c}_t, \tag{14.14d}$$
$$o_t = \sigma\big(W_o x_t + U_o h_{t-1} + b_o\big), \tag{14.14e}$$
$$h_t = o_t \odot \tanh(c_t), \tag{14.14f}$$

where $\sigma(\cdot)$ denotes the sigmoid activation, $\odot$ is element-wise multiplication, and $W_*, U_*, b_*$ are trainable parameters. As illustrated by Figure 14.4, the LSTM's operation (14.14a)–(14.14f) can be understood as a controlled flow of information:

- **Forget gate** $f_t$: Determines which components of the previous cell state $c_{t-1}$ should be retained or discarded. This selective forgetting allows the network to

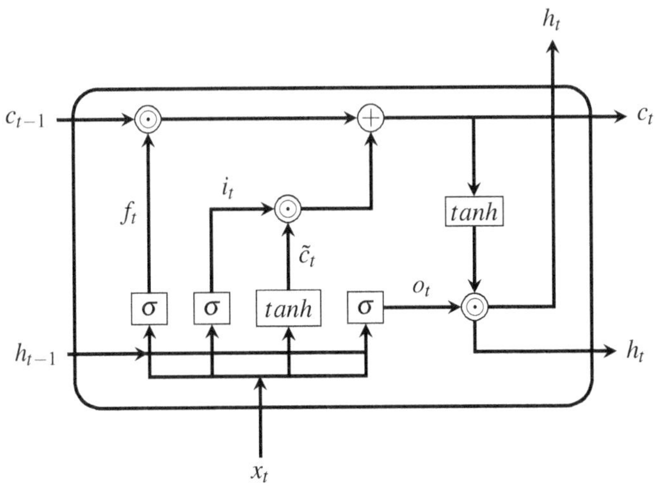


Figure 14.4 *Schematic of a single LSTM cell showing gates and cell state.*

remove irrelevant or outdated information, mitigating the problem of exploding memory over time.

- **Input gate** $i_t$: Controls the extent to which the new candidate memory $\tilde{c}_t$ is incorporated into the cell state. By scaling the candidate cell, the network selectively updates $c_t$ with relevant new information.
- **Candidate cell** $\tilde{c}_t$: Represents the new content generated from the current input $x_t$ and previous hidden state $h_{t-1}$, which is modulated by the input gate before entering the memory.
- **Memory cell** $c_t$: Acts as the long-term memory of the LSTM, combining retained old information (via $f_t \odot c_{t-1}$) and newly relevant information (via $i_t \odot \tilde{c}_t$). This linear additive update is key to preserving long-range dependencies and avoiding vanishing gradient issues.
- **Output gate** $o_t$: Determines which parts of the memory cell are exposed to the hidden state $h_t$, effectively controlling what information is sent to the next layer or time step. The nonlinearity $\tanh(c_t)$ ensures that the hidden state remains bounded, while the output gate allows selective propagation.

Through these mechanisms, the LSTM can maintain, update, and expose relevant information across many time steps, making it particularly effective for sequential data tasks such as language modelling, time-series forecasting, and speech recognition. The interplay between gates allows the network to learn *when to remember*, *when to forget*, and *when to output*, thereby addressing the limitations of traditional RNNs in capturing long-term dependencies.

LSTM networks are trained using *backpropagation through time* (BPTT), which extends standard backpropagation to sequential data by unrolling the network across time steps and computing gradients with respect to all weights. A key advantage of LSTMs over standard RNNs is their gating mechanisms – the input, forget, and output gates – which regulate information flow. These gates allow the network to retain important signals over long sequences, effectively alleviating the *vanishing* and *exploding gradient* issues common in vanilla RNNs.

To further improve training stability and convergence, several techniques are commonly employed:

- **Gradient clipping:** Limits gradients exceeding a threshold to prevent explosion.
- **Weight initialization:** Proper initialization of recurrent and input weights ensures stable signal propagation.
- **Regularization:** Dropout, $L_2$ weight decay, or layer normalization reduce overfitting and improve generalization.
- **Adaptive optimization:** Optimizers such as Adam or RMSProp, sometimes with learning rate decay, accelerate convergence.

LSTMs are widely applied in sequential tasks that require modelling long-term dependencies. In time series forecasting, they are used to capture complex temporal patterns in financial data, weather, sensor readings, and energy consumption. In *natural language processing* (NLP), LSTMs support tasks such as language modelling, machine translation, text generation, and sentiment analysis. They are also employed in speech and audio processing, including speech recognition, speaker identification, and music generation. Additionally, LSTMs are useful in video and sequential event modelling, where they help in action recognition and predictive maintenance by capturing temporal dependencies across frames or events.

The ability of LSTMs to capture both short- and long-term patterns makes them a cornerstone of sequence modelling. While Transformers are increasingly popular, LSTMs remain essential for long or streaming sequences and resource-limited settings. Variants such as bidirectional, stacked, and attention-augmented LSTMs further enhance their modelling capabilities.

*Example 14.2. LSTM time series forecasting with long-range dependence*

We demonstrate a simple LSTM for forecasting a synthetic time series exhibiting long-range dependence. The series is generated as

$$X_t = \left(1 + \frac{1}{2}\sin\left(\frac{t}{10}\right)\right)\sin(t) + Z_t, \qquad Z_t \text{ i.i.d. } N(0, \sigma_Z^2), \quad \sigma_Z = 0.5, \qquad (14.15)$$

for $t = 1, \ldots, 1000$. The resulting series combines a slowly varying amplitude modulation with a nonlinear sinusoidal component and Gaussian noise. To capture temporal dependencies, input-output pairs are created using a sliding window of 50 time steps, and the data are divided into training and testing sets.

A two-layer LSTM network is constructed with 64 and 32 hidden units, respectively, followed by a dense output layer. The model is trained with mean squared error loss using the Adam optimizer, and its performance is monitored through both training and validation loss curves. After training, the model predicts future values on the test data to assess its capability in learning long-term temporal patterns.

The results are shown in Figure 14.5. The top row shows the full time series, while the bottom row presents the training/validation loss and the predicted versus true values on the test set. The model successfully captures the underlying temporal patterns and long-range dependencies in the series, accurately forecasting the next values despite the noise and slowly varying amplitude. The training and validation loss curves indicate stable convergence, and the predicted values closely follow the true sequence, demonstrating the LSTM's effectiveness in modelling complex sequential data.

### 14.2.3 Gated recurrent units

The GRU is a streamlined variant of the LSTM designed to reduce computational complexity while still capturing long-term dependencies in sequential data. Unlike LSTMs, GRUs combine the forget and input gates into a single *update gate* $z_t$ and merge the cell state with the hidden state into a single vector $h_t$. The GRU updates

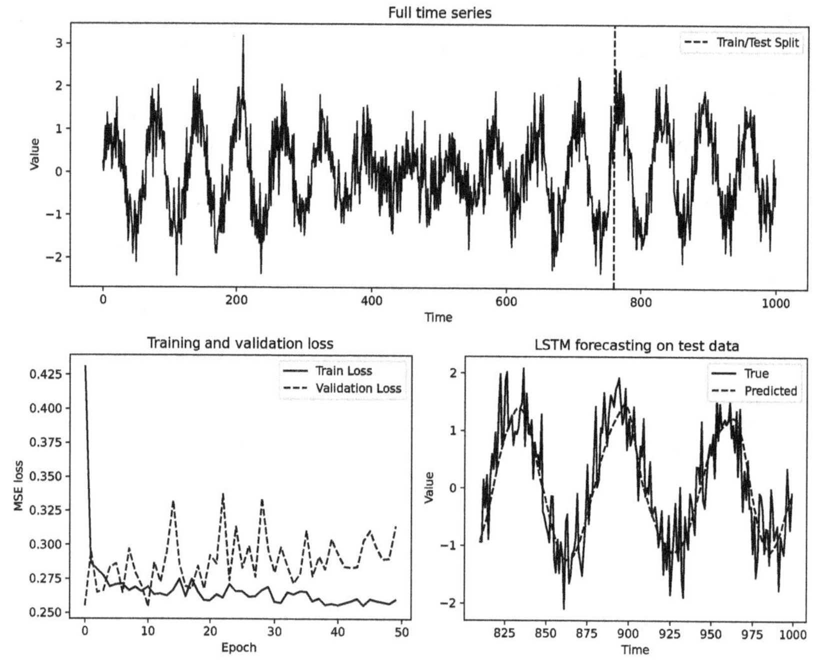


Figure 14.5 *Top panel: synthetic time series. Bottom left: training and validation loss. Bottom right: FNN time series predictions (Example 14.2).*

are defined as:

$$z_t = \sigma(W_z x_t + U_z h_{t-1} + b_z), \tag{14.16a}$$
$$r_t = \sigma(W_r x_t + U_r h_{t-1} + b_r), \tag{14.16b}$$
$$\tilde{h}_t = \tanh(W_h x_t + U_h(r_t \odot h_{t-1}) + b_h), \tag{14.16c}$$
$$h_t = (1 - z_t) \odot h_{t-1} + z_t \odot \tilde{h}_t, \tag{14.16d}$$

where $z_t$ is the update gate, $r_t$ is the reset gate, $\tilde{h}_t$ is the candidate hidden state, and $\sigma(\cdot)$ denotes the sigmoid activation function. As illustrated in Figure 14.6, the key components of the GRU operate as follows:

- **Update gate** $z_t$: Balances retention of the previous hidden state $h_{t-1}$ with incorporation of the new candidate $\tilde{h}_t$, enabling long-term dependency modelling.
- **Reset gate** $r_t$: Controls how much past information contributes to the candidate state, allowing selective forgetting.
- **Candidate hidden state** $\tilde{h}_t$: Combines current input with reset past information to form the potential new state.
- **Hidden state** $h_t$: Integrates retained past and candidate information based on $z_t$, analogous to the LSTM output gate in a simpler form.

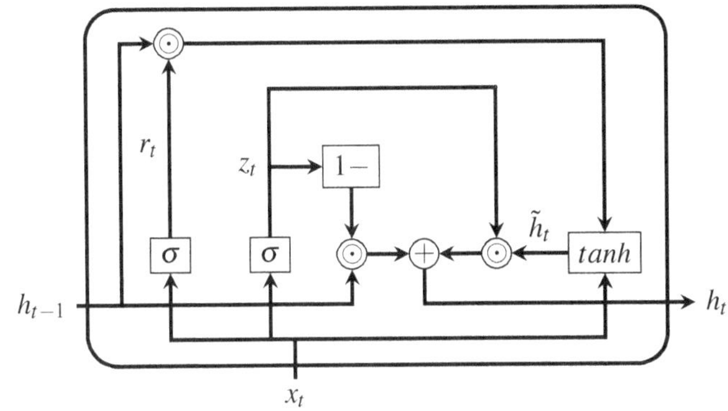


Figure 14.6 *Schematic of a single GRU cell showing gates and cell state.*

GRUs offer several advantages over LSTMs. By merging gates and combining the memory and hidden states, they have fewer parameters, making them computationally lighter, faster to train, and less prone to overfitting, especially on smaller datasets. Despite their simpler architecture, GRUs are capable of capturing long-range dependencies in sequences, often performing comparably to LSTMs in many tasks. Additionally, with fewer gates and operations, gradient flow tends to be more stable, reducing the need for extensive gradient clipping or complex regularization.

GRUs have been successfully applied to a wide range of sequential modelling tasks similar to LSTMs, including time series forecasting, speech recognition, machine translation, and natural language understanding. The choice between LSTM and GRU often depends on the dataset size, computational resources, and specific task requirements. GRUs are typically preferred for smaller datasets or real-time applications where training speed and model simplicity are important, while LSTMs may offer slightly better performance for tasks that require very long-term memory.

*Example 14.3. GRU time series forecasting*

We construct a GRU model to forecast a synthetic time series exhibiting long-range temporal dependence. The time series data are same as the one used in Example 14.2. To transform the time series into supervised learning form, we use a look-back window of size 50, constructing input-output pairs $(x_{t-50:t-1}, x_t)$ for model training.

The GRU network consists of two stacked recurrent layers with 64 and 32 units, respectively, followed by a fully connected layer that outputs a single-step forecast. The model is trained using the Adam optimizer with a learning rate of $10^{-3}$ and mean squared error (MSE) loss, for 50 epochs with a batch size of 32. Eighty percent of the samples are used for training, while the remaining 20% are reserved for testing.

Figure 14.7 displays the training and validation losses together with the true and predicted values on the test set. The GRU model effectively learns the underlying periodic structure and captures smooth transitions in the data, with minor prediction

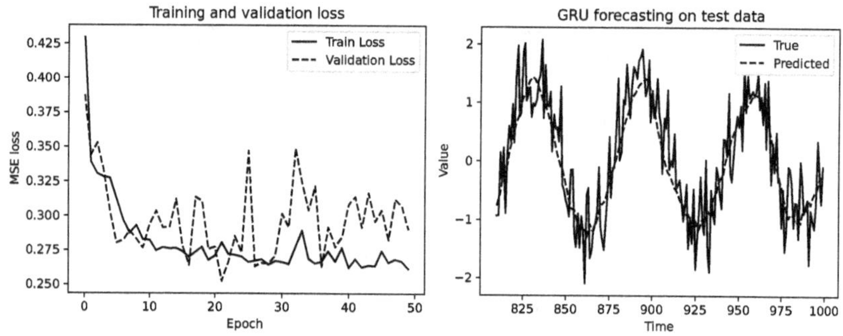


Figure 14.7 *Training and validation loss (left) and GRU time series prediction (right) in Example 14.3.*

errors occurring in regions of higher noise variance. This example illustrates the capability of GRUs to model nonlinear time series with long-term dependencies.

## 14.3 Convolutional Neural Networks (CNN)

While RNNs, including LSTMs and GRUs, are designed to capture long-term temporal dependencies, they can be computationally intensive, challenging to train on very long sequences, and still susceptible to vanishing or exploding gradients. *Convolutional Neural Networks* (CNNs), originally developed for image processing, offer an effective alternative for time series modelling by capturing local patterns, learning hierarchical features, and sharing parameters across time steps, which improves generalization and reduces overfitting. CNNs also allow parallel processing of multiple time steps and, with techniques like global pooling or dilated convolutions, can handle variable-length sequences while capturing multi-scale temporal dependencies. These advantages make CNNs well suited for tasks such as ECG classification, IoT sensor anomaly detection, speech and audio modelling, financial forecasting, and activity recognition, and hybrid CNN-RNN models can combine local feature extraction with sequential modelling for richer temporal representations and improved computational efficiency.

### *14.3.1 Convolutional layers and feature extraction*

CNNs extract meaningful local features from sequential data using convolutional layers. For a univariate time series $x = (x_1, x_2, \dots, x_T)$, a one-dimensional convolution applies a filter (or kernel) $w = (w_1, w_2, \dots, w_k)$ of length $k$. The convolution at time step $t$ is computed as

$$(y * w)_t = \sum_{i=1}^{k} w_i x_{t-i+1}. \tag{14.17}$$

Intuitively, a filter acts as a short-term pattern detector: it slides along the sequence, responding strongly when a specific local pattern, such as a spike, oscillation, or trend, appears. By using multiple filters per layer, the network can capture a variety of local temporal motifs simultaneously. Convolution is translation-invariant, meaning a learned pattern can be recognized anywhere in the sequence. Design choices such as stride, which determines how far the filter moves at each step, and padding, which preserves sequence length near boundaries, are important to balance computational efficiency and feature coverage. Nonlinear activation functions, such as ReLU or tanh, are applied after convolution to enable the network to model complex, nonlinear patterns.

Stacking multiple convolutional layers allows the network to build hierarchical representations. Lower layers capture primitive local features (e.g. spikes, edges, abrupt changes), intermediate layers combine these into higher-level motifs (e.g. repeated cycles or trends), and higher layers integrate information across time for global sequence-level representations. Dilated convolutions further increase the receptive field without adding layers:

$$(y *_d w)_t = \sum_{i=0}^{k-1} w_i x_{t-d \cdot i}, \tag{14.18}$$

where $d$ is the dilation rate, allowing the network to capture longer-range dependencies while maintaining computational efficiency.

Pooling layers complement convolutions by reducing the temporal dimension while retaining salient information. Max pooling emphasizes dominant patterns,

$$y_t = \max(x_t, x_{t+1}, \ldots, x_{t+k-1}),$$

while average pooling summarizes trends,

$$y_t = \frac{1}{k} \sum_{i=0}^{k-1} x_{t+i}.$$

Pooling reduces computation, increases robustness to small temporal shifts, and focuses the network on the most informative features.

Overall, one-dimensional CNNs function as hierarchical pattern extractors: filters at lower layers detect short-term motifs, intermediate layers combine them into more complex structures, and higher layers capture global temporal patterns suitable for downstream tasks such as classification, anomaly detection, or forecasting. By automatically learning which local patterns are most informative, CNNs provide an efficient and interpretable framework for sequential data modelling and form the basis for more advanced architectures, including hybrid CNN-RNN models and *temporal convolutional networks* (TCNs).

### *14.3.2 Hybrid CNN-RNN architectures*

Hybrid models combine convolutional and recurrent neural networks to capture both local and long-range temporal patterns in sequential data (Lin et al., 2017; Lai et al.,

2018). In these architectures, convolutional layers are applied first to extract meaningful local features from the input sequence. These features encode short-term patterns, such as spikes, oscillations, or trends, into compact representations, reducing dimensionality and highlighting the most informative motifs. By summarizing local dependencies into feature maps, CNNs provide a rich input for subsequent recurrent layers while reducing the computational burden of processing raw sequences directly with RNNs.

The CNN-extracted features are then fed into recurrent networks, such as LSTMs or GRUs, which model long-term dependencies and maintain memory across time steps. Recurrent layers integrate local features over time, capturing temporal correlations and sequential relationships that extend beyond the receptive field of the convolutional filters. This combination enables the network to leverage precise local pattern detection while incorporating global temporal context. Figure 14.8 illustrates a hybrid CNN-LSTM architecture. Convolutional and pooling layers in the top row extract spatial or local features from the input sequence, which are then fed into the LSTM layer in the bottom row to capture temporal dependencies, followed by a dense output layer for prediction.

Hybrid CNN-RNN models are particularly effective in applications where both short-term and long-term dependencies are important. For example, in ECG classification, CNNs detect characteristic waveforms like QRS complexes, while RNNs capture rhythm variations across cycles. In IoT anomaly detection, convolutional layers identify transient deviations, and recurrent layers model temporal dependencies to distinguish true anomalies from normal fluctuations. In speech recognition, CNNs extract local acoustic features, and RNNs model sequences of phonemes or words, improving transcription accuracy.

Design considerations for hybrid architectures include the number and size of convolutional filters to capture relevant local patterns, the depth and type of recurrent layers for effective long-term modelling, and whether to use pooling or residual connections to manage feature map resolution and gradient flow. Training involves backpropagation through both convolutional and recurrent components, often combined with regularization techniques such as dropout or batch normalization to enhance generalization.

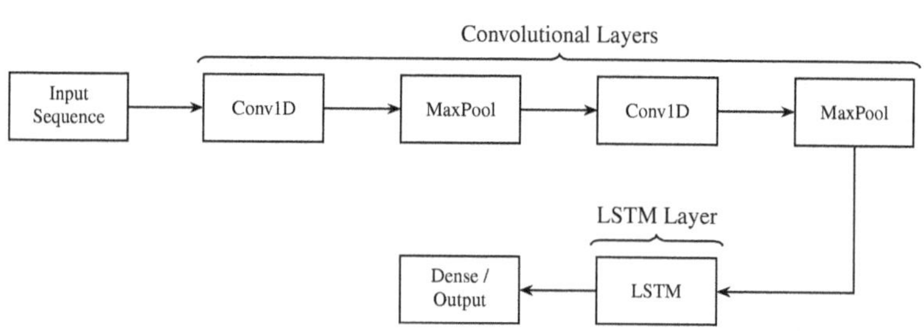


Figure 14.8 *Schematic of a hybrid CNN–LSTM network showing convolutional feature extraction and temporal modelling via an LSTM layer.*

Overall, hybrid CNN-RNN networks provide a versatile framework for sequential data modelling, integrating local feature extraction with long-term dependency learning. They have found wide application in healthcare, industrial IoT, finance, and speech processing.

*Example 14.4. Hybrid CNN-LSTM for time series forecasting*

In this example, we demonstrate a hybrid CNN-LSTM model for forecasting a synthetic time series with long-range dependencies. The synthetic data are same as the one used in Example 14.2. To prepare the data for modelling, we construct overlapping sliding windows of size 50 as input sequences and the next time step as the target, and then reshape the data for the CNN-LSTM input format (samples, timesteps, features). The dataset is split into training and test sets, with 80% of the samples used for training.

The model consists of two convolutional layers with causal padding to extract local temporal features, followed by a max-pooling layer to reduce sequence length. The extracted feature sequences are fed into an LSTM layer to capture long-term dependencies. Finally, fully connected dense layers map the LSTM outputs to a single scalar for one-step-ahead prediction. The network is trained using the mean squared error (MSE) loss with the Adam optimizer for 50 epochs, with 20% of the training set reserved for validation.

Figure 14.9 illustrates the performance of the hybrid CNN-LSTM model on the synthetic time series. The left panel shows the training and validation loss over epochs, highlighting the model's convergence and indicating whether overfitting occurs. The right panel compares the predicted values on the test set with the true series, demonstrating the model's ability to capture both short-term fluctuations and longer-term temporal patterns.

### 14.3.3 Temporal convolutional networks

*Temporal convolutional networks* (TCNs) are a class of fully convolutional architectures designed for sequence modelling that can serve as an alternative to recurrent

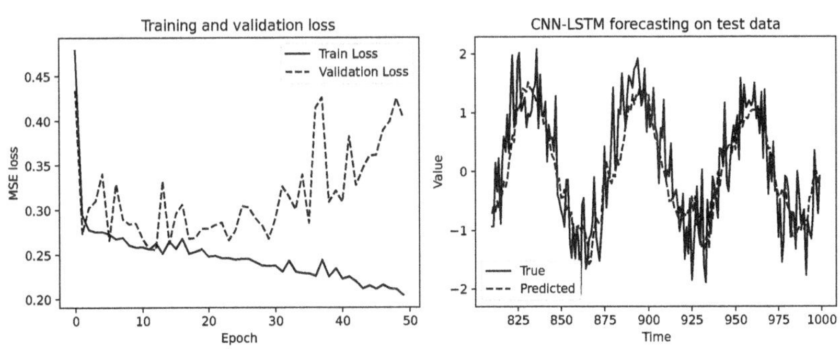


Figure 14.9 *Training and validation loss (left) and hybrid CNN-LSTM time series prediction (right) in Example 14.3.*

networks (Lea et al., 2016). Unlike standard CNNs, TCNs employ *causal convolutions*, ensuring that the output at time step $t$ depends only on inputs from time steps 1 to $t$. This causality preserves the temporal ordering of the sequence, making TCNs suitable for forecasting, control, and other tasks where future information must not be accessed.

To capture long-range dependencies efficiently, TCNs utilize *dilated convolutions*, where the convolutional filter is applied over increasingly spaced input points. For a one-dimensional input sequence $x = (x_1, x_2, \ldots, x_T)$, a dilated convolution with filter $w$ of length $k$ and dilation factor $d$ is defined as (14.18). Dilations exponentially increase the receptive field with depth, allowing TCNs to model long-term dependencies without requiring extremely deep networks.

Residual connections are commonly incorporated in TCN blocks to facilitate training by mitigating gradient vanishing and improving convergence. Each residual block typically consists of two layers of dilated causal convolutions followed by nonlinear activations, optional normalization, and a skip connection that adds the block's input to its output. These connections allow the network to preserve low-level features while learning higher-level temporal patterns. Figure 14.10 illustrates a TCN with stacked 1D convolutional layers of increasing dilation. Dashed arrows indicate residual connections that propagate earlier layer outputs to later layers, enabling better gradient flow and feature reuse.

Unlike RNNs, TCNs process sequences in parallel during training and inference, resulting in faster computation and easier gradient propagation. Empirical studies show that TCNs are competitive with, and sometimes outperform, LSTMs and GRUs on tasks requiring long-term memory, such as sequence forecasting, speech synthesis, and video modelling.

Design considerations for TCNs include selecting the filter size, dilation schedule, number of layers, and residual block configuration. Larger filters and deeper networks increase the receptive field, while dilations enable coverage of long sequences without excessive parameters. Overall, TCNs combine the efficiency of convolutions with the ability to model long-term dependencies, offering a powerful and parallelizable alternative to traditional recurrent architectures for sequential data.

*Example 14.5. TCN for time series forecasting*

In this example, we demonstrate a TCN model for forecasting a synthetic time series with long-range dependencies. The synthetic data are generated by process (14.15),

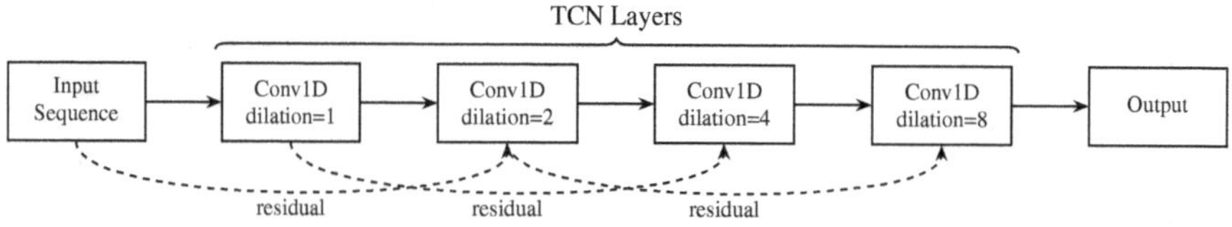


Figure 14.10 *Schematic of a TCN architecture showing dilated convolutional layers and residual connections f or sequence modelling.*

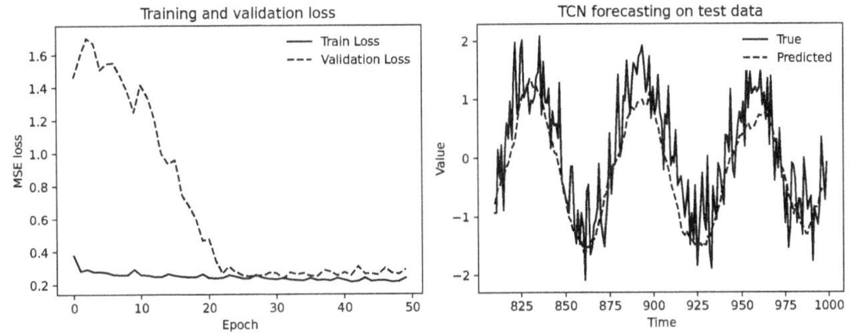


Figure 14.11 *Training and validation loss (left) and TCN time series prediction (right).*

which are same as the one used in Example 14.2. To convert the series into supervised learning samples, a look-back window of size 50 is used, producing input–output pairs $(x_{t-50:t-1}, x_t)$. The first 80% of the data are used for training and the remaining 20% for testing.

The TCN model consists of stacked causal convolutional layers with increasing dilation rates to capture long-range dependencies efficiently. Each convolutional block contains a 1D causal convolution layer with ReLU activation, followed by batch normalization and dropout for regularization. Specifically, three convolutional blocks are used with dilation rates 1, 2, and 4, respectively, followed by a global average pooling layer and dense layers to produce the next-step forecast. Causal convolutions ensure that predictions at time $t$ only depend on past inputs.

Figure 14.11 shows the training and validation losses on the left and the predicted versus true values on the test set on the right. The TCN successfully captures the long-range dependencies and nonlinear oscillatory patterns in the series, providing accurate forecasts across most regions despite the added noise. This example illustrates the effectiveness of temporal convolutional networks for modelling time series with long-range dependencies and complex temporal structures.

### *14.3.4 Multivariate time series*

In many practical applications, time series data are multivariate, meaning that multiple related signals or channels are observed simultaneously. Let $X \in \mathbb{R}^{T\times C}$ denote a multivariate sequence, where $T$ is the number of time steps and $C$ is the number of channels, such as multiple sensor readings in an IoT system or multiple ECG leads in healthcare monitoring.

One-dimensional convolutions can be applied to multivariate time series in several ways. Convolutions may operate independently on each channel, allowing the network to learn channel-specific temporal patterns. Alternatively, convolutions can jointly process all channels, enabling the network to capture inter-channel correlations and combined patterns that span multiple variables. This joint processing is particularly important when signals interact or exhibit synchronized behaviour, such as coupled sensor readings in industrial systems or correlated physiological signals in healthcare.

The network can also incorporate hierarchical feature extraction across channels. Lower layers may focus on short-term temporal patterns within individual channels, while deeper layers learn cross-channel interactions and higher-level temporal structures. Pooling, residual connections, and dilated convolutions can further enhance the model's ability to capture long-range dependencies and complex relationships across multiple variables.

Applications of multivariate time series modelling include multilead ECG classification, where relationships between leads are critical for detecting arrhythmias; sensor networks in smart homes or industrial systems, where anomalies may be revealed only through correlations among sensors; and financial forecasting, where multiple market indicators evolve simultaneously and influence each other. By effectively capturing both within-channel and cross-channel dependencies, convolutional networks for multivariate time series provide a powerful framework for modelling complex sequential data.

### *14.3.5 Interpretability and applications*

Understanding the inner workings of convolutional models is crucial in high-stakes domains such as healthcare, finance, and industrial monitoring. Visualizing learned filters can reveal the temporal patterns the network detects – spikes, oscillations, trends, or more complex motifs – while techniques like saliency maps, activation maximization, and class activation mapping highlight which regions of the input most influence predictions. Such interpretability fosters trust and transparency, ensures that decisions rely on relevant features rather than spurious correlations, and supports model diagnostics and improvement by guiding architectural adjustments, preprocessing refinements, or data augmentation. In applications like ECG classification, IoT anomaly detection, or financial forecasting, visualization can pinpoint critical waveform segments, sensor readings, or temporal patterns, enabling actionable insights and facilitating regulatory compliance or human-in-the-loop decision-making.

These interpretability principles extend naturally to more complex architectures. In hybrid CNN-RNN models, convolutional layers extract local motifs while recurrent layers integrate them over time, allowing visualization to show how short-term features contribute to longer-term predictions. Similarly, in TCNs, analysing dilated causal convolutions and residual blocks can identify the temporal ranges most influential for model outputs. By illuminating both local and long-range dependencies, CNNs provide a framework that is both effective and interpretable for sequential modelling.

CNNs have been successfully applied to tasks including ECG signal classification and arrhythmia detection, anomaly detection in IoT sensor networks, financial time series forecasting, and human activity recognition or speech/audio analysis. Through hierarchical feature extraction, weight sharing, and parallel computation, CNNs efficiently capture both short-term and long-term temporal dependencies while maintaining interpretability.

## 14.4 Attention Mechanisms and Transformer Models

Time series often exhibit long-range dependencies and complex temporal patterns that are challenging for traditional sequential models to capture. In finance, price movements may depend on events from weeks or months earlier; in climate modelling, current conditions reflect seasonal or long-term shifts. While RNNs maintain hidden states and CNNs extract local patterns, both have limitations: RNNs can struggle with vanishing or exploding gradients, and CNNs' fixed receptive fields restrict modelling of distant dependencies. Even LSTMs and GRUs, which mitigate gradient issues, may fail to capture very long-range interactions effectively.

Attention mechanisms provide a flexible solution by assigning learnable weights to past time steps, enabling the model to focus selectively on the most relevant observations. This allows the simultaneous capture of short-term patterns and long-range dependencies while offering interpretability, highlighting which inputs most influence predictions. Attention-based architectures thus overcome key limitations of RNNs and CNNs, making them well suited for complex, high-dimensional, and noisy time series data. This section introduces the principles of attention and their integration into time series models, illustrating how these architectures leverage historical information more effectively.

### *14.4.1 Attention mechanism*

The core idea of attention is to compute a weighted sum of information, where the weights reflect the relevance of each element in the sequence to the current prediction. For a sequence of length $T$, let the input be $x_1, x_2, \ldots, x_T \in \mathbb{R}^d$. Attention introduces three vectors for each time step: queries $q_t$, keys $k_t$, and values $v_t$, defined as

$$q_t = W_Q x_t, \quad k_t = W_K x_t, \quad v_t = W_V x_t,$$

where $W_Q, W_K, W_V \in \mathbb{R}^{d \times d_k}$ are learnable projection matrices that transform the input into a lower-dimensional space suitable for computing relevance scores. Intuitively, the *query* represents the information we are seeking, the *key* encodes what each time step offers, and the *value* contains the information to be aggregated.

The attention mechanism produces an output by taking a weighted sum of the value vectors, where the weights are determined by the similarity between queries and keys. In the widely used *scaled dot-product attention*, this operation is given by

$$\text{Attention}(Q, K, V) = \text{softmax}\left(\frac{QK^\top}{\sqrt{d_k}}\right) V, \tag{14.19}$$

where $Q, K, V$ are matrices of queries, keys, and values stacked across all time steps, and the scaling factor $\sqrt{d_k}$ prevents excessively large dot-product values that can lead to vanishing gradients after the softmax. The softmax ensures that attention weights sum to one, giving a probabilistic interpretation of the importance of each past observation.

For illustration, consider a univariate stock price sequence $[x_1, x_2, \dots, x_T]$. If a sudden jump occurs at $x_3$ that signals a future trend, the attention mechanism can assign greater weight to $x_3$, while downweighting other less informative points. This dynamic weighting enables the model to focus selectively on the most relevant historical information – unlike conventional RNNs, which pass all past information uniformly through hidden states.

Attention mechanisms offer several advantages for time series modelling. They capture long-range dependencies without relying on recurrent computation, mitigating issues such as vanishing or exploding gradients. By explicitly weighting the influence of each time step, attention also enhances interpretability, allowing practitioners to identify which historical events most strongly drive predictions. In addition, attention operations can be fully parallelized across time steps, enabling efficient training and scalability to long sequences. These properties make attention particularly effective for high-dimensional, noisy, or irregular time series, where dependencies may span multiple temporal scales and certain past events exert disproportionate influence.

Modern architectures extend this framework through *multi-head attention*, in which multiple sets of queries, keys, and values are learned in parallel. This enables the model to capture diverse temporal relationships – such as short-term fluctuations and long-term trends in financial data, or seasonal and non-seasonal components in climate systems. Moreover, attention layers are often combined with convolutional or recurrent modules to form hybrid models, integrating both local and global temporal representations for richer and more flexible time series analysis.

### *14.4.2 Transformer architecture for sequential data*

Transformers extend attention mechanisms by stacking multiple layers of multi-head self-attention and position-wise feedforward networks. Unlike recurrent models, transformers process sequences in parallel, making them highly efficient for long sequences. Their design enables the modelling of complex temporal dependencies while capturing both local and global patterns in the data.

In standard attention, a single head may focus on only one type of dependency at a time. *Multi-head self-attention* addresses this limitation by computing several attention operations in parallel, allowing the model to capture different aspects of temporal relationships simultaneously. Formally, if there are $h$ attention heads, each head $i \in \{1, \dots, h\}$ is computed as

$$\text{head}_i = \text{Attention}(Q^i, K^i, V^i), \quad Q^i = XW_Q^i,\ K^i = XW_K^i,\ V^i = XW_V^i,$$

where $X \in \mathbb{R}^{T \times d_{\text{model}}}$ is the input sequence, and $W_Q^i, W_K^i, W_V^i$ are learnable projection matrices for the $i$-th head. The outputs of all heads are concatenated and projected through a learnable matrix $W_O$:

$$\text{MultiHead}(Q, K, V) = \text{Concat}(\text{head}_1, \dots, \text{head}_h) W_O.$$

By attending to multiple subspaces of the input simultaneously, multi-head attention captures both short-term variations and long-term dependencies, making it effective for complex time series with multiple temporal scales.

Each transformer layer also contains a position-wise *feedforward network* (FFN) that applies non-linear transformations to each time step independently:

$$\mathrm{FFN}(z) = \mathrm{ReLU}(zW_1 + b_1)W_2 + b_2,$$

where $W_1, W_2$ and $b_1, b_2$ are learnable parameters. To ensure stable training and ease gradient flow, *residual connections* are applied around both the multi-head attention and feedforward sublayers, followed by *layer normalization*:

$$z_t' = \mathrm{LayerNorm}\left(z_t + \mathrm{MultiHead}(z_t, z_t, z_t)\right),$$
$$z_t'' = \mathrm{LayerNorm}\left(z_t' + \mathrm{FFN}(z_t')\right).$$

Residual connections prevent degradation of learned representations in deep stacks, while layer normalization stabilizes training and accelerates convergence. Figure 14.12 illustrates this Transformer block for time series. The input sequence, optionally augmented with positional encodings, is projected into queries, keys, and values. Attention weights are computed for each head, applied to the values, and the multi-head outputs are concatenated and linearly projected to produce the final output sequence.

Since transformers process sequences in parallel, they lack inherent information about the ordering of time steps. *Positional encodings* inject temporal information into the input embeddings, enabling the model to distinguish the relative and absolute

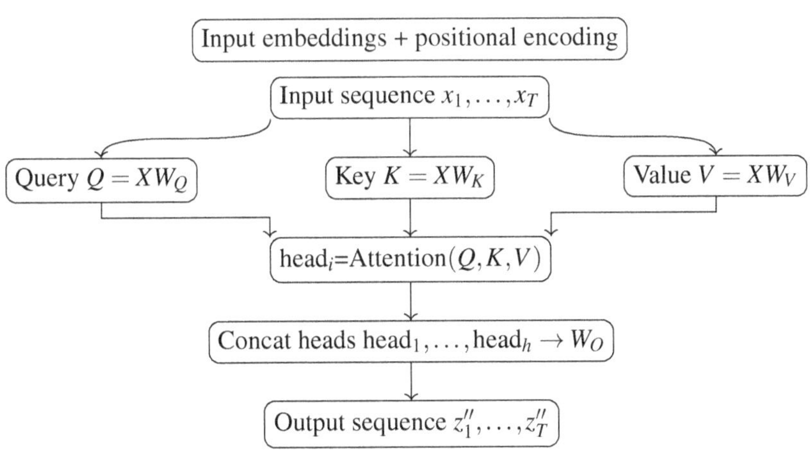


Figure 14.12 *Transformer block for time series: input sequences are projected into queries, keys, and values, attention weights are computed, and multiple heads are concatenated to produce the output.*

positions of time steps:

$$\mathrm{PE}(t,2i) = \sin\left(\frac{t}{10000^{2i/d}}\right), \quad \mathrm{PE}(t,2i+1) = \cos\left(\frac{t}{10000^{2i/d}}\right),$$

where $t$ is the time step index, $i$ indexes the embedding dimension, and $d$ is the embedding size. This choice of sinusoidal functions allows the model to generalize to longer sequences than those seen during training, as relative distances between positions can be inferred from the encodings. Variants such as learned positional embeddings or relative positional encodings have also been proposed, particularly in time series tasks, to better capture temporal patterns and dependencies.

Transformers have emerged as a powerful alternative to traditional RNNs and CNNs for modelling sequential data. By leveraging self-attention mechanisms, transformers can capture both local and global dependencies, enabling effective modelling of complex temporal patterns. Unlike recurrent models, transformers process sequences in parallel, allowing for efficient training on long sequences. Moreover, the attention weights provide interpretability, revealing which past observations most influence predictions. Hybrid architectures that combine transformers with convolutional or recurrent modules, or incorporate temporal embeddings, further enhance their capacity to model high-dimensional, heterogeneous time series.

### *14.4.3 Transformer models for time series forecasting*

Transformers have become powerful tools for time series forecasting due to their ability to capture long-range dependencies and handle heterogeneous, high-dimensional data. Several variants have been proposed to tailor the architecture for sequential tasks. The *Time Series Transformer* (TST) emphasizes temporal encoding for multi-step prediction, while *Informer* introduces sparse and probabilistic attention to process long sequences efficiently. *Autoformer* integrates trend-seasonal decomposition to enhance interpretability and forecasting accuracy, and the *Temporal Fusion Transformer* (TFT) combines attention with gating mechanisms and static covariate encoders for adaptive feature selection.

These transformer-based models have demonstrated strong performance across domains such as finance (stock and volatility prediction), healthcare (ECG and EEG analysis), and IoT systems (anomaly detection and predictive maintenance). Their advantages include effective long-range dependency modelling, parallelizable computation, and interpretable attention mechanisms. However, standard transformers remain computationally demanding and data-hungry. Efficient variants like Informer, Autoformer, and TFT mitigate these limitations, enabling scalable and accurate forecasting for long and complex time series.

*Example 14.6. Transformer for time series forecasting*

We demonstrate a Transformer-based model for forecasting a synthetic time series exhibiting long-range temporal dependence. The data are generated as

$$X_t = (1+0.5\sin(0.05t))\sin(t) + 0.3Z_t, \quad Z_t \sim \mathcal{N}(0,1), \quad t = 1,\ldots,1000. \quad (14.20)$$

The series combines slowly varying amplitude modulation, a nonlinear sinusoidal component, and Gaussian noise. To construct supervised learning samples, a look-back window of size 50 is used, producing input–output pairs $(x_{t-50:t-1}, x_t)$ for model training. The first 80% of the data are used for training and the remaining 20% for testing.

The Transformer model consists of stacked encoder blocks with multi-head self-attention and feed-forward layers. For each encoder block, the computation at time step $t$ is given by:

$$\begin{aligned}
\text{Attention}(Q,K,V) &= \text{softmax}\left(\frac{QK^\top}{\sqrt{d_k}}\right)V, \\
\text{head}_i &= \text{Attention}(XW_i^Q, XW_i^K, XW_i^V), \\
\text{MultiHead}(X) &= \text{Concat}(\text{head}_1, \ldots, \text{head}_h)W^O, \\
X' &= \text{LayerNorm}(X + \text{MultiHead}(X)), \\
X'' &= \text{LayerNorm}(X' + \text{FFN}(X')), \\
\text{FFN}(X') &= W_2\,\text{ReLU}(W_1 X' + b_1) + b_2,
\end{aligned}$$

where $X \in \mathbb{R}^{T \times d}$ is the input sequence, $d$ is the feature dimension, $h$ is the number of attention heads, and $d_k$ is the key dimension. Positional encodings are added to the input to preserve temporal order. Finally, a global average pooling over the sequence is applied, followed by a dense layer to predict the next time step.

The results are shown in Figure 14.13. The left panel presents the training and validation losses, while the right panel shows the true and predicted values on the test set. The Transformer successfully captures the long-range dependencies and nonlinear patterns in the series, producing accurate forecasts even in regions with varying amplitude. This example illustrates the effectiveness of attention-based models for modelling complex temporal dependencies in time series.

## 14.5 Bibliography

ANNs form the foundation of predictive modelling for sequential data. The seminal work by Rumelhart et al. (1986) introduced the back-propagation algorithm, enabling multi-layer networks to learn complex non-linear mappings. Goodfellow et al. (2016) provide a comprehensive treatment of deep learning architectures and training techniques, which serve as the basis for many time series models.

RNNs and their variants, particularly LSTM networks, address the challenge of modelling temporal dependencies. Hochreiter and Schmidhuber (1997) introduced LSTM networks to overcome the vanishing gradient problem, enabling the capture of long-term dependencies in sequential data. Extensions and applications to continual prediction are discussed by Gers et al. (2000). CNNs have also been adapted for sequence modelling; Bai et al. (2018) demonstrate that CNNs can effectively capture local patterns in time series and, when combined with pooling and dilation, can handle long-range dependencies.

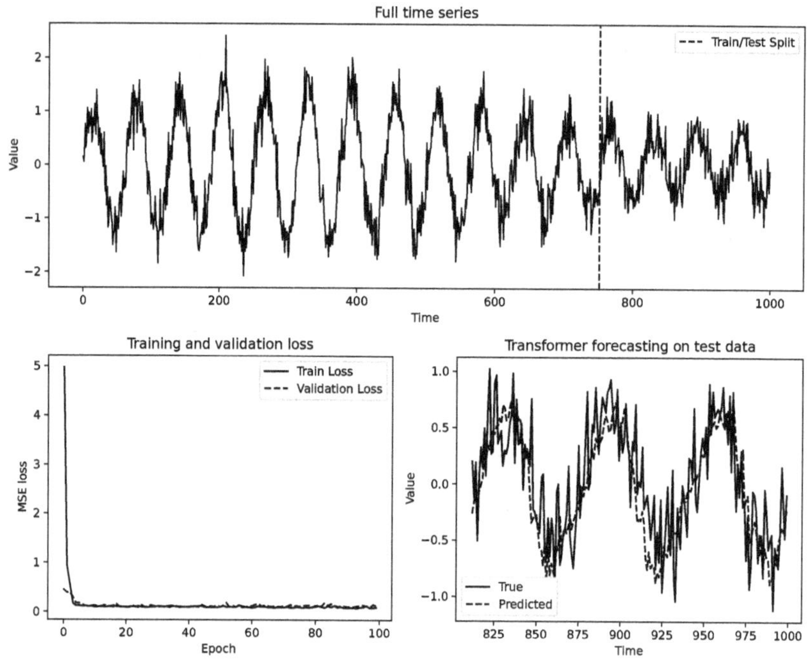


Figure 14.13 *Top panel: synthetic time series. Bottom left: training and validation loss. Bottom right: FNN time series predictions (Example 14.6).*

Attention mechanisms and Transformer architectures have further advanced time series forecasting by enabling models to focus selectively on relevant past information. Bahdanau et al. (2015) and Luong et al. (2015) laid the groundwork for attention in sequence-to-sequence tasks, while Vaswani et al. (2017) proposed the Transformer, which eliminates recurrence entirely and leverages parallel multi-head self-attention. Recent work by Li et al. (2019) adapts Transformer architectures specifically for time series forecasting, addressing challenges such as locality and long-term dependency modelling.

Hybrid architectures leverage the complementary strengths of these paradigms: CNN layers extract local temporal features, RNNs or LSTMs capture sequential dependencies, and attention or Transformer modules model global dependencies while providing interpretability. Such hybrid approaches enable predictive models to handle complex, high-dimensional time series with multiple temporal scales, offering both accuracy and flexibility across diverse forecasting tasks.

## Exercises

**14.1** Consider a simple RNN defined by

$$h_t = \tanh(W_h x_t + U_h h_{t-1} + b_h), \quad y_t = W_y h_t + b_y,$$

where $x_t \in \mathbb{R}^{n_x}$, $h_t \in \mathbb{R}^{n_h}$, $y_t \in \mathbb{R}^{n_y}$.

(a) Given $W_h, U_h, b_h, W_y, b_y$ and an input sequence $x_1, x_2$, compute $h_1, h_2$ and $y_1, y_2$.

(b) Explain how the hidden state $h_t$ encodes information from past inputs.

(c) Derive the gradient of the loss $L = \frac{1}{2}\sum_t \|y_t - \hat{y}_t\|^2$ with respect to $W_h$ using the chain rule for $t = 2$.

**14.2** Generate an AR(2) time series and prepare supervised data using a sliding window. Build and train a simple RNN with one recurrent layer and a Dense output to predict the next value. Plot the predicted versus true series and briefly discuss how well the RNN captures temporal dependencies and any limitations observed.

**14.3** Consider a 1D convolution with kernel $K = [k_1, k_2, k_3]$ and input sequence $x = [x_1, x_2, x_3, x_4, x_5]$.

(a) Compute the convolution output (no padding, stride 1) for a single channel.

(b) Apply a ReLU activation and compute the output again.

(c) Explain how increasing the kernel size affects the receptive field and the ability to capture temporal patterns.

**14.4** Generate an ARMA(2, 1) time series and prepare input-output pairs using a sliding window approach. Build a 1D CNN with several Conv1D layers followed by a Dense output layer to predict the next time step. Train the model, plot the predicted versus true series, and discuss how the CNN captures local temporal patterns compared to an RNN.

**14.5** Let $X \in \mathbb{R}^{T\times d}$ be an input sequence and consider one attention head with projections $Q = XW_Q$, $K = XW_K$, $V = XW_V$.

(a) Write the formula for the scaled dot-product attention:

$$\text{Attention}(Q,K,V) = \text{softmax}\left(\frac{QK^\top}{\sqrt{d_k}}\right)V.$$

(b) Given small matrices $X, W_Q, W_K, W_V$ (e.g. $2\times 2$), compute $Q, K, V$ and the attention output.

(c) Explain mathematically how multi-head attention captures dependencies in different subspaces of the input sequence.

**14.6** Generate an ARIMA(1, 1, 2) time series and prepare supervised data using a sliding window. Build a Transformer model with multi-head self-attention and position-wise feedforward layers to predict the next value. Train the model, visualize predicted versus true values, and discuss how attention mechanisms help capture both short-term and long-term dependencies compared to RNNs or CNNs.

Chapter 15

# Generative AI for Time Series

Predictive models for time series are commonly designed to produce *point forecasts* — estimates of expected future values. While such forecasts are useful in many applications, they provide only a limited perspective, failing to capture the full range of possible outcomes or the uncertainty inherent in complex temporal systems. In practice, understanding the *distribution* of future observations is often as important as predicting their mean. For example, financial risk management relies on evaluating the likelihood of extreme market movements, healthcare monitoring depends on anticipating rare but critical events, and industrial operations benefit from quantifying uncertainty to support preventive maintenance. These challenges motivate the need for approaches that go beyond point predictions and explicitly model the probabilistic structure of sequential data.

Generative models offer a principled solution by learning the underlying data distribution and producing synthetic sequences that reflect realistic temporal patterns. By capturing stochasticity, temporal dependencies, and nonlinear dynamics, these models enable scenario simulation, uncertainty quantification, and robust decision-making under uncertainty. In finance, generative models can simulate multiple potential asset-price trajectories for stress testing and risk assessment; in healthcare and IoT applications, they facilitate anomaly detection by identifying unusual sequences; and in data-scarce settings, they support model training through realistic data augmentation. This probabilistic framework extends the capabilities of traditional predictive approaches, providing richer insights into complex time-evolving systems.

This chapter introduces four main families of modern generative methods for time series. *Variational Autoencoders* learn latent probabilistic representations that support sampling, reconstruction, and interpolation. *Generative Adversarial Networks* employ adversarial training to produce highly realistic sequences, often used for scenario generation and data augmentation. *Diffusion models* generate samples by iteratively denoising sequences, offering strong likelihood-based performance and diverse outputs. Finally, *Transformers*, with attention mechanisms, capture long-range temporal dependencies and serve as both autoregressive generative models and components in hybrid architectures. Together, these methods cover the majority of modern probabilistic and generative techniques for time series, with other


DOI: 10.1201/9781003646112-15

approaches — such as normalizing flows or energy-based models — often integrated within these frameworks.

## 15.1 Variational Autoencoders (VAEs)

VAEs are generative models that learn a probabilistic latent representation of data. Unlike traditional autoencoders, they model the underlying distribution, enabling generation of new sequences that capture stochasticity, temporal dependencies, and complex patterns. This makes VAEs well suited for time series tasks such as probabilistic forecasting, anomaly detection, and scenario simulation.

### *15.1.1 Variational inference and ELBO*

*Variational inference* (VI) provides a powerful and general framework for approximate Bayesian inference by transforming the problem of computing intractable posterior distributions into an optimization problem (see, e.g. Jordan et al., 1999; Blei et al., 2017; Zhang et al., 2018). Instead of directly sampling from the true posterior, VI introduces a family of tractable distributions $\{q_\phi(z)\}$, parameterized by variational parameters $\phi$, and seeks to minimize the Kullback – Leibler (KL) divergence $\mathrm{KL}(q_\phi(z) \,\|\, p(z \mid x))$. This optimization is equivalent to maximizing the so-called *evidence lower bound* (ELBO), which provides both a variational approximation to the marginal likelihood and a practical learning objective for model estimation.

Recent advances such as stochastic and black-box variational inference have significantly broadened the applicability of VI to large-scale probabilistic models and streaming data (Hoffman et al., 2013; Ranganath et al., 2014). In the context of deep generative modelling, the introduction of the variational autoencoder (VAE) (Kingma and Welling, 2014; Rezende et al., 2014) has unified variational methods with deep neural networks, enabling efficient *amortized inference* via the reparameterization trick. Subsequent developments, including normalizing flows and inverse autoregressive flows, have further enhanced the flexibility of approximate posteriors, allowing them to capture complex, multimodal latent structures (Rezende and Mohamed, 2015; Kingma et al., 2016).

VAEs are fundamentally grounded in the principle of VI, in which an intractable posterior distribution is replaced by a tractable surrogate. Given observed data $x \in \mathbb{R}^d$ and latent variable $z \in \mathbb{R}^k$, the goal is to infer the true posterior

$$p_\theta(z|x) = \frac{p_\theta(x,z)}{p_\theta(x)} = \frac{p_\theta(x|z)p(z)}{p_\theta(x)}, \tag{15.1}$$

which is generally intractable. Here, $\theta$ parameterizes the generative model (**decoder**) $p_\theta(x|z)$, while $p(z)$ is the prior over latent variables. To make inference tractable, VAEs introduce an approximate posterior $q_\phi(z|x)$, parameterized by an **encoder** network with parameters $\phi$, to serve as a variational approximation.

The VAE objective maximizes the *evidence lower bound* (ELBO) on the marginal likelihood $\log p_\theta(x)$. Starting from the marginal likelihood,

$$\log p_\theta(x) = \log \int p_\theta(x,z)\,dz = \log \int p_\theta(x|z)\,p(z)\,dz,$$

we introduce the approximate posterior $q_\phi(z|x)$ and rewrite the log-likelihood as

$$\log p_\theta(x) = \log \int q_\phi(z|x) \frac{p_\theta(x,z)}{q_\phi(z|x)} dz = \log \mathbb{E}_{q_\phi(z|x)} \left[ \frac{p_\theta(x,z)}{q_\phi(z|x)} \right].$$

Applying Jensen's inequality gives a lower bound:

$$\log p_\theta(x) \geq \mathbb{E}_{q_\phi(z|x)} \left[ \log \frac{p_\theta(x,z)}{q_\phi(z|x)} \right] =: \mathscr{L}_{\text{VAE}}(\theta,\phi;x). \tag{15.2}$$

Expanding $p_\theta(x,z) = p_\theta(x|z)p(z)$ yields the standard ELBO decomposition:

$$\mathscr{L}_{\text{VAE}}(\theta,\phi;x) = \mathbb{E}_{q_\phi(z|x)}[\log p_\theta(x|z)] - D_{\text{KL}}\left(q_\phi(z|x) \,\|\, p(z)\right), \tag{15.3}$$

where the first term measures reconstruction quality under the generative model, and the second term regularizes the approximate posterior towards the prior. Maximizing this ELBO with respect to $\theta$ and $\phi$ jointly improves reconstruction while ensuring a structured latent space.

The ELBO (15.2) can also be derived without using Jensen's inequality. This is done by decomposing the marginal likelihood into the ELBO plus the KL divergence between the approximate posterior and the true posterior. Specifically, we introduce the approximate posterior $q_\phi(z|x)$ to the marginal log-likelihood,

$$\log p_\theta(x) = \mathbb{E}_{q_\phi(z|x)}\Big[\log p_\theta(x)\Big] = \mathbb{E}_{q_\phi(z|x)}\Big[\log \frac{p_\theta(x,z)}{p_\theta(z|x)}\Big].$$

We add and subtract $\log q_\phi(z|x)$ inside the expectation to rewrite the marginal likelihood as a sum of the ELBO and a KL divergence term,

$$\begin{aligned}\log p_\theta(x) &= \mathbb{E}_{q_\phi(z|x)}\Big[\log p_\theta(x,z) - \log q_\phi(z|x)\Big] + \underbrace{\mathbb{E}_{q_\phi(z|x)}\Big[\log q_\phi(z|x) - \log p_\theta(z|x)\Big]}_{D_{\text{KL}}(q_\phi(z|x)\,\|\,p_\theta(z|x))} \\ &= \mathscr{L}_{\text{VAE}}(\theta,\phi;x) + D_{\text{KL}}(q_\phi(z|x) \,\|\, p_\theta(z|x)),\end{aligned}$$

where $p_\theta(z|x)$ is the true posterior under the generative model. Since the KL divergence is always non-negative, the first term provides a lower bound on the marginal likelihood:

$$\mathscr{L}_{\text{VAE}}(\theta,\phi;x) \leq \log p_\theta(x). \tag{15.4}$$

Finally, expanding the joint distribution $p_\theta(x,z) = p_\theta(x|z)p(z)$ gives the ELBO decomposition (15.3).

### *15.1.2 Encoder-decoder architecture*

VAEs implement the variational inference framework using an *encoder-decoder* architecture. The *encoder* approximates the intractable posterior $p_\theta(z|x)$ by a tractable distribution $q_\phi(z|x)$, typically chosen as a multivariate Gaussian:

$$q_\phi(z|x) = N\big(z; \mu_\phi(x), \text{diag}(\sigma^2_\phi(x))\big), \tag{15.5}$$

where $\mu_\phi(x)$ and $\sigma_\phi(x)$ are outputs of a neural network with parameters $\phi$.

To allow gradient-based optimization through the stochastic latent variable, VAEs employ the *reparameterization trick*:

$$z = \mu_\phi(x) + \sigma_\phi(x) \odot \varepsilon, \qquad \varepsilon \sim N(0, I), \tag{15.6}$$

so that sampling from $q_\phi(z|x)$ becomes a differentiable operation with respect to $\phi$. The *decoder* $p_\theta(x|z)$ then reconstructs $x$ from $z$, parameterizing the conditional distribution of the data. For real-valued data, the decoder is often Gaussian:

$$p_\theta(x|z) = N\big(x; \mu_\theta(z), \sigma^2 I\big), \tag{15.7}$$

where $\mu_\theta(z)$ is predicted by a neural network with parameters $\theta$, and $\sigma^2$ may be fixed or learned.

The VAE training objective is the ELBO (15.3). The first term encourages accurate reconstruction of $x$ from $z$, while the second term regularizes the latent space towards the prior $p(z)$, typically $N(0,I)$. In practice, the reconstruction term is often written explicitly for Gaussian decoders:

$$\mathbb{E}_{q_\phi(z|x)}[\log p_\theta(x|z)] \approx -\frac{1}{2\sigma^2}\mathbb{E}_{q_\phi(z|x)}\|x - \mu_\theta(z)\|^2 + \text{const.} \tag{15.8}$$

If the variance $\sigma^2$ is learned — or if a full covariance matrix is used — the constant term becomes data- or parameter-dependent, taking the form $-\frac{1}{2}\log\det(2\pi\Sigma_\theta(z))$. In this case, the reconstruction term involves the Mahalanobis distance $\frac{1}{2}(x - \mu_\theta(z))^\top \Sigma_\theta(z)^{-1}(x - \mu_\theta(z))$. Moreover, when the decoder mean $\mu_\theta(z)$ is a linear function of $z$ and the approximate posterior $q_\phi(z|x)$ is Gaussian, the expectation in (15.8) can be computed analytically in closed form.

In practice the expectation $\mathbb{E}_{q_\phi(z|x)}[\|x - \mu_\theta(z)\|^2]$ is usually not available in closed form because $\mu_\theta(z)$ is a nonlinear neural network. It is therefore approximated by Monte Carlo sampling using the reparameterization trick. Draw $L$ samples

$$z^{(l)} = \mu_\phi(x) + \sigma_\phi(x) \odot \varepsilon^{(l)}, \qquad \varepsilon^{(l)} \sim N(0, I),$$

and estimate

$$\mathbb{E}_{q_\phi(z|x)}\big[\log p_\theta(x|z)\big] \approx -\frac{d}{2}\log(2\pi\sigma^2) - \frac{1}{2\sigma^2}\frac{1}{L}\sum_{l=1}^{L}\big\|x - \mu_\theta(z^{(l)})\big\|^2. \tag{15.9}$$

The Monte Carlo estimator (15.9), together with the reparameterization trick, provides a standard low-variance gradient estimator for $\theta$ and $\phi$. In practice, a single-sample estimate ($L = 1$) is typically sufficient, yielding an unbiased and computationally efficient gradient estimate.

*Example 15.1. ELBO for linear Gaussian decoder with Gaussian variational posterior*

Let the observed variable be $x \in \mathbb{R}^d$ and the latent variable be $z \in \mathbb{R}^k$. We consider the following linear-Gaussian generative model:

$$p(z) = N(z; 0, I_k), \qquad p_\theta(x \mid z) = N\big(x; Wz + b,\ \sigma^2 I_d\big),$$

where $W \in \mathbb{R}^{d\times k}$ and $b \in \mathbb{R}^d$ are model parameters, and $\sigma^2 > 0$ denotes the noise variance.

To approximate the intractable posterior $p_\theta(z \mid x)$, we introduce a Gaussian variational distribution of the form

$$q_\phi(z \mid x) = \mathcal{N}(z; \mu, S),$$

where $\mu \in \mathbb{R}^k$ is the variational mean and $S \in \mathbb{R}^{k\times k}$ is a symmetric positive-definite covariance matrix. The ELBO for a single datum $x$ is

$$\begin{aligned}\mathcal{L}(\theta,\phi;x) &= \mathbb{E}_{q_\phi(z|x)}\left[\log p_\theta(x,z) - \log q_\phi(z \mid x)\right] \\ &= \mathbb{E}_q[\log p_\theta(x \mid z)] - \mathrm{KL}\big(q_\phi(z \mid x) \,\|\, p(z)\big).\end{aligned}$$

In the ELBO, the expected log-likelihood term is given by

$$\begin{aligned}\mathbb{E}_q[\log p_\theta(x \mid z)] &= -\frac{d}{2}\log(2\pi\sigma^2) - \frac{1}{2\sigma^2}\mathbb{E}_q\left[\|x-(Wz+b)\|^2\right] \\ &= -\frac{d}{2}\log(2\pi\sigma^2) - \frac{1}{2\sigma^2}\Big(\|x-(W\mu+b)\|^2 + \mathrm{tr}(WSW^\top)\Big).\end{aligned}$$

The KL divergence between two Gaussians has the form,

$$\mathrm{KL}\big(q_\phi(z \mid x) \,\|\, p(z)\big) = \tfrac{1}{2}\Big(\mathrm{tr}(S) + \mu^\top\mu - k - \log|S|\Big),$$

in which $|S|$ is the determinant of $S$. Hence, combining the two pieces yields

$$\begin{aligned}\mathcal{L}(\theta,\phi;x) = &-\frac{d}{2}\log(2\pi\sigma^2) - \frac{1}{2\sigma^2}\Big(\|x-(W\mu+b)\|^2 + \mathrm{tr}(WSW^\top)\Big) \\ &- \tfrac{1}{2}\Big(\mathrm{tr}(S) + \mu^\top\mu - k - \log|S|\Big).\end{aligned}$$

### *15.1.3 VAE for time series*

While the standard VAE assumes i.i.d. samples, time series data exhibit strong temporal dependencies that must be explicitly modelled. Extending VAEs to handle sequential data involves introducing temporal structure into both the latent variables and the encoder-decoder networks, allowing the model to learn dynamic latent representations that evolve over time.

For a sequence of observations $x_{1:T} = \{x_1,\ldots,x_T\}$, a natural extension of the VAE introduces a sequence of latent variables $z_{1:T} = \{z_1,\ldots,z_T\}$, where each $z_t$ captures latent dynamics at time $t$. The encoder and decoder are then designed to incorporate recurrent or causal dependencies:

$$q_\phi(z_t|x_{1:t}) = N(z_t; \mu_\phi(x_{1:t}), \sigma^2_\phi(x_{1:t})), \quad p_\theta(x_t|z_{1:t}) = N(x_t; \mu_\theta(z_{1:t}), \sigma^2 I), \quad (15.10)$$

where $x_{1:t} = \{x_1,\ldots,x_t\}$ and $z_{1:t} = \{z_1,\ldots,z_t\}$. Here, the encoder captures information from past observations, while the decoder generates the current observation conditioned on the latent history.

The ELBO for sequential data generalizes to

$$\begin{aligned}\mathscr{L}_{\text{VAE}}(\theta,\phi;x_{1:T}) = &\sum_{t=1}^{T}\mathbb{E}_{q_\phi(z_{1:t}|x_{1:t})}[\log p_\theta(x_t|z_{1:t})] \\ &-\sum_{t=1}^{T} D_{\text{KL}}\left(q_\phi(z_t|x_{1:t})\,\|\,p(z_t|z_{1:t-1})\right),\end{aligned} \tag{15.11}$$

where $p(z_t|z_{1:t-1})$ defines the latent transition prior, which may be stationary (e.g. standard normal) or learned through a recurrent prior network. This formulation bridges variational autoencoding and latent state-space modelling, enabling the model to represent both observation noise and latent temporal evolution.

Sequential VAEs provide a flexible framework for time-dependent modelling, supporting probabilistic forecasting, anomaly detection, and scenario simulation (e.g. Chung et al., 2015; Fraccaro et al., 2016). They can generate multiple plausible future trajectories by sampling from the learned latent dynamics, identify segments of sequences with low likelihood under the model, and simulate counterfactual scenarios to explore the stochastic behaviour of the underlying system. By combining variational inference with recurrent or convolutional architectures, sequential VAEs capture both short- and long-term dependencies, uncertainty, and nonlinear dynamics in complex temporal data.

A representative example is the *Recurrent VAE*, which extends the encoder-decoder structure with RNNs or LSTMs to process temporal dependencies in both encoding and decoding. The encoder produces a sequence of hidden states and latent variables,

$$h_t = f_\phi(x_t, h_{t-1}), \qquad z_t = g_\phi(h_t), \tag{15.12a}$$

$$\hat{x}_t = p_\theta(x_t|z_t, h_{t-1}), \tag{15.12b}$$

where $h_t$ summarizes past information, and the decoder reconstructs each $x_t$ from the latent code $z_t$ and the context $h_{t-1}$. This architecture enables the model to capture evolving hidden representations while maintaining the stochastic expressiveness of the VAE.

Another important extension is the *Variational Recurrent Neural Network (VRNN)*, which introduces a latent variable at each time step that evolves alongside the recurrent hidden state. Its joint and variational posteriors factorize as

$$p_\theta(x_{1:T}, z_{1:T}) = \prod_{t=1}^{T} p_\theta(x_t|z_t, h_{t-1})\, p_\theta(z_t|h_{t-1}), \tag{15.13a}$$

$$q_\phi(z_{1:T}|x_{1:T}) = \prod_{t=1}^{T} q_\phi(z_t|x_t, h_{t-1}), \tag{15.13b}$$

where $h_t = f_\theta(x_t, z_t, h_{t-1})$. This design couples deterministic and stochastic dynamics: the hidden state $h_t$ acts as a recurrent memory, while $z_t$ models local uncertainty and nonlinear variability. As a result, the VRNN can represent complex, non-Markovian temporal structures and perform both inference and generation in dynamic systems.

Related approaches, such as the *Deep Kalman Filter* and *Stochastic RNNs*, interpret the sequential VAE as a nonlinear state-space model. In these formulations,

$$z_t \sim p_\theta(z_t|z_{t-1}), \qquad x_t \sim p_\theta(x_t|z_t), \tag{15.14}$$

where the transition and emission functions are parameterized by deep neural networks. The inference network approximates the filtering posterior via $q_\phi(z_t|x_{1:t}, z_{1:t-1})$, analogous to amortized inference in probabilistic state-space models. This perspective generalizes classical models such as the Kalman filter and hidden Markov model (HMM) to nonlinear and non-Gaussian regimes, offering a powerful unification of deep learning, dynamical systems, and probabilistic modelling for time series representation learning.

Training VAEs for time series involves challenges such as *posterior collapse*, where the decoder ignores latent variables, and modelling *long-range dependencies*. Remedies include KL annealing, $\beta$-VAE weighting, skip connections, hierarchical latent structures, attention-based encoders, and truncated BPTT for efficient inference. Despite these issues, VAEs are effective for *forecasting*, *anomaly detection*, and *missing data imputation*, leveraging latent dynamics to capture uncertainty and nonlinear temporal patterns. Hybrid architectures combining VAEs with Transformers, diffusion models, or Gaussian process latent dynamics further enhance prediction, representation learning, and uncertainty quantification. Overall, variational autoencoders provide a principled, flexible framework for modelling complex temporal data, bridging classical time series methods with modern deep generative models.

*Example 15.2. VAE for time series generation*

This example illustrates how a VAE can be adapted for probabilistic modelling and generation of time series data. A synthetic dataset is first generated from a noisy sine wave,

$$x_t = \sin(t) + 0.3Z_t, \qquad Z_t \sim N(0,1),$$

which is normalized to have zero mean and unit variance. To incorporate temporal dependencies, the data are segmented into overlapping windows of fixed length (e.g. 20 time steps), each used as a training sample.

The VAE framework consists of two components: an *encoder network*, which maps each input window $x_{1:T}$ to the parameters of a latent Gaussian distribution

$$q_\phi(z|x_{1:T}) = N(\mu, \sigma^2 I),$$

and a *decoder network*, which reconstructs the input window from a latent sample $z$ drawn via the reparameterization trick:

$$z = \mu + \sigma \odot \varepsilon, \qquad \varepsilon \sim N(0, I).$$

The model is trained to minimize the variational loss

$$\mathscr{L} = \mathbb{E}_{q_\phi(z|x)}[\|x - \hat{x}\|^2] + \beta D_{\mathrm{KL}}(q_\phi(z|x) \,\|\, p(z)),$$

where $p(z)$ is the standard normal prior, and $\beta$ is a weighting factor controlling the regularization strength. This objective encourages the encoder to learn a smooth, structured latent space while maintaining accurate reconstruction.

After training, the VAE can generate synthetic time series in two ways:

1. **Independent sampling:** draw $z \sim N(0,I)$ and decode it to obtain a random window that reflects the learned temporal dynamics.
2. **Autoregressive chaining:** generate a continuous sequence of arbitrary length (e.g. 1000 steps) by feeding the most recent $L$ generated points (the window length) back into the encoder – decoder loop. At each iteration, a new latent variable $z_t$ is sampled from the prior, the decoder outputs the next value $\hat{x}_{t+1}$, and the window is updated by shifting forward and appending the new point:

$$[x_{t-L+2}, \ldots, x_t, \hat{x}_{t+1}] \longrightarrow \text{decoder}(z_{t+1}).$$

This iterative process maintains temporal coherence and produces a smooth, long synthetic trajectory that mimics the statistical and dynamic properties of the original signal.

In summary, the VAE provides a probabilistic, latent-variable approach to time series generation, capable of capturing nonlinear dependencies and uncertainty. When combined with autoregressive generation, VAEs can synthesize arbitrarily long and realistic temporal sequences, suitable for forecasting, simulation, or data augmentation.

## 15.2 Generative Adversarial Networks (GANs)

GANs offer a powerful framework for modelling complex data distributions without requiring explicit likelihood evaluation. Unlike VAEs, which rely on probabilistic reconstruction, GANs employ an adversarial setup: a generator produces synthetic data, while a discriminator evaluates its authenticity. This approach has been successfully adapted to time series, enabling realistic sequence generation for forecasting, simulation, anomaly detection, and data augmentation.

### *15.2.1 Fundamentals*

GANs are trained via a game-theoretic objective in which a generator $G$ and a discriminator $D$ compete. The generator transforms a noise vector $z \sim p_z(z)$ into a synthetic sample $G(z)$, while the discriminator distinguishes between real data $x \sim p_{\text{data}}(x)$ and generated data $G(z)$. The optimization problem is formulated as

$$\min_G \max_D V(G,D) \tag{15.15}$$

where $V(G,D)$ measures how well the discriminator $D$ separates real and generated data for a given generator $G$ and is given by

$$V(G,D) = \mathbb{E}_{x \sim p_{\text{data}}}[\log D(x)] + \mathbb{E}_{z \sim p_z}[\log(1 - D(G(z)))].$$

*Derivation of the Jensen – Shannon divergence objective*

At equilibrium, the generator replicates the true data distribution $p_{\text{data}}$, and the discriminator is unable to differentiate real from synthetic samples. The optimal discriminator for (15.15) satisfies

$$D^*(x) = \frac{p_{\text{data}}(x)}{2m(x)}, \qquad m(x) = \frac{1}{2}\left(p_{\text{data}}(x) + p_G(x)\right), \tag{15.16}$$

where $m(x)$ is the mixture distribution. Substitute the optimal discriminator $D^*$ into the value function. For a fixed generator $G$, this yields

$$\begin{aligned} V(G,D^*) &= \mathbb{E}_{x\sim p_{\text{data}}}\left[\log\frac{p_{\text{data}}(x)}{2m(x)}\right] + \mathbb{E}_{x\sim p_G}\left[\log\frac{p_G(x)}{2m(x)}\right] \\ &= \int p_{\text{data}}(x)\log\frac{p_{\text{data}}(x)}{2m(x)}\,dx + \int p_G(x)\log\frac{p_G(x)}{2m(x)}\,dx \\ &= \int p_{\text{data}}(x)\log\frac{p_{\text{data}}(x)}{m(x)}\,dx + \int p_G(x)\log\frac{p_G(x)}{m(x)}\,dx - 2\log 2. \end{aligned}$$

By the definition of the Jensen – Shannon divergence,

$$\text{JS}(p_{\text{data}} \,\|\, p_G) = \frac{1}{2}\text{KL}(p_{\text{data}} \,\|\, m) + \frac{1}{2}\text{KL}(p_G \,\|\, m),$$

we conclude that

$$V(G,D^*) = 2\,\text{JS}(p_{\text{data}} \,\|\, p_G) - \log 4.$$

Therefore, solving the minimax problem (15.15) is equivalent to minimizing the Jensen – Shannon divergence between the true data distribution and the model distribution. :

$$\min_G V(G,D^*) = \min_G \left\{2\,\text{JS}(p_{\text{data}} \,\|\, p_G) - \log 4\right\}. \tag{15.17}$$

### 15.2.2 *Recurrent GAN for time series generation*

For time series, the generator and discriminator must capture temporal dependencies. Recurrent structures such as LSTMs or GRUs are commonly incorporated to model sequence evolution. Early extensions include the *Recurrent GAN* (RGAN) and its conditional variant, *RcGAN*, which leverage RNN-based generators and discriminators to produce temporally coherent sequences.

Consider a univariate time series $\{x_t\}_{t=1}^T$ generated by an unknown stochastic temporal process with joint distribution $p_{\text{data}}(x_{1:T})$. This joint distribution characterizes the underlying data-generating mechanism, capturing both the marginal behaviour of individual observations and the temporal dependence structure across time. A RGAN models this data-generating process by learning a generator $G_\theta$ that induces a distribution over entire sequences and a discriminator $D_\phi$ that evaluates complete trajectories, with both networks parameterized by recurrent neural networks.

- **Generator.** The generator takes as input a sequence of random noise vectors $z_{1:T}$, where $z_t \sim N(0, I_d)$, and produces a synthetic time series $\tilde{x}_{1:T}$ via a recurrent mapping:

$$h_t^G = f_G(h_{t-1}^G, z_t; \theta),$$
$$\tilde{x}_t = g_G(h_t^G; \theta),$$

  where $f_G(\cdot)$ and $g_G(\cdot)$ denote the recurrent state update and output functions, respectively. The generated sequence $\tilde{x}_{1:T} = G_\theta(z_{1:T})$ represents a synthetic sample from the model distribution.
- **Discriminator.** The discriminator receives either a real sequence $x_{1:T}$ or a generated sequence $\tilde{x}_{1:T}$ and outputs a scalar probability indicating whether the input sequence is real:

$$h_t^D = f_D(h_{t-1}^D, x_t; \phi),$$
$$D_\phi(x_{1:T}) = \sigma\left(w_D^\top h_T^D + b_D\right),$$

  where $f_D(\cdot)$ denotes the discriminator's recurrent update function and $\sigma(\cdot)$ is the sigmoid activation function.
- **Training Objective.** The generator and discriminator are trained adversarially by solving

$$\min_\theta \max_\phi \mathbb{E}_{x_{1:T} \sim p_{\text{data}}}[\log D_\phi(x_{1:T})] + \mathbb{E}_{z_{1:T} \sim p_z}[\log(1 - D_\phi(G_\theta(z_{1:T})))].$$

In practice, gradients are propagated through both RNNs using *backpropagation through time* (BPTT). The generator learns to produce temporally consistent sequences that the discriminator cannot distinguish from real ones.

To generate time series conditioned on auxiliary information, a conditional GAN (cGAN) formulation can be adopted. Let $c$ denote a context variable that may include past observations, lagged states, control inputs, calendar effects, or other exogenous covariates. The conditional GAN objective is given by

$$\min_\theta \max_\phi \mathbb{E}_{x_{1:T} \sim p_{\text{data}}}\left[\log D_\phi(x_{1:T} \mid c)\right] + \mathbb{E}_{z_{1:T} \sim p_z}\left[\log\left(1 - D_\phi(G_\theta(z_{1:T}, c) \mid c)\right)\right].$$

In this formulation, the generator $G_\theta$ learns the conditional distribution $p_\theta(x_{1:T} \mid c)$ by producing sequences that are consistent with the given context, while the discriminator $D_\phi$ assesses the plausibility of a sequence relative to the same conditioning information. Both networks receive $c$ as an additional input, ensuring that real and synthetic samples are evaluated under identical conditions.

Conditioning enables the model to capture context-dependent temporal dynamics and reduces mode ambiguity by restricting the support of the generated distribution. This framework is particularly well suited for time-series forecasting, counterfactual simulation, and scenario analysis, where future trajectories must be generated conditional on observed histories or external signals.

### 15.2.3 C-RNN-GAN and TimeGAN

An early attempt to extend GANs to sequential and continuous data is the *C-RNN-GAN* model (Mogren, 2016). Both the generator $G$ and discriminator $D$ are implemented using RNNs, such as LSTMs, allowing the model to capture temporal dependencies through hidden states. Given a random noise sequence $z_{1:T}$, the generator produces a sequence of outputs $G(z_{1:T}) = (x_1, \ldots, x_T)$, while the discriminator evaluates the joint probability of real and generated sequences, $D(x_{1:T})$. The standard adversarial objective is maintained:

$$\min_G \max_D \mathbb{E}_{x_{1:T} \sim p_{\text{data}}}[\log D(x_{1:T})] + \mathbb{E}_{z_{1:T} \sim p_z}[\log(1 - D(G(z_{1:T})))]. \tag{15.18}$$

To stabilize training and encourage smooth temporal generation, the C-RNN-GAN introduces feature matching and gradient penalties between real and generated hidden representations. Although primarily demonstrated for music and signal generation, it laid the foundation for subsequent time-series GANs such as TimeGAN, which incorporate supervised and embedding-based components for better temporal consistency.

A representative and influential model for time series generation is *TimeGAN* (Yoon et al., 2019), which combines the strengths of autoencoder-based reconstruction and adversarial training within a unified framework. Unlike standard GANs that rely purely on adversarial loss, TimeGAN introduces a latent embedding space that explicitly preserves temporal dynamics through supervised sequence alignment. The model consists of four components:

(i) an *embedding network* $E$ that maps observed sequences $x_{1:T}$ to latent representations $h_{1:T} = E(x_{1:T})$;

(ii) a *recovery network* $R$ that reconstructs the original data $\hat{x}_{1:T} = R(h_{1:T})$;

(iii) a *generator* $G$ that generates synthetic latent trajectories $\tilde{h}_{1:T} = G(z_{1:T})$ from random noise;

(iv) a *discriminator* $D$ that distinguishes between real latent sequences $h_{1:T}$ and generated ones $\tilde{h}_{1:T}$.

The overall training objective combines three losses:

$$\begin{aligned}
\mathcal{L}_{\text{rec}} &= \mathbb{E}\left[\|x_{1:T} - \hat{x}_{1:T}\|_2^2\right], \\
\mathcal{L}_{\text{sup}} &= \mathbb{E}\left[\|h_{t+1} - \tilde{h}_{t+1}\|_2^2\right], \\
\mathcal{L}_{\text{adv}} &= \mathbb{E}_{h_{1:T}}[\log D(h_{1:T})] + \mathbb{E}_{\tilde{h}_{1:T}}[\log(1 - D(\tilde{h}_{1:T}))].
\end{aligned}$$

The total objective is a weighted sum:

$$\min_{E,R,G} \max_D \mathcal{L}_{\text{adv}} + \lambda_{\text{rec}} \mathcal{L}_{\text{rec}} + \lambda_{\text{sup}} \mathcal{L}_{\text{sup}}, \tag{15.19}$$

where $\lambda_{\text{rec}}$ and $\lambda_{\text{sup}}$ control the balance among the competing terms. The supervised loss $\mathcal{L}_{\text{sup}}$ enforces step-to-step temporal consistency between real and generated hidden trajectories, which helps TimeGAN capture both static correlations and dynamic transitions in time series data. This hybrid structure makes TimeGAN a strong baseline for realistic synthetic data generation across diverse domains such as finance, medicine, and IoT systems.

### *15.2.4 Applications, evaluation, and challenges*

GANs have emerged as powerful tools for time-series synthesis, augmentation, and simulation across a wide range of application domains. By learning complex joint distributions over entire trajectories, GANs can generate realistic synthetic sequences in data-scarce environments, enhance downstream forecasting and classification models through data augmentation, and support privacy-preserving data sharing. In addition, discriminator scores or generator-based reconstruction errors can be leveraged for anomaly detection, while controlled generation enables scenario simulation in financial markets, healthcare monitoring, and Internet-of-Things (IoT) systems.

Evaluating time-series GANs presents unique challenges, as standard image-based metrics such as the Fréchet Inception Distance (FID) do not directly apply to sequential data. Common evaluation strategies include *statistical similarity* measures — such as marginal distributions, autocorrelation functions, power spectral density, and maximum mean discrepancy (MMD) — which assess whether synthetic data preserve key temporal and distributional properties of real sequences. Another important criterion is *predictive validity*, in which models trained on synthetic data are evaluated on real data to measure how well task-relevant information is retained. These quantitative metrics are often complemented by qualitative assessments of temporal patterns, volatility clustering, seasonality, and cross-variable correlations. More recent evaluation frameworks employ discriminative classifiers or self-supervised temporal representations to quantify the degree of realism and distinguishability between real and generated sequences.

Despite their flexibility, training GANs for sequential data remains challenging due to adversarial instability, mode collapse, and the need to preserve both short- and long-range temporal dependencies. To address these issues, time-series-specific architectures such as C-RNN-GAN and TimeGAN incorporate recurrent or attention-based structures, auxiliary reconstruction losses, and embedding networks to stabilize training and encourage faithful temporal dynamics. More recent advances combine GANs with transformers or diffusion-based models, enabling improved long-horizon modelling and richer representation of non-Markovian and highly nonlinear temporal processes. While challenges remain, GAN-based approaches continue to offer a flexible and expressive framework for modelling complex time series beyond the assumptions of classical parametric models.

## 15.3 Diffusion Models for Sequential Data

Diffusion models have recently emerged as a powerful class of generative models capable of producing high-fidelity data across various domains. Unlike GANs or VAEs, diffusion models explicitly define a stochastic process that gradually transforms data into noise (the *forward* process) and then learns to reverse this process to reconstruct the original data (the *reverse* process). For time series applications, this approach enables the modelling of complex temporal dependencies and distributions without relying on adversarial training or explicit likelihood approximations. By capturing both short-term dynamics and long-range correlations, diffusion-based sequence models have demonstrated strong performance in forecasting, simulation, and scenario generation. In this section, we discuss the core components of diffusion models, including the forward and reverse processes, training objectives, sequence modelling strategies, and practical applications to time series analysis.

### *15.3.1 Forward diffusion process*

The forward process, also known as the diffusion or noising process, defines how a clean time series is gradually transformed into a noisy sequence through a Markovian stochastic process. Given an original sequence $x_0 \in \mathbb{R}^{T\times d}$ of length $T$ with $d$ features, the forward process introduces Gaussian noise incrementally over $T$ discrete time steps. Formally, this process is defined as

$$q(x_t \mid x_{t-1}) = N\left(x_t; \sqrt{1-\beta_t}\, x_{t-1}, \beta_t I\right), \quad t = 1, \ldots, T, \tag{15.20}$$

where $\beta_t \in (0,1)$ is a predefined variance schedule controlling the magnitude of noise at each step. As $t$ increases, the sequence $x_t$ becomes progressively more corrupted, eventually approaching an isotropic Gaussian distribution when $t = T$. This formulation ensures that the forward process is analytically tractable and allows for closed-form sampling of intermediate noisy sequences.

A convenient property of the forward process is that any intermediate step $x_t$ can be sampled directly from the initial data $x_0$ without sequentially iterating through all previous steps. This is achieved through the reparameterization

$$x_t = \sqrt{\bar{\alpha}_t}\, x_0 + \sqrt{1-\bar{\alpha}_t}\, \varepsilon, \quad \varepsilon \sim N(0, I), \tag{15.21}$$

where $\alpha_t = 1 - \beta_t$ and $\bar{\alpha}_t = \prod_{s=1}^{t} \alpha_s$ represents the cumulative product of the noise reduction factors. The term $\varepsilon$ corresponds to injected Gaussian noise, and this formulation is critical for training the reverse process, as it provides a clear target for denoising.

In the context of time series, the forward process must consider temporal dependencies to ensure that the noise injection does not destroy the underlying dynamics. While the process itself is memoryless (Markovian), its design — such as the choice of $\beta_t$ schedule and feature-wise noise scaling — affects the difficulty of the denoising task and the model's ability to recover realistic sequences. Common schedules include linear, quadratic, and cosine-based increments, each controlling the rate at which temporal structure is corrupted.

Overall, the forward process provides a principled way to convert structured time series into a sequence of progressively noisier observations, forming the foundation for the reverse denoising model. By carefully designing the variance schedule and noise injection strategy, one can ensure that temporal coherence is preserved sufficiently for effective reconstruction during the reverse diffusion process.

### *15.3.2 Reverse denoising process*

The reverse denoising process is the core generative mechanism of diffusion models, responsible for transforming a noisy sequence back into a coherent time series. While the forward process gradually corrupts the original data with Gaussian noise, the reverse process learns a parameterized denoising function that sequentially removes noise at each step, reconstructing the original data distribution.

Formally, the reverse process is defined as a Markov chain:

$$p_\theta(x_{t-1} \mid x_t) = N\big(x_{t-1}; \mu_\theta(x_t, t), \Sigma_\theta(x_t, t)\big), \quad t = T, \ldots, 1, \tag{15.22}$$

where $\mu_\theta$ and $\Sigma_\theta$ are functions predicted by a neural network with parameters $\theta$. At each step, the model predicts the mean and variance of the distribution from which the previous, less noisy sequence $x_{t-1}$ is sampled. By iteratively applying this denoising step from $t = T$ down to $t = 0$, the model generates a synthetic sequence that approximates the original time series $x_0$.

A key simplification often used in practice is to fix $\Sigma_\theta(x_t, t)$ to a known function derived from the forward process, leaving the network to predict only the mean $\mu_\theta(x_t, t)$. This reduces variance and stabilizes training. Additionally, using the reparameterization trick from the forward process, the reverse step can be written as

$$x_{t-1} = \frac{1}{\sqrt{\alpha_t}} \left( x_t - \frac{1-\alpha_t}{\sqrt{1-\bar{\alpha}_t}} \varepsilon_\theta(x_t, t) \right) + \sigma_t z, \quad z \sim N(0, I), \tag{15.23}$$

where $\varepsilon_\theta$ is the network predicting the injected noise, $\bar{\alpha}_t = \prod_{s=1}^{t} \alpha_s$ is the cumulative noise factor from the forward process, and $\sigma_t$ controls stochasticity in the reverse step.

For time series, the reverse process must capture temporal dependencies to generate sequences that are both realistic and coherent. To achieve this, models typically employ RNNs, temporal convolutional networks, or transformer-based architectures. Conditioning on past observations, covariates, or control signals further enhances the fidelity of generated sequences and allows the model to produce context-aware forecasts or scenario simulations.

In summary, the reverse diffusion process serves as a learned denoising mechanism that sequentially reconstructs a realistic time series from noise. Its effectiveness depends on both the accuracy of the noise-predicting network and the careful design of temporal architectures that capture the dynamics of sequential data.

### *15.3.3 Denoising objective and model training*

Training diffusion models for time series involves teaching the network to reverse the forward diffusion process by accurately predicting the noise injected at each step. Given a noisy sequence $x_t$ obtained from the forward process

$$x_t = \sqrt{\bar{\alpha}_t} x_0 + \sqrt{1-\bar{\alpha}_t} \varepsilon, \quad \varepsilon \sim N(0, I), \tag{15.24}$$

the model learns a noise-predicting function $\varepsilon_\theta(x_t, t)$ parameterized by a neural network.

The most commonly used objective is the mean-squared error between the true noise $\varepsilon$ and the predicted noise:

$$\mathscr{L}(\theta) = \mathbb{E}_{x_0,\varepsilon,t}\left[\|\varepsilon - \varepsilon_\theta(x_t,t)\|^2\right], \tag{15.25}$$

where the expectation is taken over the data distribution, the Gaussian noise, and the diffusion time steps. Minimizing this loss trains the network to iteratively denoise the sequence, effectively learning the reverse denoising process.

This formulation has several advantages for time series modelling. First, it avoids the need to compute a complex likelihood, which is often intractable for high-dimensional sequential data. Second, it allows the use of standard stochastic gradient descent methods and backpropagation through temporal architectures such as RNNs, TCNs, or transformers. Third, the objective can be easily conditioned on past observations, covariates, or control signals, enabling conditional sequence generation and forecasting.

In practice, additional techniques—such as variance weighting, noise schedule tuning, and curriculum training—can improve stability and convergence. The network trained under this denoising objective can then be applied in the reverse process to generate realistic synthetic sequences that capture both short-term dynamics and long-range temporal dependencies.

### *15.3.4 Diffusion models for sequential data*

Diffusion models have recently emerged as a powerful class of generative models for time-series data, owing to their ability to capture complex temporal dynamics, intrinsic stochasticity, and high-dimensional multivariate dependencies. By gradually perturbing data with noise and learning an iterative denoising process, diffusion models generate diverse and coherent trajectories that approximate the full conditional distribution of future paths. This property makes them particularly well suited for probabilistic forecasting, scenario simulation, and principled uncertainty quantification in sequential settings.

Diffusion-based approaches have been successfully applied across a range of domains, including finance (market simulation, stress testing, and risk scenario generation), healthcare (synthetic ECG and EEG signal generation, patient trajectory modelling), climate science (weather and climate time-series simulation), and Internet-of-Things (IoT) systems (sensor data synthesis and anomaly detection). Compared to variational autoencoders and generative adversarial networks, diffusion models are generally more robust in capturing nonlinear, non-Gaussian, and multimodal temporal dependencies, while avoiding some of the training instabilities commonly associated with adversarial objectives.

Despite these advantages, diffusion models face practical challenges, most notably the high computational cost induced by iterative denoising during sampling. Recent methodological advances address these limitations through accelerated sampling schemes, such as deterministic or higher-order solvers, optimized noise schedules, and knowledge distillation techniques that reduce the number of diffusion steps.

In addition, conditional diffusion models — incorporating historical observations, covariates, or control signals — have further improved generation fidelity and forecasting accuracy.

Overall, diffusion models complement VAEs and GANs by offering a flexible and stable framework for high-fidelity time-series generation and analysis. Their strong theoretical foundations, combined with ongoing improvements in efficiency and scalability, position diffusion models as a promising tool for modelling complex sequential data beyond the assumptions of classical time-series models.

## 15.4 Transformers for Generative Time Series

Transformers, originally developed for natural language processing, have become a powerful framework for modelling sequential data, including time series, due to their self-attention mechanism, which efficiently captures dependencies across both short and long temporal horizons. Unlike recurrent architectures, which process sequences sequentially, transformers operate on all time steps in parallel, enabling them to model complex temporal patterns, multivariate interactions, and stochastic trajectories simultaneously. These capabilities make transformers particularly well suited for generative modelling of time series.

### *15.4.1 Transformer-based generative frameworks*

In generative applications, transformer models can be implemented in either *autoregressive* frameworks, where each time step is generated conditioned on previous outputs, or *non-autoregressive* frameworks, which generate entire sequences simultaneously, often utilizing latent variables or masking strategies to capture sequence structure. In the following sections, we review architectures specifically adapted for time-series data, discuss relevant training objectives, and highlight practical applications across domains such as finance, healthcare, and sensor networks.

In the autoregressive setting, a transformer models the conditional distribution of the next time step given all previous steps:

$$p_\theta(x_t \mid x_{1:t-1}) = f_\theta(x_{1:t-1}), \tag{15.26}$$

where $x_t$ represents the observation at time $t$, and $f_\theta$ denotes the transformer network parameterized by $\theta$. Generation proceeds sequentially: at each step, the model predicts the next value, which is then fed back as input for subsequent predictions. This approach naturally captures the temporal ordering and allows the model to generate realistic sequences that respect the underlying dynamics.

Non-autoregressive generation, by contrast, predicts multiple or all future time steps simultaneously. This is achieved by conditioning the transformer on past observations or external covariates while producing a full sequence in parallel:

$$\hat{x}_{t+1:T} = f_\theta(x_{1:t}, c), \tag{15.27}$$

where $c$ denotes optional conditioning information. Non-autoregressive transformers enable faster generation but may require additional mechanisms, such as iterative refinement or masked attention, to maintain temporal consistency.

For time series, input embeddings play a critical role. Each observation $x_t$ is mapped to a high-dimensional embedding space, often combined with *positional encodings* or *time features* to retain temporal order and periodicity information. These embeddings are then processed through stacked self-attention layers to capture interactions across all time steps.

Transformer-based generative frameworks have demonstrated strong performance in long-sequence forecasting, scenario simulation, and probabilistic sequence generation. By directly modelling complex dependencies without relying on recurrence, they provide an effective alternative to RNN-based GANs or diffusion models, particularly for high-dimensional or irregularly sampled time series.

### *15.4.2 Transformer-based time series generation*

Several transformer architectures have been adapted or specifically designed for generative modelling of time series, balancing the need for long-range dependency capture, efficiency, and probabilistic forecasting. Key approaches include:

- **Time Series Transformer (TST).** The TST extends the standard transformer to time series by incorporating time-aware embeddings, such as positional encodings, periodic features, and covariate embeddings. These embeddings allow the model to represent both the temporal location of each observation and auxiliary information such as day-of-week, seasonality indicators, or external covariates. By processing sequences through stacked self-attention layers, the TST can capture temporal dependencies, cross-feature interactions, and multivariate correlations, making it well suited for multivariate sequence generation, anomaly detection, and scenario simulation.
- **Autoregressive Transformers.** Inspired by GPT-like models, autoregressive transformers model the conditional distribution of each time step given all previous observations:

  $$p_\theta(x_t \mid x_{1:t-1}) = f_\theta(x_{1:t-1}),$$

  where $f_\theta$ is the transformer network. Generation proceeds sequentially, ensuring temporal ordering is preserved. Masked self-attention prevents information leakage from future time steps, allowing the model to generate realistic sequences one step at a time. Autoregressive models naturally support probabilistic forecasting by sampling multiple trajectories from the learned distribution.
- **Informer and Efficient Transformers.** Standard transformers scale quadratically with sequence length, which can be prohibitive for long time series or high-dimensional data. Informer and related architectures employ sparse attention, probabilistic attention, or low-rank approximations to reduce computational and memory costs while retaining the ability to model long-range dependencies. These methods enable efficient training and fast generation for long sequences,

making them practical for applications such as high-frequency finance, climate modelling, or sensor network data synthesis.

- **Conditional and Context-aware Transformers.** Many time-series applications require generation conditioned on external covariates, historical observations, or scenario parameters. Conditional transformers incorporate this information via additional embedding layers, cross-attention modules, or context vectors, allowing the model to produce sequences that are consistent with the given context. This approach is particularly useful for forecasting, intervention analysis, stress-testing scenarios, and counterfactual simulations.

Collectively, these architectures provide a flexible toolkit for generative modelling, enabling transformers to produce high-fidelity, temporally coherent sequences across diverse domains. By selecting appropriate embeddings, attention mechanisms, and conditional strategies, practitioners can tailor transformer models to the specific characteristics and requirements of their time-series data, including sequence length, dimensionality, stochasticity, and temporal complexity.

### *15.4.3 Training objectives*

Training transformer models for generative time-series tasks aims to learn the underlying conditional distributions governing sequence evolution while faithfully capturing temporal dependencies and uncertainty. Depending on the generation paradigm and architectural design, several training objectives are commonly employed.

- **Autoregressive likelihood-based training.** In autoregressive settings, transformers are trained to model the conditional distribution of each observation given its past history:
$$\mathscr{L}(\theta) = -\sum_{t=1}^{T} \log p_\theta(x_t \mid x_{1:t-1}),$$
where $p_\theta(\cdot)$ denotes the predictive distribution output by the transformer at each time step. For continuous-valued time series, this distribution is often chosen to be Gaussian or mixture-based, while discrete or discretized data are modelled using categorical distributions. Minimizing the negative log-likelihood encourages the model to generate sequences consistent with the empirical data distribution.
- **Teacher forcing and exposure mitigation.** During training, autoregressive transformers typically rely on teacher forcing, where the true previous observation $x_{t-1}$ is provided as input when predicting $x_t$. While effective for optimization, this strategy can lead to exposure bias at generation time. Scheduled sampling addresses this issue by gradually replacing ground-truth inputs with the model's own predictions, improving robustness and stability during long-horizon sequence generation.
- **Masked and sequence-to-sequence objectives.** For non-autoregressive or bidirectional transformer architectures, masked modelling and sequence-to-sequence reconstruction objectives are commonly used. In these approaches, a subset of

time steps is masked and the model is trained to reconstruct the missing values:

$$\mathscr{L}(\theta) = \mathbb{E}\left[\|x_{1:T} - \hat{x}_{1:T}\|^2\right].$$

Such objectives encourage the model to exploit global contextual information and capture long-range temporal dependencies, making them particularly suitable for long-term forecasting, interpolation, and imputation tasks.

- **Conditional training objectives.** When sequence generation is conditioned on auxiliary information, such as covariates, control variables, or historical context, the training objective explicitly incorporates this conditioning:

$$\mathscr{L}(\theta) = -\sum_{t=1}^{T} \log p_\theta(x_t \mid x_{1:t-1}, c),$$

ensuring that generated trajectories remain consistent with the given context. Conditional objectives are essential for applications such as scenario analysis, intervention modelling, and probabilistic forecasting under exogenous influences.

Together, these training paradigms enable transformer-based models to generate realistic and temporally coherent sequences, while retaining the flexibility required for uncertainty quantification, conditional generation, and modelling of complex long-range temporal dependencies.

### *15.4.4 Applications and practical considerations*

Transformer-based generative models have been successfully applied to a wide range of time series problems, owing to their ability to capture long-range temporal dependencies and complex cross-variable interactions. These properties make them particularly effective in high-dimensional and multivariate settings, including:

- **Forecasting and scenario generation:** Transformers are widely used to generate realistic future trajectories in applications such as financial markets, energy systems, and supply chain management. By leveraging attention mechanisms, they can jointly model short-term fluctuations and long-run trends, enabling scenario analysis, stress testing, and probabilistic forecasting under alternative assumptions.
- **Healthcare and physiological signals:** In biomedical applications, transformer-based generative models can synthesize realistic ECG, EEG, and other physiological signals. This capability supports data augmentation for rare clinical events, improves model robustness, and facilitates privacy-preserving data sharing without revealing sensitive patient information.
- **IoT and sensor networks:** For industrial, environmental, and smart infrastructure systems, transformers can generate multivariate sensor streams that reflect complex temporal dependencies across devices. Such synthetic data are valuable for anomaly detection, system validation, and simulation of extreme or failure scenarios.

- **Climate and environmental modelling:** The capacity of transformers to capture long-range temporal and cross-variable interactions makes them attractive for climate and environmental applications. They have been used to generate temperature, precipitation, and pollution trajectories under different policy, emission, or intervention scenarios, complementing traditional physical and statistical modelling approaches.

Despite their flexibility and expressive power, transformer-based generative models raise several practical considerations. The quadratic computational complexity of standard self-attention with respect to sequence length can limit scalability for long time series, motivating the development of efficient attention mechanisms such as sparse, local, or low-rank approximations. Moreover, transformers often require large and diverse datasets to generalize effectively; when data are limited or noisy, overfitting and unstable generation may occur. Careful architectural choices, regularization strategies, and the incorporation of domain-specific inductive biases — such as conditional inputs or structured embeddings — can help mitigate these challenges.

Overall, transformers provide a powerful and versatile framework for generative modelling of time series, complementing recurrent neural networks, GAN-based approaches, and diffusion models. Their combination of parallel sequence processing, attention-based dependency modelling, and flexibility for conditional and probabilistic generation makes them particularly well suited for high-dimensional, long-horizon, and scenario-driven time series applications.

## 15.5 Hybrid Generative Models for Time Series

While individual generative paradigms such as VAEs, GANs, diffusion models, and transformers each provide distinct advantages, no single approach fully addresses all challenges inherent in time series generation. VAEs offer structured latent representations and principled probabilistic modelling, GANs excel at producing sharp and realistic samples, diffusion models achieve strong coverage of complex data distributions, and transformers are particularly effective at capturing long-range temporal and cross-variable dependencies. Hybrid generative models seek to integrate these complementary strengths, resulting in architectures that are more expressive, flexible, and robust for sequential data. By combining probabilistic latent structures, adversarial training objectives, and attention-based sequence modelling, hybrid approaches can generate high-fidelity, temporally coherent, and context-aware time series.

*VAE – GAN hybrids*

VAE – GAN architectures integrate the latent-variable framework of VAEs with the adversarial training mechanism of GANs. In this setting, an encoder maps an observed sequence $x_{1:T}$ into a latent variable $z \sim q_\phi(z \mid x_{1:T})$, while a decoder or generator $G$ reconstructs the sequence from the latent space. A discriminator $D$ is trained simultaneously to distinguish between real and generated sequences, encouraging the generator to produce more realistic outputs. A typical objective takes the form

$$\mathscr{L}_{\text{VAE-GAN}} = \mathbb{E}_{x\sim p_{\text{data}}}\left[\|x - G(z)\|^2\right] + \beta\, D_{\text{KL}}\left(q_\phi(z \mid x) \,\|\, p(z)\right) + \lambda\, \mathbb{E}_{x\sim p_{\text{data}}}\left[\log D(x) + \log(1 - D(G(z)))\right], \qquad (15.28)$$

where $\beta$ and $\lambda$ control the trade-off between reconstruction accuracy, latent regularization, and adversarial realism. Compared to standard VAEs, this hybrid substantially improves sample sharpness and temporal realism, while preserving a meaningful latent space that supports interpolation, clustering, and conditional generation. These properties are particularly useful in applications such as financial scenario generation and biomedical signal synthesis.

*Diffusion – transformer hybrids*

Diffusion models generate data by iteratively denoising samples from a simple reference distribution, leading to excellent coverage of complex and multimodal data distributions. However, naive diffusion architectures may struggle to efficiently capture long-range temporal dependencies or scale to very long sequences. Incorporating transformer-based architectures into the denoising network addresses these limitations by leveraging self-attention to model global temporal interactions:

$$x_{t-1} = f_\theta(x_t, t), \qquad f_\theta \text{ implemented as a transformer network.} \qquad (15.29)$$

In this hybrid design, the transformer conditions on both the diffusion step and the partially denoised sequence, allowing the model to coordinate denoising decisions across distant time points and multiple variables. The resulting models combine the stability and sample diversity of diffusion processes with the representational power of attention mechanisms, making them well suited for long-horizon, high-dimensional, and multivariate time series generation.

*GAN – transformer hybrids*

GAN-based models can also benefit substantially from transformer architectures. In GAN – transformer hybrids, the generator is augmented with self-attention or full transformer blocks to explicitly model temporal dependencies, while the discriminator evaluates entire sequences for realism and consistency. The adversarial objective typically remains

$$\min_G \max_D \mathbb{E}_{x\sim p_{\text{data}}}[\log D(x)] + \mathbb{E}_{z\sim p_z}[\log(1 - D(G_\theta(z)))], \qquad (15.30)$$

but the use of attention within the generator enables the model to capture complex long-term patterns that are difficult to learn with purely convolutional or recurrent architectures. This hybrid approach is especially effective for multivariate time series, irregular sampling schemes, and sequences exhibiting strong cross-variable interactions.

*Applications and advantages*

Hybrid generative models have found growing application across domains such as financial scenario simulation, healthcare signal generation, climate and environmental modelling, and industrial sensor data synthesis. By integrating complementary modelling principles, these approaches offer several key advantages:

- **Improved fidelity and temporal coherence:** Adversarial or diffusion-based objectives enhance realism, while latent-variable and attention mechanisms preserve temporal structure.
- **Structured latent representations:** Hybrid models retain interpretable latent spaces that facilitate interpolation, conditional generation, representation learning, and uncertainty quantification.
- **Scalability to long and multivariate sequences:** Attention-based components enable efficient modelling of long-range dependencies and complex cross-series interactions.

Despite their promise, hybrid models introduce additional complexity in both architecture design and training. Balancing multiple loss components, ensuring training stability, and managing computational costs remain active challenges. Nevertheless, hybrid generative frameworks represent a powerful and flexible direction for advancing high-quality time series generation, offering a unifying perspective that bridges probabilistic modelling, adversarial learning, diffusion processes, and attention-based sequence modelling.

## 15.6 Probabilistic Forecasting and Uncertainty Quantification

A central advantage of generative models for time series is their intrinsic ability to represent uncertainty and deliver fully probabilistic forecasts. In contrast to traditional point-prediction methods, which output a single expected trajectory, generative models characterize the entire predictive distribution over future outcomes. This capability is crucial in risk-sensitive and decision-critical settings, where understanding variability, tail behaviour, and extreme scenarios is often more important than minimizing average forecast error.

Many generative models produce uncertainty-aware forecasts through repeated sampling from latent or stochastic components.

- **Variational autoencoders.** VAEs encode an observed sequence $x_{1:T}$ into a latent random variable $z \sim q_\phi(z \mid x_{1:T})$. By sampling the latent space multiple times and decoding each sample, the model generates a collection of plausible future trajectories:

$$\tilde{x}_{1:T}^{(i)} = G_\theta\left(z^{(i)}\right), \qquad z^{(i)} \sim q_\phi(z \mid x_{1:T}), \quad i = 1, \ldots, N.$$

  The resulting ensemble approximates the predictive distribution of the time series. This approach enables estimation of predictive means, variances, quantiles, and tail risks, while the structured latent space facilitates scenario interpolation and conditional analysis.

- **Diffusion models.** Diffusion models generate samples by simulating a learned reverse-time stochastic process. Starting from a noise distribution, each realization produces a distinct trajectory:

$$x_0^{(i)} \sim p_\theta(x_0 \mid x_T), \qquad i = 1, \ldots, N,$$

where $x_T$ denotes the terminal noise state. The inherent stochasticity of the reverse diffusion process naturally yields a diverse set of trajectories, providing a rich representation of forecast uncertainty. Diffusion models are particularly effective at capturing multimodal and non-Gaussian predictive distributions, which frequently arise in financial, climate, and sensor data.

- **Autoregressive and conditional transformers.** Autoregressive transformers define a predictive distribution at each time step,

$$p_\theta(x_t \mid x_{1:t-1}),$$

from which samples can be drawn sequentially to form ensemble forecasts. Conditional transformers extend this framework by incorporating covariates, control variables, or scenario parameters, yielding context-aware probabilistic predictions. This design allows uncertainty to propagate dynamically over time, reflecting both model uncertainty and intrinsic stochasticity in the data-generating process.

Generative models can represent multiple forms of uncertainty relevant to time series analysis:

- **Aleatoric uncertainty:** inherent randomness in the system, such as market fluctuations or measurement noise.
- **Epistemic uncertainty:** uncertainty due to limited data or model misspecification, often explored through ensembles or Bayesian extensions.
- **Scenario uncertainty:** variability arising from different conditioning variables, policy choices, or external shocks.

By explicitly modelling these uncertainties, generative approaches provide a more comprehensive and interpretable view of future dynamics than deterministic forecasting methods.

Assessing probabilistic forecasts requires metrics that account for both calibration and sharpness of the predictive distribution. Commonly used criteria include:

- **Continuous Ranked Probability Score (CRPS):** evaluates the discrepancy between the predicted cumulative distribution function and the observed outcome, balancing accuracy and uncertainty.
- **Prediction interval coverage and width:** measures whether empirical coverage matches nominal levels and whether uncertainty bounds are informative rather than overly conservative.
- **Maximum Mean Discrepancy (MMD) or Wasserstein distance:** compares empirical distributions of generated and observed sequences, capturing differences in higher-order moments and temporal structure.

These metrics provide complementary perspectives on forecast quality, highlighting trade-offs between sharpness, calibration, and distributional fidelity.

Overall, probabilistic forecasting provides a unifying perspective that connects generative modelling with decision-making under uncertainty. By leveraging VAEs, diffusion models, transformers, and their hybrids, practitioners can not only generate realistic time series but also quantify uncertainty in a principled manner. This capability enables richer scenario analysis, improved risk assessment, and more informed decisions, naturally concluding the chapter with an emphasis on uncertainty-aware time series modelling.

## 15.7 Bibliography

This chapter has provided a comprehensive overview of generative modelling techniques for time series, highlighting their principles, architectures, and practical applications. VAEs (Kingma and Welling, 2014; Wang et al., 2021) offer a probabilistic latent-variable framework well suited for interpolation, reconstruction, and uncertainty-aware sequence generation. GANs (Goodfellow et al., 2014; Yoon et al., 2019; Esteban et al., 2017) leverage adversarial training to produce realistic time series with strong temporal coherence, making them effective for data augmentation and simulation.

Diffusion models (Ho et al., 2020; Tashiro et al., 2021) extend generative modelling to time series through iterative stochastic denoising, enabling the capture of complex nonlinear dynamics, multiscale temporal dependencies, and non-Gaussian uncertainty. Transformer-based models (Vaswani et al., 2017; Li et al., 2019; Lim et al., 2021) provide a flexible attention-based framework that excels at modelling long-range dependencies, supporting both autoregressive and conditional sequence generation in high-dimensional and multivariate settings. Hybrid models, including VAE – GAN, GAN – Transformer, and diffusion – transformer architectures (Larsen et al., 2016; Song and Ermon, 2021) further combine the strengths of probabilistic latent representations, adversarial training, and self-attention to enhance sequence fidelity, temporal consistency, and representation quality.

Finally, probabilistic forecasting and uncertainty quantification (Gneiting and Raftery, 2007; Salinas et al., 2020) emphasize the ability of generative models to characterize full predictive distributions rather than point estimates. This capability is critical for risk-aware decision-making in domains such as finance, healthcare, climate science, and industrial systems. Collectively, these generative approaches form a versatile and powerful toolkit for time series synthesis, scenario simulation, and uncertainty-aware analysis, offering both theoretical foundations and practical solutions for complex sequential data.

### Exercises

**15.1** Consider a sequence of observations $x_{1:T} = \{x_1, \dots, x_T\}$ and a sequence of latent variables $z_{1:T} = \{z_t, \dots, z_{1:T}\}$, where each $z_t$ captures latent dynamics at time $t$. Note that $\{x_{1:T}, z_{1:T}\}$ has the following state-space (or sequential) VAE structure.

$$p_\theta(x_{1:T}, z_{1:T}) = \prod_{t=1}^{T} p_\theta(x_t|z_{1:t})p(z_t|z_{1:t-1}). \tag{15.31}$$

Suppose that $\{x_{1:T}, z_{1:T}\}$ follows a filtering variational family, i.e.

$$q_\phi(z_{1:T}|x_{1:T}) = \prod_{t=1}^{T} q_\phi(z_t|x_{1:t}). \tag{15.32}$$

Replacing $x$ and $z$ in (15.2) with $x_{1:T}$ and $z_{1:T}$, respectively, and making use of (15.31) and (15.32), show that the ELBO for sequential data is given by (15.11).

**15.2** Consider a $\beta$-Variational Autoencoder ($\beta$-VAE) with objective function

$$\mathscr{L}(\theta, \phi) = \mathbb{E}_{q_\phi(z|x)}\left[\log p_\theta(x \mid z)\right] - \beta\, D_{\mathrm{KL}}\left(q_\phi(z \mid x) \,\|\, p(z)\right), \tag{15.33}$$

where $p(z)$ is a prior distribution on the latent variable $z$ (typically $N(0,I)$), $q_\phi(z \mid x)$ is the encoder (variational posterior), and $p_\theta(x \mid z)$ is the decoder.

(a) Analyse the effect of choosing $\beta > 1$ on the learned latent representation. In particular, discuss how increasing $\beta$ influences disentanglement, information compression, and reconstruction accuracy.

(b) Consider the limiting cases, $\beta \to 0$, and $\beta \to \infty$. Describe the behaviour of the encoder and decoder in each case and explain the implications for the learned latent space.

(c) Discuss the trade-off between reconstruction fidelity and latent regularization induced by $\beta$. How does this trade-off affect the interpretability and smoothness of the latent space?

**15.3** Let $\{x_t\}_{t=0}^{T}$ denote a forward diffusion process defined by

$$x_t = \sqrt{1-\beta_t}\, x_{t-1} + \sqrt{\beta_t}\, \varepsilon_t, \qquad \varepsilon_t \sim N(0,I),$$

where $\{\beta_t\}_{t=1}^{T}$ is a predefined noise schedule with $\beta_t \in (0,1)$.

(a) Show that the distribution of $x_t$ conditioned on the original signal $x_0$ admits the closed-form representation

$$x_t = \sqrt{\bar{\alpha}_t}\, x_0 + \sqrt{1-\bar{\alpha}_t}\, \varepsilon, \qquad \varepsilon \sim N(0,I),$$

where $\alpha_t = 1 - \beta_t$ and

$$\bar{\alpha}_t = \prod_{s=1}^{t} \alpha_s.$$

(b) Using the above representation, derive the conditional distribution $q(x_t \mid x_0)$ and show that it is Gaussian.

(c) Explain intuitively why, as $t$ increases and $\bar{\alpha}_t \to 0$, the forward process gradually destroys information about the original signal $x_0$.

# Appendix A Fourier, Laplace, and z-Transforms

This appendix provides a short introduction to the Fourier transform, which is a valuable mathematical tool in time-series analysis. The related Laplace and z-transforms are also briefly introduced.

Given a (possibly complex-valued) function $h(t)$ of a real variable $t$, the **Fourier transform** of $h(t)$ is usually defined as

$$H(\omega) = \int_{-\infty}^{\infty} h(t)\,\mathrm{e}^{-\mathrm{i}\omega t}\,\mathrm{d}t \tag{34}$$

provided the integral exists for every real $\omega$. Note that $H(\omega)$ is in general complex. A sufficient condition for $H(\omega)$ to exist is that

$$\int_{-\infty}^{\infty} |h(t)|\,\mathrm{d}t < \infty.$$

If Equation (34) is regarded as an integral equation for $h(t)$ given $H(\omega)$, then a simple inversion formula exists of the form

$$h(t) = \frac{1}{2\pi}\int_{-\infty}^{\infty} H(\omega)\,\mathrm{e}^{\mathrm{i}\omega t}\,\mathrm{d}\omega. \tag{35}$$

Then $h(t)$ is called the (inverse) Fourier transform of $H(\omega)$. The two functions $h(t)$ and $H(\omega)$ are commonly called a Fourier transform pair.

The reader is warned that many authors use a slightly different definition of a Fourier transform to Equation (34). For example, some authors put a constant $1/\sqrt{(2\pi)}$ outside the integral in Equation (34), and then the inversion formula for $h(t)$ has a symmetric form. In time-series analysis many authors (e.g. Cox and Miller, 1968, p. 315) find it convenient to put a constant $1/2\pi$ outside the integral in Equation (34). In the inversion formula, the constant outside the integral is then unity.

In time-series analysis, it is often convenient to work with the variable $f = \omega/2\pi$ rather than $\omega$. The resulting Fourier transform pair is

$$G(f) = \int_{-\infty}^{\infty} h(t)\,\mathrm{e}^{-2\pi\mathrm{i}ft}\,\mathrm{d}t, \tag{36}$$

$$h(t) = \int_{-\infty}^{\infty} G(f)\,\mathrm{e}^{2\pi\mathrm{i}ft}\,\mathrm{d}f. \tag{37}$$

Note that the constant outside each integral is now unity.

DOI: 10.1201/9781003646112-A

When working with discrete-time series, we typically use the discrete form of the Fourier transform where $h(t)$ is only defined for integer values of $t$. Then

$$H(\omega) = \sum_{t=-\infty}^{\infty} h(t)\,e^{-i\omega t} \qquad \text{for } -\pi \leq \omega \leq \pi \tag{38}$$

is the discrete Fourier transform of $h(t)$. Note that $H(\omega)$ is only defined in the interval $[-\pi, \pi]$. The inverse transform is

$$h(t) = \frac{1}{2\pi}\int_{-\pi}^{\pi} H(\omega)\,e^{i\omega t}\,d\omega. \tag{39}$$

Fourier transforms have many useful properties, some of which are used during the later chapters of this book. However, we do not attempt to review them all here, as there are numerous mathematical books that cover the Fourier transform in varying depth.

One special type of Fourier transform arises when $h(t)$ is a real-valued even function such that $h(t) = h(-t)$. The autocorrelation function of a stationary time series has these properties. Then, using Equation (34) with a constant $1/\pi$ outside the integral, we find

$$\begin{aligned} H(\omega) &= \frac{1}{\pi}\int_{-\infty}^{\infty} h(t)\,e^{-i\omega t}\,dt \\ &= \frac{2}{\pi}\int_{0}^{\infty} h(t)\cos\omega t\,dt \end{aligned} \tag{40}$$

and it is easy to see that $H(\omega)$ is a real-valued even function. The inversion formula is then

$$\begin{aligned} h(t) &= \frac{1}{2}\int_{-\infty}^{\infty} H(\omega)\,e^{i\omega t}\,d\omega \\ &= \int_{0}^{\infty} H(\omega)\cos\omega t\,d\omega. \end{aligned} \tag{41}$$

Equations (40) and (41) are similar to a discrete Fourier transform pair and are useful when we only wish to define $H(\omega)$ for $\omega > 0$. We have generally adopted the latter convention in this book. When $h(t)$ is only defined for integer values of $t$, Equations (40) and (41) become

$$H(\omega) = \frac{1}{\pi}\left[h(0) + 2\sum_{t=1}^{\infty} h(t)\cos\omega t\right] \tag{42}$$

$$h(t) = \int_{0}^{\pi} H(\omega)\cos\,\omega t\,d\omega \tag{43}$$

and $H(\omega)$ is now only defined on $[0, \pi]$.

The **Laplace transform** of a function $h(t)$, which is defined for $t > 0$, is given by

$$H(s) = \int_{0}^{\infty} h(t)\,e^{-st}\,dt. \tag{44}$$

As compared with the Fourier transform, note that the lower limit of the integral is zero, and not $-\infty$, and that $s$ is a complex variable, which may have real *and* imaginary parts. The integral converges when the real part of $s$ exceeds some number called the 'abscissa of convergence'.

Given a function $h(t)$, such that

$$h(t) = 0 \qquad \text{for } t < 0 \tag{45}$$

then the Laplace and Fourier transforms of $h(t)$ are the same, provided that the real part of $s$ is zero. More generally, the Laplace transform is a generalization of the Fourier transform for functions defined on the positive real line. Control engineers often prefer to use the Laplace transform when investigating the properties of a linear system, as this will cope with physically realizable systems which are stable *or* unstable. The impulse response function of a physically realizable linear system satisfies Equation (45) and so, for such functions, the Fourier transform is a special case of the Laplace transform. More details about the Laplace transform may be found in many mathematics books.

The **z-transform** of a function $h(t)$ defined on the non-negative integers is given by

$$H(z) = \sum_{t=0}^{\infty} h(t) z^{-t}, \tag{46}$$

where $z$ is a complex variable. Comparing Equation (46) with (44) (or with (47)), we see that the z-transform can be thought of as a discrete version of the Laplace transform, on replacing $e^s$ by $z$. In discrete time, with a function satisfying Equation (45), some authors (e.g. Hayes, 1996, Section 2.2.5) prefer to use the z-transform rather than the discrete Fourier transform (i.e. Equation (38)) or the discrete form of the Laplace transform, namely

$$H(s) = \sum_{t=0}^{\infty} h(t) e^{-st}. \tag{47}$$

All three transforms have somewhat similar properties, in that a convolution in the time domain corresponds to a multiplication in the frequency domain.

The more advanced reader will observe that, when $\{h(t)\}$ is a probability function such that $h(t)$ is the probability of observing the value $t$, for $t = 0, 1, \ldots$, then Equation (46) is related to the probability generating function of the distribution, while Equations (38) and (47) are related to the moment generating and characteristic functions of the distribution.

## Exercises

**A.1** If $h(t)$ is real, show that the real and imaginary parts of its Fourier transform, as defined by Equation (34), are even and odd functions, respectively.

**A.2** If $h(t) = e^{-a|t|}$ for all real $t$, where $a$ is a positive real constant, show that its Fourier transform, as defined by Equation (34), is given by

$$H(\omega) = 2a/(a^2 + \omega^2) \qquad \text{for } -\infty < \omega < \infty$$

**A.3** Show that the Laplace transform of $h(t) = \mathrm{e}^{-at}$ for $(t > 0)$, where $a$ is a real constant, is given by

$$H(s) = 1/(s+a) \qquad \text{for } \mathrm{R}e(s) > a$$

where $\mathrm{R}e(s)$ denotes the real part of $s$.

# Appendix B Dirac Delta Function

Suppose that $\phi(t)$ is any function which is continuous at $t = 0$. Then the **Dirac delta function** $\delta(t)$ is such that

$$\int_{-\infty}^{\infty} \delta(t)\,\phi(t)\,\mathrm{d}t = \phi(0). \tag{48}$$

Because it is defined in terms of its integral properties alone, it is sometimes called the 'spotting' function since it picks out one particular value of $\phi(t)$. It is also sometimes simply called the **delta function**.

Although it is called a 'function', it is important to realize that $\delta(t)$ is *not* a function in the usual mathematical sense. Rather it is what mathematicians call a generalized function, or distribution. This maps a function, $\phi(t)$ say, into the real line, by producing the value $\phi(0)$.

Some authors define the delta function by

$$\delta(t) = \begin{cases} 0 & t \neq 0 \\ \infty & t = 0 \end{cases} \tag{49}$$

such that

$$\int_{-\infty}^{\infty} \delta(t)\,\mathrm{d}t = 1.$$

While this is often intuitively helpful, it is mathematically meaningless.

The Dirac delta function can also be regarded as the limit, as $\varepsilon \to 0$, of a pulse of width $\varepsilon$ and height $1/\varepsilon$ (i.e. having unit area) defined by

$$u(t) = \begin{cases} 1/\varepsilon & 0 < t < \varepsilon \\ 0 & \text{otherwise.} \end{cases}$$

This definition is also not mathematically rigorous, but is heuristically useful. In particular, control engineers can approximate such an impulse by an impulse with unit area whose duration is short compared with the least significant time constant of the response to the linear system being studied.

Even though $\delta(t)$ is a generalized function, it can often be handled as if it were an ordinary function except that we will be interested in the values of integrals involving $\delta(t)$ and never in the value of $\delta(t)$ by itself.

The delta function has many useful properties and we have used $\delta(t)$ in Chapter 9 to analyze particular linear systems in continuous time.

DOI: 10.1201/9781003646112-B

## Exercises

**B.1** The function $\phi(t)$ is continuous at $t = t_0$. If $a < b$, show that

$$\int_a^b \delta(t - t_0)\,\phi(t)\,\mathrm{d}t = \begin{cases} \phi(t_0) & \text{for } a < t_0 < b \\ 0 & \text{for } t_0 < a \text{ or } t_0 > b. \end{cases}$$

**B.2** The function $\phi(t)$ is continuous at $t = 0$. Show that $\phi(t)\,\delta(t) = \phi(0)\,\delta(t)$.

# Appendix C Covariance and Correlation

This book assumes knowledge of basic statistical topics such as the laws of probability, probability distributions, expectation and basic statistical inference (including estimation, significance testing and linear regression). Any reader who is unfamiliar with these topics should consult one of the numerous elementary texts covering this material.

The topics of covariance and correlation are usually studied as part of elementary probability or linear regression. However, they are not always clearly understood at first and are particularly important in the study of time series. Thus they will now be briefly reviewed.

Suppose two random variables $X$ and $Y$ have means $E(X) = \mu_X, E(Y) = \mu_Y$, respectively. Then the **covariance** of $X$ and $Y$ is defined to be

$$\text{Cov}(X,Y) = E[(X - \mu_X)(Y - \mu_Y)] \tag{50}$$

and may be denoted $\gamma_{XY}$. If $X$ and $Y$ are independent, then

$$\begin{aligned} E[(X - \mu_X)(Y - \mu_Y)] &= E(X - \mu_X)E(Y - \mu_Y) \\ &= 0 \end{aligned}$$

so that the covariance is zero. If $X$ and $Y$ are *not* independent, then the covariance may be positive or negative depending on whether 'high' values of $X$ tend to go with 'high' or 'low' values of $Y$. Here high means greater than the appropriate mean.

Covariance is a useful quantity for many mathematical purposes, but it is difficult to interpret, as it depends on the units in which $X$ and $Y$ are measured. Thus it is often useful to standardize the covariance between two random variables by dividing by the product of their respective standard deviations to give a quantity called the **correlation coefficient**. If we denote the standard deviations of $X$ and $Y$ by $\sigma_X$ and $\sigma_Y$, respectively, then the correlation of $X$ and $Y$ is defined by

$$\text{Corr}(X,Y) = \text{Cov}(X,Y)/\sigma_X\sigma_Y \tag{51}$$

and is typically denoted by $\rho_{XY}$. It can readily be shown that a correlation coefficient must lie between $\pm 1$ and is a useful measure of the linear association between two variables. $\text{Corr}(X,Y) = 1$ if $Y = aX + b$ with $a > 0$, and $\text{Corr}(X,Y) = -1$ if $Y = aX + b$ with $a < 0$.

Given $N$ pairs of observations, $\{(x_i, y_i);\ i = 1, \ldots, N\}$, the usual estimate of the covariance between two variables is given by

$$s_{xy} = \sum_{i=1}^{N} (x_i - \bar{x})(y_i - \bar{y})/(N-1).$$

DOI: 10.1201/9781003646112-C

If we denote the sample variances of the two variables by $s_x^2$ and $s_y^2$, where $s_x^2 = \sum_{i=1}^{N}(x_i - \bar{x})^2/(N-1)$, then the usual estimate of the correlation coefficient is given by the sample correlation coefficient

$$r_{xy} = s_{xy}/s_x s_y. \tag{52}$$

This is the intuitive estimate of $\rho_{XY}$ as defined above in Equation (51), and also agrees with Equation (2.3) in Chapter 2 on cancelling the denominators $(N-1)$.

The above definitions are applied to time series as described in Chapters 2 and 3. For a stochastic process $\{X_t\}$ $\mathrm{Cov}(X_t, X_{t+k})$ is a function of $t$ and $k$, and a function of only $k$ if these process is stationary. Then it is called the autocovariance coefficient function at lag $k$, denoted as $\gamma(k)$ with $\gamma(0) = \mathrm{Cov}(X_t, X_t) = \mathrm{Var}(X_t)$. Similarly for a stochastic process $\{X_t\}$,

$$\mathrm{Corr}(X_t, X_{t+k}) = \frac{\mathrm{Cov}(X_t, X_{t+k})}{\mathrm{sd}(X_t)\mathrm{sd}(X_{t+k})}$$

is a function of $t$ and $k$. For a stationary process, it becomes a function of only $k$, called the autocorrelation coefficient function at lag $k$, equal to $\gamma(k)/\gamma(0)$ and denoted as $\rho(k)$. See Section 3.2.

# Answers to Exercises

*Chapter 2*

**2.1** There are various ways of assessing the trend and seasonal effects. A simple method is to calculate the four yearly averages in 1995, 1996, 1997 and 1998, and also the average sales in each of periods I, II, ..., XIII (i.e. calculate the row and column averages). The yearly averages provide a crude estimate of trend, while the differences between the period averages and the overall average estimate the seasonal effects. With such a small downward trend, this rather crude procedure may well be adequate for most purposes. It has the advantage of being easy to understand and compute. A more sophisticated approach would be to calculate a 13-month simple moving average, moving along one period at a time. This will give trend values for each period from period 7 to period 46. The end values for periods 1 to 6 and 47 to 52 need to be found by some sort of extrapolation or by using a non-centred moving average. The differences between each observation and the corresponding trend value provide individual estimates of the seasonal effects. The average value of these differences in each of the 13 periods can then be found to estimate the overall seasonal effect, assuming that it is constant over the 4-year period.

**2.2** It is hard to guess autocorrelations even when plotting $x_t$ against $x_{t-1}$. Any software package will readily give $r_1 = -0.55$.

**2.4** The usual limits of 'significance' are at $\pm 2/\sqrt{N} = \pm 0.1$. Thus $r_7$ is just 'significant'. However, unless there is some contextual reason for an effect at lag 7, there is no real evidence of non-randomness, as one expects 1 in 20 values to be 'significant' when data really are random.

**2.5** (b) Two seasonal differences, $\nabla_{12}^2$, are needed to transform data to stationarity.

*Chapter 3*

**3.1** For example, $\rho(1) = (0.7 - 0.7 \times 0.2)/(1 + 0.7^2 + 0.2^2) = 0.56/1.53$.

**3.3** $\text{Var}(X_t)$ is not finite. Then consider $Y_t = X_t - X_{t-1} = Z_t + (C-1)Z_{t-1}$. The latter expression denotes a stationary MA(1) process, with ac.f.

$$\rho_Y(k) = \begin{cases} 1 & k = 0, \\ (C-1)/[1+(C-1)^2] & k = \pm 1, \\ 0 & \text{otherwise.} \end{cases}$$

**3.4** $\rho(k) = 0.7^{|k|}$ for $k = 0, \pm 1, \pm 2, \ldots$. Note this does not depend on $\mu$.

**3.6** Looking at the roots of the auxiliary equation (see Section 3.4.4, general-order process), the roots must satisfy $\left|[\lambda_1 \pm \sqrt{(\lambda_1^2+4\lambda_2)}]/2\right| < 1$. If $\lambda_1^2+4\lambda_2 > 0$, it can easily be shown that $\lambda_1+\lambda_2 < 1$, and $\lambda_1-\lambda_2 > -1$. If $\lambda_1^2+4\lambda_2 < 0$, the roots are complex and we find $\lambda_2 > -1$.
When $\lambda_1 = 1/3$ and $\lambda_2 = 2/9$, the Yule–Walker equations have auxiliary equation $y^2 - \frac{1}{3}y - \frac{2}{9} = 0$, which has roots $(\frac{2}{3})$ and $(-\frac{1}{3})$. Thus the general solution is $\rho(k) = A_1(\frac{2}{3})^{|k|} + A_2(-\frac{1}{3})^{|k|}$. Now use $\rho(0) = 1$ and $\rho(1) = [\frac{1}{3} + \frac{2}{9}\rho(1)]$, giving $\rho(1) = 3/7$, to evaluate $A_1 = 16/21$ and $A_2 = 5/21$.

**3.8**
$$\gamma_Y(k) = 2\gamma_X(k) - \gamma_X(k+1) - \gamma_X(k-1).$$

$$\gamma_Y(k) = \begin{cases} 2-2\lambda & k=0 \\ -\lambda^{|k|-1}(1-\lambda)^2 & k=\pm1,\pm2,\ldots. \end{cases}$$

**3.9** All three models are stationary and invertible. For model (a), we find $X_t = Z_t + 0.3Z_{t-1} + 0.3^2 Z_{t-2} + \cdots$.

**3.12** (a) $p = d = q = 1$.

(b) Model is non-stationary, but is invertible.

(c) 0.7, 0.64, 0.628 —- decreasing very slowly as process is non-stationary.

(d) 0.7, 0.15, 0.075, 0.037 — decreasing quickly towards zero as invertible.

**3.13** $\rho(k) = \frac{45}{38}(3/4)^{|k|} - \frac{7}{38}(1/4)^{|k|} \qquad k = 0, \pm1, \pm2, \ldots$.
The AR(3) process is non-stationary, as the equation $(1 - B - cB^2 + cB^3) = 0$, has a root on the unit circle, namely, $B = 1$.

**3.14** $\mathrm{Cov}[Ye^{i\omega t}, \overline{Y}e^{-i\omega(t+\tau)}] = e^{-i\omega\tau}\,\mathrm{Cov}[Y, \overline{Y}]$ does not depend on $t$.
$e^{i\theta} = \cos\theta + i\sin\theta$. When $\theta$ is uniformly distributed on $(0, 2\pi)$, then $\mathrm{E}(\cos\theta) = \mathrm{E}(\sin\theta) = 0$. Hence result.

*Chapter 4*

**4.2** The least squares normal equations are the same as the sample Yule–Walker equations in Equation (4.12) except that the constant divisor $\sum(x_t - \bar{x})^2$ is omitted and the estimated autocovariance in Equation (4.1) is effectively estimated by summing over $(N-p)$, rather than $(N-k)$ cross-product terms.

**4.3** When fitting an AR(2) process, $\pi_2$ is the coefficient $\alpha_2$. For such a process, the first two Yule–Walker equations are: $\rho(2) = \alpha_1\rho(1) + \alpha_2$ and $\rho(1) = \alpha_1 + \alpha_2\rho(-1) = \alpha_1 + \alpha_2\rho(+1)$. Solve for $\pi_2 = \alpha_2$ by eliminating $\alpha_1$.

**4.4** $\pi_j$ is zero for $j$ larger than 2, the order of the process. $\pi_2$ equals the final coefficient in the model, namely, $\alpha_2 = 2/9$; and so we only have to find $\pi_1$, which equals $\rho(1)$. This is 9/21 — see Exercise 3.6.

**4.5** The values come down quickly to zero, suggesting a stationary process. Values outside the range $\pm2/\sqrt{100} = \pm0.2$ are significantly different from zero, in this case just $r_1$, $r_2$ and $r_5$. An MA(2) process has an ac.f. of this shape, meaning non-zero coefficients at lags 1 and 2. Given the small sample size, it is, however, possible that an AR(2) process could be appropriate, as the theoretical autocorrelations beyond lag 2 would be quite small, even if not exactly zero as for the

MA(2) model. In practice, it is often difficult to distinguish between competitor models having similar ac.f.s and there may be no 'right answer'. All models are approximations anyway.

**4.6** The ac.f. of the data indicates non-stationarity in the mean, as the values of $r_k$ are only coming down to zero very slowly. So we need to take first differences (at least). The values of $r_k$ for the first differences are all 'small' and are not significantly different from zero (all are less than $\pm 2/\sqrt{60}$). This suggests the series is now stationary and is just a purely random process. Thus a possible model for the original series is an ARIMA(0, 1, 0) model, otherwise known as a random walk. Of course, in practice, we would like to see a time plot of the data, check for outliers, etc. as well as getting more relevant background information about the economic variable concerned.

*Chapter 5*

**5.4** Suppose we denote the model by

$$(1-\alpha B^{12})W_t = (1+\theta B)Z_t$$

$$\text{or} \qquad X_t = Z_t + \theta Z_{t-1} + (1+\alpha)X_{t-12} - \alpha X_{t-24}.$$

Then $\hat{x}_N(1) = (1+\hat{\alpha})x_{N-11} - \hat{\alpha}\, x_{N-23} + \hat{\theta}\, \hat{z}_N$ and $\hat{x}_N(h) = (1+\hat{\alpha})x_{N+h-12} - \hat{\alpha}\, x_{N+h-24}$ for $h = 2, 3, \ldots, 12$.

**5.5**

$$\hat{x}_N(1) = 1.2x_N - 0.2x_{N-1} - 0.5z_N$$

$$\hat{x}_N(2) = 1.2\hat{x}_N(1) - 0.2x_N$$

$\text{Var}[e_N(h)]$ equals $\sigma_Z^2$ when $h=1$, $1.49\sigma_Z^2$ when $h=2$, and $1.90\sigma_Z^2$ when $h=3$.

*Chapter 6*

**6.1** (b) $f^*(\omega) = (1-\lambda^2)/\pi(1-2\lambda\cos\omega + \lambda^2)$ for $0 < \omega < \pi$.

**6.2** (a) $f(\omega) = \sigma_Z^2[3 + 2(2\cos\omega + \cos 2\omega)]/\pi$ for $0 < \omega < \pi$.
(b) $f^*(\omega) = \sigma_Z^2[1.34 + 2(0.35\cos\omega - 0.3\cos 2\omega)]/\pi$ for $0 < \omega < \pi$.

**6.3** The non-zero mean makes no difference to the acv.f., ac.f. or spectrum.

$$\gamma(k) = \begin{cases} 1.89\sigma_Z^2 & k=0, \\ 1.2\sigma_Z^2 & k=\pm 1, \\ 0.5\sigma_Z^2 & k = \pm 2, \\ 0 & \text{otherwise.} \end{cases}$$

$$\rho(k) = \begin{cases} 1 & k=0, \\ 1.2/1.89 & k=\pm 1, \\ 0.5/1.89 & k = \pm 2, \\ 0 & \text{otherwise.} \end{cases}$$

**6.4** This triangular spectrum gives rise to a rather unusual ac.f. Dividing through Equation (6.9) by $\sigma_X^2$ (or $\gamma(0)$), we get $\rho(k) = \int_0^\pi f^*(\omega)\cos\omega k\, d\omega$ When $k=0$, we find, as expected, that $\rho(0) = \int_0^\pi f^*(\omega)\, d\omega = 1$. When $k$ is even, $\cos\omega k$ completes an integer number of cycles between 0 and $\pi$ so that $\int_0^\pi \cos\omega k\,\mathrm{d}\omega = 0$, and can further show that $\int_0^\pi \omega\cos\omega k\,\mathrm{d}\omega = 0$ after some messy algebra and integrating by parts (integrate $\cos\omega k$ and differentiate $\omega$). When $k$ is odd, $\int_0^\pi \cos\omega k\,\mathrm{d}\omega$ is still zero, but we find, on integrating by parts, that $\int_0^\pi \omega\cos\omega k\,\mathrm{d}\omega = 2/k^2$.

**6.5** Clearly $E[X(t)] = 0$ and $\mathrm{Var}[X(t)] = 1$. Thus $\rho(u) = \gamma(u) =$ $E[X(t)X(t+u)] = \mathrm{Prob}[X(t)$ and $X(t+u)$ have same sign$]$ $-$ $\mathrm{Prob}[X(t)$ and $X(t+u)$ have opposite sign$]$.
(Hint: Prob(observing an even number of changes in time $u$) =
$\mathrm{e}^{-\lambda u}\left[1+\frac{(\lambda u)^2}{2!}+\frac{(\lambda u)^4}{4!}+\cdots\right] = \mathrm{e}^{-\lambda u}(\mathrm{e}^{\lambda u}+\mathrm{e}^{-\lambda u})/2$ .)
With a unit variance, $f(\omega) = f^*(\omega)$. As process is in continuous time, use Equation (6.17). Algebra is rather messy. It may help to write $\cos\omega u = [\mathrm{e}^{\mathrm{i}\omega u} + \mathrm{e}^{-\mathrm{i}\omega u}]/2$.

**6.6** Here $f_Y = \sigma^2/\pi$ and $f_X = \sigma^2/\pi(1-2\alpha\cos\omega+\alpha^2)$ for $0<\omega<\pi$.

**6.7** Use Equation (6.16). As in the second part of the solution of Equation (6.5), algebra is a bit messy, and it may help to write $\cos\omega k$ as $[\mathrm{e}^{\mathrm{i}\omega k}+\mathrm{e}^{-\mathrm{i}\omega k}]/2$. Then you need to be able to sum a geometric progression (G.P.).

*Chapter 8*

**8.1** $f_{XY}(\omega) = \sigma_Z^2\,[\beta_{11}\beta_{21}+\beta_{12}\beta_{22}+\beta_{21}\mathrm{e}^{-\mathrm{i}\omega}+\beta_{12}\mathrm{e}^{+\mathrm{i}\omega}]\,/\,\pi$ for $0<\omega<\pi$.

**8.2** Use the fact that $\mathrm{Var}[\lambda_1 X(t)+\lambda_2 Y(t+\tau)] \geq 0$ for any constants $\lambda_1$, $\lambda_2$ (e.g. putting $\lambda_1 = \sigma_Y, \lambda_2 = \sigma_X$ gives $\rho_{XY} \geq -1$).
For the given bivariate process, we find

$$\gamma_{XY}(k) = \begin{cases} 0.84\sigma_Z^2 & k=0,\\ -0.4\sigma_Z^2 & k=1,\\ 0.4\sigma_Z^2 & k=-1,\\ 0 & \text{otherwise.}\end{cases}$$

Hence we find (for $0<\omega<\pi$)
$c(\omega) = 0.84\sigma_Z^2/\pi$, $q(\omega) = -0.8\sigma_Z^2\sin\omega\,/\,\pi$,
$\alpha_{XY}(\omega) = \sigma_Z^2\sqrt{(0.84^2+0.8^2\sin^2\omega)}\,/\,\pi$, $\tan\phi_{XY}(\omega) = 0.8\sin\omega\,/0.84$,
$C(\omega) = 1$.

*Chapter 9*

**9.1** They are all linear except (b) and (g). Note that (e) is linear but not time-invariant.

**9.2** Selected answers only: For (a), $h_k$ is $\frac{1}{2}$ for $k=\pm1$, 1 for $k=0$, and zero otherwise. Then we find

$$S_t = \begin{cases} 0 & t<-1,\\ \frac{1}{2} & t=-1,\\ 1\frac{1}{2} & t=0,\\ 2 & t\geq 1.\end{cases}$$

Then $H(\omega) = \frac{1}{2}\mathrm{e}^{-\mathrm{i}\omega}+1+\frac{1}{2}\mathrm{e}^{\mathrm{i}\omega} = 1+\cos\omega$. As this is real and non-negative, $G(\omega) = H(\omega)$, and $\phi(\omega) = 0$.
(b) $H(\omega) = \frac{1}{5}+\frac{2}{5}\cos\omega+\frac{2}{5}\cos 2\omega$

(c)

$$h_k = \begin{cases} +1 & k = 0, \\ -1 & k = +1, \\ 0 & \text{otherwise.} \end{cases}$$

Filters (a) and (b) are low pass, while (c) and (d) are high pass. The combined filter from (a) and (b) has
$H(\omega) = (1+\cos\omega)\left(\frac{1}{5} + \frac{2}{5}\cos\omega + \frac{2}{5}\cos 2\omega\right)$.

**9.3** (a) $H(\omega) = ge^{-i\omega\tau}$ for $\omega > 0$.
(b) $H(\omega) = g(1 - i\omega T)/(1+\omega^2 T^2)$ for $\omega > 0$.

**9.4** For the one-parameter AR(2) process, we find

$$\gamma(k) = \begin{cases} \alpha^{|k/2|}\sigma_Z^2/(1-\alpha^2) & k \text{ even} \\ 0 & k \text{ odd} \end{cases}$$

$f_X(\omega) = f_Z(\omega)/(1 - 2\alpha\cos 2\omega + \alpha^2)$ for $0 < \omega < \pi$.

*Chapter 10*

**10.1** It is generally sensible to restrict $\alpha$ to the range $(0,1)$, although it is possible to devise a plausible ARIMA scenario with $\theta$ positive in the range $(0,1)$, and this would extend the plausible range for $\alpha$ from $(0,1)$ to $(0,2)$.

**10.2** $X_2 = \mu_2 + n_2$ and $X_1 = \mu_1 + n_1 = \mu_2 - \beta_1 + n_1 = \mu_2 - \beta_2 + n_1 + w_2$. Thus

$$\begin{bmatrix} X_2 \\ X_1 \end{bmatrix} = \begin{bmatrix} 1 & 0 \\ 1 & -1 \end{bmatrix}\begin{bmatrix} \mu_2 \\ \beta_2 \end{bmatrix} + \begin{bmatrix} n_2 \\ n_1 + w_2 \end{bmatrix}$$

So

$$\hat{\theta}_2 = \begin{bmatrix} \hat{\mu}_2 \\ \hat{\beta}_2 \end{bmatrix} = \begin{bmatrix} 1 & 0 \\ 1 & -1 \end{bmatrix}^{-1}\begin{bmatrix} X_2 \\ X_1 \end{bmatrix} = \begin{bmatrix} 1 & 0 \\ 1 & -1 \end{bmatrix}\begin{bmatrix} X_2 \\ X_1 \end{bmatrix} = \begin{bmatrix} X_2 \\ X_2 - X_1 \end{bmatrix}$$

— see Equation (10.17).
To find $P_2$, $\text{Var}(\hat{\beta}_2)$, for example, is

$$E(X_2 - X_1 - \beta_2)^2 = E(n_2 - n_1 + \beta_2 - w_2 - \beta_2)^2 = 2\sigma_n^2 + \sigma_w^2.$$

Using the Kalman filter, we find $\hat{\theta}_{3|2} = \begin{bmatrix} 2X_2 - X_1 \\ X_2 - X_1 \end{bmatrix}$, $P_{3|2} = \sigma_n^2\begin{bmatrix} 5 & 3 \\ 3 & 2 \end{bmatrix}$, and $K_3 = \begin{bmatrix} 5/6 \\ 1/2 \end{bmatrix}$. Hence $\hat{\theta}_3$ using Equation (10.13).

**10.3** (a) For $\theta_t^T = [X_t, \beta Z_t]$, we find $h^T = [1,0]$; $G = \begin{bmatrix} 0 & 1 \\ 0 & 0 \end{bmatrix}$; $w_t^T = [1,\beta]Z_t$. Alternatively, if we take $\theta_t^T = [X_t, Z_t]$, for example, then we find $G = \begin{bmatrix} 0 & \beta \\ 0 & 0 \end{bmatrix}$.

(b) Try $\theta_t^T = [X_t, \hat{X}_t(1), \hat{X}_t(2)] = [X_t, \beta_1 Z_t + \beta_2 Z_{t-1}, \beta_2 Z_t]$ with
$G = \begin{bmatrix} 0 & 1 & 0 \\ 0 & 0 & 1 \\ 0 & 0 & 0 \end{bmatrix}$. Or try $\theta_t^T = [X_t, Z_t, Z_{t-1}]$.

*Chapter 13*

**13.1** This is a VAR(1) model with a coefficient matrix at lag one equal to $\Phi = \begin{pmatrix} 1 & 0.5 \\ 0.2 & 0.7 \end{pmatrix}$. The roots of the equation: determinant $\{I - \Phi x\} = 0$ are 1/1.2 and 1/0.5. Thus one of the roots lies 'inside the unit circle'. (When, as here, the roots are real, then one only needs to look and see if the magnitude of the root exceeds unity.) This means the process is non-stationary. This could be guessed from the first equation of the model where the coefficient of $X_{1,t-1}$ is 'large' (unity) and so, in the absence of $X_{2t}$ would give a (non-stationary) random walk for $X_{1t}$ (this is not meant to be a proper mathematical argument).

**13.2** The roots of the equation: determinant $\{I - \Phi x\} = 0$ are as follows: (a) a single root of 1/0.6; (b) a single root of 1/1.4; (c) two roots of 1/0.7 and 1/0.3; (d) two roots of 1/0.8 and 1/0.4. This means that models (a), (c) and (d) are stationary (giving roots exceeding unity in magnitude so they are 'outside the unit circle') but (b) is not.

**13.3** Model (c) of Exercise 12.2 is stationary and has a diagonal coefficient matrix. This means that the model consists of two independent AR(1) processes. Any cross-covariances are zero and so the covariance and correlation matrix functions are diagonal. We find

$$\Gamma(k) = \begin{pmatrix} 0.7^{|k|}\sigma_1^2 & 0 \\ 0 & 0.3^{|k|}\sigma_2^2 \end{pmatrix} \text{ and } P(k) = \begin{pmatrix} 0.7^{|k|} & 0 \\ 0 & 0.3^{|k|} \end{pmatrix}.$$

**13.4** All pure MA processes, whether univariate or multivariate, are stationary. The model is invertible if the roots of the equation, determinant $\{I + \Theta_1 x\} = 0$, are outside the unit circle. In this case, we find that the roots are $-1/0.8$ and $-1/0.2$, which are both smaller than $-1$ and so lie outside the unit circle. Thus the model *is* invertible.

**13.5** Denote the forecast of $X_{N+h}$ made at time $N$ by $\hat{X}_N(h)$. Then $\hat{X}_N(1) = \Phi X_N$ and $\hat{X}_N(2) = \Phi\hat{X}_N(1) = \Phi^2 X_N$. Picking out the first component of $X$, for example, we find $\hat{X}_{1,N}(1) = 0.9X_{1,N} + 0.5X_{2,N}$ while $\hat{X}_{1,N}(2) = 0.76X_{1,N} + 0.60X_{2,N}$.

# Bibliography


B. Abraham and J. Ledolter. *Statistical Methods for Forecasting*. Wiley, New York, 1983.

B. Abraham and J. Ledolter. Forecast functions implied by autoregressive integrated moving average models and other related forecast procedures. *International Statistical Review*, 54:51–66, 1986.

M. Adya, J. Armstrong, F. Collopy, and M. Kennedy. An application of rule-based forecasting to a situation lacking domain knowledge. *International Journal of Forecasting*, 16:477–484, 2000.

H. Akaike. On the use of a linear model for the identification of feedback systems. *Annals of the Institute of Statistical Mathematics*, 20:425–439, 1968.

T. Anderson. *The Statistical Analysis of Time Series*. Wiley, New York, 1971.

R. Andrews. Forecasting performance of structural time series models. *Journal of Business & Economic Statistics*, 12:129–133, 1994.

M. Aoki. *State Space Modeling of Time Series*. Springer-Verlag, Berlin, 2nd edition, 1990.

J. Armstrong. *Principles of Forecasting: A Handbook for Researchers and Practitioners*. Kluwer, Norwell, MA, 2001.

J. Armstrong and F. Collopy. Error measures for generalizing about forecasting methods: Empirical comparisons. *International Journal of Forecasting*, 8:69–80, 1992.

R. Ashley. On the relative worth of recent macroeconomic forecasts. *International Journal of Forecasting*, 4:363–376, 1988.

K. Astrom. *Introduction to Stochastic Control Theory*. Academic Press, New York, 1970.

K. Astrom and T. Bohlin. Numerical identification of linear dynamic systems from normal operating records. In P. Hammond, editor, *Theory of Self-Adaptive Control Systems*, pages 94–111. Plenum Press, New York, 1966.

D. Bahdanau, K. Cho, and Y. Bengio. Neural machine translation by jointly learning to align and translate. In *International Conference on Learning Representations (ICLR)*, 2015.

S. Bai, J. Z. Kolter, and V. Koltun. An empirical evaluation of generic convolutional and recurrent networks for sequence modeling. arXiv preprint arXiv:1803.01271, 2018.

M. Ball and A. Wood. Trend growth in post-1850 British economic history: The Kalman filter and historical judgment. *The Statistician*, 45:143–152, 1996.

A. Banerjee, J. Dolado, J. Galbraith, and D. Hendry. *Co-Integration, Error-Correction, and the Econometric Analysis of Non-Stationary Data.* Oxford University Press, 1993.

M. Bartlett. Chance or chaos (with discussion). *Journal of the Royal Statistical Society, Series A*, 153:321–347, 1990.

W. Bell and S. Hillmer. Modelling time series with calendar variation. *Journal of the American Statistical Association*, 78:526–534, 1983.

J. Bendat and A. Piersol. *Random Data: Analysis and Measurement Procedures.* Wiley, 3rd edition, 2000.

J. Beran. *Statistics for Long-Memory Processes.* Chapman & Hall, 1994.

W. Berveridge. Weather and harvest cycles. *The Economic Journal*, 31:429–452, 1921.

P. Bidarkota. The comparative forecast performance of univariate and multivariate models: An application to real interest rate forecasting. *International Journal of Forecasting*, 14:457–468, 1998.

D. M. Blei, A. Kucukelbir, and J. D. McAuliffe. Variational inference: A review for statisticians. *Journal of the American Statistical Association*, 112(518):859–877, 2017.

P. Bloomfield. *Fourier Analysis of Time Series.* Wiley, 2nd edition, 2000.

G. Boero. Comparing *ex-ante* forecasts from a sem and var model: An application to the Italian economy. *Journal of Forecasting*, 9:13–24, 1990.

T. Bollerslev, Y. Chou, and K. Kroner. Arch models in finance. *Journal of Econometrics*, 52:5–59, 1992.

T. Bollerslev, R. Engle, and D. Nelson. Arch models. In R. Engle and D. McFadden, editors, *Handbook of Econometrics*, volume IV, pages 2959–3038. Elsevier, 1994.

G. Box and G. Jenkins. *Time-Series Analysis, Forecasting and Control.* Holden-Day, revised edition, 1970.

G. Box and J. MacGregor. The analysis of closed-loop dynamic-stochastic systems. *Technometrics*, 16:391–398, 1974.

G. Box and P. Newbold. Some comments on a paper of coen, gomme and kendall. *Journal of the Royal Statistical Society: Series A*, 134:229–240, 1971.

G. Box, G. Jenkins, and G. Reinsel. *Time Series Analysis, Forecasting and Control.* Prentice-Hall, 3rd edition, 1994.

D. Brillinger. *Time Series: Data Analysis and Theory.* SIAM, classics edition, 2001.

W. Brock and S. Potter. Non-linear time series and macroeconometrics. In G. Maddala, C. Rao, and H. Vinod, editors, *Handbook of Statistics, Vol. 11, Econometrics*, pages 195–229. North-Holland, 1993.

P. Brockwell and R. Davis. *Time Series: Theory and Methods.* Springer-Verlag, 2nd edition, 1991.

P. Brockwell and R. Davis. *Introduction to Time Series and Forecasting.* Springer, 2nd edition, 2002.

R. Brown. *Smoothing, Forecasting and Prediction.* Prentice-Hall, 1963.

K. Burnham and D. Anderson. *Model Selection and Multi-Model Inference.* Springer-Verlag, 2nd edition, 2002.

F. d. Butter and M. Fase. *Seasonal Adjustment as a Practical Problem.* Elsevier, Amsterdam, 1991.

K.-S. Chan and H. Tong. *Chaos: A Statistical Perspective.* Springer, New York, 2001.

D. Chappell, J. Padmore, P. Mistry, and C. Ellis. A threshold model for the French franc/deutschmark exchange rate. *Journal of Forecasting*, 15:155–164, 1996.

C. Chatfield. Some recent developments in time-series analysis. *Journal of the Royal Statistical Society: Series A*, 140:492–510, 1977.

C. Chatfield. The holt-winters forecasting procedure. *Appl. Stat.*, 27:264–279, 1978.

C. Chatfield. Inverse autocorrelations. *Journal of the Royal Statistical Society: Series A*, 142:363–377, 1979.

C. Chatfield. What is the best method of forecasting? *Journal of Applied Statistics*, 15:19–38, 1988.

C. Chatfield. Calculating interval forecasts (with discussion). *Journal of Business & Economic Statistics*, 11:121–144, 1993.

C. Chatfield. Positive or negative? *International Journal of Forecasting*, 11:501–502, 1995.

C. Chatfield. *Time-Series Forecasting.* Chapman & Hall/CRC Press, Boca Raton, 2001.

C. Chatfield and A. Collins. *Introduction to Multivariate Analysis.* Chapman & Hall, London, 1980.

C. Chatfield and M. Yar. Holt-winters forecasting: Some practical issues. *The Statistician*, 37:129–140, 1988.

C. Chatfield, A. Koehler, J. Ord, and R. Snyder. Models for exponential smoothing: A review of recent developments. *The Statistician*, 50:147–159, 2001.

C. Chen. Robustness properties of some forecasting methods for seasonal time series: A Monte Carlo study. *International Journal of Forecasting*, 13:269–280, 1997.

B. Choi. *ARMA Model Identification.* Springer-Verlag, New York, 1992.

A. Choudhury, R. Hubata, and R. St. Louis. Understanding time-series regression estimators. *The American Statistician*, 53:342–348, 1999.

J. Chung, K. Kastner, L. Dinh, K. Goel, A. Courville, and Y. Bengio. A recurrent latent variable model for sequential data. In *Advances in Neural Information Processing Systems (NeurIPS)*, pages 2980–2988, 2015.

W. Cleveland. *The Elements of Graphing Data.* Hobart Press, Summit, NJ, 2nd edition, 1994.

P. Coen, E. Gomme, and M. Kendall. Lagged relationships in economic forecasting. *Journal of the Royal Statistical Society: Series A*, 132:133–163, 1969.

P. Cowpertwit and A. Metcalfe. *Introductory Time Series with R.* Springer, New York, 2009.

D. Cox and V. Isham. *Point Processes.* Chapman & Hall, London, 1980.

D. Cox and H. Miller. *The Theory of Stochastic Processes.* Wiley, New York, 1968.

J. Craddock. The analysis of meteorological time series for use in forecasting. *The Statistician*, 15:169–190, 1965.

N. Crato and B. Ray. Model selection and forecasting for long-range dependent processes. *Journal of Forecasting*, 15:107–125, 1996.

G. Darbellay and M. Slama. Forecasting the short-term demand for electricity: Do neural networks stand a better chance? *International Journal of Forecasting*, 16: 71–83, 2000.

N. Davies and P. Newbold. Some power studies of a portmanteau test of time series model specification. *Biometrika*, 66:153–155, 1979.

D. DeJong and C. Whiteman. Unit roots in U.S. macroeconomic time series: A survey of classical and Bayesian perspectives. In D. e. a. Brillinger, editor, *New Directions in Time Series Analysis: Part II*, pages 43–59. Springer-Verlag, New York, 1993.

P. Dhrymes. *Time Series, Unit Roots and Cointegration.* Academic Press, San Diego, 1997.

F. Diebold. *Elements of Forecasting.* South-Western, Cincinnati, 2001.

F. Diebold and L. Kilian. Unit root tests are useful for selecting forecasting models. *J. Bus. Econ. Stat.*, 18:265–273, 2000.

P. Diggle. *Time Series: A Biostatistical Introduction.* Oxford University Press, Oxford, 1990.

J. Durbin and S. Koopman. *Time Series Analysis by State Space Methods.* Oxford University Press, Oxford, 2001.

J. Durbin and M. Murphy. Seasonal adjustment based on a mixed additive-multiplicative model. *Journal of the Royal Statistical Society: Series A*, 138:385–410, 1975.

W. Enders. *Applied Econometric Time Series.* Wiley, New York, 1995.

R. Engle and C. Granger. *Long-Run Economic Relationships: Readings in Cointegration.* Oxford University Press, Oxford, 1991.

C. Esteban, S. L. Hyland, and G. Rätsch. Real-valued (medical) time series generation with recurrent conditional gans. *arXiv preprint arXiv:1706.02633*, 2017.

J. Fan and Q. Yao. *Nonlinear Time Series: Nonparametric and Parametric Methods.* Springer, New York, 2003.

J. Faraway and C. Chatfield. Time series forecasting with neural networks: A comparative study using the airline data. *Applied Statistics*, 47:231–250, 1998.

R. Fildes. An evaluation of Bayesian forecasting. *Journal of Forecasting*, 2:137–150, 1983.

R. Fildes. Quantitative forecasting: The state of the art. econometric models. *Journal of the Operational Research Society*, 36:549–580, 1985.

R. Fildes. The evaluation of extrapolative forecasting methods. *International Journal of Forecasting*, 8:81–98, 1992.

R. Fildes and S. Makridakis. The impact of empirical accuracy studies on time series analysis and forecasting. *International Statistical Review*, 63:289–308, 1995.

D. Findley, B. Monsell, W. Bell, M. Otto, and B.-C. Chen. New capabilities and methods of the x-12-arima seasonal adjustment program (with discussion and reply). *Journal of Business & Economic Statistics*, 16:127–177, 1998.

M. Fraccaro, S. K. Sønderby, U. Paquet, and O. Winther. Sequential neural models with stochastic layers. In *Advances in Neural Information Processing Systems (NeurIPS)*, pages 2199–2207, 2016.

P. Franses. *Time Series Models for Business and Economic Forecasting*. Cambridge University Press, Cambridge, 1998.

P. Franses and F. Kleibergen. Unit roots in the nelson-plosser data: Do they matter for forecasting? *International Journal of Forecasting*, 12:283–288, 1996.

W. Fuller. *Introduction to Statistical Time Series*. Wiley, New York, 2nd edition, 1996.

E. J. Gardner. Automatic monitoring of forecast errors. *Journal of Forecasting*, 2: 1–21, 1983.

E. J. Gardner. Exponential smoothing: The state of the art. *Journal of Forecasting*, 4:1–28, 1985.

E. J. Gardner and E. McKenzie. Forecasting trends in time series. *Management Science*, 31:1237–1246, 1985.

F. A. Gers, J. Schmidhuber, and F. Cummins. Learning to forget: Continual prediction with lstm. *Neural Computation*, 12(10):2451–2471, 2000.

J. Gleick. *Chaos*. Viking, 1987.

I. Glover and P. Grant. *Digital Communications*. Prentice-Hall, 1998.

T. Gneiting and A. E. Raftery. Strictly proper scoring rules, prediction, and estimation. *Journal of the American Statistical Association*, 102(477):359–378, 2007.

I. Goodfellow, J. Pouget-Abadie, M. Mirza, B. Xu, D. Warde-Farley, S. Ozair, A. Courville, and Y. Bengio. Generative adversarial nets. *Advances in Neural Information Processing Systems*, 27, 2014.

I. Goodfellow, Y. Bengio, and A. Courville. *Deep Learning*. MIT Press, Cambridge, MA, 2016.

J. d. Gooijer, B. Abraham, A. Gould, and L. Robinson. Methods for determining the order of an autoregressive-moving average process: A survey. *International Statistical Review*, 53:301–329, 1985.

C. Gourieroux. *ARCH Models and Financial Applications*. Springer-Verlag, 1997.

C. Granger. Forecasting stock market prices: Lessons for forecasters. *International Journal of Forecasting*, 8:3–13, 1992.

C. Granger and P. Newbold. Spurious regressions in econometrics. *Journal of Econometrics*, 2:111–120, 1974.

C. Granger and P. Newbold. *Forecasting Economic Time Series*. Academic Press, 2nd edition, 1986.

C. Granger and T. Teräsvirta. *Modelling Nonlinear Economic Relationships*. Oxford University Press, 1993.

G. Grimmett and D. Stirzaker. *Probability and Random Processes*. Oxford University Press, 3rd edition, 2001.

H. Grubb and A. Mason. Long lead-time forecasting of UK air passengers by holt-winters methods with damped trend. *International Journal of Forecasting*, 17: 71–82, 2001.

I. Gustavsson, L. Ljung, and T. Söderstrom. Identification of process in closed loop – identifiability and accuracy aspects. *Automatica*, 13:59–75, 1977.

V. Gómez and A. Maravall. Seasonal adjustment and signal extraction in economic time series. In D. P. na, G. Tiao, and R. Tsay, editors, *A Course in Time Series Analysis*, chapter 8. Wiley, 2001.

J. Hamilton. *Time Series Analysis*. Princeton University Press, 1994.

D. Hand, F. Daly, A. Lunn, K. McConway, and F. Ostrowski. *A Handbook of Small Data Sets*. Chapman & Hall, 1994.

E. Hannan. *Multiple Time Series*. Wiley, 1970.

R. Harris and R. Sollis. *Applied Time Series Modelling and Forecasting*. John Wiley & Sons Ltd., 2003.

P. Harrison. Short-term sales forecasting. *Applied Statistics*, 14:102–139, 1965.

A. Harvey. *Forecasting, Structural Time Series Models and the Kalman Filter*. Cambridge University Press, 1989.

A. Harvey. *The Econometric Analysis of Time Series*. Philip Allan, 2nd edition, 1990.

A. Harvey. *Time Series Models*. Harvester Wheatsheaf, 2nd edition, 1993.

M. Hayes. *Statistical Digital Signal Processing and Modeling*. Wiley, 1996.

J. Ho, A. Jain, and P. Abbeel. Denoising diffusion probabilistic models. *Advances in Neural Information Processing Systems*, 33, 2020.

S. Hochreiter and J. Schmidhuber. Long short-term memory. *Neural Computation*, 9(8):1735–1780, 1997.

M. D. Hoffman, D. M. Blei, C. Wang, and J. Paisley. Stochastic variational inference. In *Journal of Machine Learning Research (JMLR)*, volume 14, pages 1303–1347, 2013.

K. Hornik, M. Stinchcombe, and H. White. Multilayer feedforward networks are universal approximators. *Neural Networks*, 2(5):359–366, 1989.

S. Hylleberg, editor. *Modelling Seasonality*. Oxford University Press, 1992.

V. Isham. Statistical aspects of chaos. In O. Barndorff-Nielsen et al., editors, *Networks and Chaos - Statistical and Probabilistic Aspects*, pages 124–200. Chapman & Hall, 1993.

G. Janacek and L. Swift. *Time Series: Forecasting, Simulation, Applications*. Ellis Horwood, Chichester, U.K., 1993.

G. Jenkins. *Practical Experiences with Modelling and Forecasting Time Series*. Gwilym Jenkins and Partners (Overseas) Ltd., Jersey, 1979.

G. Jenkins and G. McLeod. *Case Studies in Time Series Analysis*, volume 1. Gwilym Jenkins and Partners Ltd., 1982.

G. Jenkins and D. Watts. *Spectral Analysis and its Applications*. Holden-Day, San Francisco, 1968.

S. Johansen. Cointegration in the var model. In D. P. na, G. Tiao, and R. Tsay, editors, *A Course in Time Series Analysis*, chapter 15. Wiley, New York, 2001.

M. I. Jordan, Z. Ghahramani, T. S. Jaakkola, and L. K. Saul. An introduction to variational methods for graphical models. *Machine Learning*, 37(2):183–233, 1999.

K. Kadiyala and S. Karlsson. Forecasting with generalized Bayesian vector autoregressions. *Journal of Forecasting*, 12:365–378, 1993.

H. Kantz and T. Schreiber. *Nonlinear Time Series Analysis*. Cambridge University Press, 1997.

M. Kendall and J. Ord. *Time Series*. Arnold, Sevenoaks, U.K., 3rd edition, 1990.

M. Kendall, A. Stuart, and J. Ord. *The Advanced Theory of Statistics*, volume 3. Griffin, 4th edition, 1983.

P. Kenny and J. Durbin. Local trend estimation and seasonal adjustment of economic and social time series (with discussion). *Journal of the Royal Statistical Society: Series A*, 145:1–41, 1982.

D. P. Kingma and M. Welling. Auto-encoding variational bayes. In *Proceedings of the 2nd International Conference on Learning Representations (ICLR)*, 2014.

D. P. Kingma, T. Salimans, and M. Welling. Improving variational inference with inverse autoregressive flow. In *Advances in Neural Information Processing Systems (NeurIPS)*, pages 4743–4751, 2016.

R. Kohn and C. Ansley. Estimation, prediction and interpolation for arima models with missing data. *Journal of the American Statistical Association*, 81:751–761, 1986.

L. Koopmans. *The Spectral Analysis of Time Series*. Academic Press, 2nd edition, 1995.

G. Lai, Y. Chang, Y. Yang, and H. Liu. Modeling long- and short-term temporal patterns with deep neural networks. In *Proceedings of the 31st AAAI Conference on Artificial Intelligence (AAAI)*, 2018. arXiv:1703.07015.

T. Lai and H. Xing. *Statistical Models and Methods for Financial Markets*. Springer, New York, 2008.

A. B. L. Larsen, S. K. Sønderby, H. Larochelle, and O. Winther. Autoencoding beyond pixels using a learned similarity metric. *International Conference on Machine Learning*, 2016.

A. Lawrance. Directionality and reversibility in time series. *The International Statistical Review*, 59:67–79, 1991.

C. Lea, R. Vidal, A. Reiter, and G. D. Hager. Temporal convolutional networks: A unified approach to action segmentation. In *European Conference on Computer Vision (ECCV) Workshops*, pages 47–54. Springer, 2016.

S. Li, X. Jin, H. Xiong, X. Zhou, W. Chen, and Y. Wang. Enhancing the locality and breaking the memory bottleneck of transformer on time series forecasting. In *Advances in Neural Information Processing Systems*, volume 32, pages 1–13, 2019.

B. Lim, S. Ö. Arık, N. Loeff, and T. Pfister. Temporal fusion transformers for interpretable multi-horizon time series forecasting. *International Journal of Forecasting*, 37(4):1748–1764, 2021.

J.-L. Lin and C. Granger. Forecasting from non-linear models in practice. *Journal of Forecasting*, 13:1–9, 1994.

T. Lin, T. Guo, and K. Aberer. Hybrid neural networks for learning the trend in time series. In *Proceedings of the International Joint Conference on Artificial Intelligence (IJCAI)*, 2017.

M.-T. Luong, H. Pham, and C. D. Manning. Effective approaches to attention-based neural machine translation. In *Proceedings of the 2015 Conference on Empirical Methods in Natural Language Processing (EMNLP)*, pages 1412–1421, 2015.

H. Lütkepohl. *Introduction to Multiple Time Series Analysis*. Springer-Verlag, New York, 2nd edition, 1993.

S. Makridakis and M. Hibon. Accuracy of forecasting: An empirical investigation (with discussion). *Journal of the Royal Statistical Society: Series A*, 142:97–145, 1979.

S. Makridakis and M. Hibon. The m3 competition: results, conclusions and implications. *International Journal of Forecasting*, 16:451–476, 2000.

S. Makridakis, A. Anderson, R. Carbone, R. Fildes, M. Hibon, R. Lewandowski, J. Newton, E. Parzen, and R. Winkler. *The Forecasting Accuracy of Major Time Series Methods*. Wiley, New York, 1984.

S. Makridakis, C. Chatfield, M. Hibon, M. Lawrence, T. Mills, K. Ord, and L. Simmons. The m2-competition: A real-time judgmentally based forecasting study (with commentary). *International Journal of Forecasting*, 9:5–29, 1993.

R. M. May. Chaos and the dynamics of biological populations. *Proceedings of the Royal Society of London. Series A*, 413:27–44, 1987.

J. McLain. Dominant tracking signals. *International Journal of Forecasting*, 4:563–572, 1988.

N. Meade. The use of growth curves in forecasting market development – a review and appraisal. *Journal of Forecasting*, 3:429–451, 1984.

N. Meade and I. Smith. Ararma vs. arima – a study of the benefits of a new approach to forecasting. *International Journal of Management Science*, 13:519–534, 1985.

R. Meinhold and N. Singpurwalla. Understanding the Kalman filter. *The American Statistician*, 32:123–127, 1983.

T. Mills. *Time Series Techniques for Economists.* Cambridge University Press, Cambridge, 1990.

T. Mills. *The Econometric Modelling of Financial Time Series, 2nd edition* Cambridge University Press, Cambridge, 1999.

G. Mizon. A simple message for autocorrelation correctors: Don't. *Journal of Econometrics*, 69:267–288, 1995.

O. Mogren. C-rnn-gan: Continuous recurrent neural networks with adversarial training. *arXiv preprint arXiv:1611.09904*, 2016.

D. Montgomery. *Introduction to Statistical Quality Control, 3rd edition* Wiley, New York, 1996.

D. Montgomery, L. Johnson, and J. Gardiner. *Forecasting and Time Series Analysis, 2nd edition* McGraw-Hill, New York, 1990.

M. Murray. A drunk and her dog: An illustration of cointegration and error correction. *The American Statistician*, 48:37–39, 1994.

H. Neave. A comparison of lag window generators. *Journal of the American Statistical Association*, 67:152–158, 1972.

H. Nelson and C. Granger. Experience with using the box-cox transformation when forecasting economic time series. *Journal of Econometrics*, 10:57–69, 1979.

P. Newbold. Some recent developments in time-series analysis, iii. *International Statistical Review*, 56:17–29, 1988.

P. Newbold and C. Granger. Experience with forecasting univariate time-series and the combination of forecasts (with discussion). *Journal of the Royal Statistical Society: Series A*, 137:131–165, 1974.

P. Newbold, C. Agiakloglou, and J. Miller. Long-term inference based on short-term forecasting models. In T. S. Rao, editor, *Time Series Analysis*, pages 9–25. Chapman & Hall, London, 1993.

J. Ord, A. Koehler, and R. Snyder. Estimation and prediction for a class of dynamic nonlinear statistical models. *Journal of the American Statistical Association*, 92: 1621–1629, 1997.

T. Otomo, T. Nakagawa, and H. Akaike. Statistical approach to computer control of cement rotary kilns. *Automatica*, 8:35–48, 1972.

A. Pankratz. *Forecasting with Dynamic Regression Models*. Wiley, New York, 1991.

A. Papoulis. *Probability, Random Variables and Stochastic Processes*. McGraw-Hill, New York, 2nd edition, 1984.

E. Parzen. Ararma models for time series analysis and forecasting. *Journal of Forecasting*, 1:67–82, 1982.

D. Peña and G. Box. Identifying a simplifying structure in time series. *Journal of the American Statistical Association*, 82:836–843, 1987.

D. Percival. Three curious properties of the sample variance and autocovariance for stationary processes with unknown mean. *The American Statistician*, 47:274–276, 1993.

P. Phillips. Understanding spurious regressions in econometrics. *Journal of Econometrics*, 33:311–340, 1986.

A. Pole, M. West, and J. Harrison. *Applied Bayesian Forecasting and Time Series Analysis*. Chapman & Hall, New York, 1994.

M. Priestley. *Spectral Analysis and Time Series*, volume 1 and 2. Academic Press, London, 1981.

M. Priestley. The frequency domain approach to the analysis of closed-loop systems. In D. Brillinger and P. Krishnaiah, editors, *Handbook of Statistics*, volume 3, pages 275–291. North-Holland, Amsterdam, 1983.

M. Priestley. *Non-linear and Non-stationary Time Series Analysis*. Academic Press, London, 1988.

R. Ranganath, S. Gerrish, and D. M. Blei. Black box variational inference. In *Proceedings of the 17th International Conference on Artificial Intelligence and Statistics (AISTATS)*, pages 814–822, 2014.

G. Reinsel. *Elements of Multivariate Time Series Analysis*. Springer-Verlag, New York, 2nd edition, 1997.

D. J. Rezende and S. Mohamed. Variational inference with normalizing flows. In *Proceedings of the 32nd International Conference on Machine Learning (ICML)*, pages 1530–1538, 2015.

D. J. Rezende, S. Mohamed, and D. Wierstra. Stochastic backpropagation and approximate inference in deep generative models. In *Proceedings of the 31st International Conference on Machine Learning (ICML)*, pages 1278–1286, 2014.

P. Robinson and P. Zaffaroni. Nonlinear time series with long memory: A model for stochastic volatility. *Journal of Statistical Planning and Inference*, 68:359–371, 1998.

L. Rogers and D. Williams. *Diffusions, Markov Processes and Martingales*. Cambridge University Press, Cambridge, 2nd edition, 1994.

S. Ross. *Introduction to Probability Models*. Academic Press, San Diego, 6th edition, 1997.

G. Rowe and G. Wright. The Delphi technique as a forecasting tool: Issues and analysis. *International Journal of Forecasting*, 15:353–375, 1999.

D. E. Rumelhart, G. E. Hinton, and R. J. Williams. Learning representations by back-propagating errors. *Nature*, 323:533–536, 1986.

D. Salinas, V. Flunkert, J. Gasthaus, and T. Januschowski. Deepar: Probabilistic forecasting with autoregressive recurrent networks. *International Journal of Forecasting*, 36(3):1181–1191, 2020.

N. Shephard. Statistical aspects of arch and stochastic volatility. In D. Cox, D. Hinkley, and O. Barndorff-Nielsen, editors, *Time Series Models*, pages 1–67. Chapman & Hall, London, 1996.

R. Shumway and D. Stoffer. *Time Series Analysis and Its Applications: With R Examples*. Springer, New York, 3rd edition, 2010.

J. Smith and S. Yadav. Forecasting costs incurred from unit differencing fractionally integrated processes. *International Journal of Forecasting*, 10:507–514, 1994.

F. Snodgrass, G. Groves, K. Hasselmann, G. Miller, W. Munk, and W. Powers. Propagation of ocean swells across the pacific. *Philosophical Transactions of the Royal Society of London, Series A*, 259:431–497, 1966.

Y. Song and S. Ermon. Score-based generative modeling through stochastic differential equations. *International Conference on Learning Representations*, 2021.

D. Spencer. Developing a Bayesian vector autoregression forecasting model. *International Journal of Forecasting*, 9:407–421, 1993.

A. Sutcliffe. Time-series forecasting using fractional differencing. *Journal of Forecasting*, 13:383–393, 1994.

Y. Tashiro, Y. Song, and S. Ermon. Csdi: Conditional score-based diffusion models for probabilistic time series imputation. *Advances in Neural Information Processing Systems*, 34, 2021.

L. Tashman and J. Kruk. The use of protocols to select exponential smoothing procedures: A reconsideration of forecasting competitions. *International Journal of Forecasting*, 12:235–253, 1996.

A. Tay and K. Wallis. Density forecasting: A survey. *Journal of Forecasting*, 19: 235–254, 2000.

S. Taylor. Modeling stochastic volatility: A review and comparative study. *Mathematical Finance*, 4:183–204, 1994.

L. Tee and S. Wu. An application of stochastic and dynamic models in the control of a papermaking process. *Technometrics*, 14:481–496, 1972.

G. Tiao. Vector arma models. In D. P. na, G. Tiao, and R. Tsay, editors, *A Course in Time Series Analysis*, chapter 14. Wiley, New York, 2001.

G. Tiao and R. Tsay. Some advances in non-linear and adaptive modelling in time series. *Journal of Forecasting*, 13:109–131, 1994.

H. Tong. *Non-linear Time Series.* Oxford University Press, Oxford, 1990.

H. Tong. A personal overview of non-linear time series analysis from a chaos perspective. *Scandinavian Journal of Statistics*, 22:399–446, 1995.

R. Tsay. Testing and modelling multivariate threshold models. *Journal of the American Statistical Association*, 93:1188–1202, 1998.

R. Tsay. Nonlinear time series models: Testing and applications. In D. P. na, G. Tiao, and R. Tsay, editors, *A Course in Time Series Analysis*, chapter 10. Wiley, New York, 2001.

R. Tsay. *Analysis of Financial Time Series.* Wiley, New York, 3rd edition, 2010.

R. Tsay. *Multivariate Time Series Analysis With R and Financial Applications.* Wiley, New York, 2015.

W. Vandaele. *Applied Time Series and Box-Jenkins Models.* Academic Press, New York, 1983.

A. Vaswani, N. Shazeer, N. Parmar, J. Uszkoreit, L. Jones, A. N. Gomez, L. Kaiser, and I. Polosukhin. Attention is all you need. In *Advances in Neural Information Processing Systems (NeurIPS)*, pages 5998–6008, 2017.

K. Wallis. Asymmetric density forecasts of inflation and the bank of England's fan chart. *National Institute Economic Review*, 167:106–112, 1999.

Z. Wang, W. Yan, and T. Oates. Time series generative modeling with variational autoencoders. *arXiv preprint arXiv:2106.01355*, 2021.

R. Webby and M. O'Connor. Judgemental and statistical time series forecasting: A review of the literature. *International Journal of Forecasting*, 12:91–118, 1996.

W. Wei. *Time Series Analysis: Univariate and Multivariate Methods.* Addison-Wesley, Redwood City, CA, 1990.

M. West and J. Harrison. *Bayesian Forecasting and Dynamic Models.* Springer-Verlag, New York, 2nd edition, 1997.

P. Whittle. *Prediction and Regulation.* University of Minnesota Press, Minneapolis, 2nd edition, 1983.

N. Wiener. *Extrapolation, Interpolation, and Smoothing of Stationary Time-Series.* MIT Press, Cambridge, MA, 1949.

A. Yaglom. *An Introduction to the Theory of Stationary Random Functions.* Prentice-Hall, Englewood Cliffs, NJ, 1962.

Q. Yao and H. Tong. Quantifying the influence of initial values on non-linear prediction. *Journal of the Royal Statistical Society, Series B*, 56:701–725, 1994.

J. Yoon, D. Jarrett, and M. van der Schaar. Time-series generative adversarial networks. In *Advances in Neural Information Processing Systems (NeurIPS) 32*, pages 8789–8800, 2019.

P. Young. *Recursive Estimation and Time-Series Analysis.* Springer-Verlag, Berlin, 1984.

C. Zhang, J. Butepage, H. Kjellstrom, and S. Mandt. Advances in variational inference. *IEEE Transactions on Pattern Analysis and Machine Intelligence (TPAMI)*, 41(8):2008–2026, 2018.

# Index

For Product Safety Concerns and Information please contact our EU representative GPSR@taylorandfrancis.com
Taylor & Francis Verlag GmbH, Kaufingerstraße 24, 80331 München, Germany

www.ingramcontent.com/pod-product-compliance
Lightning Source LLC
Chambersburg PA
CBHW070247230726
48635CB00030B/336

* 9 7 8 1 0 4 1 0 2 6 3 3 4 *